Advances in Wireless Communication Networks

Advances in Wireless Communication Networks

Edited by Phoebe Hill

MURPHY & MOORE
www.murphy-moorepublishing.com

Murphy & Moore Publishing,
1 Rockefeller Plaza,
New York City, NY 10020, USA

ISBN: 978-1-63987-024-0

Cataloging-in-Publication Data

Advances in wireless communication networks / edited by Phoebe Hill.
p. cm.
Includes bibliographical references and index.
ISBN 978-1-63987-024-0
1. Wireless communication systems. 2. Telecommunication systems. I. Hill, Phoebe.
TK5103.2 .A38 2022
621.384--dc23

For information on all Murphy & Moore Publications
visit our website at www.murphy-moorepublishing.com

MURPHY & MOORE

Contents

Preface

This book aims to highlight the current researches and provides a platform to further the scope of innovations in this area. This book is a product of the combined efforts of many researchers and scientists, after going through thorough studies and analysis from different parts of the world. The objective of this book is to provide the readers with the latest information of the field.

The electromagnetic transfer of information between two or more points that do not use an electrical conductor as a medium to perform the transfer is known as wireless communication. Radio waves are the most common wireless technology used for wireless communication. It comprises various types of fixed, mobile, and portable applications such as two-way radios, cellular telephones, wireless networking, and personal digital assistants. A wireless network refers to a computer network that uses wireless data connections between network nodes. Some other methods of achieving wireless communication are the use of sound and the use of electromagnetic wireless technologies such as light, magnetic and electric fields. As this field is emerging at a rapid pace, the contents of this book will help the readers understand the modern concepts and applications of the subject. It is a valuable compilation of topics, ranging from the basic to the most complex advancements in the field of wireless communication. This book will provide comprehensive knowledge to the readers.

I would like to express my sincere thanks to the authors for their dedicated efforts in the completion of this book. I acknowledge the efforts of the publisher for providing constant support. Lastly, I would like to thank my family for their support in all academic endeavors.

Editor

1

Cross layer resource allocation for fault-tolerant topology control in wireless mesh networks based on genetic algorithm

Esmaeil Nik Maleki and Ghasem Mirjalily*

Abstract

Optimal topology control is an essential factor for efficient development of wireless mesh networks. For this purpose, a set of available tools can be exploited including power control, rate adaptation, channel assignment, channel selection, scheduling and routing. In most recent studies, only some of these tools are applied for throughput maximization. In this paper, we first propose a comprehensive cross-layer resource allocation model for topology control in which a complete set of available tools are exploited in order to guarantee the fairness, balancing and robustness, in addition to throughput maximization. This leads to an NP-complete problem; therefore, we propose a four steps heuristic method based on problem decomposition to reduce the computational complexity. In first step, the best K potential paths with disjoint vertices are extracted between each pair of nodes. In second step, a method based on the genetic algorithm is proposed in order to assign frequency channels to the links of these paths. This assignment procedure must preserve the essential links and must reduce the potential interference of the network. In third step, best compatible configurations are extracted on each frequency channel using power control and rate adaptation. It must be performed such that minimizes the power consumption, maximizes the transmission rate and provides the transmission rate balancing on the links. In last step, a cross-layer method is proposed for selecting the best path between each pair of nodes such that throughput maximization, fairness, and balancing on nodes and frequency channels are met. Validation in terms of numerical results demonstrates the efficiency of our proposed method for topology control in wireless mesh networks.

Keywords: Multi-radio multi-channel wireless mesh network, Resource allocation, Topology control, Genetic algorithm, Robustness against failures, Fairness, Balancing

1 Introduction

Recently, there is a growing interest in providing multimedia and wideband services on wireless networks. In this regard, wireless mesh networks have attracted much attention. These networks are classified as multi-hop networks and provide significant benefits including low-cost deployment, robustness and simple configuration [1–3]. In order to guarantee the quality of service in wireless mesh networks, the optimal resource allocation and topology control is essential [4, 5]. To this end, extensive research has been performed to modify the existing algorithms and to develop the novel algorithms.

* Correspondence: mirjalily@yazd.ac.ir
Department of Electrical Engineering, Yazd University, Yazd, Iran

1.1 Motivation

A wireless mesh network is composed of wireless mesh routers connected to each other by wideband connections. Mesh routers are fixed nodes that play both the role of data routing and providing access points for network clients. With the development of technology, wireless mesh routers are equipped with several radio interfaces capable of rate and power adjustment. Although, increasing the transmission power leads to the increasing transmission rate, it increases the interference and therefore, reduces the number of simultaneous transmissions [3]. Moreover, several non-overlapping channels are available in these networks where the efficient assignment of these channels to the radio interfaces results in throughput improvement [6, 7]. Consequently, the problem of trade-off between power control and rate adaptation and efficient channel assignment must be addressed in wireless mesh networks. This procedure may result in more than one common channel between two nodes producing a multi-graph topology for the network. Therefore, the proper selection of frequency channels is essential for increasing the network performance [8].

Since the number of frequency channels is limited, the simultaneous transmission on some links with the same channel is impossible. In order to reduce the co-channel interference, the power control and rate adjustment must be used. Moreover, the interfering transmissions can be scheduled on different time slots using a time-based scheduling algorithm. As a result, the optimal allocation of transmissions to the time slots along with the power control and rate adaptation is another problem that should be considered. Finally, the cross-layer routing together with the layer-2 tools including channel assignment and selection, scheduling, power control and rate adaptation are essential for performance improvement. In most studies [4, 5, 9–17], only some of the available tools are exploited for efficient allocation of resources. Our proposed method includes a comprehensive set of tools to improve the network operation.

Based on the application of wireless mesh network, the objectives like throughput maximization, fairness between traffic demands, balancing of frequency channels utilization, and node utilization balancing must be considered. However, only throughput maximization has been investigated in most researches [9–17]. This can result in some serious problems in the network. Firstly, because of the imbalance in resource utilization, the network congestion is occurred in some areas while the other areas carry low volume of traffic. Secondly, the fairness between traffic demands is not established and therefore, some traffic demands are not satisfied. Thirdly, the density of mesh routers is usually low and the traffic volumes are high; therefore, the failure of a router, even for a moment, greatly increases the packet losses. For this reason, in addition to the above-mentioned objectives, the robustness of the network against failures should also be considered. According to the definition provided in [18], if the network graph is K-Connected, the network will remain connected even after the failure of fewer than K nodes. Therefore, wireless mesh networks must be designed such that the K-Connectivity feature of the graph is guaranteed.

1.2 Methods

In this paper, we first introduce the problem of Fault-Tolerant Topology Control with Throughput Maximization, Balancing and Fairness (FTTC-TMBF) in which a comprehensive cross-layer resource allocation problem is modeled for topology control. In this model, a complete set of topology control tools are used including power control, rate adaptation, channel assignment, channel selection, scheduling and routing. Here, we define the objective function with the aims of throughput maximization, frequency channel utilization balancing, node utilization balancing, fairness and robustness against failures.

As it has been pointed out in [13, 14], the power control in wireless mesh network is NP-complete. This issue is a subset of the topology control problem that has been studied in this paper; therefore, FTTC-TMBF is NP-complete and the computational complexity of FTTC-TMBF increases exponentially with any increase in the network size. In this paper, we propose the Heuristic FTTC-TMBF (HFTTC-TMBF) method that is composed of four steps. In first step, we suggest an appropriate objective function to select the best K disjoint paths between each pair of nodes. This function minimizes the number of hops and the consumed power and also takes into account the node utilization balancing. In next step, the Genetic-Based Link Channel Assignment (GB-LCA) algorithm is proposed for assigning channels to links such that potential interference is minimized. In third step, we use the Genetic-Based Compatible Set Formation (GB-CSF) algorithm for extracting the set of common-channel links with minimum interference. For this purpose, power control and rate adaptation tools are employed and the cost function is minimized by using the genetic algorithm for each compatible set of links. Finally, we introduce the Genetic Based Cross Layer Paths Selection (GB-CLPS) algorithm, in which the best paths between each pair of nodes are selected by considering throughput maximization, fairness and balancing on frequency channels and nodes. The proposed algorithms of this paper is implemented in a centralized manner. Since the topology of wireless mesh networks are nearly fixed, the centralized implementation is a proper choice.

The rest of the paper is organized as follows. Related studies are reviewed in section 2. Section 3 defines the model and assumptions of the study. The cross-layer topology optimization problem is formulated in section 4. The details of HFTTC-TMBF solution are provided in section 5. The simulation results are analyzed in section 6. Finally, the paper is concluded in Section 7.

2 Related work

In recent years, extensive research has been performed on cross-layer design of wireless mesh networks. In [9], the authors exploited the Mixed-Integer Linear Programming (MILP) to model throughput optimization in wireless mesh networks based on all available tools including power control, rate adaptation, channel assignment and routing. For this reason, it is a suitable choice to be compared with our proposed approach. In [9], the optimization process is divided in two phases. In the first phase, the authors find an optimum time-slot assignment for a scenario with a single shared channel, taking into consideration the routing. In the second one, starting from the first phase solution, a complete solution is built up for the multi-channel multi-radio scenario. Since this problem is NP-complete, a heuristic method based on the compatible configurations is proposed in which all the links that are feasible to transmit on the same channel at the same time are extracted. The advantage of the model presented in [9] is its completeness because it considers all available tools and is the first complete model introduced in this field. However, its disadvantage is that its objective is only throughput maximization and it does not consider the balancing, fairness and robustness against failures that are very important in WMNs. In our proposed model FTTC-TMBF, we have added the balancing, fairness and K-connectivity feature to themodel introduced in [9]. In addition, we presented a four-step solution where in first step, a heuristic method and in second to forth ones, a solution based on the genetic algorithm are used. In summary, we have proposed a comprehensive model in FTTC-TMBF to optimize the wireless mesh networks in which the objectives of throughput maximization, balancing, fairness and K-connectivity is considered, in addition to using all available tools; while, the model introduced in [9], only have considered the objective of throughput maximization. Moreover, in comparison to [9], we have proposed HFTTC-TMBF in which a heuristic method and genetic algorithm is used for solving the problem.

In 2010, Luo et al. [10] developed two computational tools for joint optimization of rate control, power control, scheduling and routing. Moreover, they studied the relationship between frequency reuse and network performance as well as the benefits of multi-hop against single-hop. The tools developed by Luo et al. are based on the column generation and series solution of the problem, which produces the suboptimal solution in an acceptable time. In [9, 10], objectives such as fairness, balancing and robustness against failures are not considered.

In [11, 12], the authors exploited the MILP and formulated the problem of gateway nodes selection along with the power control, routing and time slot assignment to maximize the service level of the nodes. They proposed a heuristic method to solve the problem in a serial manner, which is composed of several steps. The issues of rate adaptation and channel assignment/selection are not considered in these references.

Hedayati et al. [13], proposed a centralized approach to optimize the power consumption and rate adaptation with the aim of throughput maximization. They proposed a distributed version of their algorithm in [14] where useful tools such as channel assignment and routing are not used. In [15], the problem of robust topology control in multi-channel multi-radio wireless mesh networks is investigated. The authors formulated the problem in the form of a Mixed-Integer Non-Linear Program (MINLP) to maximize the end-to-end rate by considering the constraints of routing, co-channel interference and frequency channel switching. In order to reduce the computational complexity of the problem, the authors proposed a heuristic method, which decomposes the problem, and then a binary search algorithm is used to find the sub-problem solution. However, some objectives like fairness and balancing are not considered in any of the discussed references.

In [16, 17], the fairness objective is investigated in addition to the throughput maximization. The authors in [16], developed a model and then solved it by using the serialization method. In this reference, the network is assumed to be single-radio single-channel and single-rate; hence, the channel assignment and rate adaptation problems are ignored. In [17], differentiation among traffic flows is investigated such that the minimum fairness among different flows is guaranteed. This leads to the efficient distribution of bandwidth between clients. However, the effect of power control, rate adaptation and channel assignment are not considered in this work.

In [19], we proposed a cross layer framework for the optimal topology control where different tools such as power control, rate adaptation, scheduling, channel assignment and routing tools have been used to achieve the goals such as throughput improvement and balancing. In this paper, robustness against failure and node utilization have not been modeled appropriately. Moreover, the issue of fairness is not considered. In [20], we have represented a heuristic method for topology control. However, some objectives like fairness and balancing have not been considered implicitly in this reference.

In [21, 22], the throughput maximization and balancing objectives are investigated. The authors in [21] proposed a heuristic method by using the channel selection and rate adaptation tools. In [22], the authors indicated that the ad hoc routing methods are not appropriate for wireless mesh networks. They proposed a novel routing method with the aim of balancing improvement. However, none of these references has considered the complete set of tools and objectives.

Due to the computational complexity of the cross-layer optimization problem, various meta-heuristic methods have been proposed in recent studies [23–25]. In [23], a new routing method named MNSGA-II is proposed in which a Genetic Algorithm (GA) procedure is used to extract the best paths with the aims of minimizing the number of transmissions and delay. However, the other objectives and tools are not considered. In [24], we proposed a GA-based method for power and rate control along with scheduling in the wireless mesh networks. The objective function of this method only includes the minimization of the number of time slots. Moreover, the routing and channel assignment are not considered. The authors in [25] investigated multicast routing and channel assignment problems simultaneously using a GA-based method in which power control and rate adaptation tools are ignored. In this method, the multicast trees are determined at first using differential evolution technique and then the channel assignment is performed by using a GA-based method. While the throughput maximization and fairness are considered in this reference, the balancing factor is not considered.

In [26–28], various methods of topology modification have been proposed in order to preserve the network robustness against failures. The authors in [26] proposed a new protocol for topology control in wireless mesh network of hand-held devices. They selected a dominant set of interconnected nodes where the routing function is active in these nodes. This protocol results in the reduction of collision, overhead, interference and energy consumption. However, the network is assumed single-channel, in which each node has a simple radio interface with no rate and power adjustment capability. In [27], Peng et al. proposed a linear network coding based fault-tolerant routing, which can recover the lost packets by the source. This method, by using multi-path routing and random linear network coding improves the conventional node selection methods. Another topology control method is investigated in [28], where the authors have created a *K*-Connected graph based on the channel assignment and routing. In these references, the authors have studied only some of the objectives which does not include the fairness and balancing.

Reviewing the previous references, it seems that proposing a comprehensive model for optimum resource allocation and topology control is still an open issue in wireless mesh networks. In this paper, we investigate the problem of topology control in wireless mesh networks which includes the following contributions,

- A comprehensive cross-layer model for topology control problem is developed in which tools including power control, rate adaptation, channel assignment and selection, scheduling and routing are used.
- A complete set of objectives such as throughput maximization, balancing and fairness is considered.
- A decomposition-based heuristic method is proposed in order to reduce the computational complexity of the problem.
- A new routing metric is introduced to determine the best *K* potential individual paths between node pairs based on the *K*-Connectivity feature
- A channel assignment method based on the genetic algorithm is proposed to reduce the potential interference while preserves the *K*-Connectivity feature of the network
- Genetic algorithm is used for power control and rate adaptation in order to determine the compatible links.
- A heuristic method based on genetic algorithm is introduced for selecting the best path in order to provide fairness and balancing.

3 Network model and assumptions

In this paper, we model the network with a directed graph $G = (V, E)$ where $\mathbf{V} = \{v_1, v_2, v_3, \dots, v_n\}$represents n nodes (wireless mesh routers) placed in a given area and $\boldsymbol{E}$ denotes the set of links between the nodes. In order to determine $\boldsymbol{E}$, the interference model should be defined. In [29], two different interference models are introduced including the protocol interference model and physical interference model. The first model is specified with transmission range and interference range parameters in which all nodes in the transmission range of a given node can receive its messages correctly. In this model, if all messages are transmitted on the same frequency channel, the transmission on a given link can lead to interference on all links placed in the interference range of the transmitting node. The sufficient condition for the existence of a link between nodes i and j in the physical interference model can be written as $SINR_{ij} \geq \gamma(\rho)$ in which $\gamma(\rho)$ represents a threshold level dependent on transmission rate and $SINR$ that is defined as,

$$SINR_{ij} = \frac{p_{ij} G_{ij}}{N_0 + \sum\limits_{\forall e_{mn} \in E \setminus e_{ij}} p_{mn} G_{mj}} \tag{1}$$

Here N_0 is the thermal noise power, $p_{ij}$is the transmission power from node i to j, and G_{ij} is the propagation gain, such that $G_{ij} = (1/d_{ij})^{\varepsilon}$, where ε is a parameter dependent on the shadowing and fading phenomena and d_{ij} is the geometric distance between two nodes i and j. Moreover, e_{ij} represents the link between two nodes i and j. The second term of the denominator represents the interference resulting from simultaneous transmission of other links with the transmission from i to j over the same frequency channel. It is clear that for successful reception at receiver node j with rate ρ, the $SINR_{ij}$

must be greater than or equal to $\gamma(\rho)$. In order to calculate the denominator of $SINR_{ij}$, we ignore the effect of links that the receiver j is placed outside the interference range of their transmitters [20].

Here, we assume that node i is equipped with In_iradio interfaces each has an omnidirectional antenna. Each node transmits with one of the available rates $\mathbf{R} = \{\rho_1, \rho_2, ..., \rho_M | \rho_1 < \rho_2 <. ... < \rho_M\}$, and selects its transmission power continuously from $[0, P^{\max}]$. Moreover, one of the available non-overlapping frequency channels $\mathbf{\Omega} = \{\omega_i | 1 \le i \le H\}$is assigned to each radio interface. In order to prevent the intra-node interference, different channels must be assigned to radio interfaces of each node; this means that each node can simultaneously transmit on all of its radio interfaces. In this paper, we assume a schedule-based MAC protocol according to the TDMA with the maximum number of time slots in a time frame is represented by $T_{\max}$. If $\mathbf{Q}$ is the set of node pairs which have the traffic request for transmission, then each traffic flow associated to the source and destination pair $(u, v) \in \mathbf{Q}$ is assumed to be unicast, where TD_{uv} represents the amount of traffic demand.

As mentioned before, robustness against failures requires the network graph to be K-Connected. In [30], Penrose proved that if the minimum degree of a network graph is K, the network is K-Connected with high probability. Therefore, the minimum transmission power of each node i with degree $deg_i$is,

$$P_{\mathrm{i}}^{min} = min\left(P\in[0, P^{\max}] \mid deg_i \ge K\right) \quad ;\forall i\in V \tag{2}$$

In this paper, it is assumed that the network graph is potentially K-Connected in the condition of minimum transmission rate and single frequency channel. Menger's theory says that there should be K paths with distinct vertices between each pair of the nodes to guarantee the K-Connectivity feature of the network [31].

4 Fault-tolerant topology control with throughput maximization, balancing and fairness (FTTC-TMBF)

In this section, a comprehensive cross-layer model for topology control named FTTC-TMBF is introduced. In order to model the problem, some variables are defined as follows:

- $X_{ijk}^{\omega t}$ is a binary variable which is equal to 1 if at least one packet in time slot t and frequency channel ω is transmitted from node i to node j with rate ρ_k.
- P_{ij}^{t} is a real number from $[0, P^{\max}]$ which shows the transmission power of link e_{ij}in time slot t .
- f_{ijt}^{uv}represents the amount of traffic on link e_{ij}that belongs to the traffic session $(u, v) \in \mathbf{Q}$ in time slot t.

Table 1 summarizes all notations that will be used in this section. By defining these variables, problem is formulated as follows.

In FTTC-TMBF, the objectives are throughput maximization, fairness, balancing and robustness against failures; therefore, the objective function is:

$$\textbf{Minimize} \quad \alpha_1 \frac{\sum\limits_{\forall e_{ij}\in E}\sum\limits_{\forall \omega\in\Omega}\sum\limits_{k=1}^{M}\sum\limits_{t=1}^{T_{\max}} X_{ijk}^{\omega t}}{TN_a^{\max}} + \alpha_2\left(\frac{\sigma_{SF}^2}{\sigma_{SF}^{2,\max}}\right) + \alpha_3\left(\frac{\sigma_n^2}{\sigma_n^{2,\max}}\right) + \alpha_4\left(\frac{\sigma_c^2}{\sigma_c^{2,\max}}\right) \tag{3}$$

In the objective function (3), throughput maximization, fairness, nodes utilization balancing and frequency channel utilization balancing are represented by four terms.

A- Throughput maximization - Since the matrix of traffic demands is specified, the throughput maximization is equivalent to minimizing the number of time slots in which the links are transmitting the data and is considered in first term of (3). In this equation, $TN_a^{\max}$is the maximum number of time slots in which the links are transmitting data.

B- Fairness between traffic flows - If only maximizing the total throughput is considered in the objective function, some sessions may receive the maximum service rate while the others get much less. In this paper, to provide fairness between traffic sessions, a satisfaction factor for each traffic session *(u,v)* is defined as (4),

$$SF_{uv} = \frac{C_{uv}}{TD_{uv}} \quad ;\forall(u, v)\in\mathbf{Q} \tag{4}$$

In order to provide fairness, the variance of satisfaction factor should be minimized. In (4), C_{uv} is the average capacity per time slot for the flow from u to v that is defined as (5),

$$C_{uv} = \frac{1}{TN_{uv}}\sum_{t=1}^{T_{\max}}\sum_{\forall e_{ij}\in E}\sum_{\forall\omega\in\Omega}\sum_{k=1}^{M} Z_{ijt}^{uv} \times X_{ijk}^{\omega t} \times \upsilon(\rho_k) \quad ;\forall(u, v)\in\mathbf{Q} \tag{5}$$

when TN_{uv} is the maximum number of time slots in which the flow from u to v is active. In (3), parameter $\sigma_{SF}^{2,\max}$ is the maximum variance of satisfaction factor and σ_{SF}^2 is the variance of satisfaction factor as defined by (6),

$$\sigma_{SF}^2 = \frac{1}{|\mathbf{Q}|}\sum_{\forall(u,v)\in\mathbf{Q}} (SF_{uv} - \bar{SF})^2 \tag{6}$$

where, $|\mathbf{Q}|$ represents the number of traffic demands in the network and $\bar{SF}$ is the average of satisfaction factor

Table 1 List of some notations used in this paper

Notation	Description
In_i	Number of radio interfaces of node i
$\mathbf{R}$	Set of M available transmission rates
p^{max}	Maximum transmission power
$\boldsymbol{\Omega}$	Set of non-overlapping channels
$\mathbf{ch}_i$	Set of channels assigned to node i
G	Directed network graph
V	Set of n mesh routers
E	Set of directed links
e_{ij}	Directed link from node i to node j
G_{ij}	Propagation gain
d_{ij}	Geometric distance between two nodes i and j
$SINR_{ij}$	Signal to Interference and Noise Ratio in the receiver of link e_{ij}
$\gamma(\rho)$	SINR threshold corresponding to the rate ρ
N_0	Thermal noise power
P_i^{min}	Minimum transmission power of node i to satisfy K-degree requirement
T_{max}	Maximum number of time slots in a time frame
P_{ij}	Transmission power from node i to node j
deg_i	Degree of node i
TD_{uv}	The amount of traffic demand from source node u to destination node v
$X_{ijk}^{\omega t}$	A binary variable which is equal to 1 if the transmission from node i to node j is scheduled in time slot t with frequency channel ω and transmission rate ρ_k
P_{ij}^t	Transmission power of link e_{ij} scheduled in time slot t
f_{ijt}^{uv}	The amount of traffic belongs to traffic session $(u, v) \in \mathbf{Q}$ passed on link e_{ij} in time slot t
Z_{ijt}^{uv}	A binary parameter which is equal to 1 if $f_{ijt}^{uv} > 0$
$\upsilon(\rho_k)$	The amount of transmitted traffic in one time slot at rate ρ_k
C_{uv}	End-to-end average throughput of traffic session $(u, v) \in \mathbf{Q}$ in each time slot
SF_{uv}	Satisfaction factor for each traffic session $(u, v) \in \mathbf{Q}$
σ_{SF}^2	Variance of satisfaction factor
$\overline{SF}$	Average satisfaction factor of all sessions
$\sigma_{SF}^{2,\,max}$	Maximum variance of satisfaction factor
U_n^i	Utilization of node i
σ_n^2	Variance of U_n^i
$\overline{U}_n$	Average utilization of all nodes
$\sigma_n^{2,\,max}$	Maximum variance of node utilization
U_c^ω	Utilization of channel ω
σ_c^2	Variance of U_c^ω
$\overline{U}_c$	Average utilization of all channels
$\sigma_c^{2,\,max}$	Maximum variance of channel utilization

between different traffic demands resulting from $\overline{SF} = \sum_{\forall (u,v) \in Q} SF_{uv} / |\mathbf{Q}|$.

C- Channel utilization balancing - When the objective function only includes the fairness and throughput factors, the balancing in frequency channel utilization is severely affected. In other words, congestion may occur in one frequency channel, while the others carry a low volume of traffic. Therefore, minimizing the variance of channel utilization (σ_c^2) must also be considered in the objective function. For this purpose, we first define the amount of channel utilization as follows,

$$U_c^\omega = \sum_{t=1}^{T_{max}} \sum_{\forall e_{ij} \in E} \sum_{k=1}^{M} \upsilon(\rho_k) \times X_{ijk}^{\omega t} \qquad ; \forall \omega \in \boldsymbol{\Omega} \tag{7}$$

The variance of U_c^ω is calculated from (8),

$$\sigma_c^2 = \frac{1}{|\boldsymbol{\Omega}|} \sum_{\forall \omega \in \boldsymbol{\Omega}} \left(U_c^\omega - \overline{U}_c\right)^2 \tag{8}$$

where, $\overline{U}_c = \sum_{\forall \omega \in \boldsymbol{\Omega}} U_c^\omega / |\boldsymbol{\Omega}|$ is the average of frequency channel utilization.

D- Node utilization balancing - Similar to the balancing of frequency channel utilization, minimizing the variance of node utilization (σ_n^2) is also considered in the objective function. The purpose of node utilization balancing is distributing traffic eventually across different nodes, but channel balancing means distributing traffic on different frequency channels. By defining the node utilization as (9), its variance is calculated from (10),

$$U_n^i = \frac{1}{In_i} \sum_{t=1}^{T_{max}} \sum_{\forall \omega \in \boldsymbol{\Omega}} \left(\sum_{\forall e_{ij} \in E} \sum_{k=1}^{M} \upsilon(\rho_k) \times X_{ijk}^{\omega t} + \sum_{\forall e_{qi} \in E} \sum_{k=1}^{M} \upsilon(\rho_k) \times X_{qik}^{\omega t} \right) \qquad ; \forall i \in V \tag{9}$$

$$\sigma_n^2 = \frac{1}{n} \sum_{\forall i \in V} \left(U_n^i - \overline{U}_n\right)^2 \tag{10}$$

where, $\overline{U}_n = \sum_{\forall i \in V} U_n^i / n$ shows the average of node utilization for all nodes in the network. To adjust the importance of each of the four objective factors in (3), we use the coefficients α_i; $i = 1, \ldots, 4$, where $\sum_{i=1,\ldots,4} \alpha_i = 1$. It is important to note that all of these objective factors must be normalized before summation.

The constraints of FTTC-TMBF are shown below in eqs. (11) to (23). These constraints can be categorized in three classes: second layer constraints, third layer constraints and cross-layer constraints. The second layer constraints are provided in (11) to (14). Eq. (11) shows the constraint of SINR for transmission on link e_{ij} in frequency channel ω and time slot t given that some other links are active at the same frequency channel and the same time slot. The lower bound of transmission power on each link is shown in (12). This quantity should be higher than the minimum power required for transmitter node to be K-degree. Moreover, it should be higher than the minimum power required for a successful reception assumed no other link is active with the same frequency channel. The half-duplex property of each radio interface is represented in (13). According to (14), the maximum number of concurrent connections in a node is restricted to the number of its radio interfaces.

$$P_{ij}^{t} G_{ij} \geq \gamma(\rho_k) \times X_{ijk}^{\omega t} \times \left(N_0 + \sum_{\forall e_{mn} \in E \setminus e_{ij}} \sum_{h=1}^{M} P_{mn}^{t} G_{mj} X_{mnh}^{\omega t} \right) \quad ; \forall e_{ij} \in \boldsymbol{E}, \forall t \in \{1, 2, ..., T_{\max}\}, \forall k = 1, 2, ..., M, \forall \omega \in \boldsymbol{\Omega} \tag{11}$$

$$P_{ij}^{t} \geq \max\left(\left(\frac{\gamma(\rho_k) \times N_0 \times X_{ijk}^{\omega t}}{G_{ij}} \right), P_i^{\min} \right) \quad ; \forall e_{ij} \in \boldsymbol{E}, \forall t \in \{1, 2, ..., T_{\max}\}, \forall k = 1, 2, ..., M, \forall \omega \in \boldsymbol{\Omega} \tag{12}$$

$$\sum_{\forall e_{ij} \in \boldsymbol{E}} \sum_{k=1}^{M} X_{ijk}^{\omega t} + \sum_{\forall e_{jq} \in \boldsymbol{E}} \sum_{k=1}^{M} X_{jqk}^{\omega t} \leq 1 \quad ; \forall t \in \{1, 2, ..., T_{\max}\}, \forall j \in \boldsymbol{V}, \forall \omega \in \boldsymbol{\Omega} \tag{13}$$

$$\sum_{\forall \omega \in \boldsymbol{\Omega}} \left(\sum_{\forall e_{ij} \in \boldsymbol{E}} \sum_{k=1}^{M} X_{ijk}^{\omega t} + \sum_{\forall e_{jq} \in \boldsymbol{E}} \sum_{k=1}^{M} X_{jqk}^{\omega t} \right) \leq In_j \quad ; \forall j \in \boldsymbol{V}, \forall t \in \{1, 2, ..., T_{\max}\} \tag{14}$$

For third layer, the constraints of packet delivery in end-to-end traffic flows are considered by (15)–(18). Constraint (15) represents that the total traffic flow exported from the source node (or imported to the destination node) in every session must be equal to the amount of traffic demand at that session. As mentioned before, there should be K disjoint paths between each node pair of set Q to guarantee the K-Connectivity feature of the network. This constraint is showed in (16)–(18), where Z_{ijt}^{uv} is a binary parameter which is equal to 1 if $f_{ijt}^{uv} > 0$. Constraint (16) shows that the output traffic of each source node routes to the K individual paths. Similarly, (17) shows that the input traffic to each destination node is delivered from K distinct paths. Finally, (18) describes that the traffic between u and v passes from a middle node no more than once. These constraints guarantee the distinction of paths of each traffic flow.

$$\sum_{t=1}^{T_{\max}} \left(\sum_{\forall j \in V} f_{ijt}^{uv} - \sum_{\forall q \in V} f_{qit}^{uv} \right) = \begin{cases} TD_{uv} & ; i = u \\ -TD_{uv} & ; i = v \\ 0 & ; otherwise \end{cases} \quad ; \forall i \in V, \forall (u, v) \in \boldsymbol{Q} \tag{15}$$

$$\sum_{\forall j \in \boldsymbol{V}} Z_{ujt}^{uv} \geq K \quad ; \forall (u, v) \in \boldsymbol{Q}, \tag{16}$$

$$\sum_{\forall i \in V} Z_{ivt}^{uv} \geq K \quad ; \forall (u, v) \in \boldsymbol{Q}, \tag{17}$$

$$\sum_{\forall i \in V \setminus u,v} Z_{ijt}^{uv} \leq 1, \sum_{\forall m \in V \setminus u,v} Z_{jmt}^{uv} \leq 1 \quad ; \forall j \in V \setminus u, v, \tag{18}$$

The cross-layer constraints between network and MAC layers are expressed in (19) and (20). Constraint (19) represents the relationship between variables of second and third layers. It implies that the total traffic transmitted on a link cannot be more than the maximum capacity of that link regarding the transmission rate, where $\upsilon(\rho_k)$ represents the amount of transmitted traffic in one time slot at rate ρ_k. Constraint (20) indicates that if some traffic demands pass through a link, it should be active on at least one time slot with the assigned frequency channel and transmission rate. Finally, the variations of decision variables are given in (21)–(23).

$$\sum_{\forall (u,v) \in \boldsymbol{Q}} f_{ijt}^{uv} \leq \sum_{\forall \omega \in \boldsymbol{\Omega}} \sum_{k=1}^{M} \upsilon(\rho_k) \times X_{ijk}^{\omega t} \quad ; \forall e_{ij} \in \boldsymbol{E}, \forall t \in \{1, 2, ..., T_{\max}\} \tag{19}$$

$$Z_{ijt}^{uv} \leq \sum_{\forall \omega \in \boldsymbol{\Omega}} \sum_{k=1}^{M} X_{ijk}^{\omega t} \quad ; \forall e_{ij} \in \boldsymbol{E}, \forall (u, v) \in \boldsymbol{Q}, \forall t \in \{1, 2, ..., T_{\max}\} \tag{20}$$

$$P_{ij}^{t} \leq P^{\max} \quad ; \forall e_{ij} \in \boldsymbol{E}, \forall t \in \{1, 2, ..., T_{\max}\} \tag{21}$$

$$X_{ijk}^{\omega t} \in \{0, 1\}; \ \forall e_{ij} \in \boldsymbol{E}, \forall t \in \{1, 2, ..., T_{\max}\} \quad ; \forall k = 1, 2, ..., M, \forall \omega \in \boldsymbol{\Omega} \tag{22}$$

$$f_{ijt}^{uv} \geq 0 \quad ; \forall t \in \{1, 2, ..., T_{\max}\}, \forall (u, v) \in \boldsymbol{Q}, \forall e_{ij} \in \boldsymbol{E} \tag{23}$$

5 The proposed heuristic solution for FTTC-TMBF problem

As mentioned before, the computational complexity of the proposed topology control problem is high. In order to reduce this complexity, we introduce HFTTC-TMBF which is based on decomposing the problem to four sub-problems including KPBP, GB-LCA, GB-CSF and GB-CLPS. The implementation of HFTTC-TMBF algorithm is shown in Fig. 1. In this section, the implementation of these sub-problems is presented. Table 2 summarizes all notations that will be used in this section.

5.1 First step- K-potential best paths selection (KPBP)

With respect to the Menger theory, if a network graph is K-Connected, then the graph should have K paths with distinct vertices between each pair of nodes. To this end,

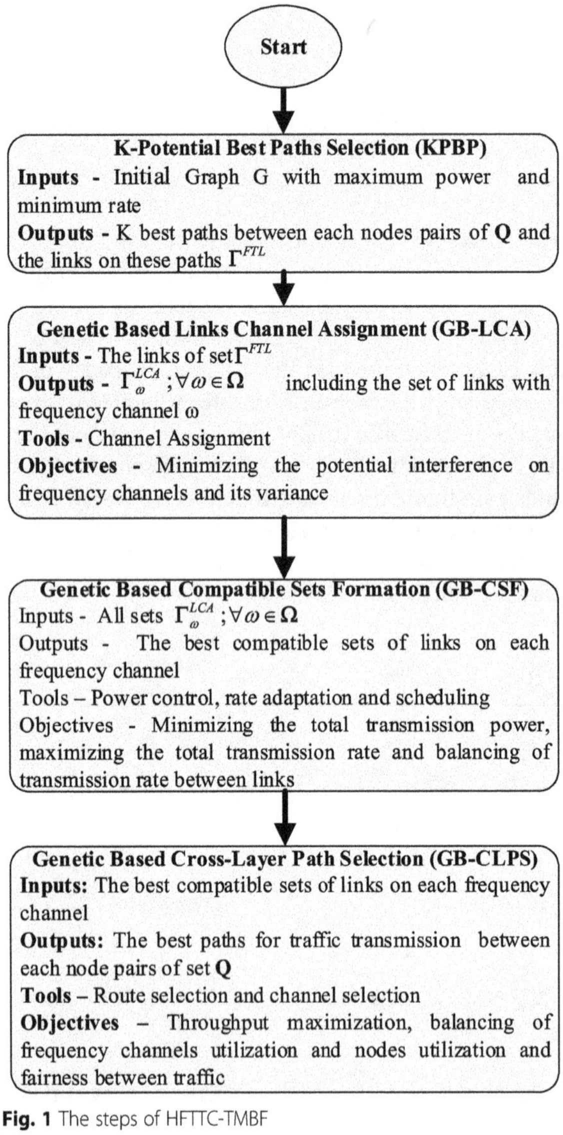

Fig. 1 The steps of HFTTC-TMBF

we have proposed KPBP in [20] which finds the best K disjoint paths between each node pair of set $\boldsymbol{Q}$ such that the traffic delivery is guaranteed in the case of K-1 failures. In brief, the initial graph is formed based on the transmission with maximum power and minimum rate. Next, all of the vertices-disjoint paths between each node pair of set $\boldsymbol{Q}$ are extracted. To this end, the shortest path between the two nodes is obtained and all of the vertices and links of this path are removed. Then the second path is determined and removed from the graph. This procedure is continued until all paths are extracted. Finally using eq. (24) for each path, the best K paths between nodes are obtained. Since in this step, no traffic demand is on the links, the proposed metric has potentially find K disjoint paths. This metric is a combination of the hop counts, power consumption for transmission on the links of the path and amount of usage from each node in different paths. If the number of disjoint paths and indices of nodes on the lth path between two nodes u and v are shown by NoP_{uv} and $\boldsymbol{Path}_{uv,\ l}$; $l = 1, \ldots, NoP_{uv}$, respectively, then proposed cost function for choosing the best K potential paths is formulated in (24),

$$RCF_{uv,l} = \alpha_1 \hat{H}_{uv,l} + \alpha_2 \hat{P}_{uv,l} + \alpha_3 \hat{B}_{uv,l}\ , \alpha_1 + \alpha_2 + \alpha_3 = 1 \qquad (24)$$

$$;\forall u, v \in \mathbf{Q}, l = 1, 2, \ldots, NoP_{uv}$$

The first term of the cost function represents a generic routing measure which lead to the minimization of the number of hop counts. In (24), $\hat{H}_{uv,l}$ is the normalized value of the number of hops on the lth path between nodes u and v as defined in (25),

$$\hat{H}_{uv,l} = \frac{|\boldsymbol{Path}_{uv,l}| - 1}{Max_{\forall 1' = 1,\ldots,NoP_{uv}}\left(H_{uv,l'}\right)} \quad ;\forall u, v \in \mathbf{Q}, l = 1, \ldots, NoP_{uv} \qquad (25)$$

Here, $|\boldsymbol{Path}_{uv,\ l}|$ indicates the dimension of the set $\boldsymbol{Path}_{uv,\ l}$, and the denominator shows the maximum number of hops on different paths between two nodes used to normalize the number of hops.

If only the minimization of hop counts is considered, the paths with long hops are selected for data transmission, which lead to increasing the power consumption of transmitters and consequently more interference. Therefore, power minimization must be considered in defining routing measure. If we show the total potential power consumption of all links on the lth path between u and v with $P^{sum}_{uv,l}$, and the maximum potential power consumption of these links with $P^{max}_{uv,l}$, then $\hat{P}_{uv,l}$ is defined according to (26),

$$\hat{P}_{uv,l} = \frac{1}{2} \times \frac{P^{sum}_{uv,l}}{Max_{\forall l' \in \{1,2,\ldots,NoP_{uv}\}}\left(P^{sum}_{uv,l'}\right)} + \frac{1}{2} \times \frac{P^{max}_{uv,l}}{Max_{\forall l' \in \{1,2,\ldots,NoP_{uv}\}}\left(P^{max}_{uv,l'}\right)} \qquad (26)$$

$$;\forall u, v \in \mathbf{Q}, l = 1, \ldots, NoP_{uv}$$

In order to determine $P^{sum}_{uv,l}$ and $P^{max}_{uv,l}$, the minimum transmission power on each link must be calculated from (1), assuming no other link is active simultaneously and the transmission is performed using the minimum rate. In eq. (26), the denominators of the first and second terms show the maximum value of total power consumption for different available paths and the maximum power consumption of links for different paths, respectively.

In order to balance the node utilization, we prefer to use the paths that their vertices are less used on other paths. In this way, the normalized balance factor $\hat{B}_{uv,l}$ is defined in (27),

Table 2 List of notations used in HFTTC-TMBF

Notation	Description
K-Potential Best Paths Selection (KPBP)	
NoP_{uv}	Number of paths between two nodes u and v
$\mathbf{Path}_{uv,l}$	Indices of nodes on the lth path between two nodes u and v
$\hat{H}_{uv,l}$	Normalized value of the number of hops on the lth path between two nodes u and v
$B_{uv,l}^{n_x}$	Number of usage of node n_x located on the lth path between nodes u and v on other paths
$P_{uv,l}^{sum}$	Total power consumption of links on the lth path between two nodes u and v
$P_{uv,l}^{max}$	Maximum power consumption of links on the lth path between two nodes u and v
$\hat{P}_{uv,l}$	Normalize power factor on the lth path between two nodes u and v
$B_{uv,l}^{sum}$	Total number of the usages of the nodes on the lth path between the two nodes u and v on other paths
$B_{uv,l}^{max}$	Maximum number of the usages of the nodes on the lth path between the two nodes u and v on other paths
$\hat{B}_{uv,l}$	Normalized balance factor on the lth path between the two nodes u and v
$RCF_{uv,\ell}$	Cost function of the lth path between nodes u and v
$\mathbf{\Gamma}^{FTL}$	The set of required links to achieve a K-connected graph
$\lvert\mathbf{\Gamma}^{FTL}\rvert$	Number of members in set $\mathbf{\Gamma}^{FTL}$
Genetic Based Links Channel Assignment (GB-LCA)	
$X_{ij\omega}^{g\tau}$	Gene in GB-LCA algorithm that is 1 if channel ω is assigned to link e_{ij} in generation g and chromosome τ
$\mathbf{C}^{g\tau}$	Chromosome in GB-LCA algorithm
G^g	Generation population
$PI^{g\tau}$	Potential interference on each chromosome $\mathbf{C}^{g\tau}$ belonging to generation g
$\sigma_{PI}^{2,g\tau}$	Variance of potential interference in each chromosome
$PI_{\omega}^{g\tau}$	Potential interference on each chromosome $\mathbf{C}^{g\tau}$belonging to generation g in frequency channel ω
$PI^{g\max}$	Maximum potential interference among all chromosomes of generation g
$\sigma_{PI}^{2,g\ \max}$	Maximum variance of interference among all chromosomes of generation g
$COST(\mathbf{C}^{g\tau})$	Cost function of chromosome $\mathbf{C}^{g\tau}$
$\mathbf{C}^{gfa}$	First parent (Father)
$\mathbf{C}^{gma}$	Second parent (Mother)
$\mathbf{C}_{c}^{g\tau}$	Children chromosome
$\widehat{X}_{ij\omega}^{g\tau_1}$	Children gene
$\mathbf{C}_{m}^{g\tau}$	Mutation chromosome
μ	Mutation rate
n_m	Number of chromosomes resulting from mutation
n_{μ}	Number of mutated genes
n_p	Number of chromosomes in current generation
n_c	Number of children chromosomes
$\mathbf{\Gamma}_{\omega}^{LCA}$	Set of links with frequency channel $\omega \in \mathbf{\Omega}$
Genetic Based Compatible Sets Formation (GB-CSF)	
$P_{ij}^{g\tau}$	Gene in GB-CSF that is the transmission power of each link e_{ij}belonging to $\mathbf{\Gamma}_{\omega}^{LCA}$
$\mathbf{S}^{g\tau}$	Chromosome in GB-CSF algorithm
$\rho_{ij}^{g\tau}$	Transmission rate correspondent to gene $P_{ij}^{g\tau}$
$\sigma_{\rho}^{2,g\tau}$	Transmission rate variance of links in the chromosome τ
$\overline{\rho}^{g\tau}$	Transmission rate average of all links of the chromosome τ
$P^{g\max}$	Maximum value of transmission power of the chromosome τ
$\rho^{g\max}$	Maximum value of transmission rate of the chromosome τ
$\sigma_{\rho}^{2,g\ \max}$	Maximum variance of transmission rate

Table 2 List of notations used in HFTTC-TMBF *(Continued)*

Notation	Description
$COST(\mathbf{S}^{g\tau})$	Cost function of chromosome $\mathbf{S}^{g\tau}$
$\mathbf{S}^{gfa}$	First parent (Father)
$\mathbf{S}^{gma}$	Second parent (Mother)
$\mathbf{S}_c^{g\tau}$	Children chromosome
$\widehat{P}_{ij}^{g\tau_1}$	Children gene
$\mathbf{S}_m^{g\tau}$	Mutation chromosome
$\tilde{P}_{ij}^{g\tau}$	Mutation gene
Genetic Based Cross Layer Path Selection (GB-CLPS)	
$X_{uvk}^{g\tau}$	This gene is 1 if *k*th path is selected for traffic flow of *(u,v)*
$\mathbf{P}^{g\tau}$	Chromosome in GB-CLPS algorithm
LT_{ij}^{τ}	Traffic on link e_{ij} when all paths are selected based on chromosome τ
$X_{uvl'}^{ij}$	A binary parameter that is 1 if link $e_{ij}$is on *l'*th path between nodes *u* and *v*
$TN_{\omega m}^{g\tau}$	Number of time slots when *m*th configuration set is active on frequency channel ωin chromosome τ
$\mathbf{\Gamma}_m^{\omega}$	Set of links belong to *m*th configuration set on frequency channel ω
$\sigma_{SF}^{2,g\tau}$	Variance of satisfaction factor of different traffic flows
SF_{uv}^{τ}	Satisfaction factor of each traffic flow *(u,v)*
$TN_S^{g\ max}$	Maximum number of total time slots of all chromosomes
C_{uv}^{τ}	Transmission rate of the link on the path from *u* to *v* which requires maximum number of time slots for transmission
$path_{uv}^{\tau}$	Set of links on the path selected between nodes *u* and *v* in chromosome τ
$\sigma_n^{2,g\tau}$	Variance of node utilization
$\sigma_c^{2,g\tau}$	Variance of frequency channel utilization
V_a^{τ}	Set of active nodes in chromosome τ
n_a^{τ}	Number of active nodes in chromosome τ
U_i^{τ}	Utilization of node *i* in chromosome τ
$\overline{U}_n^{\tau}$	Average of node utilization in chromosome τ
$\mathbf{\Omega}_a^{\tau}$	Set of active frequency channels in chromosome τ
$\lvert\mathbf{\Omega}_a^{\tau}\rvert$	Number of active frequency channels in chromosome τ
U_{ω}^{τ}	Utilization of frequency channel ω in chromosome τ
$\overline{U}_c^{\tau}$	Average of frequency channel utilization in chromosome τ
$\mathbf{P}^{gfa}$	First parent (Father)
$\mathbf{P}^{gma}$	Second parent (Mother)
$\mathbf{P}_c^{g\tau}$	Children chromosome
$\widehat{X}_{uvk}^{g\tau}$	Children gene
$\mathbf{P}_m^{g\tau}$	Mutation chromosome

$$\hat{B}_{uv,l} = \frac{1}{2} \times \frac{B_{uv,l}^{max}}{Max_{\forall l' \in \{1,2,\dots,NoP_{uv}\}}\left(B_{uv,l'}^{max}\right)} + \frac{1}{2} \times \frac{B_{uv,l}^{sum}}{Max_{\forall l' \in \{1,2,\dots,NoP_{uv}\}}\left(B_{uv,l'}^{sum}\right)} \quad ; \forall u, v \in \mathbf{Q}, l = 1, \dots, NoP_{uv} \tag{27}$$

For node n_x on the *l*th path between two nodes *u* and *v*, if we show the number of its usage on other paths by $B_{uv,l}^{n_x}$, then $B_{uv,l}^{max}$ and $B_{uv,l}^{sum}$ are defined as (28) and (29),

$$B_{uv,l}^{max} = Max_{\forall n_x \in \mathbf{path}_{uv,l}}\left(B_{uv,l}^{n_x}\right) \quad ; \forall u, v \in \mathbf{Q}, l = 1, \dots, NoP_{uv} \tag{28}$$

$$B_{uv,l}^{sum} = \sum_{\forall n_x \in \mathbf{path}_{uv,l}} B_{uv,l}^{n_x} \quad ; \forall u, v \in \mathbf{Q}, l = 1, \dots, NoP_{uv} \tag{29}$$

By applying the cost function (24) and extracting the *K* paths between each pair of nodes, the set of required links are achieved and stored in $\mathbf{\Gamma}^{FTL}$.

5.2 Second step- genetic based links channel assignment (GB-LCA)

In section 5–1, the set of essential links for preserving the K-connectivity feature of the graph is extracted and stored in $\mathbf{\Gamma}^{FTL}$. In order to establish the transmission/reception on each link of the set $\mathbf{\Gamma}^{FTL}$, there should be a common channel between two end nodes of the link. In this section, the genetic algorithm is used for assigning a channel to these links considering the limited number of radio interfaces in each node. In other words, if the end node i belongs to several links from set $\mathbf{\Gamma}^{FTL}$and the number of radio interfaces of i are In_i, the total number of frequency channels assigned to this node cannot be greater than In_i. In the following, required elements for implementing GB-LCA are presented.

5.2.1 Genes, chromosome and population

In GB-LCA, the genes are assumed to be binary and represented by $X^{g\tau}_{ij\omega}$. In each chromosome $\tau$from generation g, a gene takes the value of 1 if channelω is assigned to the link e_{ij}. Chromosome is a set of genes that is defined as$\mathbf{C}^{g\tau} = \{X^{g\tau}_{ij\omega} | \forall e_{ij} \in \mathbf{\Gamma}^{FTL}, \forall \omega \in \mathbf{\Omega}\}$. Each chromosome has $n_v = |\mathbf{\Gamma}^{FTL}| \times |\mathbf{\Omega}|$ genes in which $|\mathbf{\Gamma}^{FTL}|$ and $|\mathbf{\Omega}|$ are dimension of set $\mathbf{\Gamma}^{FTL}$and number of available frequency channels, respectively. Moreover, population is the set of n_p chromosomes that are produced in each generation of the algorithm implementation. The population of generation g is shown by $G^g = \langle \mathbf{C}^{g\tau} | \tau = 1, \ldots, n_p \rangle$.

5.2.2 Cost function of GB-LCA

For each chromosome, the fitness function $F(\mathbf{C}^{g\tau})$ is,

$$F(\mathbf{C}^{g\tau}) = \alpha_1 \frac{PI^{g\tau}}{PI^{g\ \max}} + \alpha_2 \frac{\sigma^{2,g\tau}_{PI}}{\sigma^{2,g\ \max}_{PI}} \qquad ; \forall \tau = 1, 2, \ldots, n_p \tag{30}$$

where, $PI^{g\tau}$is the potential interference on each chromosome $\tau$from generation g, which is defined as,

$$PI^{g\tau} = \sum_{\forall e_{ij} \in \mathbf{\Gamma}^{FTL}} \sum_{\forall e_{mn} \in \mathbf{\Gamma}^{FTL}_{e_{ij}}} \sum_{\forall \omega \in \mathbf{\Omega}} p^{\max} G_{mj} X^{g\tau}_{ij\omega} X^{g\tau}_{mn\omega} \tag{31}$$

This function is the total potential interference on receivers of all links in $\mathbf{\Gamma}^{FTL}$. Moreover, $\sigma^{2,g\tau}_{PI}$ in (30) represents the variance of potential interference on frequency channels in each chromosome. If potential interference on each frequency channel is defined as (32),

$$PI^{g\tau}_{\omega} = \sum_{\forall e_{ij} \in \mathbf{\Gamma}^{FTL}} \sum_{\forall e_{mn} \in \mathbf{\Gamma}^{FTL}_{e_{ij}}} p^{\max} G_{mj} X^{g\tau}_{ij\omega} X^{g\tau}_{mn\omega} \qquad ; \forall \omega \in \mathbf{\Omega}, \tag{32}$$

then the variance of potential interference would be,

$$\sigma^{2,g\tau}_{PI} = \frac{1}{|\mathbf{\Omega}|} \sum_{\forall \omega \in \mathbf{\Omega}} \left(PI^{g\tau}_{\omega} - \overline{PI}^{g\tau}\right)^2 \qquad ; \forall \tau = 1, 2, \ldots, n_p \tag{33}$$

where $\overline{PI}^{g\tau} = \sum_{\forall \omega \in \mathbf{\Omega}} PI^{g\tau}_{\omega} / |\mathbf{\Omega}|$ is the average potential interference on all frequency channels. In (30), the terms are normalized by using appropriate maximum values. In other words, denominators of first and second terms show maximum potential interference and maximum variance of interference between all chromosomes of a generation, which can be obtained using (34) and (35), respectively.

$$PI^{g\ \max} = \max_{\forall \tau' = 1, \ldots, n_p} \left(PI^{g\tau'}\right) \tag{34}$$

$$\sigma^{2,g\ \max}_{PI} = \max_{\forall \tau' = 1, \ldots, n_p} \left(\sigma^{2,g\tau'}_{PI}\right) \tag{35}$$

To adjust the effect of each of the two objectives, we use the coefficients α_i; $i = 1, 2$ where$\sum_{i=1,2} \alpha_i = 1$.

The constraints of GB_LCA are defined in (36) and (37). Equation (36) shows that a channel should be assigned to each link of set. $\forall e_{ij} \in \mathbf{\Gamma}^{FTL}$. Limitation on the number of radio interfaces are given in (37).

$$\sum_{\forall \omega \in \mathbf{\Omega}} X^{g\tau}_{ij\omega} = 1 \qquad ; \forall e_{ij} \in \mathbf{\Gamma}^{FTL}, \forall \tau = 1, \ldots, n_p \tag{36}$$

$$\sum_{\forall e_{ij} \in \mathbf{\Gamma}^{FTL}} \sum_{\forall \omega \in \mathbf{\Omega}} X^{g\tau}_{ij\omega} + \sum_{\forall e_{qi} \in \mathbf{\Gamma}^{FTL}} \sum_{\forall \omega \in \mathbf{\Omega}} X^{g\tau}_{qi\omega} \le In_i \qquad ; \forall i \in V, \forall \tau = 1, 2, \ldots, n_p \tag{37}$$

According to the definition of objective function (30) and constraints (36), (37), the cost function is defined as (38),

$$COST(\mathbf{C}^{g\tau}) = \begin{cases} F(\mathbf{C}^{g\tau}) & ; \text{if constraints (36) and (37) are met} \\ \infty & ; \text{otherwise} \end{cases} \tag{38}$$

If the constraints (36) and (37) are established for each chromosome, $COST(\mathbf{C}^{g\tau})$becomes equal to$F(\mathbf{C}^{g\tau})$; otherwise the cost of the assumed chromosome becomes infinity. After introducing the concept of gene, population and generation, now the implementation steps of GB-LCA are described below. These steps include the production of initial population, production and selection of next generation population, and the termination condition.

5.2.3 Production of the initial population

In order to produce the first-generation population G^1, Genes of each chromosome should be initialized with respect to constraints (36) and (37). For this purpose, a common channel is assigned to one of the radio interfaces of each node. We assign channels randomly to the remaining radio interfaces such that different radios on each node have different channels. By assigning frequency channels to the end nodes of each link of set $\mathbf{\Gamma}^{FTL}$, the genes of all chromosomes belonging to the first generation are initialized.

5.2.4 Next generation population

The potential population is produced by combining the chromosomes of the current generation, the children chromosomes, and the mutated chromosomes.

5.2.4.1 Children chromosomes An important part of the genetic algorithm is to create a new solution called children. In order to produce the children chromosomes, the chromosomes of the parents are extracted through tournament selection mechanism. In this method, a set of n_{fa} chromosomes from the current generation are chosen and the best one is selected as the first parent [32]. This parent is called father and denoted by $\mathbf{C}^{gfa}$. The second parent ($\mathbf{C}^{gma}$) is called mother and is obtained similarly. If we show the two chromosomes of ensuing children with $\mathbf{C}_c^{g\tau_1}$ and $\mathbf{C}_c^{g\tau_2}$, then each gene of these chromosomes is produced using crossover operator according to (39),

$$\begin{aligned} \widehat{X}_{ij\omega}^{g\tau_1} &= (1-\vartheta_{ij})X_{ij\omega}^{gfa} + \vartheta_{ij}X_{ij\omega}^{gma} \\ \widehat{X}_{ij\omega}^{g\tau_2} &= \vartheta_{ij}X_{ij\omega}^{gfa} + (1-\vartheta_{ij})X_{ij\omega}^{gma} \end{aligned} \quad ;\forall e_{ij}\in\mathbf{\Gamma}^{FTL}, \forall\omega\in\mathbf{\Omega} \tag{39}$$

The coefficient ϑ_{ij} is randomly selected as zero or one. In addition, $X_{ij\omega}^{gfa}$ and $X_{ij\omega}^{gma}$ are genes of the selected parents $\mathbf{C}^{gfa}$and $\mathbf{C}^{gma}$, respectively. This equation describes that the channel assigned to link e_{ij}(gene $\widehat{X}_{ij\omega}^{g\tau_1}$) is selected randomly among channels assigned to this link in parent chromosomes. Each time the crossover operator is run, two chromosomes are produced. In this problem, the number of children chromosomes is considered equal to n_c.

5.2.4.2 Chromosomes resulting from mutation Random variations of the chromosomes can decrease the time for achieving optimal solution. The mutation operator implements these random variations in GB-LCA. The resulting chromosome from the mutation is shown by $\mathbf{C}_m^{g\tau}$. If the number of given chromosomes for mutation is denoted by n_m, these chromosomes are selected randomly among the chromosomes of the current generation.

In order to implement the mutation operator in each selected chromosome, the genes are first categorized based on the corresponding links. The set of genes related to link $e_{ij}$are denoted as $X_{ij\omega}^{g\tau}$ $;\forall\omega\in\mathbf{\Omega}$. Therefore, the number of sets of genes is equal to $|\mathbf{\Gamma}^{FTL}|$. If we denote the required mutation rate with μ, $n_\mu = \lceil \mu \times |\mathbf{\Gamma}^{FTL}| \rceil$ sets must be selected randomly. Then in each set and for each link e_{ij}, one of the genes $X_{ij\omega}^{g\tau}$ $;\forall\omega\in\mathbf{\Omega}$ is randomly set to 1 such that constraint (37) is satisfied.

5.2.4.3 Choosing the population of next generation The next generation population includes three sets of current generation, the children chromosomes and the mutated chromosomes. Figure 2 shows how population of next generation is selected. If n_p, n_c and n_m represent the number of current generation, children and mutants, respectively, then the number of potential members of the next generation population is equal to$n_p + n_c + n_m$. The number of chromosomes in each generation is equal to n_p; therefore, some parts of the potential population should be removed. To select the population of next generation, the described three sets of

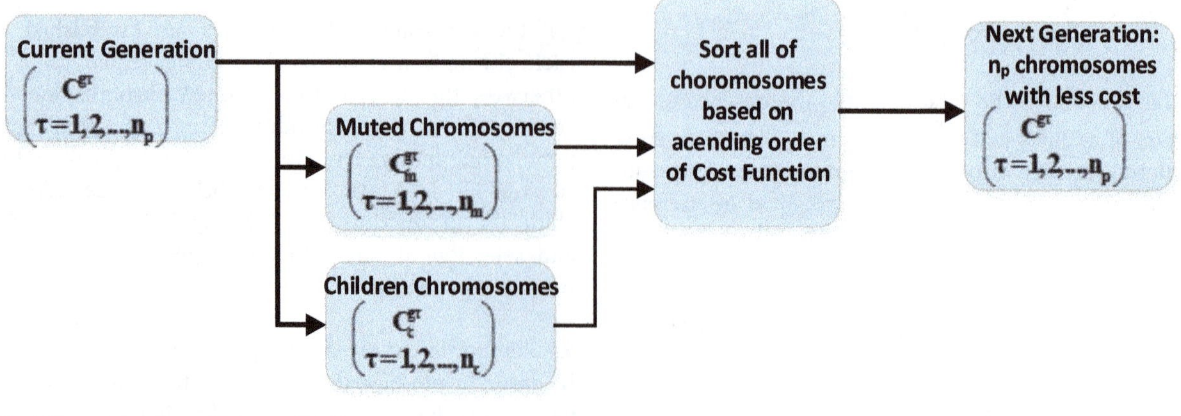

Fig. 2 Next generation selection in GB-LCA

chromosomes are merged, and then n_p members with less cost are selected according to the cost function (38).

5.2.5 Termination condition

We propose the following condition to stop execution,

$$\left|\left(\min_{\forall\tau=1,\dots,n_p}\left(COST\left(\mathbf{C}_p^{g\tau}\right)\right)-\min_{\forall\tau=1,\dots,n_p}\left(COST\left(\mathbf{C}_p^{(g-1)\tau}\right)\right)\right)\right|\leq\lambda \tag{40}$$

According to this equation, if the absolute value of the difference between minimum costs of the current and previous generations is less than λ, the algorithm is stopped. Then, the chromosomes with lower costs are chosen as the solution of the channel assignment problem. Thus, by executing the GB-LCA algorithm, frequency channels are assigned to the links of set $\mathbf{\Gamma}^{FTL}$. This process can minimize the potential interference on each frequency channel, while at the same time satisfies the constraints (36) and (37).

5.3 Third step- genetic based compatible sets formation (GB-CSF)

In this section, we introduce the GB-CSF method for extracting all compatible sets. In each compatible set, the links can be active simultaneously without having interference on each other, such that minimizing the total transmission power and maximizing the total transmission rate is guaranteed. Moreover, the balancing of transmission rate between links is considered. GB-CSF algorithm is only implemented on the links with similar frequency channels; because other links with different frequency channels can be activated simultaneously without interfering with each other. If the links having frequency channel $\omega\in\mathbf{\Omega}$ are represented with $\mathbf{\Gamma}_\omega^{LCA}$, GB-CSF is executed on each set of $\mathbf{\Gamma}_\omega^{LCA}$ iteratively. In each execution on $\mathbf{\Gamma}_\omega^{LCA}$, a set of links with lower cost are determined. Then these links are eliminated from $\mathbf{\Gamma}_\omega^{LCA}$ and the algorithm is executed again on the remained links. This procedure continues until no other link is remained in $\mathbf{\Gamma}_\omega^{LCA}$. In the following, the execution procedure of GB-CSF for a specific set $\mathbf{\Gamma}_\omega^{LCA}$ is described. For other frequency channels, the same procedure is repeated.

5.3.1 Genes, chromosome and population

In this problem, genes are represented with $P_{ij}^{g\tau}$ which shows the transmission power of each link e_{ij} belonging to $\mathbf{\Gamma}_\omega^{LCA}$. Therefore, gene is a continues variable that can take a value in the interval $[0, P^{\max}]$. As mentioned before, a chromosome is a set of n_v genes which is defined as $S^{g\tau}=\{P_{ij}^{g\tau}|\forall e_{ij}\in\mathbf{\Gamma}_\omega^{LCA}\}$. The number of members of each chromosome is equal to $n_v=|\mathbf{\Gamma}_\omega^{LCA}|$ where $|\mathbf{\Gamma}_\omega^{LCA}|$ is the dimension of set $\mathbf{\Gamma}_\omega^{LCA}$. Moreover, the set of n_p chromosomes is produced in each generation g of the algorithm implementation. This set is shown by $G^g=\langle S^{g\tau}|\tau=1,\dots,n_p\rangle$.

5.3.2 Cost function of GB-CSF

In this problem, the fitness function $F(\mathbf{S}^{g\tau})$ is defined as (41),

$$F(\mathbf{S}^{g\tau})=\alpha_1\left(\sum_{\forall e_{ij}\in\mathbf{\Gamma}_\omega^{LCA}}\frac{P_{ij}^{g\tau}}{P^{g\max}}\right)^{-1}+\alpha_2\sum_{\forall e_{ij}\in\mathbf{\Gamma}_\omega^{LCA}}\frac{\rho_{ij}^{g\tau}}{\rho^{g\max}}+\alpha_3\left(\frac{\sigma_\rho^{2,g\tau}}{\sigma_\rho^{2,g\max}}\right)^{-1}$$
$$;\forall\tau=1,2,\dots,n_p \tag{41}$$

In this equation, the first to third terms are related to the total transmission power, the total transmission rate on the links, and the transmission rate balancing in a chromosome, respectively. For every gene of a chromosome, the corresponding transmission rate is determined using (42),

$$\rho_{ij}^{g\tau}=\max_{\forall k=1,\dots,M}\left(\rho_k\;\middle|\;SINR_{ij}^{g\tau}=\frac{P_{ij}^{g\tau}G_{ij}}{\left(N_0+\sum_{\forall e_{mn}\in\mathbf{\Gamma}_\omega^{LCA}\backslash e_{ij}}P_{mn}^{g\tau}G_{mj}\right)}\geq\gamma(\rho_k)\right)$$
$$;\forall e_{ij}\in\mathbf{\Gamma}_\omega^{LCA} \tag{42}$$

which is the maximum rate of each link e_{ij} to satisfy the SINR constraint. If this constraint is not satisfied for minimum transmission rate ρ_1, the given link cannot be activated simultaneously along with other links. If the only objective is to maximize the transmission rate, some links might have the maximum transmission rate, while the others receive the minimum rate. For providing the balancing, variance of transmission rate should be as small as possible. In (41), $\sigma_\rho^{2,g\tau}$ is the transmission rate variance of links in the chromosome $S^{g\tau}$ of generation g defined as,

$$\sigma_\rho^{2,g\tau}=\frac{1}{|\mathbf{\Gamma}_\omega^{LCA}|}\sum_{\forall e_{ij}\in\mathbf{\Gamma}_\omega^{LCA}}\left(\rho_{ij}^{g\tau}-\overline{\rho}^{g\tau}\right)^2 \tag{43}$$

in which, $\overline{\rho}^{g\tau}=\sum_{\forall e_{ij}\in\mathbf{\Gamma}_\omega^{LCA}}\rho_{ij}^{g\tau}/|\mathbf{\Gamma}_\omega^{LCA}|$ is the average transmission rate for all links of the chromosome τ. All terms of (41) are normalized and then linearly combined using appropriate coefficients α_i ; $i=1,\dots,3$ such that $\sum_{i=1,\dots,3}\alpha_i=1$. The values $P^{g\max}$, $\rho^{g\max}$ and $\sigma_\rho^{2,g\max}$ represent the maximum values of transmission power, transmission rate and variance of transmission rate, respectively. These values are obtained using (44) and (45), respectively,

$$\begin{aligned} P^{g\ \max} &= \max_{\forall \tau'=1,\ldots,n_p}\left(\max_{\forall e_{ij}\in \Gamma_{\omega}^{LCA}}\left(P_{ij}^{g\tau'}\right)\right) \\ \rho^{g\ \max} &= \max_{\forall \tau'=1,\ldots,n_p}\left(\max_{\forall e_{ij}\in \Gamma_{\omega}^{LCA}}\left(\rho_{ij}^{g\tau'}\right)\right) \end{aligned} \tag{44}$$

$$\sigma_{\rho}^{2,g\ \max} = \max_{\forall \tau'=1,\ldots,n_p}\left(\sigma_{\rho}^{2,g\tau'}\right) \tag{45}$$

According to the definition of $F(\mathbf{S}^{g\tau})$, the cost function is defined in (46),

$$COST(\mathbf{S}^{g\tau}) = \frac{1}{F(\mathbf{S}^{g\tau})} \tag{46}$$

The implementation steps of GB-CSF include the production of initial population, production and selection of next generation population, and the termination condition.

5.3.3 Production of the initial population

In order to initialize the first generation, the transmission power of genes that are available in all chromosomes of the first generation should be determined. This power is selected randomly in the range $[P_{ij}^{g\tau,\ \min}, P^{\max}]$, where $P_{ij}^{g\tau,\ \min}$ represents the minimum power required for transmission with the minimum possible rate on link e_{ij} assumed no other link is active. This value is obtained as (47),

$$P_{ij}^{g\tau,\ \min} = \frac{N_0 \times \gamma(\rho_1)}{G_{ij}} \qquad ;\forall e_{ij}\in \Gamma_{\omega}^{LCA}, \forall \tau = 1,\ldots,n_p \tag{47}$$

5.3.4 Next generation population

As mentioned before, the potential population is produced by combining the chromosomes of the current generation, the children chromosomes, and the mutated chromosomes. In the following, production of children chromosomes and mutated chromosomes are presented which are represented by $\mathbf{S}_c^{g\tau}$ and $\mathbf{S}_m^{g\tau}$, respectively.

5.3.4.1 Children chromosomes In order to produce the children chromosomes, parent chromosomes $\mathbf{S}^{gfa}$ and $\mathbf{S}^{gma}$should be selected first. To have a better search in the solution space, the parent chromosomes must be selected randomly [33]. For this purpose, we use a roulette wheel mechanism based on the normalized fitness function,

$$P(\mathbf{S}^{g\tau}) = \frac{F(\mathbf{S}^{g\tau})}{\sum_{\tau=1}^{n_p} F(\mathbf{S}^{g\tau})} \qquad ;\forall \tau = 1,2,\ldots,n_p \tag{48}$$

In this method, a line with unit length is considered, and each segment of this line is assigned to each current chromosome according to the value of $P(\mathbf{S}^{g\tau})$. In order to select each parent chromosome, a random number with uniform distribution is produced in the range [0, 1]. Then the chromosome corresponding to the generated random number is selected as the parent chromosome. For instance, Fig. 3 shows how parent chromosome is selected. In this figure, the values of $P(\mathbf{S}^{g\tau})$ for chromosomes 1 to 10 is considered as 0.19, 0.16, 0.13, 0.12, 0.10, 0.09, 0.07, 0.06, 0.05, and 0.03, respectively. Based on these values, the length occupied by each chromosome is determined on the line. For selecting a chromosome as a parent, the generated random number should be located in the region that is occupied by the corresponding chromosome. It is observed that in this method, a chromosome with higher fitness is more probable to be selected as a parent chromosome. Moreover, the chromosomes with negligible probability are still possible to be selected; therefore, the variety of solutions is preserved.

If we show the two chromosomes of the ensuing children with $\mathbf{S}_c^{g\tau_1}$ and $\mathbf{S}_c^{g\tau_2}$; then, each gene of these chromosomes is calculated according to (49),

$$\begin{aligned} \widehat{P}_{ij}^{g\tau_1} &= (1-\eta)P_{ij}^{gfa} + \eta P_{ij}^{gma} \\ \widehat{P}_{ij}^{g\tau_2} &= \eta P_{ij}^{gma} + (1-\eta)P_{ij}^{gfa} \end{aligned}. \tag{49}$$

in which P_{ij}^{gfa} and P_{ij}^{gma} are the genes of the parents. According to (49), if $\eta$is randomly selected in the range [0, 1], the children gens (transmission power levels) can never be outside the range of their parents. Therefore, we define η in the interval $-\delta \le \eta \le 1 + \delta$ in which δ is selected randomly using the uniform distribution in the interval [0, 1].

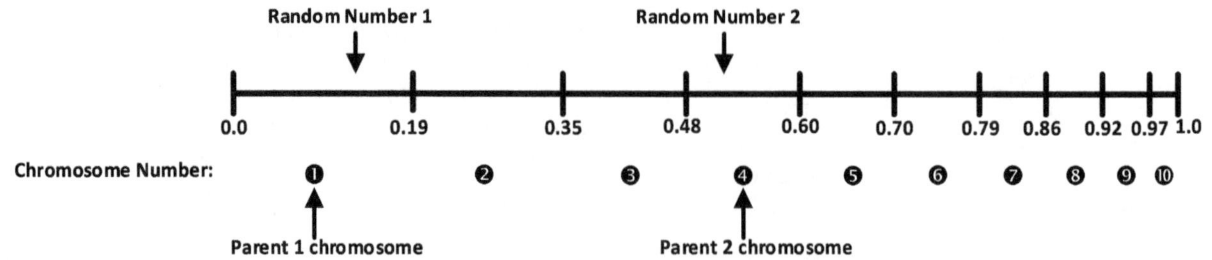

Fig. 3 Selection of parent chromosomes in GB-CSF

5.3.4.2 Chromosomes resulting from mutation If the resulting chromosome from the mutation is shown by $\mathbf{S}_m^{g\tau}$, the number of mutated genes in this chromosome is determined based on the mutation rate μ which reflects the percentage of variation in the genes. In order to preserve the historical memory of the algorithm, the new power levels (mutated genes) must be produced in a way that they are placed in the neighborhood of the selected chromosomes for mutation. Therefore, we use the normal distribution for the mutation operator. There are a lot of references such as [34–36] in which the normal distribution is used for mutation operator. Without loss of generality, we assume that the gene $P_{ij}^{g\tau}$ from chromosome $\mathbf{S}^{g\tau}$is randomly selected for mutation. The mutated gene is defined according to (50),

$$\tilde{P}_{ij}^{g\tau} = P_{ij}^{g\tau} + \varepsilon N(0,1) \tag{50}$$

in which $N(0,1)$ is the normal distribution function with zero mean and unit variance. At the beginning, ε is considered in the range$[0, P^{\max}]$. After applying the mutation operator, this parameter is modified according to (51),

$$\varepsilon_{new} = \begin{cases} \min(P^{\max}, \varepsilon_{old} + \Delta\varepsilon) & ;\textit{if successful percent of mutation} \geq \wp \\ \max(0, \varepsilon_{old} - \Delta\varepsilon) & ;\textit{otherwise} \end{cases} \tag{51}$$

The successful percentage of mutation is the number of mutant chromosomes that their cost is reduced compared to the original chromosome. In the above equation,$\wp$ is an arbitrary threshold.

5.3.4.3 Choosing the population of next generation Similar to the proposed method in 5–2-4 for GB-LCA, at first all chromosomes of current population, children chromosomes and mutation chromosomes must be merged and then $n_p$members with lower cost are selected as population of the next generation. Therefore, in this method, the best chromosomes of each generation are transferred to the next generation and the historical memory of the algorithm is preserved.

5.3.5 Termination condition

In order to stop the algorithm, the absolute value of the difference between minimum costs of the current and previous generations should not be higher than the threshold. In other words, the termination condition is similar to (40) in which the name of current generation chromosomes and previous chromosomes are replaced with $\mathbf{S}^{g\tau}$and $\mathbf{S}^{(g-1)\tau}$, respectively. Here, the chromosomes with lower costs are chosen among the current generation chromosomes. Then, the links of this chromosome are eliminated from the set Γ_ω^{LCA} and the procedure of extracting the compatible sets continues until set $\Gamma_\omega^{LCA}$is emptied. The algorithm is repeated for other sets.

5.4 Fourth step- genetic based cross-layer path selection (GB-CLPS)

In this section, we introduce a Genetic Based Cross-Layer Path Selection (GB-CLPS) algorithm that uses path and channel selection tools to fulfill the goals such as balancing and fairness. As explained earlier in sections 5–1 to 5–3, first, the K best potential paths between node pairs were extracted in step 1 using KPBP; then in step 2, the best channels were assigned to links of these paths using GB-LCA. In step 3, the best compatible sets of links on each frequency channel were obtained using GB-CSF. GB-CLPS algorithm selects the best path for transmission among K selected paths between each node pair such that not only throughput is maximized, but also balancing of frequency channels utilization and nodes utilization and fairness between traffic flows are provided. For this purpose, path and channel selection tools based on genetic algorithm are used.

5.4.1 Genes, chromosome and population

A gene is a binary value denoted by $X_{uvk}^{g\tau}$ that is equal to 1 if the kth path is selected for traffic flow (u, v). A chromosome is the set of genes that can be defined as $\mathbf{P}^{g\tau} = \{X_{uvk}^{g\tau} | (u, v) \in \mathbf{Q}, k = 1, 2, ..., K\}$, where the number of genes in the chromosome is equal to $n_v = |\mathbf{Q}| \times K$. Population is the set of n_p chromosomes of the problem.

5.4.2 Cost function of GB-CLPS

By determining each chromosome, paths between each node pair u and v belong to set $\mathbf{Q}$ are determined. Traffic on each link $e_{ij}$of the network is calculated as (52),

$$LT_{ij}^{\tau} = \sum_{\forall(u,v)\in\mathbf{Q}} \sum_{l'=1}^{K} TD_{uv} X_{uvl'}^{g\tau} \chi_{uvl'}^{ij} \quad ;\forall\tau = 1, 2, ..., n_p \tag{52}$$

where, $\chi_{uvl'}^{ij}$is a binary parameter that is equal to 1 if link e_{ij} is on the l'th path between nodes u and v. For each chromosome, the cost function $COST(\mathbf{P}^{g\tau})$ is defined with (53),

$$COST(\mathbf{P}^{g\tau}) = \alpha_1 \frac{\sum_{\forall\omega\in\Omega} \sum_{m=1}^{NoS_\omega} TN_{\omega m}^{g\tau}}{TN_s^{g\ \max}} + \alpha_2 \frac{\sigma_{SF}^{2,g\tau}}{\sigma_{SF}^{2,g\ \max}} + \alpha_3 \frac{\sigma_n^{2,g\tau}}{\sigma_n^{2,g\ \max}} + \alpha_4 \frac{\sigma_c^{2,g\tau}}{\sigma_c^{2,g\ \max}} \quad ;\forall\tau = 1, 2, ..., n_p \tag{53}$$

In above equation, throughput maximization, fairness and balancing of nodes utilization and channels utilization are considered by four terms. To adjust the effect of each objective, coefficients α_i ; $i = 1, .., 4$ are used, where $\sum_{i=1,..,4}\alpha_i = 1$. The first term of the cost function shows the total time slots required for traffic transmission. Since the transmission rate of each link is obtained using GB-CSF, the maximum throughput is achieved by minimizing the number of time slots required for traffic transmission. In (53), NoS_ω is the number of available configuration sets for frequency channel ω, which is obtained from GB-CSF. In addition, $TN^{g\tau}_{\omega m}$ that is the maximum number of used time slots when the mth configuration set is active on frequency channel ω, is obtained using (54),

$$TN^{g\tau}_{\omega m} = \max_{\forall e_{ij}\in\Gamma^{\omega}_{m}}\left(\frac{LT^{\tau}_{ij}}{\rho_{ij}\times T}\right) \quad ;\forall m = 1,\ldots,NoS_\omega, \forall\omega\in\mathbf{\Omega} \tag{54}$$

where $\mathbf{\Gamma}^{\omega}_{m}$ is the set of links of mth configuration set on frequency channel ω. In addition, ρ_{ij} is the transmission rate of link e_{ij} that is obtained from GB-CSF, and T represents the length of each time slot. In order to calculate $TN^{g\tau}_{\omega m}$, the maximum number of time slots required for transmitting traffic flows on each link of a configuration set should be obtained.

For fairness, the variance of satisfaction factors of different traffic flows ($\sigma^{2,g\tau}_{SF}$) defined in (55) should be minimized,

$$\sigma^{2,g\tau}_{SF} = \frac{1}{|\mathbf{Q}|}\sum_{\forall(u,v)\in\mathbf{Q}}\left(SF^{\tau}_{uv} - \overline{SF^{\tau}}\right)^2 \quad ;\forall\tau = 1,2,\ldots,n_p \tag{55}$$

In this equation, $SF^{\tau}_{uv} = C^{\tau}_{uv}/TD_{uv}$ is the satisfaction factor of service for each traffic flow *(u,v)* where C^{τ}_{uv} is the transmission rate of the link on the path that requires the maximum number of time slots for transmission. The amount of C^{τ}_{uv} is calculated from (56),

$$C^{\tau}_{uv} = \arg\max_{\rho_{ij}|\forall e_{ij}\in path^{\tau}_{uv}}\left(\frac{LT^{\tau}_{ij}}{\rho_{ij}\times T}\right) \quad ;\forall\tau = 1,2,\ldots,n_p, \forall(u,v)\in\mathbf{Q} \tag{56}$$

in which $path^{\tau}_{uv}$ is the set of links on the selected path between nodes u and v in chromosome τ. In addition, in (55), $\overline{SF^{\tau}} = \sum_{\forall(u,v)\in\mathbf{Q}} SF^{\tau}_{uv}/|\mathbf{Q}|$ is the average of satisfaction factors of all traffic flows.

In order to provide balancing, variances of node utilization ($\sigma^{2,g\tau}_{n}$) and frequency channel utilization ($\sigma^{2,g\tau}_{c}$) should be minimized. $\sigma^{2,g\tau}_{n}$ is defined as follows,

$$\sigma^{2,g\tau}_{n} = \frac{1}{n^{\tau}_{a}}\sum_{\forall i\in V^{\tau}_{a}}\left(U^{\tau}_{i} - \overline{U}^{\tau}_{n}\right)^2 \quad ;\forall\tau = 1,2,\ldots,n_p \tag{57}$$

where, V^{τ}_{a} and n^{τ}_{a} are the sets of active nodes and the number of active nodes in chromosome τth. An active node is a node with non-zero traffic. U^{τ}_{i} and $\overline{U}^{\tau}_{n}$ are the amount of node i utilization and average of node utilization in chromosome τ which are defined in (58) and (59),

$$U^{\tau}_{i} = \frac{1}{In_i}\left(\sum_{\forall\omega\in\mathbf{\Omega}}\sum_{m=1}^{NOS_\omega}\left(\sum_{\forall e_{ij}\in\mathbf{S}^{\omega}_{m}} LT^{\tau}_{ij} + \sum_{\forall e_{qi}\in\mathbf{S}^{\omega}_{m}} LT^{\tau}_{qi}\right)\right) \quad ;\forall\tau = 1,2,\ldots,n_p, \forall i\in V^{\tau}_{a} \tag{58}$$

$$\overline{U}^{\tau}_{n} = \frac{1}{n^{\tau}_{a}}\sum_{\forall i\in V^{\tau}_{a}} U^{\tau}_{i} \quad ;\forall\tau = 1,2,\ldots,n_p \tag{59}$$

Similarly, the variance of channel utilization ($\sigma^{2,g\tau}_{c}$) is obtained using (60),

$$\sigma^{2,g\tau}_{c} = \frac{1}{|\mathbf{\Omega}^{\tau}_{a}|}\sum_{\forall\omega\in\mathbf{\Omega}^{\tau}_{a}}\left(U^{\tau}_{\omega} - \overline{U}^{\tau}_{c}\right)^2 \quad ;\forall\tau = 1,2,\ldots,n_p \tag{60}$$

where $\mathbf{\Omega}^{\tau}_{a}$ and $|\mathbf{\Omega}^{\tau}_{a}|$ represent the set of active frequency channels and the number of active frequency channels in chromosome τ. An active channel is a channel in which its traffic load is not zero. In addition, the utilization and average utilization of each frequency channel ω are defined in (61) and (62),

$$U^{\tau}_{\omega} = \sum_{m=1}^{NOS_\omega}\left(\sum_{\forall e_{ij}\in\Gamma^{\omega}_{m}} LT^{\tau}_{ij}\right) \quad ;\forall\omega\in\mathbf{\Omega}, \forall\tau = 1,2,\ldots,n_p \tag{61}$$

$$\overline{U}^{\tau}_{c} = \frac{1}{|\mathbf{\Omega}^{\tau}_{a}|}\sum_{\forall\omega\in\mathbf{\Omega}^{\tau}_{a}} U^{\tau}_{\omega} \quad ;\forall\tau = 1,2,\ldots,n_p \tag{62}$$

In eq. (53), $TN^{g\ \max}_{s}$, $\sigma^{2,g\ \max}_{SF}$, $\sigma^{2,g\ \max}_{n}$ and $\sigma^{2,g\ \max}_{c}$ represent the maximum values of total numbers of time slots, variance of satisfaction factor, variance of node utilization and variance of channel utilization between all chromosomes, respectively. These values can be obtained using (63) to (66), respectively,

$$TN^{g\ \max}_{s} = \max_{\tau'=1,2,\ldots,n_p}\left(\sum_{\forall\omega\in\mathbf{\Omega}}\sum_{m=1}^{NoS_\omega} TN^{g\tau'}_{\omega m}\right) \tag{63}$$

$$\sigma^{2,g\ \max}_{SF} = \max_{\forall\tau'=1,\ldots,n_p}\left(\sigma^{2,g\tau'}_{SF}\right) \tag{64}$$

$$\sigma^{2,g\ \max}_{n} = \max_{\forall\tau'=1,\ldots,n_p}\left(\sigma^{2,g\tau'}_{n}\right) \tag{65}$$

$$\sigma^{2,g\ \max}_{c} = \max_{\forall\tau'=1,\ldots,n_p}\left(\sigma^{2,g\tau'}_{c}\right) \tag{66}$$

5.4.3 Implementation of GB-CLPS

GB-CLPS is based on the implementation of a binary genetic algorithm, and includes the following implementation steps: production of the initial population, production and selection of next generation population, and the termination condition. In order to produce the first-generation population, we randomly assign 0 or 1 to each gene in a chromosome. In each chromosome, one path between $\forall(u, v) \in \boldsymbol{Q}$ should be selected. The potential population of the next generation is formed from the population of chromosomes in the current generation i.e. $\mathbf{P}^{g\tau}$; $\tau = 1, \ldots, n_p$, mutation chromosomes and children chromosomes.

In order to produce the children chromosomes, the chromosomes of the parents are extracted through tournament selection mechanism. In this method, n_{fa} chromosomes are selected from the current generation for the first parent (father). These chromosomes are selected among current generation chromosomes according to the cost function. The second parent ($\mathbf{P}^{gma}$) is obtained similarly. Each of the genes in the two children chromosomes obtained from the crossover operator by using eq. (67),

$$\begin{aligned}\widehat{X}_{uvk}^{g\tau_1} &= (1-\vartheta_{uv})X_{uvk}^{gfa} + \vartheta_{uv}X_{uvk}^{gma} \qquad ;\forall(u,v)\in\boldsymbol{Q}, \forall k = 1, \ldots, K\\ \widehat{X}_{uvk}^{g\tau_2} &= \vartheta_{uv}X_{uvk}^{gfa} + (1-\vartheta_{uv})X_{uvk}^{gma}\end{aligned} \tag{67}$$

where the coefficients ϑ_{uv} are randomly selected as zero or one. Therefore, each time the crossover operator is run, two chromosomes ($\mathbf{P}_c^{g\tau_1}$ and $\mathbf{P}_c^{g\tau_2}$) are produced. In this problem, the number of children chromosomes is considered equal to n_c. As mentioned before, random variations of chromosomes are performed by using mutation operator. A chromosome produced by mutation is shown by$\mathbf{P}_m^{g\tau}$. The number of chromosomes selected for mutation is denoted by n_m. In order to implement the mutation operator, we define the set of gens corresponding to each traffic flow (u, v)as $X_{uvk}^{g\tau}$ $;\forall k = 1, \ldots, K$. If the required mutation rate is equal to μ, $n_\mu = \lceil \mu \times |\boldsymbol{Q}| \rceil$set of gens is randomly selected at first. Then in each selected set and for every traffic flow(u, v), one of the genes $X_{uvk}^{g\tau}$ $;\forall k = 1, \ldots, K$is randomly set to 1.

In order to select the next generation population, the three sets of chromosomes including current population, children chromosomes and mutation chromosomes are combined and n_p members are selected according to their costs. The termination condition is similar to the (40) where the chromosomes are replaced by $\mathbf{P}^{g\tau}$, and the cost function is similar to (53). After the termination condition is met, the chromosomes with lower cost will be selected among the current generation of chromosomes. Thus, the best paths for traffic transmission are selected by considering throughput, balancing and fairness factors.

6 Experimental results and discussion

In this section, we evaluate the performance of FTTC-TMBF and HFFTC-TMBF methods using some different scenarios. In each scenario, a number of wireless mesh routers is randomly distributed in a given area. Each router is equipped with at most three advanced radio interfaces that can adjust the transmission power in the range $[0, P^{\max}]$ with $P^{\max}$ equal to 20dbm. They also can use different transmission rates using different coding and modulation schemes. In order to have a successful reception, the SINR level of the receiver must be higher than the threshold value. According to IEEE 802.11a standard, different values of transmission rates including 6, 9, 12, 18, 24, 36, 48 and 54 Mbps are available for each interface and the SINR threshold corresponding to these values are 6.02, 7.78, 9.03, 10.79, 17.04, 18.8, 24.05 and 24.56 dB, respectively [37].

It is assumed that the maximum number of available non-overlapping frequency channels that can be assigned to radio interfaces is 12. Moreover, according to [13, 14], the interference range and the parameterε of the propagation gain are set to 350 m and 2.5 respectively, unless otherwise specified. Here, the noise power is set to $N_0 = -90$ *dBm*, and the connectivity number K is set to 2. In addition, the weight coefficients $\alpha_i$of the objective and cost functions are assumed equal, unless otherwise specified. Each point on the result graphs is the average obtained from 100 simulation runs. In this paper, we used MATLAB as simulation tool and GAMS as optimization solver.

6.1 Simulation and analysis of FTTC-TMBF

In this section, we compare the solution of the FTTC-TMBF model with the algorithm introduced in [9]. We assume that the nodes are randomly distributed in a square area of 1km^2. Also, ten traffic demands is considered for the network where the volume of each traffic flow is randomly selected between 15 and 30 MB. In Fig. 4, the performance of the FTTC-TMBF solution with objective function (3) and coefficients $\alpha_1 = 2/5$, $\alpha_i = 1/5$; $i = 2, 3, 4$ is compared with the solution introduced in [9]. As seen from Fig. 4a, the network throughput of FTTC-TMBF is lower as compared to the [9]. This is because in [9], the fairness and balancing are not considered in the objective function and the optimization is done only with the aim of throughput maximization; while, our proposed model considers a complete set of objectives including throughput maximization, fairness, balancing and robustness against failures. Moreover, by increasing the number of nodes, throughput is increased. This is because the links length is decreased and according to the power control ability of the algorithms, less power is required for successful transmission; therefore, the interference is decreased and throughput is improved. This is obvious from the figure in interval $n =$

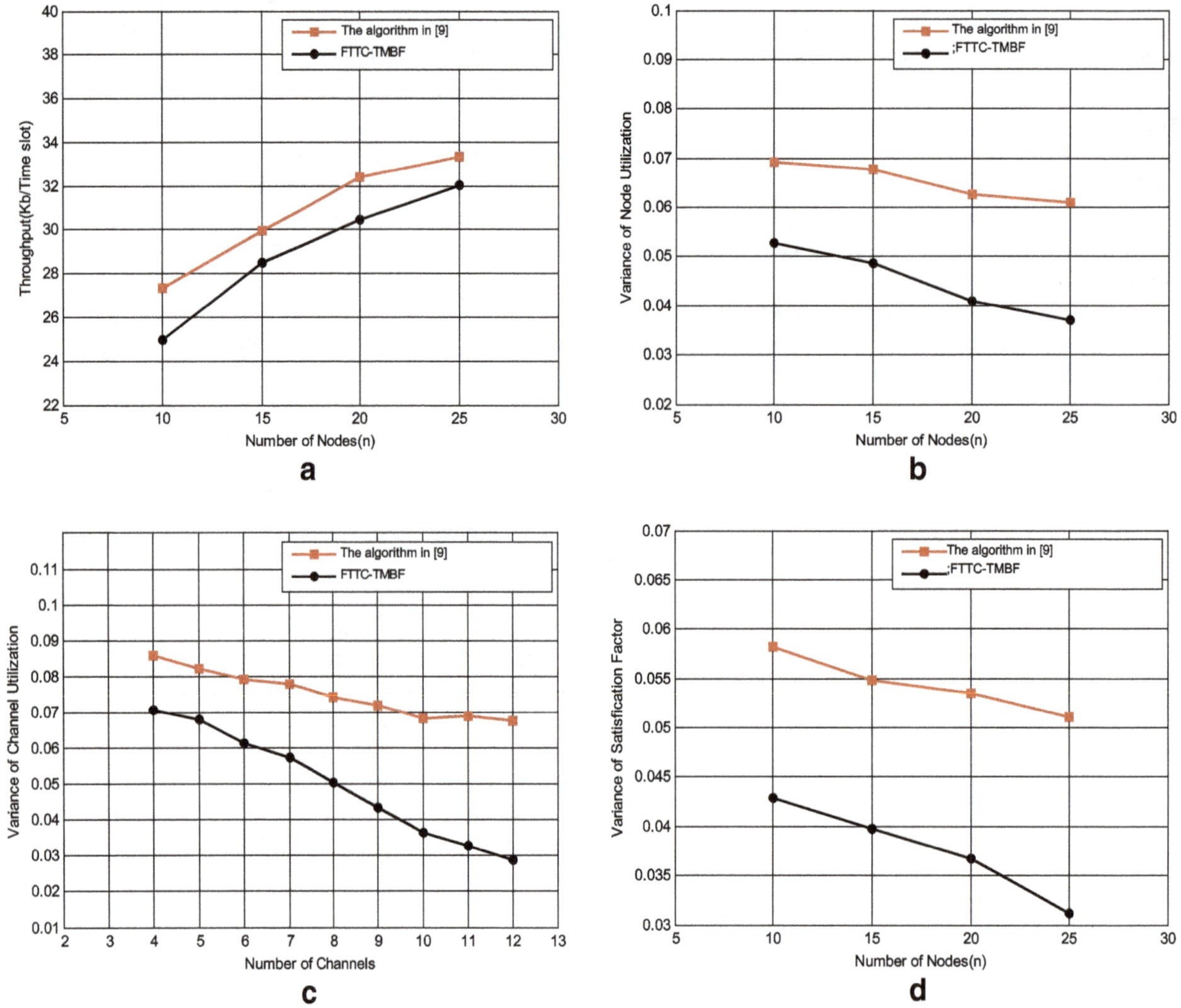

Fig. 4 The comparison of FTTC-TMBF performance and [9] (**a**) throughput (**b**) variance of node utilization (**c**) variance of channel utilization (**d**) variance of satisfaction factor

10–20 for the algorithm in [9] and in interval n = 10–15 for FTTC-TMBF. With further increase in the number of nodes, the throughput continues increasing as explained above, but because the nodes become very close to each other, we reach the minimum transmission power and from this point, the power control stops reducing the transmission power. As a result, by increasing the number of nodes beyond a threshold, the interference is increased significantly. In this case, although the received power is also increased, because of the dense nodes and high interference, the throughput increase slope is reduced.

Figure 4b represents the variance of node utilization in FTTC-TMBF, which is decreased compared to [9]. This parameter is decreased as the number of nodes is increased, because the fair distribution of traffic is more possible by having more nodes in the network. The comparison between the variance of channel utilization is depicted in Fig. 4c when the number of nodes is set to 20. It shows that the frequency channels are used more uniformly in FTTC-TMBF. Moreover, the utilization of the channels is more uniformly as the number of frequency channels is increased, which is due to the better traffic distribution among different channels. Finally, Fig. 4d shows that the variance of satisfaction factor is decreased in FTTC-TMBF, which means the fairness between different traffic flows is improved.

The impact of balancing and fairness factors on the throughput has been investigated in Fig. 5. It is shown that the variance of node utilization is decreased by increasing the effect of this factor in objective function (increasing coefficient α_3 in (3)). Also, by increasing the number of nodes, the throughput is

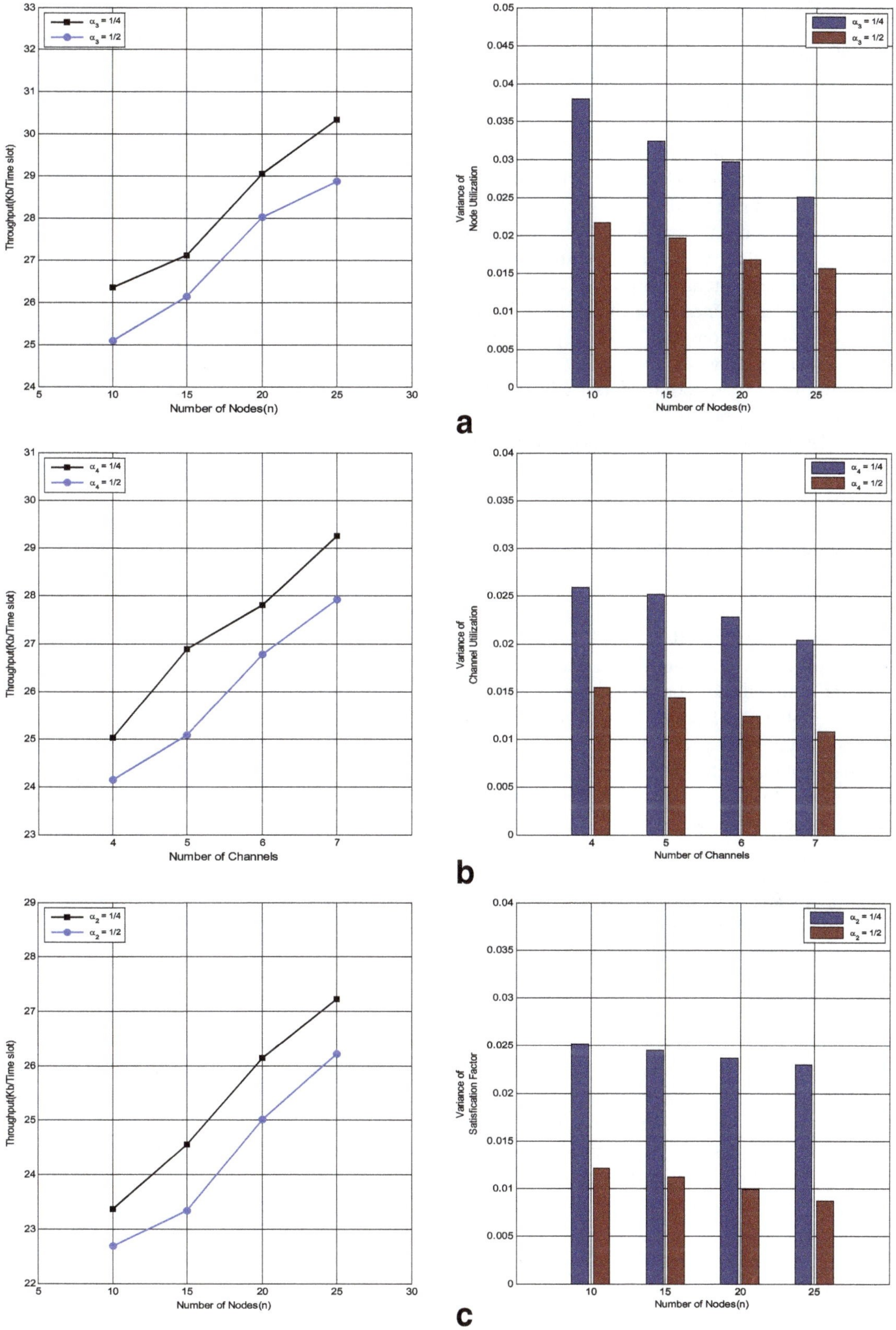

Fig. 5 Analysis of the impact of the fairness and balancing factors on the throughput in FTTC-TMBF (**a**) impact of the node utilization balancing factor (**b**) impact of the channel utilization balancing factor (**c**) impact of the fairness factor

increased because due to the reduced length of links and lower transmission power for data transmission, more links are active simultaneously. It is also evident from Fig. 5a that by increasing the number of nodes, traffic is more uniformly distributed on the nodes that results in better balancing on nodes utilization. Figure 5b shows that with any increase in the number of channels, the variance of channel utilization is decreased while the throughput is increased. This is because any increase in the number of channels results in better distribution of traffic between different channels and therefore the possibility of simultaneous transmissions is increased. This figure also clarifies that by increasing the coefficient $\alpha_4$in (3), a better balancing on the frequency channels utilization is provided. However, the throughput is decreased slightly that is due to the reduced effect of the throughput factor in the objective function. In Fig. 5c, the impact of the fairness coefficient is analyzed. By any increase in the amount ofα_2 in (3), the fairness among traffic flows is increased. However as mentioned before, the throughput is slightly decreased that is due to the reduced effect of the throughput factor in the objective function.

6.1.1 Remark

In addition to our proposed method, some other cross-layer solutions are suggested in the literature that consider throughput, balancing and fairness jointly (for example [10, 11]). However, these solutions consider only some of the available tools of power control, rate adaptation, channel assignment, scheduling and routing. Moreover, although references such as [17, 19] consider the *K*-Connectivity of the network graph, they do not consider other important objectives such as fairness and balancing. Therefore, the mentioned references are not comparable with our proposed methods that consider all available tools and a complete set of metrics including throughput, balancing, fairness, and *K*-Connectivity.

6.2 Simulation and analysis of HFTTC-TMBF

In this section, the performance of HFTTC-TMBF is studied. Here, it is assumed that the nodes have been distributed in a square area of 2km^2. At first, the network graph is specified using KPBP algorithm. In this algorithm, the three best paths between each two nodes u and v are selected using $RCF_{uv,\ l}$criteria with $\alpha_i = 1/3$, $i = 1, \ldots, 3$in (24).

As mentioned before, the implementation of steps 2 to 4 of HFTTC-TMBF are based on the genetic algorithm. In [32, 33], some recommendations are provided regarding the selection of proper parameters of genetic algorithm. Considering these recommendations, we adjust the population parameters of each generation on

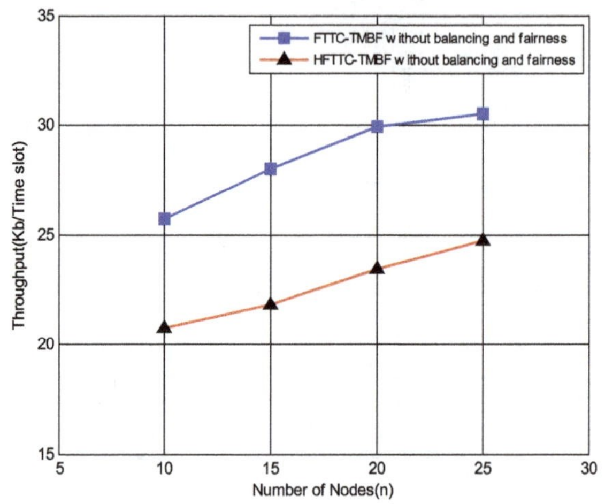

Fig. 6 The comparison of throughput in FTTC-TMBF and HFTTC-TMBF

$n_p = 40, n_c = 0.8n_p, n_m = 0.1n_p, \mu = 0.04, \lambda = 0.001.$ In second step, the channel assignment is performed using GB-LCA algorithm. In third step, we use GB-CSF algorithm to extract the compatible sets based on the parameter $\varepsilon = 0.1dBm$. Finally, we exploit the GB-CLPS algorithm for path selection.

For comparing HFTTC-TMBF with FTTC-TMBF, it is assumed that balancing and fairness factors are neglected. In other words, the coefficients α_i ; $i = 2, \ldots, 4$ in objective function (3), α_2 in (30), $\alpha_3$in (41) and α_i ; $i = 2, \ldots, 4$in (53) are considered zero.

Figure 6 shows that the amount of throughput is reduced by about 5 kb/slot, i.e., about 20%, in HFTTC-TMBF in comparison to the FTTC-TMBF. Due to the computational complexity reduction in HFTTC-TMBF, this throughput reduction is acceptable. Moreover, Fig. 7a represents the effect of increasing α_3 in (41) and α_3 in (53) on node utilization balancing. It is seen from this figure that by increasing these coefficients, the variance of node utilization and throughput is decreased. Throughput reduction is the result of reducing the throughput coefficient in the objective function. Moreover, by increasing the number of nodes, the links length is decreased which is due to the fixed network area. This reduces the interference according to the power control and rate adaptation and therefore, throughput is increased. However, with further increase in the number of nodes, they become closer to each other that increase the interference and therefore, the throughput-increasing slope is decreased. In addition, increasing the number of nodes leads to better distribution of traffic among nodes such that it improves the amount of node utilization balancing.

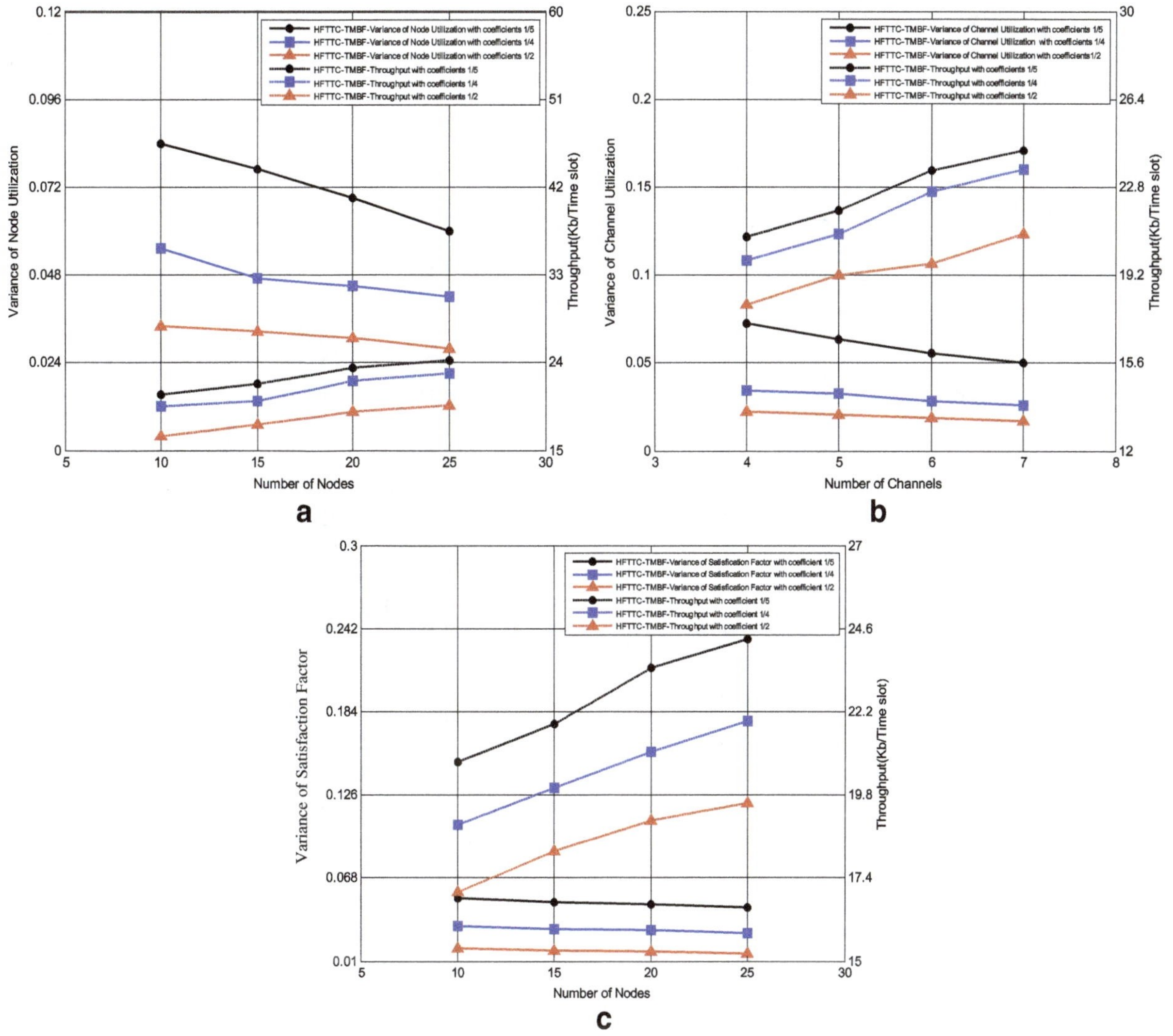

Fig. 7 Analysis of the impact of the fairness and balancing factors on the throughput in HFTTC-TMBF (**a**) impact of the node utilization balancing factor (**b**) impact of the channel utilization balancing factor (**c**) impact of the fairness factor

Figure 7b shows the performance of HFTTC-TMBF in achieving frequency channel balancing. It is observed from this figure that any increment in the amount of α_4 and α_2 in (53) and (30) respectively, results in decreasing the variance of frequency channel utilization and enhancing the balancing. Moreover, it is seen that any increase in the number of channels improves the throughput and frequency channel balancing which is due to the better traffic distribution between frequency channels. Moreover, Fig. 7c shows that the fairness is improved with the increase in the number of nodes, which is due to the selection of paths with less volume of traffic. Moreover, the variance of satisfaction factor is decreased by increasing the α_2 in (53) from 1/5 to 1/2. In other words, fairness is improved.

In the following, the effect of the coefficients $\alpha_i$from the routing cost function (24) on throughput is investigated. In the present simulation, the amount of $\alpha_2$in (30), $\alpha_3$in (41) and α_i ; $i = 2, \ldots, 4$ in (53) have been set to 1/5. As it can be seen from Fig. 8, increasing α_1 and α_2 in (24) results in increasing the throughput. This is mainly due to the higher effect of hop counts on the routing measure. Therefore, the probability of selecting shorter paths and decreasing the transmission power has been increased which in turn results in the interference reduction and throughput improvement. When α_2 in (24) is increased, the paths with lower amount of sum and maximum potential transmission power have been selected, and consequently, the throughput has been improved due to the reduction of potential interference. Figure 8also shows the effect

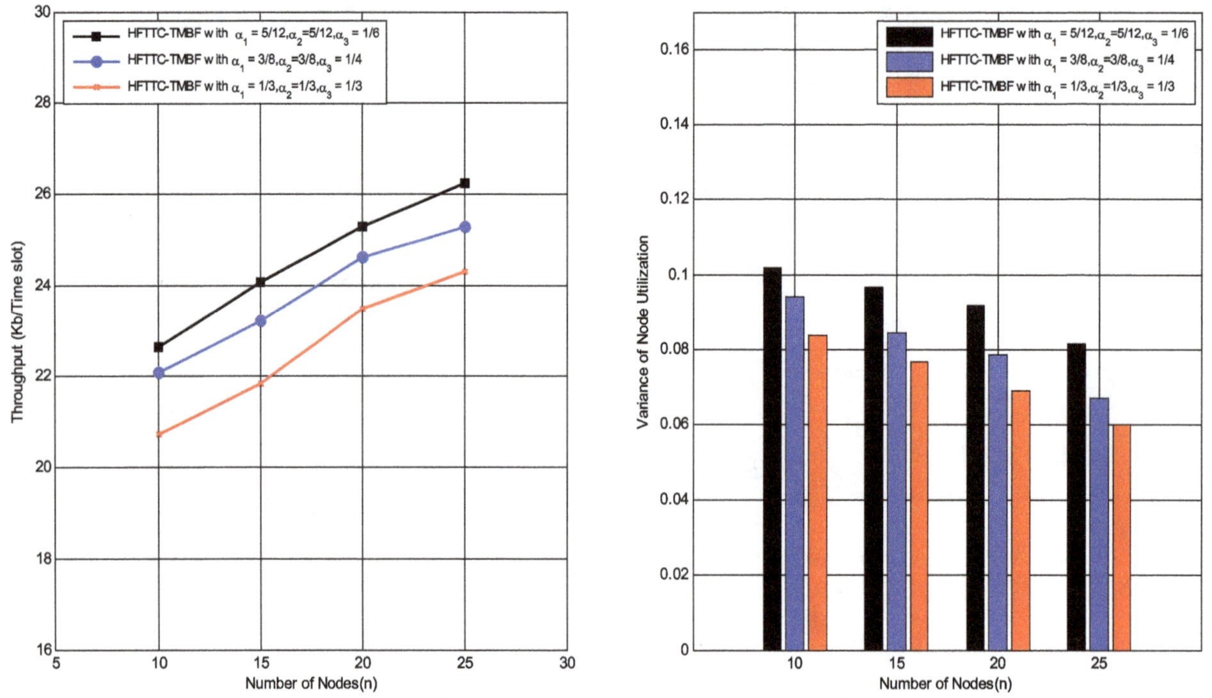

Fig. 8 Analysis of the effect of the coefficients α_i from the Cost function (24) in HFTTC-TMBF on the throughput and balancing

of increasing α_3 in (24) on balancing. Here, due to the selection of paths with smaller number of repetitive nodes, the balancing in the use of nodes resource has slightly been improved.

6.3 Complexity analysis

According to [6], the problem decomposition decreases the computational complexity considerably. Moreover, using meta-heuristic algorithms like Genetic algorithm is a common solution for decreasing the complexity of optimization problems. Therefore, the HFTTC-TMBF method which uses both the problem decomposition idea and the Genetic algorithm, definitely reduces the computational complexity as compared to the FTTC-TMBF.

As the number of variables of the FTTC-TMBF and HFTTC-TMBF methods is very high and their algorithm structures are very complex, comparing their complexities analytically is very difficult (if not impossible). So, we preferred to measure the run time as an indication of the complexity degree of the algorithms. In order to measure the run times of the algorithms, both methods FTTC-TMBF and HFTTC-TMBF are again simulated. In a certain scenario with 10 nodes and 5 non-overlapping channels, the run time needed for convergence was 356 min for FTTC-TMBF, while in HFTTC-TMBF, the run time reduced to 15.5 min. For different scenarios, the run time of FTTC-TMBF is roughly 20–30 times of the HFTTC-TMBF.

7 Conclusion and future work

In this paper, we considered the problem of resource allocation and topology control in wireless mesh networks, comprehensively. To this end, we first proposed a comprehensive cross layer model in which a complete set of available tools including power control, rate adaptation, channel assignment and selection, scheduling and routing are used simultaneously in order to maximize the throughput. Moreover, a complete set of objectives including channel utilization balancing, node utilization balancing, fairness and robustness against failures in addition to the throughput maximization are included in the proposed model.

In order to reduce the computational complexity of the comprehensive model, we have proposed four step HFTTC-TMBF heuristic method in which first the K best disjoint paths were chosen between each node pair of the network. Then GB-LCA method is proposed for assigning channels to links of these paths by using the genetic algorithm such that the potential interference is minimized. In next step, GB-CSF is employed to extract the sets of compatible links on each frequency channel that can be activated simultaneously without any interference. These sets are selected with regard to the maximization of transmission rates and providing balancing between transmission rates of the links. Finally, for selecting the best paths for transmitting the traffic flows, GB-CLPS algorithm is proposed in which objectives such as node utilization balancing, frequency channel balancing and fairness are obtained employing channel selection and path selection tools. In order to verify the performance of the proposed methods, extensive simulations were done. The analysis of simulation results showed the efficiency of the proposed methods in terms of throughput, balancing and fairness.

The proposed algorithms of this paper are implemented in a centralized manner. The centralized implementation is a proper choice, because:

1. The topology of the wireless mesh networks is nearly fixed. In addition, there are usually one or more gateways for connecting WMN to the Internet or other fixed networks. Fixed and powerful gateways are proper places for implementing centralized algorithms.
2. Today, the concept of Software Defined Networking (SDN) is accepting widely by the researches for implementing the centralized control modules for the networks. In SDN-WMN architecture, the algorithms are run on a logical centralized control device.

However, implementation of the proposed algorithms in a distributed manner is straightforward and is leaved as a future work. In this regard, we must design some control packets for collecting and distributing information locally at each node.

In addition, besides the variance criterion that is used in this paper to establish the fairness, balancing and the uniform distribution of rate and transmission power between different nodes, other metrics such as max-min and proportional fairness are also applicable in future works.

Abbreviations

FTTC-TMBF: Fault-Tolerant Topology Control with Throughput Maximization, Balancing and Fairness; GB-CLPS: Genetic Based Cross Layer Paths Selection; GB-CSF: Genetic-Based Compatible Set Formation; GB-LCA: Genetic-Based Link Channel Assignment; HFTTC-TMBF: Heuristic FTTC-TMBF; KPBP: K-Potential Best Paths Selection

Funding

The authors report no sources of funding for the research..

Authors' contributions

ENM and GM designed the algorithms, experiments and wrote the paper. Both authors read and approved the final manuscript.

Authors' information

Esmaeil Nik Maleki received his B.Sc. degree in electrical engineering from Isfahan University, Iran and the M.Sc. and Ph.D. degrees in telecommunications engineering from Yazd University, Iran in 2010 and 2017. Esmaeil is currently an assistant professor at Sheikhbahaee University, Iran. Her research interests are in the area of wireless networks.
Ghasem Mirjalily received his Ph.D. degree in telecommunication engineering in 2000. Since then, he has been with Yazd University, Iran, where he is a professor. He was a visiting researcher at McMaster University, Canada, in 1998 and a visiting research scientist at Shenzhen Research Institute of Big Data (SRIBD), Chinese University of Hong Kong, Shenzhen, in summers 2017 and 2018. Prof. Mirjalily is a senior member of IEEE, and his current research interests include Data Communication Networks, Wireless Networks, Network Virtualization and Service Chaining.

Competing interests

The authors declare that they have no competing interests.

References

1. I.F. Akyildiz, X. Wang, W. Wang, Wireless mesh networks: A survey. Comput. Netw. **47**(4), 445–487 (2005)
2. D. Benyamina, A. Hafid, M. Gendreau, Wireless mesh networks design—A survey. IEEE Commun. Surv. Tutorials **14**(2), 299–310 (2012)
3. P.H. Pathak, R. Dutta, A survey of network design problems and joint design approaches in wireless mesh networks. IEEE Commun. Surv. Tutorials **13**(3), 396–428 (2011)
4. T.-S. Kim et al., Resource allocation for QoS support in wireless mesh networks. IEEE Trans. Wirel. Commun. **12**(5), 2046–2054 (2013)
5. F. Martignon et al., Efficient and truthful bandwidth allocation in wireless mesh community networks. IEEE/ACM Trans. Networking **23**(1), 161–174 (2015)
6. I. Al, A.B.M. Alim, et al., Channel assignment techniques for multi-radio wireless mesh networks: A survey. IEEE Commun. Surv. Tutorials **18**(2), 988–1017 (2016)
7. Y. Qu, B. Ng, W. Seah, A survey of routing and channel assignment in multi-channel multi-radio WMNs. J. Netw. Comput. Appl. **65**, 120–130 (2016)
8. M. Shojafar, S. Abolfazli, H. Mostafaei, M. Singhal, Improving channel assignment in multi-radio wireless mesh networks with learning automata. Wireless Pers. Commun. **82**(1), 61–80 (2015)
9. A. Capone et al., Routing, scheduling and channel assignment in wireless mesh networks: optimization models and algorithms. Ad Hoc Netw. **8**(6), 545–563 (2010)
10. J. Luo, C. Rosenberg, A. Girard, Engineering wireless mesh networks: Joint scheduling, routing, power control, and rate adaptation. IEEE/ACM Trans. Networking **18**(5), 1387–1400 (2010)
11. K. Gokbayra, E.A. Yıldırım, Joint gateway selection, transmission slot assignment, routing and power control for wireless mesh networks. Comput. Oper. Res. **40**(7), 1671–1679 (2013)
12. K. Gokbayrak, E. Alper Yıldırım, Exact and heuristic approaches based on noninterfering transmissions for joint gateway selection, time slot allocation, routing and power control for wireless mesh networks. Comput. Oper. Res. **81**, 102–118 (2017)
13. K. Hedayati, I. Rubin, A. Behzad, Integrated power controlled rate adaptation and medium access control in wireless mesh networks. IEEE Trans. Wirel. Commun. **9**(7), 2362–2370 (2010)
14. K. Hedayati, I. Rubin, A robust distributive approach to adaptive power and adaptive rate link scheduling in wireless mesh networks. IEEE Trans. Wirel. Commun. **11**(1), 275–283 (2012)
15. X. Shao, C. Hua, A. Huang, Robust resource allocation for multi-hop wireless mesh networks with end-to-end traffic specifications. Ad Hoc Netw. **13**, 123–133 (2014)
16. J. Tang et al., Link scheduling with power control for throughput enhancement in multihop wireless networks. IEEE Trans. Veh. Technol. **55**(3), 733–742 (2006)
17. S. Chakraborty, S. Nandi, Distributed service level flow control and fairness in wireless mesh networks. IEEE Trans. Mob. Comput. **14**(11), 2229–2243 (2015)
18. R. Diestel, *Graph Theory*, 4th edn. (Springer-Verlag, Berlin, 2010)
19. E.N. Maleki, G. Mirjalily, R. Saadat, *Cross layer optimization framework for fault-tolerant topology control in wireless mesh networks.*" In Electrical Engineering (ICEE), 2016 24th Iranian Conference on (IEEE, Shiraz, 2016), pp. 288–293
20. E.N. Maleki, G. Mirjalily, Fault-tolerant interference-aware topology control in multi-radio multi-channel wireless mesh networks. Comput. Netw. **110**, 206–222 (2016)

21. A. Avokh, G. Mirjalily, Interference-aware multicast and broadcast routing in wireless mesh networks using both rate and channel diversity. Comput. Electr. Eng. **40**(2), 624–640 (2014)
22. C. Zhang, Z. Fang, *A Hybrid Routing Protocol Based on Load Balancing in Wireless Mesh Network* (Wireless Communications, Networking and Applications, Springer India, 2016), pp. 273–283
23. R. Murugeswari, S. Radhakrishnan, D. Devaraj, A multi-objective evolutionary algorithm based QoS routing in wireless mesh networks. Appl. Soft Comput. **40**, 517–525 (2016)
24. E.N. Maleki, G. Mirjalily, E.N. Maleki, R. Saadat, *A hybrid genetic-based scheduling optimization in wireless mesh network joint with power and rate control."* In 2014 22nd Iranian Conference on Electrical Engineering (ICEE) (IEEE, Tehran, 2014), pp. 937–942
25. D. Chakraborty, I-QCA: An intelligent framework for quality of service multicast routing in multichannel multiradio wireless mesh networks. Ad Hoc Netw. **33**, 221–232 (2015)
26. A. Vázquez-Rodas, J. Luis, A centrality-based topology control protocol for wireless mesh networks. Ad Hoc Netw. **24**, 34–54 (2015)
27. Y. Peng et al., Fault-tolerant routing mechanism based on network coding in wireless mesh networks. J. Netw. Comput. Appl. **37**, 259–272 (2014)
28. X. Bao et al., Design of logical topology with K-connected constraints and channel assignment for multi-radio wireless mesh networks. Int. J. Commun. Syst. 30(1), 1-18 (2017)
29. P. Gupta, P.R. Kumar, The capacity of wireless networks. IEEE Trans. Inf. Theory **46**(2), 388–404 (2000)
30. M.D. Penrose, On k-connectivity for a geometric random graph. Random Struct. Algoritm. **15**(2), 145–164 (1999)
31. D.B. West, *Introduction to Graph Theory*, vol 2 (Prentice hall, Upper Saddle River, 2001)
32. C. Reeves, in *Handbook of metaheuristics*. Genetic algorithms (Springer, Boston, 2003), pp. 55–82
33. R.L. Haupt, S.E. Haupt, *Practical genetic algorithms* (John Wiley & Sons, New Jersy, 2004)
34. H. Mühlenbein, D. Schlierkamp-Voosen, Predictive models for the breeder genetic algorithm i. continuous parameter optimization. Evol. Comput. **1**(1), 25–49 (1993)
35. M. Gen, R. Cheng, *Genetic algorithms and engineering optimization*, vol 7 (John Wiley & Sons, 2000)
36. K. Deb, D. Deb, Analysing mutation schemes for real-parameter genetic algorithms. Int. J. Artif. Intell. Soft Comp. **4**(1), 1–28 (2014)
37. IEEE 802.11a WG Part 11, "Wireless LAN Medium Access Control (MAC) and Physical Layer (PHY) Specifications: High-speed Physical Layer in the 5 GHz Band", 1999

2

Precoding design of NOMA-enabled D2D communication system with low latency

Ping Deng[1,2*], Wei Wu[1], Xinmeng Shen[1], Pei Li[1] and Baoyun Wang[3]

Abstract

This letter investigated a device-to-device (D2D) communication underlaying cellular system, where one wireless device (WD) can directly transmit information to two inner users under non-orthogonal multiple access (NOMA) protocol with low latency. Considering the user fairness and reliability, both users can adopt successive interference cancellation (SIC) to decode information by the given decoding order. Subject to predefined quality-of-service (QoS) requirement and the constraint of WD's transmit power, the maximal rate of latter decoded information and the corresponding precoding vectors are obtained by applying the technique of semidefinite relaxation (SDR) and the Langrangian duality method. Then, the suboptimal scheme based on singular value decomposition (SVD) is also proposed with a lower computational complexity. Numerical simulation shows that with the design of proper precoding vectors D2D communication system assisted by NOMA has a better performance than orthogonal multiple access (OMA).

Keywords: Non-orthogonal multiple access (NOMA), Low latency, Semidefinite relaxation, Langrangian duality, Singular value decomposition (SVD)

1 Introduction

Non-orthogonal multiple access (NOMA) is one of the promising multiple access to realize the challenging requirements of 5G [1, 2], such as massive connectivity, high data rate, and low latency. It has proved to be a viable solution for future dense networks and Internet of Things (IoT) devices. Unlike conventional orthogonal multiple access, NOMA uses the power domain to serve multiple users at different power levels at the same time, code, and frequency [3], in which superposition coding and successive interference cancellation (SIC) are employed [4]. Many various NOMA designs combined with multiple-input multiple-output (MIMO) [5], cooperative relaying [6] and millimeter-wave communications [7], have appeared in recent researches. In [8], the random opportunistic beamforming, which is a signal processing technique used in various wireless systems for directional communications [9], is first proposed for the MIMO NOMA systems, and the transmitter generated multiple beams and superposed multiple users within each beam. In [10], a beamforming design based on zero-forcing and user pairing scheme are proposed for the downlink multi-user NOMA system, assuming the perfect channel state information (CSI) is available at the transmitter. The integration of NOMA and multi-user beamforming thus has the potential to capture the benefits of both NOMA and beamforming.

Device-to-device (D2D) communication makes it possible for users in proximity to communicate with each other directly rather than relying on base stations (BSs) [11], and thus, it is an available way for reliable and low-latency communication. In [12] and [14], mode selection in underlay D2D networks is studied, while [13] investigates an efficient way of reusing the downlink resources for cellular and D2D mode communication. A step further from D2D pairs, [15] studies D2D groups that use NOMA as their transmission technique to serve multiple D2D receivers. Zhao et al. [16] consider the setting of an uplink single-cell cellular network communications. In order to further improve the outage performance of the NOMA-weak user in a user pair and reduce cooperative

*Correspondence: dengp@njupt.edu.cn
[1]College of Communication and Information, Nanjing University of Posts and Telecommunications, Xinmofan Road 66, Nanjing 210003, China
[2]College of Automation and College Of Artificial Intelligence, Nanjing University of Posts and Telecommunications, Xinmofan Road 66, Nanjing 210003, China
Full list of author information is available at the end of the article

delay, [17] focuses on full-duplex D2D-aided cooperative NOMA.

2 Method

This paper considers the D2D communication underlaying cellular system in a multiple cellular networks, which takes advantages of NOMA and D2D systems to increase the available throughput of the wireless networks. The superposed signals are sent from the multi-antenna wireless device (WD) to two single-antenna users under NOMA protocol. In particular, taking the user fairness into account, we study the case in which both users can adopt SIC to decode information by the given decoding order. To guarantee predefined quality-of-service (QoS) requirement and the constraint of WD's transmit power, the maximal rate of latter decoded information and the corresponding precoding vectors are obtained by applying the technique of semidefinite relaxation and the Langrangian duality method. The suboptimal solution based on single value decomposition (SVD) is proposed with lower computational complexity. Then, simulations are also provided to verify the performance of the proposed NOMA-enabled D2D schemes.

The rest of this letter is organized as follows. In Section 3, we introduce the system model and formulate the problem . In Section 4, we derive the optimal solution to this optimization problem. In Section 5, we propose the suboptimal solution based on SVD. In Section 6, we show simulation results that justify the performance of the proposed approaches. Our conclusions are included in Section 7.

Notations: Scalars are denoted by lowercase letters, vectors are denoted by boldface lowercase letters, and matrices are denoted by boldface uppercase letters. For a square matrix $\mathbf{A}$, $tr(\mathbf{A})$, $rank(\mathbf{A})$, and $\mathbf{A}^H$ denote its trace, rank, and conjugate, respectively. $\mathbf{A} \succeq 0$ and $\mathbf{A} \preceq 0$ represent that $\mathbf{A}$ is a positive semidefinite matrix and a negative semidefinite matrix, respectively. $\|\mathbf{x}\|$ denotes the Euclidean norm of a complex vector $\mathbf{x}$. $E[\cdot]$ denotes the statistical expectation. $[\cdot]^+$ means $\max(0, \cdot)$. $\mathbb{P}(\cdot)$ defines an outage probability event. The distribution of a circularly symmetric complex Gaussian (CSCG) random vector with mean vector $\mathbf{x}$ and covariance matrix Σ is denoted by $CN(0, \Sigma)$, and $\sim$ stands for "distributed as". $\mathbb{C}^{x \times y}$ denotes the space of $x \times y$ complex matrices.

3 System model and problem formulation

3.1 System model

In this paper, one D2D communication underlaying cellular system in a multiple cellular networks is considered, as illustrated in Fig. 1. One BS can serve a set of cellular users by WDs. Different cellular networks are allocated with orthogonal resource blocks, such as time, frequency, code, and space, in order to eliminate the inter interference between different cellular networks. Actually, WD is also regarded as a special user with relaying function and has a higher priority to decode its own message compared to s_1 and s_2, which are transmitted to two ordinary users. So WD has direct links with two ordinary users, respectively, while no direct link between the BS and each ordinary user is assumed due to significant path loss [18]. We focus on a single-cell downlink transmission scenario, where a WD can receive the signals from the BS and then transmit the superposed signals using the well-known amplify-and-forward (AF) protocol [19] to corresponding users under NOMA protocol. The WD is equipped N antennas,

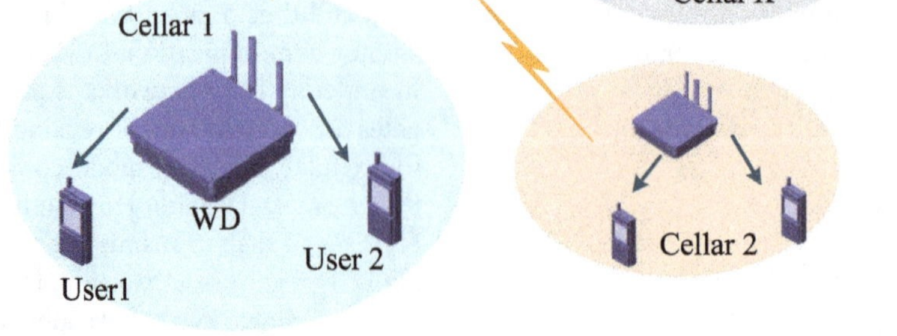

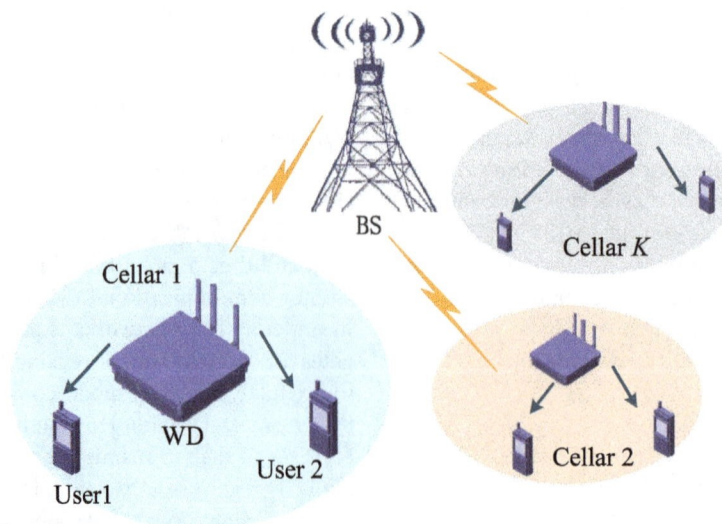

Fig. 1 System model. A D2D communication networks including one BS and K underlaying systems, in which a WD can transmit information to two users under NOMA protocol

and each user has a single antenna for the facility cost. Considering the complexity of the system, only two users are served at the same resource block. Note that it is of practical significance to choose two users to perform NOMA since NOMA systems are strongly interference-limited [20]. It is often more appropriate to group two users together to perform NOMA with user pairing [21] to realize reliable and low latency communication. It is assumed that the users' channel state information is perfectly available at the WD. We denote two types of information to both users by s_1 and s_2, respectively. It is assumed that s_1 and s_2 are independent and identically distributed (i.i.d.) circularly symmetric complex Gaussian (CSCG) random signals with unit average power [22], i.e., $E[|s_1|^2] = E[|s_2|^2] = 1$. Therefore, the complex baseband transmitted signal of WD can be expressed as:

$$\mathbf{x} = \mathbf{w}_1 s_1 + \mathbf{w}_2 s_2 \tag{1}$$

where the $\mathbf{w}_1, \mathbf{w}_2 \in \mathbb{C}^{N\times 1}$ are the precoding vectors of s_1 and s_2, respectively [23]. And the observations at two users are given by:

$$y_1 = \mathbf{h}_1^H \mathbf{x} + n_1 = \mathbf{h}_1^H(\mathbf{w}_1 s_1 + \mathbf{w}_2 s_2) + n_1 \tag{2}$$

$$y_2 = \mathbf{h}_2^H \mathbf{x} + n_2 = \mathbf{h}_2^H(\mathbf{w}_1 s_1 + \mathbf{w}_2 s_2) + n_2 \tag{3}$$

where $\mathbf{h}_1, \mathbf{h}_2 \in \mathbb{C}^{N\times 1}$ denote the complex Gaussian channel vector of user 1 and user 2, respectively, which are independent and identically distributed (*i.i.d*) fading channels. n_1 and n_2 are additive Gaussian noise (AGN), satisfying $n_1, n_2 \sim CN(0, \sigma^2)$.

3.2 Problem formulation

First, by taking the fairness of users and the decoding order into consideration, it is assumed that both users can use SIC to decode s_2 first, then subtract it from the observation before s_1 is decoded. The signal-to-interference-plus-noise ratio (SINR) of two users to decode s_1 and s_2 are respectively given by:

$$\mathrm{SINR}_{i,1} = \frac{|\mathbf{h}_i^H \mathbf{w}_1|^2}{\sigma^2}, i = 1, 2 \tag{4}$$

$$\mathrm{SINR}_{i,2} = \frac{|\mathbf{h}_i^H \mathbf{w}_2|^2}{|\mathbf{h}_i^H \mathbf{w}_1|^2 + \sigma^2}, i = 1, 2 \tag{5}$$

Then, according to the decoding order, s_2 has a higher priority to be decoded. Actually, the cognitive radio concept is used here [24]. So the achievable rate of s_2 is dependent on the minimal SINR of s_2 which is decoded by each user, and its achievable rate can be expressed as:

$$R_2 = \min[\log_2(1 + \mathrm{SINR}_{1,2}), \log_2(1 + \mathrm{SINR}_{2,2})] \tag{6}$$

In fact, s_1, which is intend for user 1, is inevitably decoded by user 2, and the achievable rate of s_1 is considered to subtract the interception by user 2. So the achievable rate of s_1 is written by [25]:

$$R_1 = \left[\log_2(1 + \mathrm{SINR}_{1,1}) - \log_2(1 + \mathrm{SINR}_{2,1})\right]^+ \tag{7}$$

It can be verified that there is an effective rate of s_1 only if the channel condition of user 1 is no worse than that of use 2.

Next, we will discuss the transmit power of WD. Suppose that the energy for receiving and amplifying the information in WD is supplied by extra power, and the transmit power P of WD is almost used to forward two types of information. So the transmit power must be satisfied [26]:

$$E\left[\mathbf{x}^H \mathbf{x}\right] = \|\mathbf{w}_1\|^2 + \|\mathbf{w}_2\|^2 \le P \tag{8}$$

At last, in this paper, we aim to maximize the achievable rate of s_1 subject to the predefined QoS of user 2 and the given transmit power of the WD. The optimization problem can be formulated as:

$$\max_{\mathbf{w}_1, \mathbf{w}_2} R_1 \tag{9a}$$

$$s.t. \; R_2 \ge \gamma_M \tag{9b}$$

$$\|\mathbf{w}_1\|^2 + \|\mathbf{w}_2\|^2 \le P \tag{9c}$$

where $\gamma_M = 2^{2R_M - 1}$, and R_M is the target rate of s_2 to satisfy the corresponding requirement of QoS.

Note that due to the non-convex nature of (9a) and (9b), problem (9) is undoubtedly non-convex in its current form. In the following subsection, we will find the optimal solution based on the analysis and transformation of problem (9).

4 The optimal solution

In order to solve the above non-convex problem, we consider the nontrivial case of the problem (9), in which the objective function is positive and can be rewritten as:

$$\begin{aligned}&\log_2(1 + \mathrm{SINR}_{1,1}) - \log_2(1 + \mathrm{SINR}_{2,1})\\ &= \log_2 \frac{1 + \mathrm{SINR}_{1,1}}{1 + \mathrm{SINR}_{2,1}} = \log_2 \frac{\sigma^2 + |\mathbf{h}_1^H \mathbf{w}_1|^2}{\sigma^2 + |\mathbf{h}_2^H \mathbf{w}_1|^2}\end{aligned} \tag{10}$$

It is obvious that with the same constraints, $\log_2 \frac{\sigma^2 + |\mathbf{h}_1^H \mathbf{w}_1|^2}{\sigma^2 + |\mathbf{h}_2^H \mathbf{w}_1|^2}$ and $\frac{\sigma^2 + |\mathbf{h}_1^H \mathbf{w}_1|^2}{\sigma^2 + |\mathbf{h}_2^H \mathbf{w}_1|^2}$ have the same optimal solution.

Meanwhile, (9b) is divided into two constraints about the SINR of s_2. So the problem (9) has the same optimal solution with the following problem:

$$\max_{\mathbf{w}_1,\mathbf{w}_2} \frac{\sigma^2+\left|\mathbf{h}_1^H\mathbf{w}_1\right|^2}{\sigma^2+\left|\mathbf{h}_2^H\mathbf{w}_1\right|^2} \tag{11a}$$

$$s.t. \quad \frac{\left|\mathbf{h}_1^H\mathbf{w}_2\right|^2}{\left|\mathbf{h}_1^H\mathbf{w}_1\right|^2+\sigma^2} \geq \gamma_M \tag{11b}$$

$$\frac{\left|\mathbf{h}_2^H\mathbf{w}_2\right|^2}{\left|\mathbf{h}_2^H\mathbf{w}_1\right|^2+\sigma^2} \geq \gamma_M \tag{11c}$$

$$\|\mathbf{w}_1\|^2+\|\mathbf{w}_2\|^2 \leq P \tag{11d}$$

The semidefinite relaxation (SDR) [27] is applied to obtain the optimal solution of (9). Define $\mathbf{W}_1 = \mathbf{w}_1\mathbf{w}_1^H, \mathbf{W}_2 = \mathbf{w}_2\mathbf{w}_2^H, \mathbf{H}_1 = \mathbf{h}_1\mathbf{h}_1^H, \mathbf{H}_2 = \mathbf{h}_2\mathbf{h}_2^H$, and ignore the constraint of $rank(\mathbf{W}_1) = rank(\mathbf{W}_2) = 1$, we can obtain:

$$\max_{\mathbf{W}_1,\mathbf{W}_2} \frac{\sigma^2+tr(\mathbf{H}_1\mathbf{W}_1)}{\sigma^2+tr(\mathbf{H}_2\mathbf{W}_1)} \tag{12a}$$

$$s.t. \quad \frac{tr(\mathbf{H}_1\mathbf{W}_2)}{tr(\mathbf{H}_1\mathbf{W}_1)+\sigma^2} \geq \gamma_M \tag{12b}$$

$$\frac{tr(\mathbf{H}_2\mathbf{W}_2)}{tr(\mathbf{H}_2\mathbf{W}_1)+\sigma^2} \geq \gamma_M \tag{12c}$$

$$tr(\mathbf{W}_2)+tr(\mathbf{W}_1) \leq P \tag{12d}$$

The problem (12) we proposed is a fractional programming obviously. The Dinkelbach method is widely adopted in solving the fractional programming. Thus, we use this method to transform and solve the problem. With a continuous auxiliary variable t [28], we reformulate the objective function in (12a) as:

$$F(t) = \max_{\{\mathbf{W}_1,\mathbf{W}_2\}\in\Psi} \sigma^2+tr(\mathbf{H}_1\mathbf{W}_1)-t[\sigma^2+tr(\mathbf{H}_2\mathbf{W}_1)] \tag{13}$$

where $\left\{\Psi \mid \frac{tr(\mathbf{H}_1\mathbf{W}_2)}{tr(\mathbf{H}_1\mathbf{W}_1)+\sigma^2} \geq \gamma_M, \frac{tr(\mathbf{H}_2\mathbf{W}_2)}{tr(\mathbf{H}_2\mathbf{W}_1)+\sigma^2} \geq \gamma_M, tr(\mathbf{W}_1)+ tr(\mathbf{W}_2) \leq P\right\}$ is the feasible set in (12). We have the following lemma, whose proof can refer to [29]:

Lemma 1 *$F(t)$ is a strictly decreasing and continuous function, and it has a unique zero solution, which is denoted as t^*. Then, the optimal value of the objective function in (12a) is t^*.*

From Lemma 1, we can see that the optimal solution can be obtained by solving (12) if we know t^* in advance. Though t^* is unknown at first, Lemma 1 tells us that the Dinkelbach method by round search [28] can be used to find the root of $F(t)$ efficiently. In each iteration, the non-negative variable t will update its value and finally reach the optimal denoting as t^*.

Therefore, in the following, we will optimize (12) for a given t at first. For convenience, we rewrite (13) into the following form:

$$F(t) = \max_{\mathbf{W}_1,\mathbf{W}_2} \sigma^2+tr(\mathbf{H}_1\mathbf{W}_1)-t\left[\sigma^2+tr(\mathbf{H}_2\mathbf{W}_1)\right] \tag{14a}$$

$$s.t. \quad tr(\mathbf{H}_1\mathbf{W}_2) \geq \gamma_M\left[tr(\mathbf{H}_1\mathbf{W}_1)+\sigma^2\right] \tag{14b}$$

$$tr(\mathbf{H}_2\mathbf{W}_2) \geq \gamma_M\left[tr(\mathbf{H}_2\mathbf{W}_1)+\sigma^2\right] \tag{14c}$$

$$tr(\mathbf{W}_1)+tr(\mathbf{W}_2) \leq P \tag{14d}$$

Meanwhile, problem (14) is a convex semidefinite problem (SDP) and can be efficiently solved by convex optimization solvers, e.g., CVX [30].

Proposition 1 *The optimal solution to problem (14) satisfies $rank(\mathbf{W}_1^*) = rank(\mathbf{W}_2^*) = 1$.*

Proof Obviously, the problem (14) is a separate SDP with three generalized constraints. According to [31], the optimal solution $(\mathbf{W}_1^*, \mathbf{W}_2^*)$ to (14) always satisfies $rank^2(\mathbf{W}_1^*) + rank^2(\mathbf{W}_2^*) \leq 3$. Here, we consider the non-trivial case where $\mathbf{W}_1^* \neq 0$ and $\mathbf{W}_2^* \neq 0$, then $rank(\mathbf{W}_1^*) = rank(\mathbf{W}_2^*) = 1$ can be obtained. Proposition 1 is proved, and this implies that the SDR here is tight. □

Let α_1, α_2, and α_3 denote the dual variables associated with constraints (14b), (14c), and (14d), respectively. The dual problem of (14) is expanded as:

$$d_{\min} = \min_{\alpha_1,\alpha_2,\alpha_3} \alpha_3 P - (\alpha_1+\alpha_2)\gamma_M\sigma^2 + (1-t)\sigma^2 \tag{15a}$$

$$s.t. \quad \mathbf{A} \leq 0, \mathbf{B} \leq 0, \tag{15b}$$

$$\alpha_1 \geq 0, \alpha_2 \geq 0, \alpha_3 \geq 0 \tag{15c}$$

where

$$\mathbf{A} = -\alpha_3\mathbf{I} + (1-\alpha_1\gamma_M)\mathbf{H}_1 - (t+\alpha_2\gamma_M)\mathbf{H}_2, \tag{16}$$

$$\mathbf{B} = -\alpha_3\mathbf{I} + \alpha_1\mathbf{H}_1 + \alpha_2\mathbf{H}_2, \tag{17}$$

and $d_{\min}$ is the value of problem (15).

Proposition 2 *The optimal dual solution α_3^* to problem (15) satisfies $\alpha_3^* > 0$.*

Proof We show that the optimal solution $\alpha_3^* > 0$ by contradiction. Assume that $\alpha_3^* = 0$, since the Lagrangian dual variables are all non-negative. The matrix $\mathbf{B}^*$ is negative semidefinite, i.e., $\alpha_1^*\mathbf{H}_1 + \alpha_2^*\mathbf{H}_2 \preceq 0$. Moreover, (14) is convex and satisfies Slater's condition, and the duality gap between (14) and (15) is zero. Thus $F(t) = (1-t)\sigma^2$. According to Lemma 1, the optimal solution to problem (14) is the same with the problem (12) when $F(t^*) = 0$. So $t^* = 1$, and the maximal rate of R_1 is $\log_2 t^* = 0$. It is not reasonable, and then $\alpha_3^* > 0$ must be true. Proposition 2 is proved. □

With the optimal dual solution $(\alpha_1^*, \alpha_2^*, \alpha_3^*)$ and optimal value $d_{\min}^*$ obtained by solving problem (15), we can derive $\mathbf{A}^*$ and $\mathbf{B}^*$, respectively, by substituting $(\alpha_1^*, \alpha_2^*, \alpha_3^*)$ into (16) and (17). Moreover, the complementary slackness condition of (15b) yields to $\mathbf{A}^*\mathbf{W}_1^* = 0, \mathbf{B}^*\mathbf{W}_2^* = 0$. Since $rank(\mathbf{W}_1^*) = 1$ and $rank(\mathbf{W}_2^*) = 1$, we have $rank(\mathbf{A}^*) = N-1$ and $rank(\mathbf{B}^*) = N-1$. Let $\mathbf{u}_1$ and $\mathbf{u}_2$ be the basis of the null space of $\mathbf{A}^*$ and $\mathbf{B}^*$, respectively, and define $\hat{\mathbf{W}}_1 = \mathbf{u}_1\mathbf{u}_1^H, \hat{\mathbf{W}}_2 = \mathbf{u}_2\mathbf{u}_2^H$, then we have:

$$\begin{cases} (1-t)\sigma^2 + \tau_1^2 tr[(\mathbf{H}_1 - t\mathbf{H}_2)\hat{\mathbf{W}}_1] = d_{\min}^* \\ \tau_2^2 tr(\hat{\mathbf{W}}_2) + \tau_1^2 tr(\hat{\mathbf{W}}_1) = P \end{cases} \tag{18}$$

where τ_1 and τ_2 are the power allocation coefficients for transmitting information s_1 and s_2, respectively. And

$$\begin{cases} \tau_1 = \sqrt{\dfrac{d_{\min}^* - (1-t)\sigma^2}{tr\left[(\mathbf{H}_1 - t\mathbf{H}_2)\hat{\mathbf{W}}_1\right]}}, \\ \tau_2 = \sqrt{P - \tau_1^2}. \end{cases} \tag{19}$$

Thus, the optimal precoding vectors are $\mathbf{w}_1^* = \tau_1\mathbf{u}_1, \mathbf{w}_2^* = \tau_2\mathbf{u}_2$ with given t.

Note that $2N$ complex variables are to be optimized for problem (12), while only three real variables for problem (15). So problem (15) has the lower computational complexity than problem (12). Actually, the complexity reduction is significant as the number of antennas at WD increases. Detailed steps of proposed optimal algorithm are summarized as Algorithm 1.

Algorithm 1 The optimal solution to problem (11)

Initialize t satisfying $F(t) \geq 0$ and tolerance ε
while ($|F(t)| > \varepsilon$) **do**
  Solve problem (15) to obtain $(\alpha_1^*, \alpha_2^*, \alpha_3^*)$
  Calculate $\mathbf{A}^*$ and $\mathbf{B}^*$ according to (16) and (17), respectively.
  Calculate $\mathbf{w}_1^*$, and $\mathbf{w}_2^*$ according to (19)
  $t \leftarrow \frac{\sigma^2 + |\mathbf{h}_1^H\mathbf{w}_1^*|^2}{\sigma^2 + |\mathbf{h}_2^H\mathbf{w}_1^*|^2}$
end while
return $\mathbf{w}_1^*$ and $\mathbf{w}_2^*$

5 The SVD-based suboptimal solution

As described in the previous section, we can derive the optimal solution to problem (11) by using fractional programming and solving dual problem. But the round search for finding optimal t^* reduces the feasibility of the optimal solution to a certain extent in practice. In this section, we propose a suboptimal solution based on SVD to further reduce the computational complexity.

When $N \geq 2$, the SVD-based precoding scheme can be used to eliminate the interference caused by s_1 at the WD by restricting precoding vector $\mathbf{w}_1$ to satisfy $\mathbf{h}_2^H\mathbf{w}_1 = 0$ [32], which simplifies the precoding vector design. With SVD-based precoding vector, user 2 cannot decode the information s_1. It implies that the precoding vector $\mathbf{w}_1$ must lie in the null space of $\mathbf{h}_2$. Let the SVD of $\mathbf{h}_2$ be expressed as $\mathbf{h}_2^H = \mathbf{u}\Lambda\mathbf{v}^H = \mathbf{u}\Lambda[\mathbf{v}_1\mathbf{v}_0]^H$, where $\mathbf{u} \in \mathbb{C}^{1\times 1}, \mathbf{v} \in \mathbb{C}^{N\times N}$ are orthogonal left and right singular vectors of $\mathbf{h}_2$, respectively, and $\Lambda \in \mathbb{C}^{1\times N}$ contains one positive singular value of $\mathbf{h}_2$. $\mathbf{v}_0 \in \mathbb{C}^{N\times(N-1)}$, which satisfies $\mathbf{v}_0^H\mathbf{v}_0 = \mathbf{I}_{N-1}$, is the last $N-1$ columns of $\mathbf{v}$ and forms an orthogonal basis for the null space of $\mathbf{h}_2^H$. The SVD-based precoding vector $\mathbf{w}_1$ can be expressed as $\mathbf{w}_1 = \mathbf{v}_0\tilde{\mathbf{w}}_1$, where $\tilde{\mathbf{w}}_1$ denotes the new vector to be designed, and the corresponding precoding vector of s_2 to be designed is $\tilde{\mathbf{w}}_2$. It is obvious to observe that in order for the SVD-based solution to be feasible, we must have $N \geq 2$. Problem (11) is consequently formulated as:

$$\max_{\tilde{\mathbf{w}}_1, \tilde{\mathbf{w}}_2} 1 + \frac{1}{\sigma^2}|\mathbf{h}_1^H\mathbf{v}_0\tilde{\mathbf{w}}_1|^2 \tag{20a}$$

$$s.t.\ |\mathbf{h}_1^H\tilde{\mathbf{w}}_2|^2 \geq \gamma_M\left(|\mathbf{h}_1^H\mathbf{v}_0\tilde{\mathbf{w}}_1|^2 + \sigma^2\right) \tag{20b}$$

$$|\mathbf{h}_2^H\tilde{\mathbf{w}}_2|^2 \geq \gamma_M\sigma^2 \tag{20c}$$

$$||\tilde{\mathbf{w}}_2||^2 + ||\tilde{\mathbf{w}}_1||^2 \leq P \tag{20d}$$

Define $\tilde{\mathbf{W}}_1 = \tilde{\mathbf{w}}_1\tilde{\mathbf{w}}_1^H$, $\tilde{\mathbf{W}}_2 = \tilde{\mathbf{w}}_2\tilde{\mathbf{w}}_2^H$, $\tilde{\mathbf{H}}_1 = \mathbf{v}_0^H\mathbf{h}_1\mathbf{h}_1^H\mathbf{v}_0$, we have the SVD-based SDP:

$$\max_{\tilde{\mathbf{W}}_1, \tilde{\mathbf{W}}_2} 1 + \frac{1}{\sigma^2}tr(\tilde{\mathbf{H}}_1\tilde{\mathbf{W}}_1) \tag{21a}$$

$$s.t.\ tr(\mathbf{H}_1\tilde{\mathbf{W}}_2) \geq \gamma_M[tr(\tilde{\mathbf{H}}_1\tilde{\mathbf{W}}_1) + \sigma^2] \tag{21b}$$

$$tr(\mathbf{H}_2\tilde{\mathbf{W}}_2) \geq \gamma_M\sigma^2 \tag{21c}$$

$$tr(\tilde{\mathbf{W}}_1) + tr(\tilde{\mathbf{W}}_2) \leq P \tag{21d}$$

Obviously, the achieved optimal solution also satisfies the rank-one constraint. Let $\tilde{\alpha}_1$, $\tilde{\alpha}_2$, and $\tilde{\alpha}_3$ denote dual variables, and its dual problem is given by:

$$d^* = \min_{\tilde{\alpha}_1, \tilde{\alpha}_2, \tilde{\alpha}_3} 1 + \tilde{\alpha}_3 P - (\tilde{\alpha}_1 + \tilde{\alpha}_2)\gamma_M\sigma^2 \tag{22a}$$

$$s.t.\ \tilde{\mathbf{A}} \preceq 0, \tilde{\mathbf{B}} \preceq 0, \tilde{\alpha}_1 \geq 0, \tilde{\alpha}_2 \geq 0, \tilde{\alpha}_3 > 0 \tag{22b}$$

where

$$\tilde{\mathbf{A}} = -\tilde{\alpha}_3\mathbf{I} + \frac{1}{\sigma^2}\tilde{\mathbf{H}}_1 - \tilde{\alpha}_1\tilde{\mathbf{H}}_1\gamma_M \quad (23)$$

$$\tilde{\mathbf{B}} = -\tilde{\alpha}_3\mathbf{I} + \tilde{\alpha}_1\mathbf{H}_1 + \tilde{\alpha}_2\mathbf{H}_2 \quad (24)$$

Different from problem (12), problem (20) is convex and the dual gap between (21) and (22) is also zero. In the same way as the previous section, we can also solve the problem (20) by its Lagrangian dual problem (22) for complexity reduction. With the SVD-based solution $(\tilde{\alpha}_1^*, \tilde{\alpha}_2^*, \tilde{\alpha}_3^*)$ achieved by problem (22), we can derive $\tilde{\mathbf{A}}^*$ and $\tilde{\mathbf{B}}^*$ according to (23) and (24). Let $\tilde{\mathbf{u}}_1$ and $\tilde{\mathbf{u}}_2$ be the basis of the null space of $\tilde{\mathbf{A}}^*$ and $\tilde{\mathbf{B}}^*$, respectively, and define $\hat{\mathbf{W}}_1=\tilde{\mathbf{u}}_1\tilde{\mathbf{u}}_1^H, \hat{\mathbf{W}}_2 = \tilde{\mathbf{u}}_2\tilde{\mathbf{u}}_2^H$. Similar to (18), we have the precoding vectors based on SVD as $\tilde{\mathbf{w}}_1^*=\tilde{\tau}_1\tilde{\mathbf{u}}_1, \tilde{\mathbf{w}}_2^*=\tilde{\tau}_2\tilde{\mathbf{u}}_2$, where

$$\begin{cases} \tilde{\tau}_1 = \sqrt{\dfrac{(d^* - 1)\sigma^2}{tr\left(\mathbf{H}_1\hat{\mathbf{W}}_1\right)}}, \\ \tilde{\tau}_2 = \sqrt{P - \tilde{\tau}_1^2} \end{cases} \quad (25)$$

And the detailed steps of proposed SVD-based suboptimal algorithm are presented as Algorithm 2. Compared with Algorithm 1, the proposed SVD-based suboptimal scheme in Algorithm 2 further reduces the computational complexity without the Dinkelbach method.

Algorithm 2 The SVD-based suboptimal solution to problem (11)

1: Set $\mathbf{v}_0 = null(\mathbf{h}_2)$, where '$null(\cdot)$' is a MATLAB function which computes the orthonormal basis for the null space of a matrix using SVD.
2: Set $\tilde{\mathbf{H}}_1 = \mathbf{v}_0^H\mathbf{h}_1\mathbf{h}_1^H\mathbf{v}_0$.
3: Solve problem (22) to obtain $(\tilde{\alpha}_1^*, \tilde{\alpha}_2^*, \tilde{\alpha}_3^*)$
4: Calculate $\tilde{\mathbf{A}}^*$ and $\tilde{\mathbf{B}}^*$ according to (23) and (24), respectively.
5: Calculate $\tilde{\mathbf{w}}_1^*$, and $\tilde{\mathbf{w}}_2^*$ according to (25)
6: **return** $\tilde{\mathbf{w}}_1^*$ and $\tilde{\mathbf{w}}_2^*$

6 Simulation results and discussions

In this section, we numerically evaluate the performance of the proposed optimal and suboptimal schemes. The simulation parameters are listed in Table 1. It is assumed that the channels from WD to two ordinary users are deep fading channels and the information signal attenuations are 55 dB and 60 dB, respectively, corresponding to an identical distance of 15 m and 20 m. We set the parameter in the Dinkelbach method $\varepsilon = 10^{-4}$ and the power of noise $\sigma^2 = -50$ dBm. The channels $\mathbf{h}_1$ and $\mathbf{h}_2$ are assumed to be quasi-static flat Rayleigh fading, and each element of them follows an independent complex Gaussian distribution $CN(0, 1)$. All the simulation results are averaged over 1000 channel realizations. The optimal scheme and SVD-based scheme in this section, respectively, mean the optimal precoding vector scheme and the SVD-based precoding vector scheme of WD.

Table 1 Simulation parameters

Parameter	Value
Attenuation of $\mathbf{h}_1$	55 dB
Attenuation of $\mathbf{h}_2$	60 dB
Power of noise (σ^2)	− 50 dBm
Parameter in Dinkelbach (ε)	10^{-4}
Number of antennas at WD (N)	4, 6
Transmit power of WD (P)	20, 25 dBm
Target rate R_M	4, 6 bps/Hz
Number of channel realizations	1000

Figure 2 shows the maximal rate performance of different schemes versus the transmit power P under the number of antennas at WD $N = 4$ and $R_M = 4$ bps/Hz. Without loss of generality, time division multiple access (TDMA) is used to a representative of orthogonal multiple access (OMA), in which WD only serves single user in one time slot with joint power allocation among two time slots for two users. It can be observed that the proposed optimal and suboptimal scheme outperform traditional TDMA in terms of maximal rate of R_1, and this performance advantage is more obvious in the high transmit power region, though in TDMA scheme the s_1 is not interrupted by s_2. And it is noted that the proposed SVD-based suboptimal scheme only has a slight performance loss compared to the optimal scheme.

Figure 3 compares the maximal rate of R_1 versus the number of antennas at WD for different schemes under the transmit power of WD $P = 25$ dBm, when the target rate of s_2, i.e., R_M, is 4 bps/Hz and 6 bps/Hz, respectively. It is obviously noted that the maximal rate of R_1 is enhanced as the number of antennas grows. However, the growth trend gradually becomes slow. Besides, the gap between proposed optimal scheme and SVD-based schemes in terms of the maximal rate of R_1 is reducing with the increasing R_M.

In Fig. 4, the maximal rate region of R_1 versus R_M is characterized for different schemes with the number of antennas at WD $N = 4$, when the transmit power of WD is 25 dBm and 20 dBm, respectively. It is also noted that the optimal scheme achieves better rate regions than the SVD-based scheme. Furthermore, the higher target rate of s_2 requires, the smaller gap between the optimal and suboptimal scheme is. Meanwhile, the impact of transmission power at WD on the achieved rate regions for different

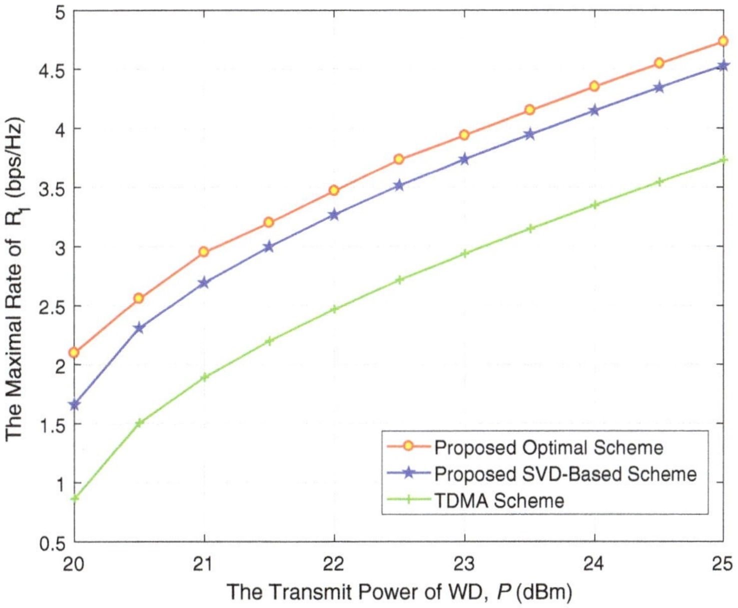

Fig. 2 Simulation 1. The maximal rate of R_1 versus the transmit power at WD for different schemes with $N = 4$ and $R_M = 4$ bps / Hz

schemes is also shown in Fig. 4. We can find that under certain transmit power of WD, when R_M is large, the rate of R_1 may be zero. The reason is that all power should be allocated to precoding vector $\mathbf{w}_2$ to first satisfy the target rate demand of s_2. So the precoding vector $\mathbf{w}_1$ has little effect on the system performance no matter it is designed in optimal solution or suboptimal solution. And the rate of R_1 becomes zero almost at the same value of target rate R_M for the optimal and SVD-based schemes.

Finally, Fig. 5 presents the outage performance of R_1 when R_M varies from 2 to 11 bps/Hz with the transmit power of WD P = 25 dBm and the number of antennas at WD N = 4. Given the transmit power of WD P, the number of antennas N, and the target rate R_M, the outage probability is $p_{\text{out}}(P, R_M, N) \triangleq \mathbb{P}(R_1 = 0)$. Especially, we set P = 25 dBm and N = 4. It is observed that the proposed optimal and suboptimal solutions achieve a similar performance to each other, and our proposed two

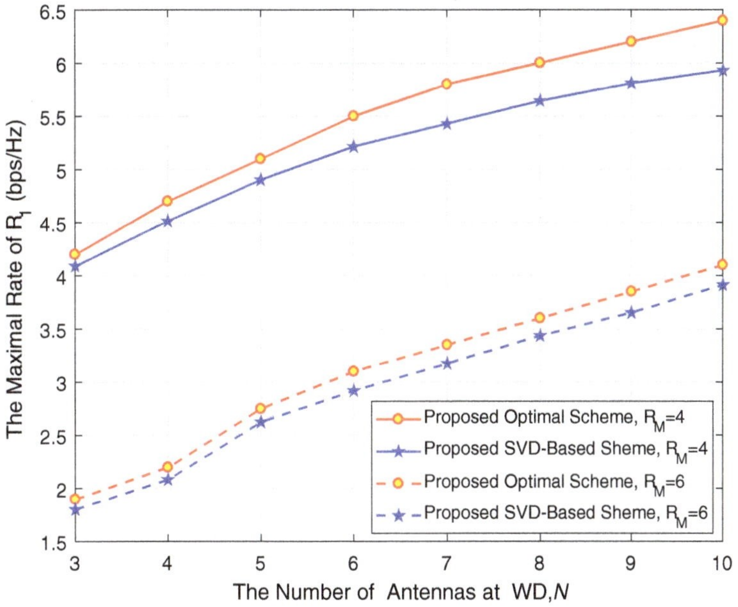

Fig. 3 Simulation 2. The maximal rate of R_1 versus the number of antennas at WD for different schemes with P = 25 dBm

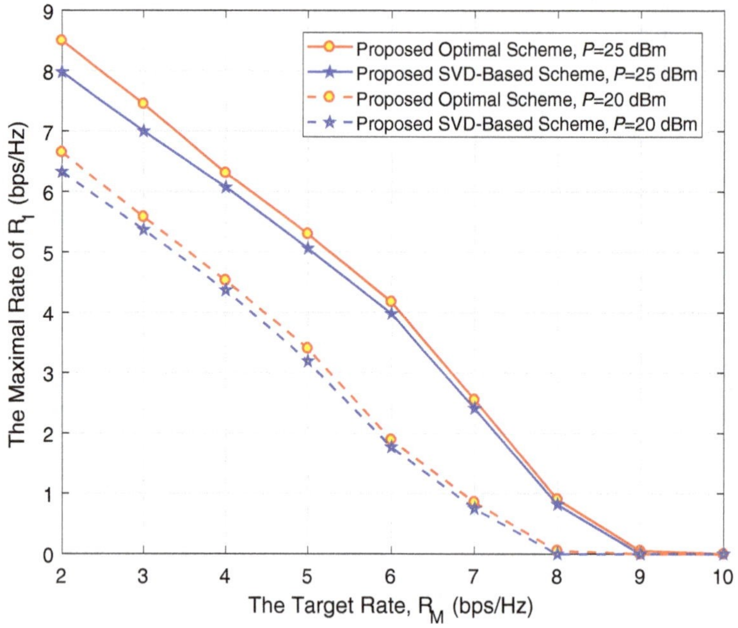

Fig. 4 Simulation 3. The maximal rate of R_1 versus R_M for different schemes with $N = 4$

schemes significantly decrease the outage probability of rate of R_1 compared with the TDMA scheme.

7 Conclusion

In this paper, an optimization problem of precoding vectors for two-user D2D communication underlaying system enabled by NOMA is investigated. Given the target rate R_M and the transmit power of WD P, the maximal rate of R_1 and corresponding optimal precoding vectors have been obtained. Then, the suboptimal solution based on SVD is proposed for complexity reduction. Finally, simulation results are provided to show the proposed optimal

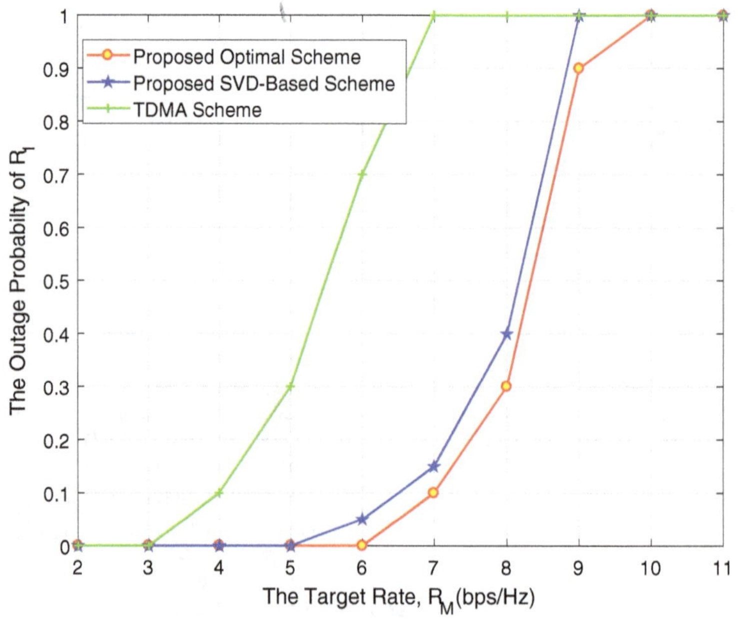

Fig. 5 Simulation 4. The outage probability of R_1 versus R_M for different schemes with N=4 and $P = 25$ dBm

and suboptimal precoding algorithms can outperform OMA scheme, such as TDMA.

Abbreviations

AF: Amplify-and-forward; AGN: Additive Gaussian noise; BS: Base station; CSI: Channel state information; CVX: Matlab software for disciplined convex programming; D2D: Device-to-device; IoT: Internet of Things; MIMO: Multiple-input multiple-output; NOMA: Non-orthogonal multiple access; OMA: Orthogonal multiple access; QoS: Quality of service; SDP: Semidefinite problem; SDP: Semidefinite relaxation; SIC: Successive interference cancellation; SINR: Signal-to-interference-plus-noise-ratio; SVD: Singular value decomposition; TDMA: Time division multiple access; WD: Wireless device

Acknowledgements

Not applicable.

Authors' contributions

PD is the main author of the current paper. PD contributed to the development of the ideas, design of the study, theory, result analysis, and article writing. WW contributed to the development of the ideas, design of the study, theory, and article writing. PD and PL conceived and designed the experiments. XS and PL performed the experiments. BW undertook revision works of the paper. All authors read and approved the final manuscript.

Funding

This paper was supported in part by the National Natural Science Foundation of China under Grant 61271232, the Natural Science Foundation of Jiangsu Province of China under BK20180757, the Project of Educational Commission of Jiangsu Province of China under 18KJB510028, the Introducing Talent Research Start-Up Fund of Nanjing University of Posts and Telecommunications under NY218100, and the Postgraduate Research & Practice Innovation Program of Jiangsu Province under Grant CXZZ13_0487.

Competing interests

The authors declare that they have no competing interests.

Author details

[1]College of Communication and Information, Nanjing University of Posts and Telecommunications, Xinmofan Road 66, Nanjing 210003, China. [2]College of Automation and College Of Artificial Intelligence, Nanjing University of Posts and Telecommunications, Xinmofan Road 66, Nanjing 210003, China. [3]College of Overseas Education, Nanjing University of Posts and Telecommunications, Xinmofan Road 66, Nanjing 210003, China.

References

1. Z. Ding, et al., Application of non-orthogonal multiple access in LTE and 5G networks. IEEE Commun. Mag. **55**(2), 185-191 (2017)
2. R. Hu, Y. Qian, An energy efficient and spectrum efficient wireless heterogeneous network framework for 5G systems. IEEE Commun. Mag. **52**(5), 94–101 (2014)
3. Y. Saito, Y. Kishiyama, A. Benjebbour, T. Nakamura, A.LK. Higuchi, in *Proc. IEEE 77th Vehicular Technology Conference (VTC Spring)*. Non-orthogonal multiple access (NOMA) for cellular future radio access. (IEEE, Dresden, 2013), 1–5
4. K. Higuchi, A. Benjebbour, Non-orthogonal multiple access (NOMA) with successive interference cancellation. IEICE Trans. Commum. **E98-B**(3), 403–414 (2015)
5. Z. Ding, F. Adachi, H Poor.The application of MIMO to non-orthogonal multiple access. IEEE Trans. Wirel. Commum. Lett. **15**(1), 537–552 (2016)
6. D. Wan, M. Wen, F. Ji, Y. Liu, Y. Huang, Cooperative NOMA systems with partial channel state information over Nakagami- m fading channels. IEEE Trans. Commun. **66**(3), 947–958 (2018)
7. B. Wang, L. Dai, M. Xiao, *Millimeter Wave NOMA. Encyclopedia of Wireless Networks*. (Springer, Cham, 2018)
8. A. Benjebbour, Y. Saito, Y. Kishiyama, A. Li, A. Harada, T. Nakamura, in *Proc. 2013 International Symposium on Intelligent Signal Processing and Communication Systems (ISPACS)*. Concept and practical considerations of non-orthogonal multiple access (NOMA) for future radio access. (IEEE, Naha, 2013), 770–774
9. W. Wu, B. Wang, Robust secrecy beamforming for wireless information and power transfer in multiuser MISO communication system. EURASIP J Wireless Com Network. **2015**(1), 161 (2015)
10. B. Kim, S. Lim, H. Kim, S. Suh, J. Kwun, S. Choi, C. Lee, S. Lee, D. Hong, in *Proc. MILCOM 2013-2013 IEEE Military Communications Conference*. Non-orthogonal multiple access in a downlink multiuser beamforming system (IEEE, San Diego, 2013), pp. 1278-1283
11. H. ElSawy, E. Hossain, M. Alouini, Analytical modeling of mode selection and power control for underlay D2D communication in cellular networks. IEEE Trans. Commun. **62**(11), 4147–4161 (2014)
12. K. Doppler, C. Yu, C. Ribeiro, P. Janis, in *Proc. 2010 IEEE Wireless Communication and Networking Conference*. Mode selection for device-to-device communication underlaying an LTE-advanced network. (IEEE, Sydney, 2010), 1–6
13. K. Akkarajitsakul, P. Phunchongharn, E. Hossain, V.K. Bhargava, in *Proc 2012 IEEE International Conference on Communication Systems (ICCS)*. Mode selection for energy-efficient D2D communications in LTE-advanced networks: A coalitional game approach (IEEE, Singapore, 2012), pp. 488–492
14. D. Zhu, J. Wang, A. Swindlehurst, C. Zhao, Downlink resource reuse for device-to-device communications underlaying cellular networks. IEEE Signal Process. Lett. **21**(5), 531–534 (2014)
15. J. Zhao, Y. Liu, K. Chai, Y. Chen, M. Elkashlan, J. Alonso-Zarate, in *Proc. IEEE Global Commun. Conf. (GLOBECOM)*. NOMA-based D2D communications: towards 5G. (IEEE, Washington, 2016), 1–6
16. J. Zhao, Y. Liu, K. Chai, Y. Chen, M. Elkashlan, Joint subchannel and power allocation for NOMA enhanced D2D communications. IEEE Trans. Commun. **65**(11), 5081–5094 (2017)
17. Z. Zhang, Z. Ma, M. Xiao, Z. Ding, P. Fan, Full-duplex device-to-device-aided cooperative nonorthogonal multiple access. IEEE Trans. Veh. Technol. **66**(5), 4467–4471 (2017)
18. C. Li, H. Yang, F. Sun, J. Cioffi, L. Yang, Multiuser overhearing for cooperative two-way multiantenna relays. IEEE Trans. Veh. Technol. **65**(5), 3796–3802 (2016)
19. C. Li, S. Zhang, P. Liu, F. Sun, J. Cioffi, L. Yang, Overhearing protocol design exploiting inter-cell interference in cooperative green networks. IEEE Trans. Veh. Technol. **65**(1), 441–446 (2016)
20. Z. Ding, X. Lei, G. Karagiannidis, R. Schober, J. Yuan, V. Bhargava, A survey on non-orthogonal multiple access for 5G networks: research challenges and future trends. IEEE J. Select. Areas Commun. **35**(10), 2181–2195 (2017)
21. Z. Ding, P. Fan, H. Poor, Impact of user pairing on 5G non-orthogonal multiple access downlink transmissions. IEEE Trans. Veh. Technol. **65**(8), 6010–6023 (2016)
22. C. Li, P. Liu, C. Zou, F. Sun, J. Cioffi, L. Yang, Spectral-efficient cellular communications with coexistent one- and two-hop transmissions. IEEE Trans. Veh. Technol. **65**(8), 6010–6023 (2016)
23. C. Li, H. Yang, F. Sun, J. Cioffi, L. Yang, Adaptive overhearing in two-way multi-antenna relay channels. IEEE Signal Process. Lett. **23**(1), 117–120 (2016)
24. F. Zhou, Y. Wu, R. Hu, Y. Wang, K.-K. Wong, Energy-efficient NOMA heterogeneous cloud radio access networks. IEEE Netw. **32**(2), 152–160 (2017)
25. W. Wu, F. Zhou, P.ei. Li, P. Deng, B. Wang, V.CM. Leung, in *Proc IEEE Inter. Conf. Commun. (ICC)*. Energy-efficient secure NOMA-enabled mobile edge computing networks (IEEE, Shanghai, 2019)
26. P. Gover, A. Sahai, in *Proc. IEEE International Symposium on Information Theory*. Shannon meets Tesla: wireless information and power transfer. (IEEE, Austin, 2010), 2363–2367
27. Z. Luo, W. Ma, A.M. So, Y. Ye, S. Zhang, Semidefinite relaxation of quadratic optimization problems. IEEE Signal Process. Mag. **27**(3), 20–34 (2010)
28. H. Zhang, Y. Huang, S. Li, L. Yang, Energy-efficient precoder design for MIMO wiretap channels. IEEE Commun. Lett. **18**(9), 1559–1562 (2014)
29. J. Xu, L. Qiu, Energy efficiency optimization for MIMO broadcast channels. IEEE Trans. Wirel. Commun. **12**(2), 690–701 (2013)
30. M. Grant, S. Boyd, CVX: Matlab software for disciplined convex programming, version 2.0 beta (2013). http://cvxr.com/cvx
31. Y. Huang, D. Palomar, Rank-constrained separable semidefinite programming with applications to optimal beamforming. IEEE Trans. Signal Proc. **58**(2), 664–678 (2010)
32. C. Li, F. Sun, J. Cioffi, L. Yang, Energy efficient MIMO relay transmissions via joint power allocations. IEEE Trans. Circ. Syst. **61**(7), 531–535 (2014)

3

Quantitative social relations based on trust routing algorithm in opportunistic social network

Genghua Yu, Zhi Gang Chen*, Jia Wu* and Jian Wu

Abstract

The trust model is widely used in the opportunistic social network to solve the problem of malicious nodes and information flooding. The previous method judges whether the node is a cooperative node through the identity authentication, forwarding capability, or common social attribute of the destination node. In real applications, this information does not have integrity and does not take into account the characteristics and dynamic adaptability of nodes, network structures, and the transitivity of social relationships between nodes. Therefore, it may not be effective in solving node non-cooperation problems and improving transmission success rate. To address this problem, the proposed node social features relationship evaluation algorithm (NSFRE) establishes a fuzzy similarity matrix based on various features of nodes. Each node continuously and iteratively deletes the filtered feature attributes to form a multidimensional similarity matrix according to the confidence level and determines the weights under different feature attributes. Then, the social relations of nodes are further quantified. The experimental results show that, compared with the traditional routing algorithm, NSFRE algorithm can effectively improve the transmission success rate, reduce transmission delay, ensure the safe and reliable transmission of information in the network, and require low buffer space and computing capacity.

Keywords: Opportunistic social network, Malicious nodes, Information flooding, Cooperative node, Feature attributes, Social relationship

1 Introduction

In recent years, as wireless networks have penetrated into our daily lives, the application scale of the network has been increasing. As a new type of self-organizing network, it has attracted the attention of researchers at home and abroad [1, 2]. In order to get rid of the restriction of establishing the end-to-end communication path to achieve network communication, the concept of the opportunistic social network is proposed. This concept has been widely used in animal tracking, vehicle network, and other fields [3, 4]. Opportunistic social networks belong to intermittent connectivity networks. Opportunistic social network nodes are characterized by typical mobility, openness, and sparseness. Nodes have low encounter rates and lack fixed and secure connectivity links. Generally, the "Storage-Carrying-Forwarding" mechanism [5] relies on the opportunity brought by node mobility to realize routing. This model requires that all nodes cooperate to forward the routing messages of other nodes in a coordinated manner and realize communication hop by hop through the chances of encounters caused by node movement.

Due to the limitations of energy, computational capacity, network bandwidth, and buffer space of the nodes in the opportunistic network [6, 7], as well as the instability and uncertainty of node connections, existing trust modeling schemes are difficult to directly applied to the opportunistic social networks. These may lead to the following problems: (1) It is difficult to collect the evidence of direct trust accurately and timely. Because of the dynamic nature of the node, it is possible to leave the connected domain after delivering the message to the next hop node. Therefore, the evidence of successful forwarding cannot be collected by using neighbor node monitoring methods, and there is no credible authorization center. (2) The node's

* Correspondence: czg@csu.edu.cn; jiawu5110@163.com
School of Computer Science and Engineering, Central South University, Changsha 410083, China

uncooperative behavior results in the inability of a trusted authority to verify whether the next-hop node is a trusted node. Because in the process of transmitting information, the node may mask or not forward the received message for some reason. If the message is passed to more uncooperative nodes, the node's transmission success rate will be reduced. (3) The computational power and cache space of the nodes are limited. Existing trust modeling schemes [7] need to spend a lot of money on trust relationship acquisition, trust relationship maintenance and evaluation, and cache space. If they are forwarded to all nodes unconditionally, they will consume network resources [6–8]. Therefore, nodes in the resource-constrained opportunistic social network need to pay as little cost as possible to realize reliable message delivery.

In this paper, to address the challenges above, we propose a secure routing method named Node Social Features Relationship Evaluation (NSFRE) algorithm for screening trusted nodes based on social relations. To prevent the packet forwarding performance caused by packet forwarding in a flooding manner and thus causing network congestion, we introduce the relevant eigenvalues in the routing algorithm. NSFRE uses interaction records to establish feature information and network structure information of mobile nodes such as the number of connections, geographical location, and transitivity of relationships. NSFRE also establishes a fuzzy similarity matrix based on fuzzy feature vectors of nodes, and then iteratively computes social relationship values between nodes. According to the calculation results, the trusted nodes are selected, and it is considered that the same threshold is a trusted node, and a method is provided for a message source node to select a next-hop node with higher trustworthiness. This forwarding method easily finds the best path to the destination node. In this paper, through in-depth study of the internal relations between node activity rules and social relations, a trust routing table for node message forwarding is established, and the cooperative nodes that can forward the messages to the destination node are discovered and selected. Finally, it solves the problem of information flooding and node non-cooperation in the opportunistic social network and improves the real-time and reliability of information transfer between nodes. The algorithm performs experiments on real data sets and uses Stanford University's road structure as a network topology for simulation experiments. The experimental results show that our algorithm is superior to the four classic routing algorithms.

Specifically, the main contribution of this paper can be summarized as the following three aspects:

1. This paper studies the application of trust mechanism based on social relations in the network. In the network of opportunities, we proposed a method for filtering information by computing trust scores based on the value of social relations. This promotes that data packets in the network are always transmitted along trusted nodes, which reduces the blindness of information transmitted by other routing methods. The method minimizes the negative impact of uncooperative nodes on the network and improves the overall network performance
2. In the calculation of the value of a node's social relations, we are no longer only aimed at a single social attribute or adopting a subjective method such as the average weight method. In this paper, we calculate the value of social relations by first screening out the available nodes according to the characteristics of nodes, calculating feature weights according to the characteristics of node characteristics and social relations transitivity. Then, the social relationship values of the filtered nodes are calculated by the weight of each feature and the nodes that satisfy the characteristics. It reduces the subjective elements that give weights to features and increases the feasibility of screening mobile nodes for transmitting information.
3. In calculating the value of social relations, we consider that social relations change dynamically with time and increase the flexibility of social relations. In the new round of message information transmission process, according to the dynamic changes of social relations at different times, we will regenerate a new routing table, so that the social relationship calculation model has sufficient adaptability to the dynamic changes of the network and improve the accuracy of the model.

2 Related works

In order to reduce the harm of uncooperative nodes to the network, there have been many researches on the trust-based mechanism in the network at home and abroad, but the research on the trust mechanism in the opportunistic social network is still in its infancy. In the network, it is difficult to establish an end-to-end communication path between source and destination. Therefore, ad hoc routing protocols cannot be directly applied to opportunistic social networks. Instead, it uses a store-and-forward mechanism to communicate. However, selfish nodes will delete some data obtained from other nodes and thus seriously affect network performance. People have designed a trust-based routing protocol [9] (TRP) that combines various practical algorithms to reduce the negative impact of malicious nodes.

For many uncooperative nodes in the opportunistic social network, the uncooperative nodes are mainly divided into two types: (1) The received information is not forwarded. (2) The information is rejected and not forwarded. These uncooperative nodes have been able to detect packet loss behavior through some uncooperative node detection algorithms such as LARS [10]. However, the uncooperative nodes could use more concealed uncooperative behaviors to conceal themselves. For example, the probability of losing packets is controlled to be less than the threshold so as to avoid being punished. Therefore, many researchers propose to reduce the impact of uncooperative nodes on network resources based on the social trust model. Among them, a social trust model is proposed for secure routing in the opportunistic network [11]. This algorithm uses the node state forwarding capabilities and common attributes to evaluate social trust values.

For malicious nodes, in the harsh environment where node density is sparse, slow-moving nodes do not have the opportunity to join network routes because they cannot effectively use the opportunities encountered to achieve self-organized authentication. There is no need to establish a complete mutual authentication for each dialog. People are aware of the inadequacy of the idea of "authenticating" nodes to determine trust relationships. Therefore, people proposed a new trust management scheme based on behavior feedback information [12]. By using a certificate chain based on social attributes, the mobile node gradually establishes a local certificate graph and implements an "identity authentication" trust relationship.

For the network based on model trust proposed by people, you first need to use the proposed method to detect malicious nodes. When quantifying the trust of node, most existing methods rely on the number of final ACK messages received by the node. If the final ACK message cannot be reliably received, it will affect the reliability of the malicious node's judgment. Therefore, the researchers proposed a "double-hop feedback method" [13] to design a dynamic trust framework. It promotes the node to obtain the trust value of another node based on the behavior of the latter node to detect the selfish malicious node.

Furthermore, people apply social relations to networks. The analysis of social relationships is generally based on the structure obtained by social networks and uses the user's information flow to analyze the strength of the user's relationship, and then uses the weights on the edges of the graph [7] to establish trust model. In the opportunistic network, it has been proven that the user's social profiles are useful for finding suitable forwarding nodes in a delay tolerant network. A Social Relationship Opportunistic Routing Algorithm (SROR) [14] was proposed for mobile social networks. Social relations and profiles between nodes were used as the key indicators for calculating the optimal forwarding node in the route to maximize the packet transmission probability.

In addition to using social relationships to select the optimal forwarding node, people also propose a method extracting information from people's social interactions to quantify the trust relationship between nodes. Some researchers believe that acquiring trust from real-world social interactions can play an important role in understanding social behavior. Therefore, an opportunistic sensing system [15] is proposed, which can detect social interactions based on the real world and acquire and quantify trust relationships among people through smartphones.

Through the above analysis, we can see that the existing work is mainly focused on the limited communication radius and the nodes can be trusted. The information is transmitted accurately through multiple hops. The working method is similar to the WSN, except that the original static node is extended to the dynamic mobile node. As long as the communication between nodes is reachable, the data can be transmitted, but without considering the trust relationship delivery between nodes. There have been social relations calculations that start from a real scenario to analyze a specific attribute, or treat social relationships between nodes as static, without taking into account the dynamic changes in node social relations, and the decision feature that affects the quantification of social relationships is not enough considered. Therefore, this paper uses the research results of trust model and social computing in social networks to determine the trust relationship between nodes based on the dynamic characteristics and social relationships of mobile nodes based on social network theory. Through analyzing the characteristics of the social network of mobile nodes in the opportunistic social network, the corresponding feature information such as interactive quality feature Q, position feature P, trust quality characteristics T, and social relation feedback feature S are extracted to study the social relationships among mobile nodes. Then, based on the theory of information entropy, rough set, and so on, a mobile node social relations computing model is proposed. The model is used as the quantification of social relations between nodes and the weight distribution of decision features, so that the trustworthiness between nodes can be calculated to filter out the next hop node set for information forwarding.

3 Social characteristics analysis of nodes

In the opportunistic social network scene, cognition of social relationships is through the use of various sensing devices (mobile phones, PDAs, etc.) attached to mobile nodes [4]. The real-time information such as the activity rules and interaction records of the nodes can be obtained in real time to analyze the key feature influencing the social relations and explore the internal relations among them [16, 17]. Quantifying social relations can more objectively and accurately reflect the changes in the relationship between mobile nodes. Through the analysis of social networks, it shows that the social relations between mobile nodes have the following characteristics:

(1) Diversity: The information transmission of opportunistic social networks mainly relies on mobile nodes. Social relationship is the tie of the mobile node and also affects the activity rule of the mobile node. Due to the spatiotemporal characteristics of the mobile node, the calculation of the relationship involves many factors such as behavior and environment, and it is difficult to accurately quantify and predict [3].
(2) Inconsistencies: It refers to the directionality of social relationships in the interaction process of mobile nodes. This directionality causes the relative social relationships between the two parties to differ due to their internal and external factors [4, 6], namely the different perceptions of the same event and information. For example, the social relations between nodes u and v are $M(u, v) = 0.8$, whereas the relations between v and u may be $M(v, u) = 0.6$.
(3) Mobility: The value of social relations is a variable over time. With the change of decision features such as the law of activity and degree of interaction between each other, the social relations change dynamically. For example, the social relationship between nodes u and v may be $M(u, v) = 0.6$ for a period of time, and $M(u, v)$ may be 0.3 at the next moment as various network features change.
(4) Transitivity: When calculating social relations among mobile agents indirectly through judgments made by other nodes or environmental information rather than based on direct contact with each other, we call this process the transitivity of social relations. For example, nodes u and v, v, and w have social relations, then nodes u and w can establish social relations through v under certain conditions.
(5) Sociality: In the scene, the node behavior is not disordered, but is influenced by the characteristics [18] of individual consciousness, social role, demand, etc., and has certain social characteristics. For example, the daily activities of office workers are driven by events such as scheduling.

Through the above analysis, it can be seen that social relations are inherent manifestations of different modes of interaction among nodes and involve a variety of factors. Considering the complex and diverse characteristics [16, 17, 19] of the interaction patterns among nodes in the information delivery service for opportunistic social networks, we introduce various types of decision features to describe the spatial and temporal characteristics of social relations of nodes. Through the collection, analysis of the history record of connections between nodes and calculating interactive quality feature Q between each other, the interaction rules between each other are found out. Position feature P reflects the trajectory characteristics of nodes in the time period. It statistically analyzes the trajectory characteristics of nodes based on different geographical position and studies the frequency of different mobile nodes reaching the same sensing area within a certain period of time. The defined trust quality characteristics T represent the mobile node's evaluation of historical interaction information records, which reflects the satisfaction of service requesters with service providers in former information interactions. Similarly, the social relation feedback feature S is defined to reflect the transitivity of social relations between nodes [17, 19] and further improve the accuracy of quantitative social relations.

It can be seen that the social relationship cognitive model building process includes the following features:

(1) Real-time perception of mobile node information.

Mobile node through its own terminal equipment can obtain various behavior information of the user in real time [20–22], such as location information, network information, current status of the active node, etc. Through the classification and preprocessing of the original information of the nodes, the perceived service center extracts the trajectory information [22–24], connection records, and historical information interaction records of the nodes to provide data support for the next decision feature calculation and social relations quantification.

(2) Calculation of mobile node decision feature.

In the opportunistic social network, the forwarding of message is based on the social relations between nodes. The dynamic changes of social relations are determined by both space and time. Therefore, we introduce Q, P, T and S to describe the dynamic social relations of nodes.

(3) Mobile node decision feature weight assignment.

Social relationships of mobile nodes change over time, and at different times, states are interrelated. In this paper, the decision feature knowledge base is obtained by using rough set theory. Based on the information entropy, the decision feature is dynamically and rationally weighted. Finally, the nodes' next-hop nodes are screened by its social relation quantification algorithm to form a node forwarding domain, which provides the decision-making basis for the realization of the opportunistic social network information transmission.

From the above analysis, we can see that the quantitative modeling of social relations of mobile nodes is of great significance for expanding the scope of information transmission, reducing transmission delays and improving the quality of information transmission in opportunistic social networks. It is embodied in (1) using the main perceptive devices of the network to form a virtual social network, analyzing and calculating the social relations. In addition, a new intelligent network of object-matter, human-object, and human-human interaction is truly realized [19], so that when the next hop node is filtered by the source node, we can better integrate the dynamic characteristics of the network (2) in the social relation calculation, giving full consideration to the characteristics of the mobile node attributes and network structure characteristics. From different perspectives for feature analysis and node screening modeling, social relationship calculation is more comprehensive, objective, and reasonable; (3) social relations is the basis of this node to choose the next hop node to transfer information, and the follow-up work is based on this expands.

4 Quantitative model of social relations

In this paper, we consider various features that affect social relations and introduce decision features such as *P*, *Q*, *T* and *S* to describe the complexity, transitivity, and uncertainty of social relations from different perspectives.

4.1 Multi-dimensional decision feature of node calculation

Definition 1 The quantified value of social relations $M(u, v)$ between node u and v is:

$$\begin{aligned} M(u,v) &= \sum_{i=1}^{m} w_i f_i(u,v) \\ s.t.\ & 0 \le w_i \le 1 \\ & \sum_{i=1}^{m} w_i = 1 \\ & u, v \in N \end{aligned} \tag{1}$$

where f_i represents different types of decision feature and ω_i represents the weights of different decision feature; m is decision characteristic number, and N is the set of nodes in the network. When $M(u, v) = 0$, there is no social relationship between nodes u and v. On the other hand, when $M(u, v) = 1$, u and v are the same nodes, that is, $u = v$.

As mobile smart terminals accelerate people's information exchange and the evolution of their social relationships, analyzing the model of connection between mobile nodes can reflect the social relations and interaction rules among different nodes. For example, the interaction model between friends may be different to strangers.

Definition 2 The total time for the mobile node u to establish a connection in the period and the total number of connections are denoted as T_u^K and N_u^K, respectively. The connection time and the number of connections between nodes u and v are denoted as $T_{u,v}^K$ and $N_{u,v}^K$, then the interactive quality feature $Q(u, v)$ of nodes u and v can be expressed as:

$$Q(u,v) = \sqrt{\sum_{k \in \{\text{req},\text{res}\}} \left(\frac{T_{u,v}^k}{T_u^k} \cdot \frac{N_{u,v}^k}{N_u^k} \right)} \tag{2}$$

where $k \in \{\text{req}, \text{res}\}$ represents the request and response record.

We can divide the sensing area into areas of different radius according to the different types of services perceived by the mobile terminal. There may be a deviation in the final stop position of the mobile node entering the same area multiple times. Therefore, by clustering the staying positions in the trajectory of the node, the staying positions of the same area are divided into the same cluster. In this way, the trace of the mobile node can be expressed as a time sequence reaching different areas.

Definition 3 The trajectory information of mobile node u in the time period is expressed as $L = \{U(q_i, c_i, p_i, \gamma)\}$, where q_i, c_i are the time sets when the node arrives and leaves the region respectively, p_i is position information of the sensing area of the ith track information, and γ is the time threshold, which is used to control the time interval between different nodes reaching the target area. Then, the location feature $P(u, v)$ of nodes u and v are expressed as:

$$P(u,v) = \frac{\sum_{i=1}^{n} B(L_u^i, L_v^i)}{T} \tag{3}$$

In the above formula, T is the time period, $B(L_u^i, L_v^i)$ is a similarity function between mobile nodes u and v in position p_i, which reflects the duration of the encounter between different nodes at the same location by the result which can be calculated by Eq. (4). Among them,

n is the total number of trajectory information in the T time period, and γ is the time threshold used to control the time interval between different nodes reaching the target area.

$$B(L_u^i, L_v^i) = \max\{q_u^i, q_v^i\} - \min\{c_u^i, c_v^i\} \quad s.t. \ |q_u^i - q_v^i| \leq \gamma \tag{4}$$

In the process of information transmission of nodes, the evaluation of the quality of connections established between nodes indicates the degree of stability of the information transmitted from the requesting connector to the receiver. Therefore, establishing the connection between the two parties to the stable evaluation of information transmission will change the social relationship between each other, while the trust quality factor reflects the property that the social relationship dynamically changes with the change of node connection stability.

Definition 4 Assume that the evaluation of mobile node u for v in the last n connections is recorded as $R(u, v) = \{t_{u,v}^1, t_{u,v}^2, \ldots, t_{u,v}^n\}$, $0 \leq t_{u,v}^i \leq 1, i \in [1, n]$, where the elements are arranged according to the historical interaction time, n is the historical interaction record threshold. The quality of trust characteristics between nodes can be expressed as

$$T(u, v) = \begin{cases} \sum_{i=1}^{n} \frac{t_{u,v}^i \cdot \sigma(i)}{n}, & n \neq 0 \\ 0, & n = 0 \end{cases} \tag{5}$$

In the formula above, $\sigma(i)$ is the attenuation function used to weight the interaction feedback evaluation that occurs at different times. Note that we give a higher weight to the latest interaction records evaluation. The attenuation function $\sigma(i)$ is calculated as:

$$\sigma(i) = \begin{cases} 1, & i = n \\ \sigma(i-1) = \sigma(i) - 1/n, & 1 \leq i \leq n \end{cases} \tag{6}$$

In calculating the social relations between nodes u and v, we should considering the transitivity of social relations; in addition to directly calculating the social relations between u and v, u can indirectly obtain the social relations values about v from other nodes. As shown in Fig. 1, nodes A and D are indirectly connected, and the social relations among them can be transmitted through other peers in the path. Therefore, we can use the feedback aggregation process which calculates the social relationship values of different information feedback nodes to the target node according to the transitivity between social relations. Due to different relationship information feedbacks having different number of hops from the source node, the reliability of the feedback information is different, and so, a simple arithmetic average calculation cannot be adopted. This paper uses the aggregation algorithm [25, 26] to calculate the social relation feedback feature between different mobile nodes.

Definition 5 Suppose the node that receives the information and feedback set as $\{b_1, b_2, \ldots, b_n\}$ and $M(b_i, v)$ represents the social relationship value between the ith information feedback and the mobile node v. Then, u and v $(u, v \in N)$ social relation feedback feature is:

$$S(u, v) = \begin{cases} \frac{\sum_{i=1}^{n} (w(b_i) \cdot M(b_i, V))}{\sum_{i=1}^{n} w(b_i)}, & n \neq 0 \\ 0, & n = 0 \end{cases} \tag{7}$$

In the above formula, n is the number of relationship information feedback nodes, and $S(u, v) = 0$ and $w(b_i)$ are

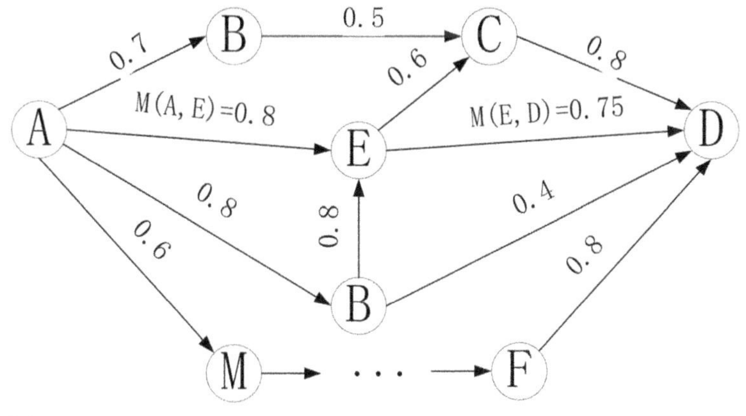

Fig. 1 Transitivity of social relations

the feedback weighting functions when there are no feedback nodes providing information in the opportunistic social network.

$$w(b_i) = \begin{cases} \prod_{i=0}^{l-1} M(a_i, a_{\text{next}}), & l > 1 \\ 1, & l = 1 \end{cases} \quad (8)$$

Among them, $M(a_i, a_{\text{next}})$ represents the social relationship value between the node a_i and its next node in the social relation transfer path from the source node u to the destination node v. And l represents the distance between the information feedback node and the source node.

4.2 Decision feature weight distribution

In the process of quantification of social relations, the size of weight reflects the status of each attribute index in the quantification of social relations decision-making, which directly affects the quality of service of the subsequent node information delivery. Therefore, an important precondition for solving the quantitative problem of social relations is designed—a reasonable and effective weight distribution method.

Rough set theory is a tool to deal with the uncertainty of knowledge, and the information entropy [26] is often used to describe the knowledge uncertainty.

Definition 6 The system uncertainty can be expressed as entropy [26] $E(X^*)$, which is

$$E(X^*) = -\sum_{i=1}^{n} P(A_i) \log_2^{P(A_i)} \quad (9)$$

where X^* is the partition of X on domain U, $X^* = U/X = \{A_1, A_2, \dots, A_n\}$.

Definition 7 Let Y be another kind of equivalence relation on domain U, $Y^* = \{B_1, B_2, \dots B_m\}$, then X^* is known, the conditional entropy of Y^* is:

$$E(Y^*|X^*) = -\sum_{i=1}^{n} P(A_i) \sum_{j=1}^{m} P(B_j|A_i) \log_2^{P(B_j|A_i)} \quad (10)$$

Definition 8 The amount of mutual information of knowledge reflects the amount of information that Y gets from X and can be expressed as

$$I(X^*; Y^*) = E(Y^*) - E(Y^*|X^*) = E(X^*) - H(X^*|Y^*) \quad (11)$$

4.3 Node social features relationship evaluation algorithm

The quantitative model of social relations in the process of information transmission of opportunistic network nodes is based on the analysis of social characteristics of nodes and calculate decision feature that affects the change of node relationships. The decision feature and the entropy theory are combined to determine the weight distribution of different decision feature. Finally, for the social relationship between mobile nodes to make a reasonable quantification, the following is given the overall realization of this model process.

Algorithm 1 Node Social Features Relationship Evaluation Algorithm (NSFRE).

Input: characteristic information of node N

STEPS:

1. First of all, input the characteristic information of the node, including the location information, connection information, interaction information, feedback information, time information, and other features, and then select the nodes that contain the relevant features in turn, and remove the nodes with less features.

2. Calculate the characteristic attribute of mobile node according to Eqs. (2)–(8).

3. There are n sample objects, each of which has m feature vectors due to a total of m decision features. Then some sample objects can be expressed in matrix as:

$$A = \begin{bmatrix} a_{11} & a_{12} & \cdots & a_{1n} \\ a_{21} & a_{22} & \cdots & a_{2n} \\ \vdots & \vdots & \ddots & \vdots \\ a_{m1} & a_{m2} & \cdots & a_{mn} \end{bmatrix}$$

4. We establishing a fuzzy similarity matrix [26] between objects and object $R = (\lambda_{ij})_{n \times n}$. The method is as follows:

$$\lambda_{ij} = \sum_{k=1}^{m} (a_{ki} \wedge a_{kj}) / \sum_{k=1}^{m} (a_{ki} \vee a_{kj}) \quad (12)$$

5. We find its transitive closure matrix by fuzzy similarity matrix, that is

$$t(R) = R^{2^k} \leftarrow \cdots \leftarrow R^4 \leftarrow R^2 R^{2^k} = R^{2^{\frac{k}{2}}} \circ R^{2^{\frac{k}{2}}} \quad (13)$$

We can classify by fuzzy similarity matrix. Similar classes on similar relations can be merged into equivalence classes about their transitive closures. Mergers corresponding to high-threshold equivalence classes can directly obtain equivalence classes corresponding to low thresholds, so the principle of merging is: If $\lambda_{ij} = \delta$, the equivalence class $[a_i]_{t(R)}$ is merged with $[a_j]_{t(R)}$, where $[a_i]_{t(R)}$ represents the equivalence class containing the element $a_i \in U$ in the transitive closure relation $t(R)$, and δ is the threshold. In this way, we can choose the threshold δ from large to small and realize the classification of different needs.

6. Sort λ_{ij} from large to small as the basis for selecting threshold δ.

7. Select the maximum value δ_1, and take a_i and $a_j$satisfying $\lambda_{ij} = \delta_1$ as a class. If $\lambda_{ij} = 1$, then a_i and a_j satisfying $\lambda_{ij} = \delta_1$ at this time are exactly the elements in the same equivalence class in the rough set.

8. Taking the second largest value δ_2 in λ_{ij}, directly find the element pair (a_i, a_j) whose similarity is equal to δ_2 from $(\lambda_{ij})_{nxn}$, and correspondingly $[a_i]_{t(R)}$ and $[a_j]_{t(R)}$ merge.

9. Repeat step (7) until the selected value is less than the predetermined threshold δ_0.

10. The last merge will be undone and form the final classification $C_i (i = 1, 2, \ldots, \zeta)$.

We classify the domain U by sorting λ_{ij}, the classification results are recorded as $C_i (i = 1, 2, \ldots, \zeta)$; ζ is the classification number. Second, after deleting each attribute from all attributes in turn, repeat steps 4–9 to determine the number of categories within the same threshold range, denoted as C_j, and the like, examining the impact of each attribute on the classification and storing the result into the decision feature classification knowledge base.

11. Calculate the amount of mutual information $C_{i,\ j}$ of nodes that have been selected as feature factors of screening condition within the same threshold range by C_i, C_j, and Eqs. (9)–(11), and calculate the trustworthiness of nodes under different characteristic attributes by the following formula:

$$\theta_j = \sum \frac{1}{\gamma} \{C_{i,j} | i, j \in [1, \zeta]\} \tag{14}$$

12. Calculate the weight of each characteristic attribute according to the value of the trustworthiness of the characteristic attribute. The weight distribution formula of the decision feature is as follows:

$$w_j = \frac{\theta_j}{\sum_{j=1}^{n} \theta_j}, (j = 1, 2, \cdots, m) \tag{15}$$

13. According to Eq. (1) to calculate the mobile node social relations quantitative value V, and finally output social relations value.

14. Output social relations value, the end.

Algorithm 1: Node Social Features Relationship Evaluation Algorithm (NSFRE)

Input: m:feature number;
n:node number;
ζ :Classification numbe;
τ :Classification confidence level number;

Output: M

```
1:  Initialize Matrix A,X   //Assign initial value ;
2:  Initialize M[][],k[],w []
3:  n=selectFeatureNode(m,n)   //Select n nodes with m feature attributes;
4:  For i=1 to n and j=i to n do
5:      Calculate the Q(i,j) base on (2)   //calculate interactive quality feature value;
6:      Calculate the P(i,j) base on (3)(4)   //calculate location feature value;
7:      Calculate the T(i,j) base on (5)(6)   //calculate the quality of trust characteristics between nodes;
8:      Calculate the S(i,j) base on (7)(8)   //calculate nodes social relation feedback feature value;
9:  End for
10:   a_i ← getFeatureVector()   //Get the feature vector for each node , i∈[0,n);
11:   A ← getMatrix( a_i ,m,n)   //The feature matrix is obtained from the feature vector;
12: For z=1 to m do
13:     For i=1 to n and j=1 to n do
14:       λ_ij ← getVagueValue(X.getValue(k,i),X.getValue(k,j))
15:         X.add( λ_ij ,i,j)   //the similarity between different objects,establish a fuzzy similar matrix R;
16:     End for
17:     For k=1 to τ do
18:         s(X)= X^(2^k)   //Using squared self-synthesis method to find the fuzzy equivalent closure matrix
19:     End for
20:     Calculate C_i when the attribute is not deleted, calculate C_j base on (9)(10) (11) when each attribute is removed from all atribures in turn //the number of categories within the same threshold range
21: If C_j is not NULL
22:    C_i,j =getMutualInf( C_i ,C_j );
23:    For i=1 to ζ and j=i to m do
24:         k_j = k_j + (1/r) C_i,j   //calculate the trustworthiness of nodes under different characteristic attributes;
25:       k.add( k_j )
26:    End for
27: End IF
28:    X=X.remove(z,*)   //delete the filtered feature attributes and repeat steps 15-27;
29: End for
30: For j=1 to m do
31:       w_j ← getWeightFeature(k.getValue(j))   //Calculate the weight distribution of feature attributes;
32:       w.add( w_j )
33: End for
34: M ← getSocialValue(w, Q, P, T, S, m, n)   //Calculate social relations value;
35: Output M
```

5 Experiment analysis

In this paper, The One Simulator is used to simulate the proposed algorithm, and some opportunistic network classical routing algorithms are compared. The performance of the NSFRE algorithm is evaluated from the aspects of transmission success rate, routing overhead, and transmission delay.

5.1 Simulation tools and scenes

In the experiment, the Stanford University Topology was used as a simulation scenario. The simulation scenario is set as follows (Fig. 2):

The data we use in the simulation scenario is a real data set of Stanford University. We design different numbers of pedestrians, cars, and electric tracks to simulate the effect of the number of nodes, simulation time, and node caching on simulation results. Among them, the experimental parameters are set as follows: We choose Stanford University real map area is 1070 m × 810 m. The simulation time is from 1 to 12 h. The simulation node is set to 120–900, and the node cache is set to 5–40 M. The speeds of pedestrians, cars, and trams are 5 km/h, 100 km/h, and 60 km/h, respectively, the channel bandwidth is 250 kb/s, and the bandwidth of high-speed transmission interfaces is 10 m/s. In addition, the node's mobility model is Shortest Path Map Based Movement [1]. The node's transmission mode is a social model. The default number of nodes is 300. Each node's

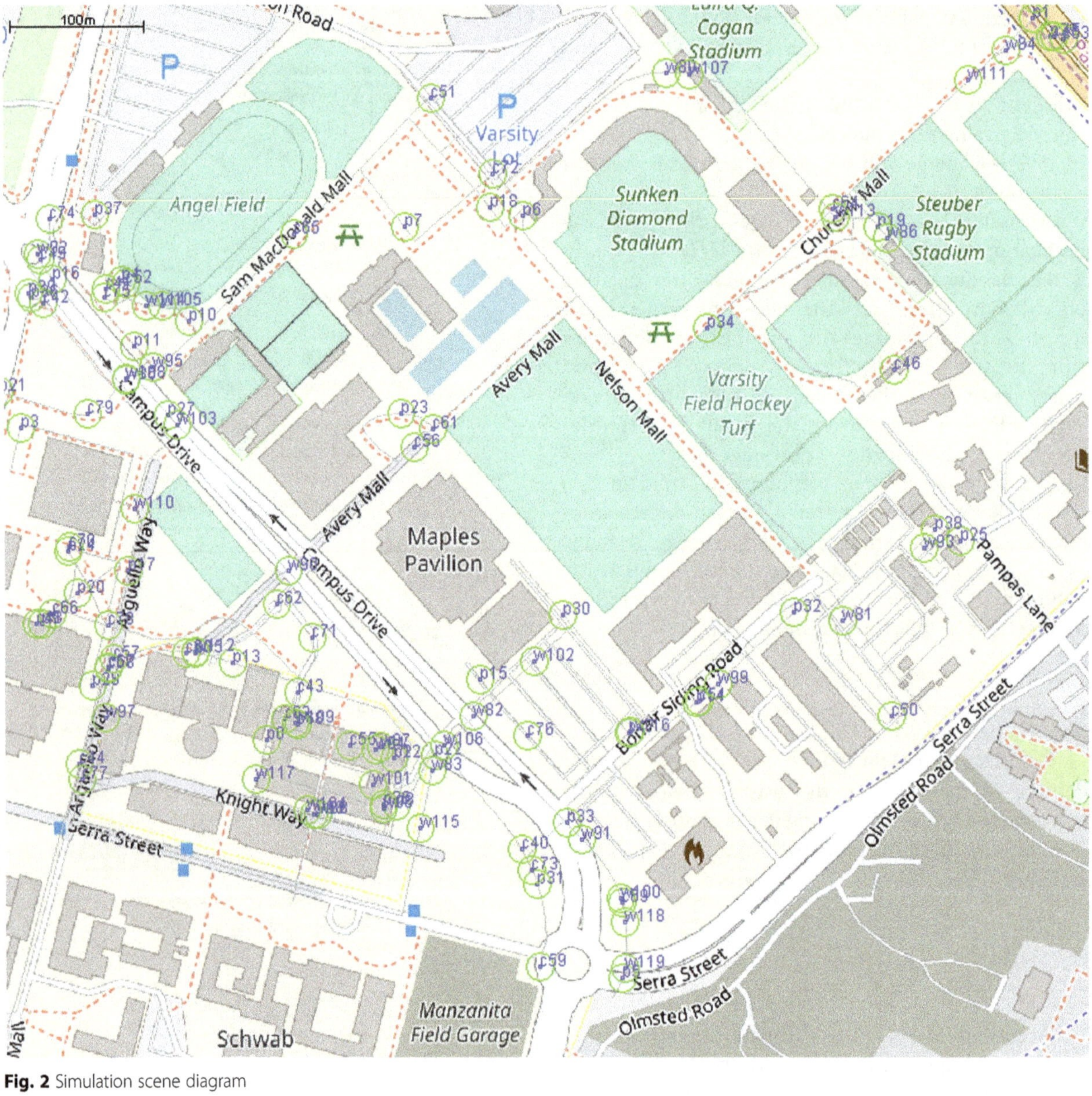

Fig. 2 Simulation scene diagram

cache is 8M. The maximum transmission area of each node is $10\,m^2$. The frequency ranges from 25 to 35 HZ and the packet type is a random array. The topology of the simulation experiment is shown in Fig. 3.

In the model simulation experiment proposed in this paper, we simulate the real data set in The One, in which the wired connection node is the trusted node selected by the source node. It can be seen that the trust node is filtered by the model. With the increase of time, the transmission connection established between the nodes is reduced, which means that the number of cooperative nodes is reduced by the model selection, and the nodes that can cooperate are changed from high density to low density. The screening of trusted nodes is more and more accurate. It not only avoids the insecurity of data flooding, but also improves the transmission efficiency of the network, reduces the probability of network congestion, and enhances the robustness of the network structure.

5.2 Experimental results

The experiment mainly analyzes the algorithm's performance in transmission success rate, transmission delay, and routing cost by adjusting parameters. The model is mainly evaluated from two aspects: (1) effectiveness analysis: experiments were conducted to compare the difference

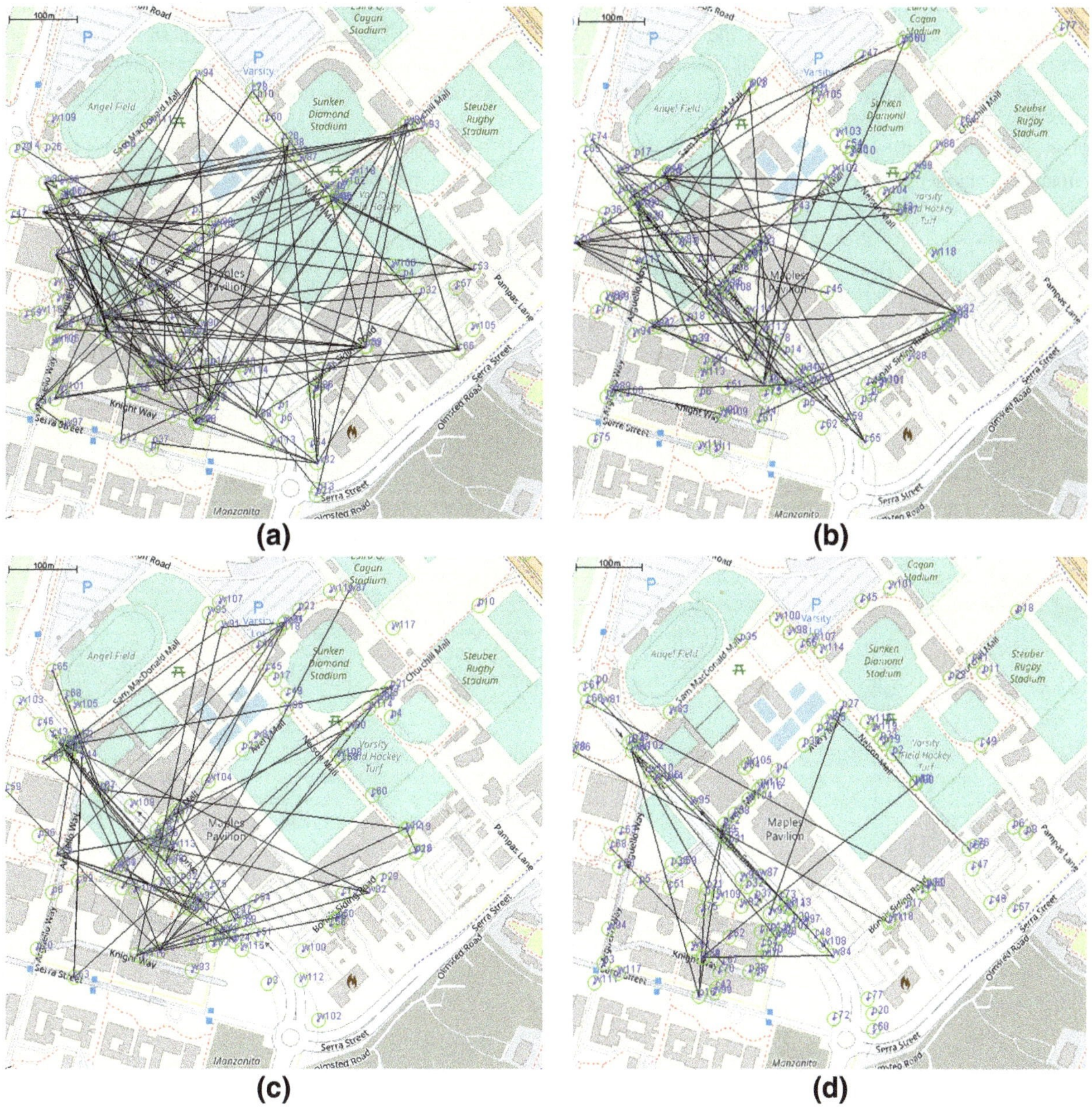

Fig. 3 Simulation experiment topology

between the model and other existing models in optimizing the network structure and increasing the success rate of information transmission and (2) adaptability analysis: through the process of dynamic changes in various uncertainties, the trusted nodes are selected to send data, which reduces the amount of information on the network and improves information transmission capabilities. Because our information transmission concept is still the classic way of probabilistic routing. The difference is that we form a quantitative probability value based on the social characteristics of the node, and the nodes are filtered by this value. Furthermore, we have made great improvements to the method of filtering nodes of the Prophet algorithm. Therefore, by comparing with other classical transmission methods applied in opportunistic social networks, we demonstrate the advantages and importance of node social characteristics and probabilistic transmission. As a reference, the algorithm of this paper is compared with Epidemic [27], Spray and wait [28], First Contact [29], and MaxProp [30] routing algorithm to analyze and compare the characteristics of each algorithm. It is proved that the proposed algorithm is more effective.

Figure 4 shows the relationship between the transmission success rate and simulation time. We can see that the algorithm's transmission success rate gradually increases as the simulation time increases. First Contact and Epidemic routing algorithm have the lowest transmission success rates, only 0.2 and 0.18, respectively. The reason is that Epidemic algorithm uses flooding to transmit information in nodes. Each node has too many message information. When the node cache is small, it is easy to cause a large amount of information data to be lost. However, First Contact is based on the forwarding strategy. It does not duplicate the information in the node and only transmits one copy of the message in the network. Therefore, the transmission success rate is low. Spray and wait routing algorithm improves information transmission success rate while reducing the number of information copies. However, the node of the MaxProp algorithm maintains a packet queue according to the transmission cost to the target node and determines the replication order according to the priority of the packet, which not only avoids network congestion, but also avoids waste of resources by blindly copying message information between nodes. Therefore, the transmission success rate of the spray and wait routing algorithm is lower than that of the MaxProp algorithm. The NSFRE algorithm has the highest transmission success rate, reaching 0.68. Precisely, it uses a combination of features to calculate the social network trust relationship to select trusted nodes, which reduces network congestion and information replication and improves the efficiency of the selected node and the reachability of the destination node. It also effectively improves the algorithm's transmission success rate.

Figure 5 shows the relationship between routing overhead and time. From the figure, we can see that the routing overhead of the NSFRE algorithm is not affected by the time. In the early stage, like the MaxProp routing algorithm, the routing overhead has a sharp downward trend, and the cost of the later routing is maintained between 12 and 26. The reason is that, as time increases, the number of nodes filtered by the algorithm will gradually decrease. Because the selection of nodes for

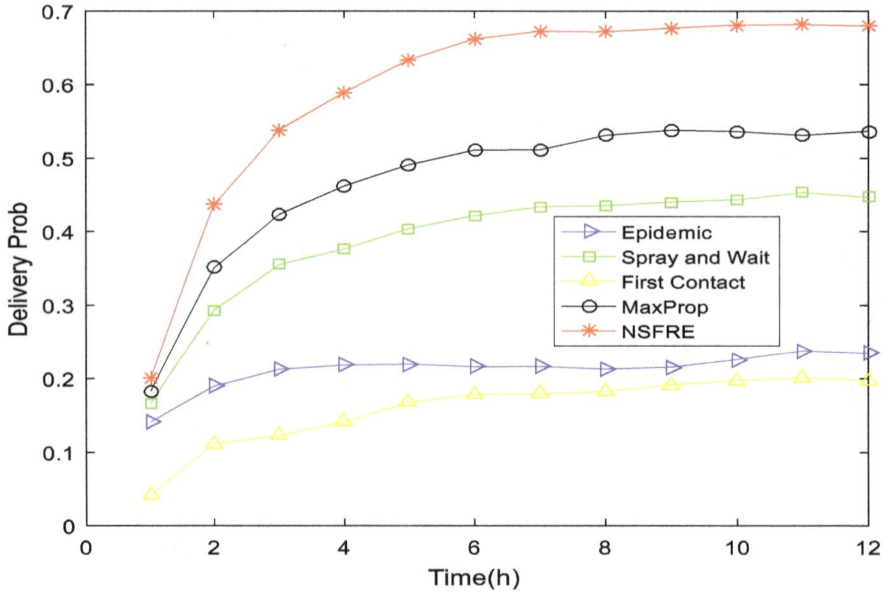

Fig. 4 The relationship between transmission success rate and simulation time

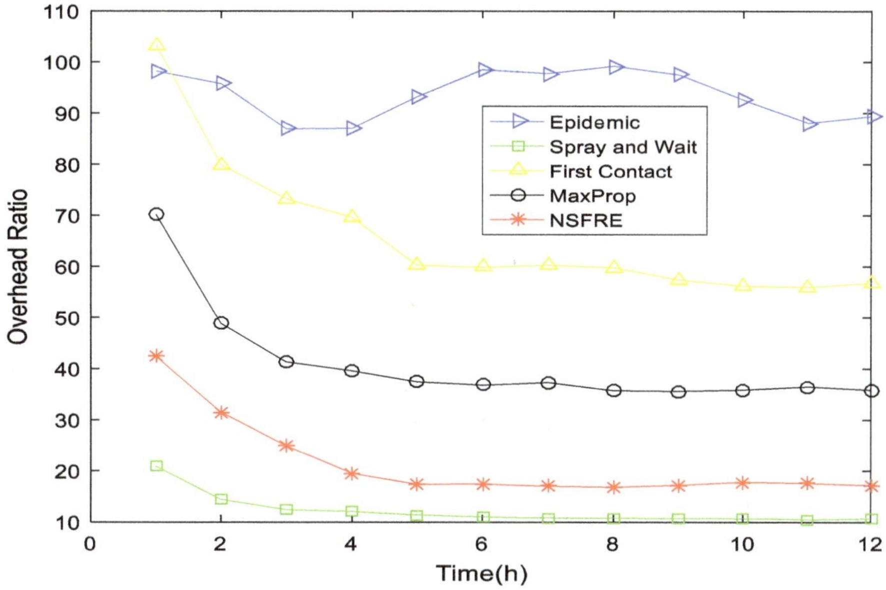

Fig. 5 The relationship between routing overhead and time

information transmission will be more accurate and the number of nodes sharing information transmission tends to be stable, the routing overhead can be kept stable. Spray and wait routing algorithm is similar to the algorithm in this paper. Compared with other algorithms, the routing overhead is relatively small, and as the time increases, the routing overhead decreases. This is because the Spray and wait routing algorithm reduces the amount of data transmission in the network and has good scalability. However, the route overhead of the Epidemic routing algorithm is relatively large, and the fluctuation is relatively large. This is because the algorithm maximizes the success rate of packet transmission. Each node carries a copied packet, and a large number of packet copies exist in the network. The network performance is degraded and the network structure is unstable.

Figure 6 shows the relationship between the average transmission. From the figure, the MaxProp algorithm determines the packet priority based on the transmission

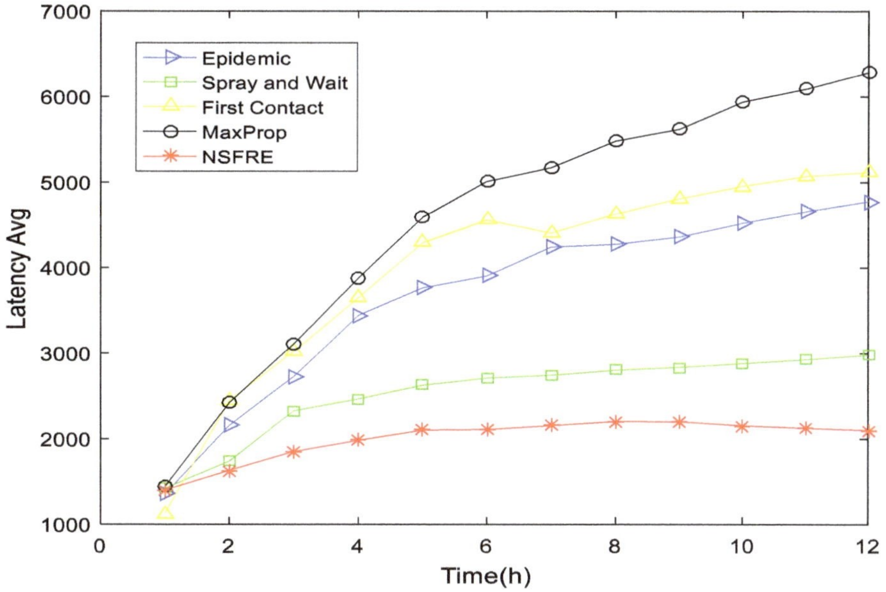

Fig. 6 The relationship between average transmission delay and time

cost, and the transmission cost is an estimate of the probability that the packet is successfully transmitted to the target node, which is estimated by the incremental averaging method. Because each transmission message must be calculated, the algorithm has the highest transmission delay. Because the First Contact algorithm only sends a copy of the message to the node that met for the first time, and the probability of the first node meeting with the destination node is very small, the algorithm has a higher propagation delay. The Epidemic routing algorithm uses flooding to deliver packets. As time increases, more and more packets are transmitted on the network. The resources in the network are consumed in large amounts. It is easy to cause network congestion, resulting in a high transmission delay. The average delay of the Epidemic algorithm reaches 4000. The NSFRE algorithm's delay is similar to the Spray and wait routing algorithm. This is because the algorithm uses resources to filter the nodes according to different factors, causing delays. However, as the NSFRE algorithm increases with time, the transmission delay tends to be flat, which proves that the algorithm has good stability and the information transmission delay does not increase exponentially with time.

Figure 7 shows the relationship between the transmission success rate and the number of nodes. We can see that the transmission success rates of the First Contact routing algorithm and the Epidemic routing algorithm are the lowest. With the increase in the number of nodes, the success rate of these two algorithms does not change much, only 0.24 and 0.18, respectively. The reason is that the Epidemic routing algorithm uses the form of flooding to carry out the information transmission to the nodes, causing a great deal of information data loss. The First Contact routing algorithm only transmits one replication message. With the increase of the number of nodes, the transmission success rate is stable at 0.24. The Spray and wait routing algorithm copies a certain amount of information for each message. With the increase of the number of nodes, the number of nodes receiving the copied message information increases, and the probability of encountering the node and the target node also increases, so the transmission success rate has increased, reaching 0.8. The transmission success rate of the MaxProp routing algorithm and the NSFRE algorithm is high, exceeding 50%. The MaxProp routing algorithm sends packets based on the calculated routing overhead, which improves the transmission and reception of valid information. Therefore, the transmission success rate reaches 0.51–0.82. The NSFRE algorithm has the highest transmission success rate, reaching 0.53–0.93. This is because the NSFRE algorithm takes into account the reachability of the destination node to calculate and filter the transmission of trusted mobile nodes. With the increase of nodes, the nodes are more closely connected and the reachability between nodes is higher, which effectively improves the transmission success rate of the algorithm.

Figure 8 shows the routing overhead and the number of nodes. In the figure, the routing overhead of the NSFRE

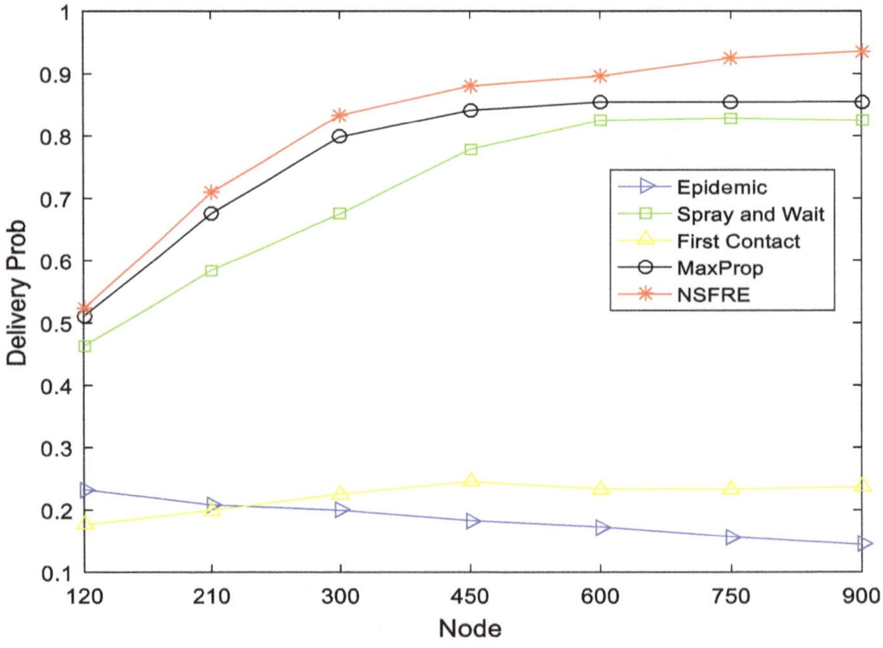

Fig. 7 The relationship between transmission success rate and number of nodes

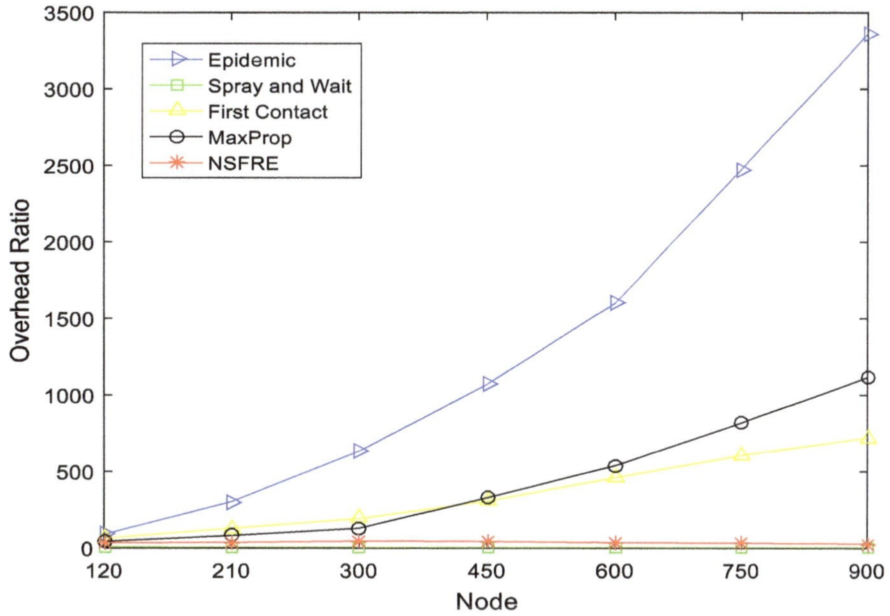

Fig. 8 The routing overhead and the number of nodes

and Spray and Wait algorithms is basically independent of node density. For the NSFRE algorithm, as the number of nodes increases, the connections between different communities will become closer. The energy consumption of the nodes filtered by the algorithm will not affect the information transmission overhead of the entire network, and the algorithm has good stability. For the other three algorithms, the routing overhead increases sharply when the node density reaches a certain level, which will lead to a significant increase in node energy consumption, thus limiting the scope of application of these routing algorithms. Among them, based on the flooding method of the Epidemic routing algorithm, as the number of nodes increases, the routing overhead approaches exponential growth, indicating that the Epidemic algorithm consumes network resources as the number of nodes increases, so the routing overhead of this algorithm is the largest. For First Contact algorithm and MaxProp algorithm, as the number of nodes increases, the number of hops the node takes to reach the destination node also increases. Therefore, the increase in the number of nodes increases the routing overhead of the network, but the increase is not significant.

Figure 9 shows the relationship between the average transmission delay and the number of nodes. We can see that the transmission delay of the NSFRE algorithm is low, and the trend of the relationship between the transmission delay of the algorithm and the node density is obviously different from other algorithms. As with the propagation delay trend of the MaxProp algorithm, it does not increase significantly with the increase in the density of nodes but decrease. This is because the NSFRE algorithm is not limited to a single social relationship. As the number of nodes increases, the number of social relationships existing between nodes increases. Therefore, the increase of social relations will make the calculation of trust values between nodes more accurate. Consequently, as the number of nodes increases, the transmission delay tends to decrease. Among them, the First Contact routing algorithm has the highest transmission delay. Because the number of nodes increases, the number of nodes from the source node to the target node increases. The Spray and wait routing algorithm has a low transmission delay. The initial phase is similar to the NSFRE. However, as the number of nodes increases, the reachability of randomly dispersed packets to the destination node decreases, and the transmission delay increases. The main reason why the NSFRE algorithm is superior to other algorithms is that the nodes that transmit the message information are filtered, which can reduce some nodes with low cooperation, reduce the network overhead, improve the reachability of the destination node, and reduce the transmission delay.

Figure 10 shows the relationship between the transmission success rate and the node cache. Node cache has different effects on the transmission success rate of each routing algorithm, and the effect is relatively significant when the node cache is relatively small. From the trend point of view, increasing the node cache can improve the transmission success rate. As can be seen in the figure, the NSFRE routing algorithm has the highest transmission success rate. When the number of nodes is set to 210 and the node cache reaches 8, the transmission success rate is gradually stable. The trend of the

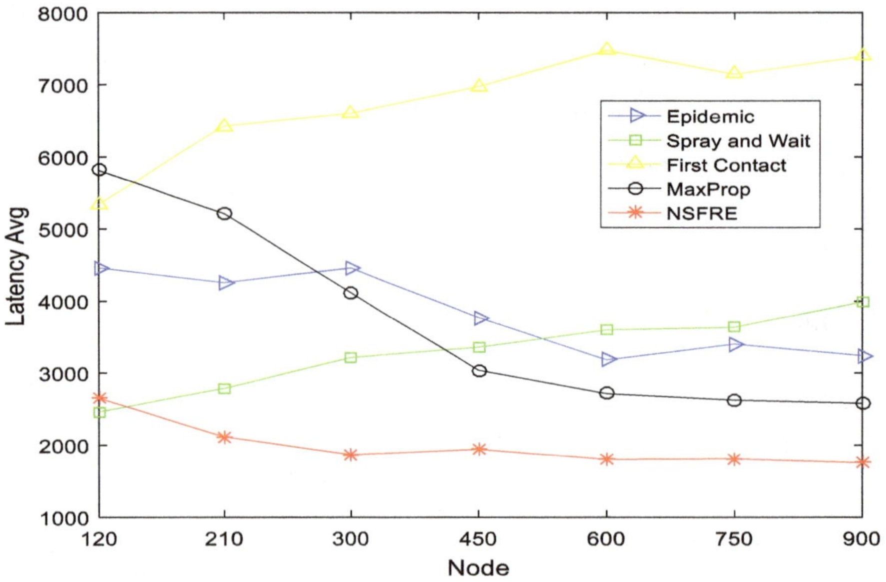

Fig. 9 The relationship between the average transmission delay and the number of nodes

algorithm curve is the same as Spray and wait and MaxProp routing algorithm. This is because the feature considered by the nodes selected by the NSFRE algorithm is the characteristic information between the nodes and the nodes themselves, regardless of the node cache size. When the node cache reaches a certain value, it will not have a great impact on the transmission success rate. The Epidemic algorithm uses a large amount of cache for the flooded message information. As node caching increases, more information can be delivered. Therefore, as the node cache increases, the transmission success rate increases. The First Contact algorithm's transmission success rate is still small, and the node cache is stable at 15. Because only the information is passed to the node that meets the first time, the encounter node has randomness, which will reduce the reachability of the destination node. Therefore, the increase in node cache does not significantly increase the transmission success rate.

Figure 11 shows the relationship between routing overhead and node caching. In the figure, the node cache has a large impact on the routing overhead of the

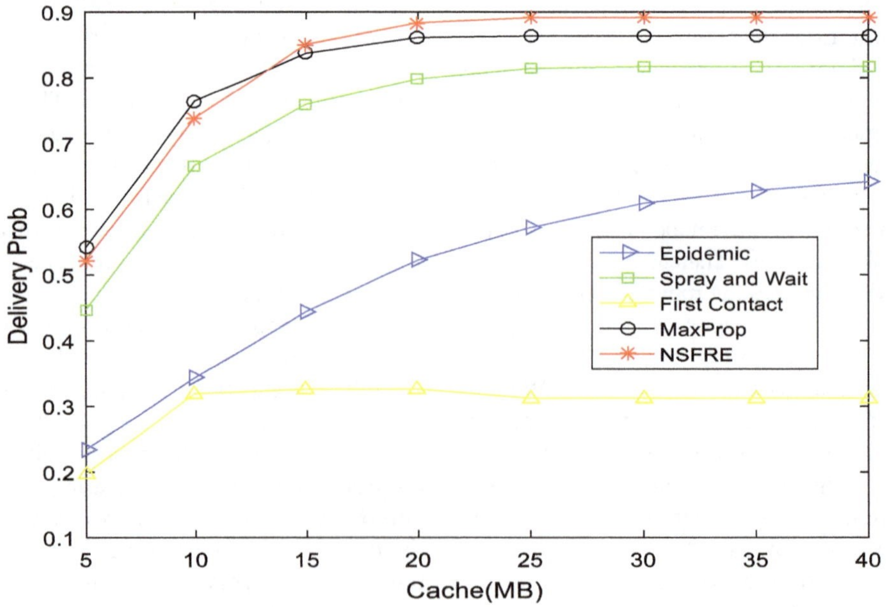

Fig. 10 The relationship between transmission success rate and the node cache

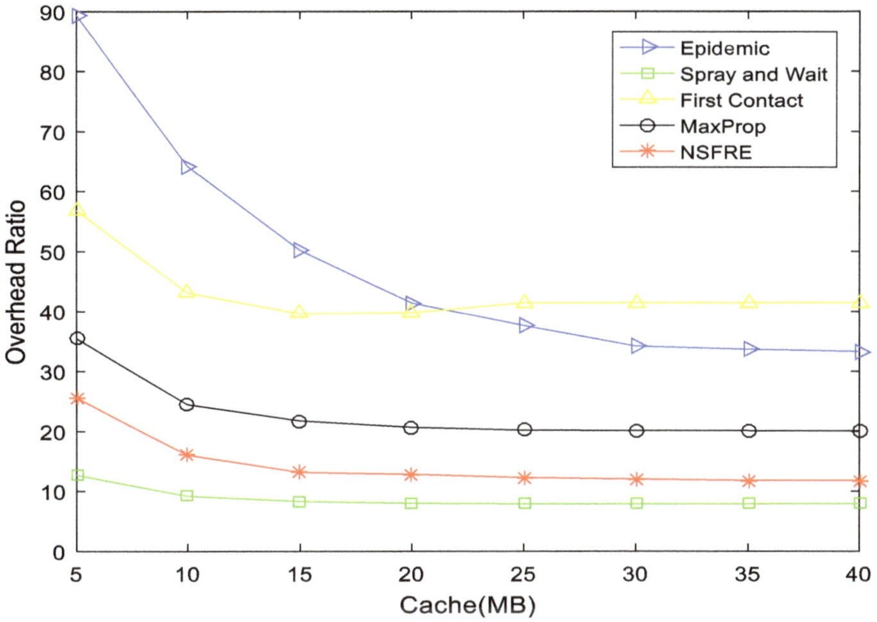

Fig. 11 The relationship between routing overhead and node caching

epidemic routing algorithm. The impact on each routing algorithm is relatively significant when the node cache is relatively small. From a trend point of view, increasing the node cache can reduce the routing overhead. When the node cache is small, the epidemic algorithm has the highest routing overhead. With the increase of node caches, the information that can be carried by a node increases, which makes it difficult to cause network congestion and information loss. Therefore, the routing overhead is significantly reduced. Because the First Contact algorithm only transmits one copy of the message at a time in the network, the routing overhead increases with the node cache and decreases significantly when there is less node cache. After that, the routing overhead does not change much, and it ranges between 40 and 57. As can be seen from the figure, the routing overhead of the NSFRE algorithm is larger when the node cache is smaller. As with the Spray and wait routing algorithm, as the node cache is increased, the routing overhead is gradually reduced, and finally, it becomes stable. The NSFRE algorithm's routing overhead is stable in the 11–13 range. This is because both algorithms are copying information to a part of the node, so the trends of the two algorithms are the same. However, because the NSFRE algorithm filters out some cooperating nodes to send message information according to the feature information, there is a certain consumption in the screening process, and the expenditure in the process of information transmission is small. Therefore, the routing overhead of the NSFRE algorithm is much less than that of the Spray and wait routing algorithm.

Figure 12 shows the relationship between the average transmission delay and the node cache. In the figure, in addition to the MaxProp algorithm, increasing the cache of other nodes to a certain extent will increase the transmission delay, especially when the cache is relatively small. It can be seen in the figure that the first contact algorithm is still the highest transmission delay. When the node cache is small, the propagation delay of the NSFRE algorithm and the Spray and wait routing algorithm is relatively small. With the increase of node cache, the trend of the curve tends to be stable, and the transmission delay of the NSFRE algorithm is still minimal. This is because the nodes selected to transmit information are filtered out according to the network structure and node characteristics, and the access to the destination node is more accessible than the random filtering of nodes by the spray and wait routing algorithm. Therefore, in this process, the transmission delay is smaller. The Epidemic algorithm has a smaller transmission delay when the node cache is small. With the increase of node cache, the amount of information in the network increases dramatically, which makes the transmission delay increase exponentially. Therefore, there is still not much advantage under other algorithms.

The relationship between the three indicators and the time, the node cache, and the number of nodes can be concluded. The NSFRE algorithm is superior to other algorithms in terms of transmission success rate, transmission delay, and energy consumption. And it has less advantage in routing overhead than the Spray and wait routing algorithm. In the real environment, NSFRE

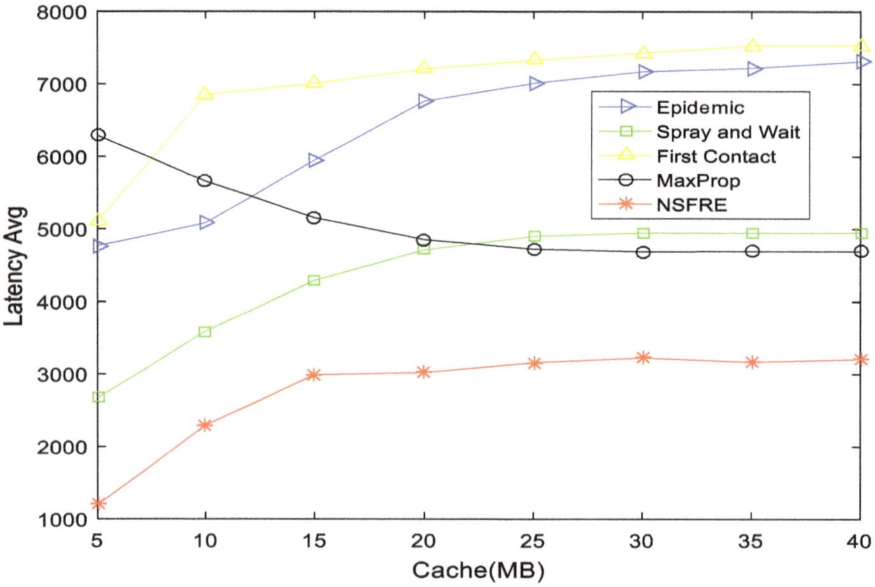

Fig. 12 The relationship between the average transmission delay and the node cache

algorithm is superior to other algorithms for long time information transmission.

6 Conclusions

In this paper, a decision-making method of opportunistic social network routing based on trust mechanism social relations is proposed, NSFRE. The algorithm calculates the trust degree of the forwarding node according to the forwarding path of trust message and message delay time collected by the destination node. Firstly, it analyzes the social elements of the impact of mobile node social relations and the characteristics of nodes and combine the features of network structure to extract location characteristics, interaction quality characteristics, trust quality characteristics, and social relationship feedback characteristics as the decision feature of social relations quantified. Secondly, through the introduction of rough set and information entropy theory, the different attributes of the mobile node are deeply studied, and the law of social attribute changes is excavated to dynamically and adaptively allocate the weights of different attributes. Finally, experimental verification of the proposed social relationship quantification model to screen trusted nodes as the next-hop nodes for data transmission has good results. The model allows data to be transmitted along the trusted cooperative nodes in the network, and at the same time, the uncooperative nodes gradually active participate in the data forwarding in the network. In the process of information transmission, the model has better dynamic adaptability, higher transmission success rate, and lower average transmission delay, so that the negative effect of uncooperative nodes on the network is minimized and the overall performance of the network is improved.

In the future, we will conduct in-depth research on the node's trusted interactions based on the characteristics of nodes through machine learning. According to the selected nodes build trust routing tables, the timestamp mechanism is used to prevent the routing table from being tampered with by malicious nodes in the feedback process, thereby further improving the stability and security of information transmission.

Abbreviation

NSFRE: Node Social Features Relationship Evaluation

Funding

This work was supported in The National Natural Science Foundation of China (61672540); Hunan Provincial Natural Science Foundation of China (2018JJ3299, 2018JJ3682); China Postdoctoral Science Foundation funded project (2017M612586); Foundation of Central South University (185684); Major Program of National Natural Science Foundation of China (71633006). Also, this work was supported partially by "Mobile Health" Ministry of Education-China Mobile Joint Laboratory.

About the authors

GengHua YU received the Master Degree Candidate in School of Computer science and engineering at Central South University, Chang-sha, Hunan, P.R.China, in 2017. She is the 2017 outstanding graduate of Nanchang University. Her research interests include wireless communications and networking, big data research, wireless network, data mining.

Zhigang Chen received the BE, the MS and PhD from Central South University in China in 1984, 1987 and 1998. He is currently a Professor, Supervisor of PhD and chair in School of Computer Science and Engineering, Central South University. He is also director and advanced member of China Computer Federation (CCF), and member of pervasive computing committee of CCF. His research interests cover the general area of cluster computing, parallel and distributed system, computer security, wireless networks.

Jia WU received the Ph.D. Degrees in software engineering Central South

University, Chang-sha, Hunan, P.R.China, in 2016. He is engineer in "Mobile Health" Ministry of Education-China Mobile Joint Laboratory and assocate professor in School of information science and engineering Central South University. Since 2010, he has been Algorithm engineer in IBM company in Seoul, Republic of Korea and in Shang-hai, P.R.China. He is a senior member of CCF(China Computer Federation), a member of IEEE and ACM. His research interests include wireless communications and networking, wireless network, big data research, mobile health in network communication.
Jian Wu Currently he is a student in School of Computer Science and Engineering of Central South University, China. His research interest is network computing.

Authors' contributions
GY, ZC, and JW conceived the idea of the paper. GY, ZC, JW, and JW drafted the manuscript and collected the data, wrote the code, and performed the analysis. ZC contributed reagents/materials/analysis tools. GY wrote and revised the paper. All authors read and approved the final manuscript.

Competing interests
The authors declare that they have no competing interests.

References

1. W.U. Jia, Z. Chen, Z. Ming, Effective information transmission based on socialization nodes in opportunistic networks. Comput. Netw. **129**, 297–305 (2017).
2. W. Jia, Z. Chen, Z. Ming, Information cache management and data transmission algorithm in opportunistic social networks. Wirel. Netw. (8), 1–12 (2018).
3. H. Zhou, V.C.M. Leung, C. Zhu, S. Xu, J. Fan, Predicting temporal social contact patterns for data forwarding in opportunistic mobile networks. IEEE Trans. Veh. Technol. PP (99), 1–1 (2017).
4. H. Zhou, H. Wang, X. Chen, X. Li, S. Xu, *Data offloading techniques through vehicular ad hoc networks: a survey, IEEE Access* (2018).
5. W. Jia, Z. Ming, Z. Chen, Small data: effective data based on big communication research in social networks. Wirel. Pers. Commun. **99**(3), 1391–1404 (2018).
6. C. Huang, Y. Chen, S. Xu, H. Zhou, The vehicular social network (VSN)-based sharing of downloaded geo data using the credit-based clustering scheme. IEEE Access (2018).
7. G. Yu, Z. Chen, J. Wu, J. Wu, A transmission prediction neighbor mechanism based on a mixed probability model in an opportunistic complex social network. Symmetry (2018).
8. R. Ju, G. Hui, C. Xu, Y. Zhang, Serving at the edge: a scalable IoT architecture based on transparent computing. IEEE Netw. **31**(5), 96–105 (2017).
9. X.J. Wu, J.K. Xiao, J.B. Shao, *Trust-based protocol for securing routing in opportunistic networks* (2015), pp. 434–439.
10. S. Bansal, M. Baker, Observation-based cooperation enforcement in ad hoc network. Comput. Sci. (2003).
11. L. Li, X. Zhong, Q. Yang, in *IEEE International Conference on Communication Systems*. A secure routing based on social trust in opportunistic networks (2016).
12. L. Chen, J. Zhuo, A trust management scheme based on behavior feedback for opportunistic networks. China Commun. **12**(4), 117–129 (2015).
13. E.K. Wang, Y. Li, Y. Ye, S.M. Yiu, L.C.K. Hui, A dynamic trust framework for opportunistic mobile social networks. IEEE Trans. Netw. Serv Manag. PP (99), 1–1 (2017).
14. G.K. Wong, Y. Chang, X. Jia, W.Y. Hui, in *International conference on computing*. Performance evaluation of social relation opportunistic routing in dynamic social networks (2015).
15. N. Palaghias, N. Loumis, S. Georgoulas, K. Moessner, in *IEEE International Conference on Communications*. Quantifying trust relationships based on real-world social interactions (2016).
16. R. Atat, L. Liu, J. Wu, G. Li, C. Ye, Y. Yi, *Big data meet cyber-physical systems: a panoramic survey* (2018).
17. H. Chen, L. Wei, Z. Wang, X. Feng, On achieving asynchronous energy-efficient neighbor discovery for mobile sensor networks. IEEE Trans. Emerg. Top. Comput. **6**(4), 553–565 (2018).
18. C. Zhu, V.C.M. Leung, J.J.P.C. Rodrigues, L. Shu, L. Wang, H. Zhou, *Social sensor cloud: framework, greenness, issues, and outlook. IEEE Network* (2018).
19. H. Chen, W. Lou, Z. Wang, Q. Wang, A secure credit-based incentive mechanism for message forwarding in noncooperative DTNs. IEEE Trans. Veh. Technol..
20. J. Wu, G. Song, H. Huang, W. Liu, X. Yong, Information and communications technologies for sustainable development goals: state-of-the-art, needs and perspectives. IEEE Commun. Surv. Tutorials (99), 1–1 (2018).
21. J. Wu, G. Song, L. Jie, D. Zeng, Big data meet green challenges: big data toward green applications. IEEE Syst. J. **10**(3), 888–900 (2016).
22. X. Peng, R. Ju, S. Liang, D. Zhang, L. Jie, Y. Zhang, BOAT: a block-streaming app execution scheme for lightweight IoT devices. IEEE Internet Things J. (99), 1–1 (2018).
23. W. Tang, K. Zhang, R. Ju, Y. Zhang, X.S. Shen, Flexible and efficient authenticated key agreement scheme for BANs based on physiological features. IEEE Trans. Mob. Comput. (99), 1–1 (2018).
24. R. Ju, Y. Zhang, R. Deng, Z. Ning, X. Shen, Joint channel access and sampling rate control in energy harvesting cognitive radio sensor networks. IEEE Trans. Emerg. Topics Comput. (99), 1–1 (2016).
25. Z. Chen, Y. Song, X. Huang, L. Yang, in *International Conference on Cyber-Enabled Distributed Computing & Knowledge Discovery*. An adaptive feedback timing control algorithm of delay-constrained data aggregation in wireless sensor networks (2009).
26. J. Yi, M. Yin, Y. Zhang, X. Zhao, in *IEEE international conference on computational intelligence and applications*. A novel recommender algorithm using information entropy and secondary-clustering (2017), pp. 128–132.
27. A.Y. Lokhov, M. Mézard, H. Ohta, L. Zdeborová, Inferring the origin of an epidemic with a dynamic message-passing algorithm. Phys. Rev. E Stat. Nonlinear Soft Matter Phys. **90**(1), 012801 (2014).
28. G. Wang, B. Wang, Y. Gao, in *International conference on communications & mobile computing*. Dynamic spray and wait routing algorithm with quality of node in delay tolerant network (2010).
29. S. Jain, K. Fall, R. Patra, in *SIGCOMM'04: Proceedings of the 2004 Conference on applications, Technologies, Architectures and Protocols for Computer Communications*. Routing in a delay tolerant network (ACM, New York, 2004), pp. 145–158.
30. J. Burgess, B. Gallagher, D. Jensen, B.N. Levine, in *IEEE International Conference on Computer Communications*. MaxProp: routing for vehicle-based disruption-tolerant networks (2006).

4

Study on communication channel estimation by improved SOMP based on distributed compressed sensing

Biao Wang*, Yufeng Ge, Cheng He, You Wu and Zhiyu Zhu

Abstract

Wireless communication channel usually show the feature of time-varying; however, the time-varying channel has the characteristic that the channel structure within the adjacent time slots having serious time correlation. Therefore, how to use the time slow-changing characteristics of the channel to design the suitable channel state information acquisition method is of great significance to further improve communication performance with low communication bit error rate (BER) for OFDM communication system. The distributed compressed sensing (DCS) is proposed for the phenomenon that multiple sparse signals with time correlation. Based on the DCS theory framework, this article will re-build a time-domain channel estimation method with joint structure by improving the synchronous orthogonal matching pursuit (simultaneous orthogonal matching pursuit, SOMP) algorithm to get better channel information acquisition performance. Simulation results demonstrate the effectiveness of the proposed channel estimation method. Compared with the conventional compressed sensing-based channel estimations which perform at each time separately, the method proposed has better performance in terms of BER.

Keywords: Channel estimation, Compressed sensing, Joint sparse model, Orthogonal frequency division multiplexing (OFDM)

1 Introduction

Compressive sensing (CS) [1–3] as a new signal processing theory, when the signal is sparse in an orthogonal transformation domain, it can be under-sampled at a frequency far lower than Nyquist sampling rate, then the original sparse signal is recovered by nonlinear reconstruction algorithm with high probability. CS theory provides a new solution for traditional channel estimation, which has been applied to the channel estimation of a communication system by many scholars. Cotter and Rao proposed sparse channel estimation based on matching pursuit (MP) algorithm [4] for single-carrier communication system [5, 6]. Dongming and Bing extended the above sparse channel estimation method based on MP to the MIMO-OFDM multicarrier communication system [7]. Based on literature [8], C. R, Berger, S. Zhou, and others deeply studied the sparse channel estimation method based on CS, proposed an OFDM frequency domain channel estimation algorithm and analyzed the advantages and disadvantages of various sparse reconstruction algorithm, the simulation and experiment show that the bit error rate of BP channel estimation algorithm can reach lower bit error rate, which is far less than that of LS channel estimation [9]. In addition, the team mentioned above also conducted other CS-based channel estimation studies in 2010 and 2011, which all verified the superiority of CS-based algorithm over traditional algorithm [10, 11]. Compared with the traditional pilot aided method, CS-based method greatly reduces the number of insertion pilot frequencies and improves the communication efficiency when the performance of the channel estimation remains unchanged. But in the study of the CS theory applied to the channel, the CS reconstruction algorithm is based on the static sparse signal, only considering the sparse reconstruction of the single observable signal at a certain point of time. However, the high-speed mobile communication channel is often regarded as a coherent multipath channel with slow time-varying, the traditional channel estimation method based on static CS will seriously reduce the efficiency of channel estimation.

* Correspondence: mail-wb@163.com
School of Electronic and Information, Jiangsu University of Science and Technology, Zhenjiang 212003, China

At present, some scholars have begun to study the application of compressed sensing for the processing of time-varying sparse signals. The existing research results show that for the dynamic characteristics of time-varying sparse signals, if the random signal processing method is used to conduct a dynamic reconstruction, a better sparse reconstruction effect will be obtained.

To sum up, the traditional channel estimation method has the disadvantages of high pilot cost and low spectral efficiency, the classic static CS channel estimation method has good performance of channel estimation, but due to the high complexity of the calculations, especially poor performance for tine-varying channel estimation. In this paper, based on this background, considering high-speed mobile communication channel is time-varying sparse and statistical dynamic, therefore, based on OFDM technology as the basic framework, a dynamic CS channel estimation method for high-speed vehicular communication channels is studied to further improve the performance of channel estimation and reduce the computational complexity of the algorithm.

2 Methodology

For DCS theory, the most typical application scenario is that multiple sensor nodes observe signals at the same time. These observed signals are usually non-sparse, but can be sparse represented under some sparse basis. More importantly, there is a certain correlation between the signals. When the corresponding signal is independently observed by each sensor, the original signal set can be recovered according to the observation results of each sensor node, and the reconstruction effect is better than that of each sensor node. Thus, it can be seen that the research object of DCS theory is the signal set with joint sparse characteristics. By analyzing the correlation structure between each sparse signal and adopting the strategy of joint signal recovery, the better sparse signal reconstruction effect is obtained.

This section will focus on JSM-1 and JSM-2 models commonly used in joint sparsity models (JSM):

(a) JSM-1
In the JSM-1 model, each signal in the observed signal set is composed of a common sparse part and a separate sparse part. Among them, all the signals contain a common sparse part, and the separate sparse part in each are not identical, meanwhile, the common sparse part and separate sparse part can be sparsely represented through the same sparse basis. The following is the mathematical description of this model:

$$\mathbf{X}_j = \mathbf{Z}_c + \mathbf{Z}_j, \quad j \in \{1, 2, \cdots, J\} \tag{1}$$

Where $\mathbf{Z}_c = \mathbf{\Psi}\boldsymbol{\theta}_c$ represents the common sparse part of the signal, $\| \boldsymbol{\theta}_c \|_0 = K_c$ represents $\mathbf{Z}_c$ is the sparse signal of K_c, $\mathbf{Z}_j = \mathbf{\Psi}\boldsymbol{\theta}_j$ represents the separate sparse part of the signal, and $\| \boldsymbol{\theta}_j \|_0 = K_j$ represents $\mathbf{Z}_j$ is the sparse signal of K_j. In addition, the common sparse part only means that the coordinate of non-zero coefficient of the signal is the same, that is, the support set is the same.

(b) JSM-2

Unlike the JSM-1 model, all signals in the JSM-2 model have the same sparse support set but different non-zero coefficients. In this model, all the original signals can be constructed by the same sparse basis, and the mathematical representation is as follows:

$$\mathbf{X}_j = \mathbf{\Psi}\boldsymbol{\theta}_j, \quad j \in \{1, 2, \cdots, J\} \tag{2}$$

Where $\|\boldsymbol{\theta}_j\|_0 = K$ denotes that the sparsity of all signals is K. For the JSM-2 model, its typical application scenarios include MIMO communication, sound localization, etc. SOMP algorithm is also based on this model.

2.1 The system model

The OFDM channel estimation method based on CS is to transform the problem of OFDM channel estimation into the problem of sparse signal reconstruction, which is solved by the CS reconstruction algorithm. Specifically expressed as

$$\mathbf{Y}_P = \mathbf{X}_P \mathbf{F}_P \mathbf{h} + \mathbf{W}_P = \mathbf{A}\mathbf{h} + \mathbf{W}_P \tag{3}$$

In the above equation, for the receiver, $\mathbf{Y}_P$, $\mathbf{X}_P$, $\mathbf{F}_P$ are all known and h satisfies the sparse characteristic. Therefore, it is a typical CS theoretical model, which can reconstruct the time-domain channel response h by using the l_1 minimum norm method or greedy algorithm with high probability.

On the other hand, high-speed mobile communication channel is considered to be slowly time-varying, and the channel coherence time is greater than the symbol period of OFDM. Within each OFDM symbol, the sparse multipath structure of communication channels remains unchanged. Therefore, the channel estimation method based on CS above is often carried out in a way of symbol-by-symbol, that is, the guide frequency is inserted into each OFDM symbol, and the impulse response of the channel is estimated by the receiver symbol-by-symbol. Although this method is simple and practical, it does not take into account the slowly varying characteristics of the high-speed mobile communication channel. Due to the channel change rate is slower than the OFDM symbol rate, channel impulse response has strong correlation within the duration of several consecutive OFDM symbols, namely, the corresponding channel response of each symbol have a common sparse part. The performance of channel estimation in OFDM system can be further improved if the characteristic of common sparse of channel response at each time can be fully utilized.

Based on this, the relevant literature, under the framework of DCS theory, converts the channel estimation under the slow time-varying channel into the problem of joint sparse recovery under the JSM-2 model, and carries out the joint sparse recovery through the SOMP algorithm to obtain channel response, which is introduced as follows:

According to Eq. (3), if the channel estimation within the time of continuous T OFDM symbols is taken into account, then

$$\begin{cases} \mathbf{Y}_P^{(1)} = \mathbf{A}\mathbf{h}^{(1)} + \mathbf{W}_P^{(1)} \\ \mathbf{Y}_P^{(2)} = \mathbf{A}\mathbf{h}^{(2)} + \mathbf{W}_P^{(2)} \\ \vdots \\ \mathbf{Y}_P^{(T)} = \mathbf{A}\mathbf{h}^{(T)} + \mathbf{W}_P^{(T)} \end{cases} \tag{4}$$

Where $\mathbf{Y}_P^{(t)}, \mathbf{h}^{(t)}, \mathbf{W}_P^{(t)}$ respectively represents the receiving signal at the pilot frequency, the impulse response of the time domain channel and the noise in the frequency domain of the tth consecutive OFDM symbol, $1 \le t \le T$. When the channel changes slowly, the channel response $\mathbf{h}^{(t)}$ corresponding to T consecutive OFDM symbols owns time correlation and common sparsity. By making full use of this common sparsity between signals, the channel estimation of multiple consecutive symbols can be considered as a whole, then the channel response $\mathbf{h}^{(t)}$ can be restored by joint sparse reconstruction instead of independent channel estimation of single OFDM symbol.

For the joint sparse model in Eq. (4), the following problems of optimization are constructed to solve the joint channel estimation:

$$\hat{\mathbf{H}} = \arg\min \sum_{t=1}^{T} \left\|\mathbf{h}^{(t)}\right\|_1 \quad s.t. \quad \sum_{t=1}^{T} \left\|\mathbf{Y}_P^{(t)} - \mathbf{A}\mathbf{h}^{(t)}\right\|_2^2 \le \varepsilon \tag{5}$$

In Eq. (5), ε is the parameter related to noise. The optimization problem above can be obtained by a joint reconstruction algorithm corresponding to JSM-2 model. This paper mainly discusses SOMP algorithm.

2.2 Time domain joint channel estimation based on SOMP algorithm

Signal sets constructed by different JSM models require different joint recovery algorithms to recover the signal. SOMP algorithm is a classic joint recovery algorithm in JSM-2 model, which was proposed by Tropp and Gilbert and then introduced into DCS environment by Baron [12]. As an improved greedy algorithm based on OMP algorithm, SOMP algorithm selects the atoms that best matched with the residual error to update the support set in each iteration and completes joint recovery of the signal set through multiple rapid iterations. However, the biggest difference between SOMP algorithm and OMP algorithm is that OMP algorithm is to select the atomic set that best matches the residual error of a single signal, while SOMP algorithm selects the atomic set that best matches the residual error of the whole signal group.

Combined with the basic principles of SOMP algorithm and the OFDM channel estimation model derived above, the time domain joint channel estimation based on SOMP algorithm is given. The specific steps are as follows:

1. Input: receive pilot signal $\mathbf{Y}_P = [\mathbf{Y}_P^{(1)}, \mathbf{Y}_P{}^{(2)}, \cdots, \mathbf{Y}_P{}^{(T)}]$; the perception matrix $\boldsymbol{A}$; maximum sparsity K of high-speed mobile communication channel.
2. Initialization: make the signal residual $\mathbf{r}_0^{(i)} = \mathbf{Y}^{(i)}, i = 1, 2, \cdots, T$, the upper label represents the ith OFDM symbol, and the lower index represents the number of iterations. The index set $\Lambda_0 =$; reconstructed atomic set $\Phi_0 =$; number of iterations $t = 1$.
3. Iteration process: the tth iteration.
   Step 1: Take the inner product of the perceived matrix A and the residual $\mathbf{r}_{t-1}^{(i)}$, and find the inner product sum corresponding to T OFDM symbols, then calculate the maximum position λ_t of the inner product sum:

$$\lambda_t = \arg\max_j \sum_{i=1}^{T} \left| \left\langle \mathbf{A}_j, \mathbf{r}_{t-1}^{(i)} \right\rangle \right| \tag{6}$$

   Where $\mathbf{A}_j$ represents the jth column atom of the perceptive matrix A;
   Step 2: Update the index set and the reconstructed atomic set:

$$\begin{aligned} \Lambda_t &= \Lambda_{t-1} \cup \lambda_t \\ \Phi_t &= \Phi_{t-1} \cup \mathbf{A}_{\lambda_t} \end{aligned} \tag{7}$$

   Step 3: Using LS algorithm to calculate the multipath coefficient corresponding to each symbol:

$$\hat{\mathbf{h}}_t^{(i)} = \arg\min \left\|\mathbf{Y}_P{}^{(i)} - \boldsymbol{\Phi}_t \mathbf{h}^{(i)}\right\|_2, i = 1, 2, \cdots, T \tag{8}$$

   Step 4: Update the residual according to the estimated value of channel impulse response:

$$\mathbf{r}_t^{(i)} = \mathbf{Y}_P{}^{(i)} - \boldsymbol{\Phi}_t \hat{\mathbf{h}}_t^{(i)}, i = 1, 2, \cdots, T \tag{9}$$

   Step 5: Judge whether the condition $t = K$ is satisfied. If it satisfied, then stop iteration. If not, let $t = t + 1$, execute step 1.
4. Output: channel impulse response estimated value $\hat{\mathbf{h}} = [\hat{\mathbf{h}}^{(1)}, \hat{\mathbf{h}}^{(2)}, \cdots, \hat{\mathbf{h}}^{(T)}]$ corresponding to each OFDN symbol.

2.3 Analysis on the disadvantages of SOMP algorithm

For the phenomenon that multiple sparse signals are correlated in distribution, DCS theory can further improve the recovery performance by utilizing the common sparsity between signals on the basis of classical CS theory, which has been widely studied and applied in wireless sensor network and MIMO communication.

However, the path delay of the actual high-speed mobile communication channel may change in multiple OFDM data symbols, and even the path generation and death may occur. The above time domain joint channel estimation based on SOMP algorithm assumes that the channel shares the same path time delay set in several continuous data symbols, which is obviously difficult to be completely satisfied and inconsistent with the actual situation! In addition, from the above description of the channel time correlation, it can be seen that the support sets for the channel sparse paths in adjacent OFDM symbols are not identical, in other words, there is a separate sparse multiple paths between the symbols, and therefore the JSM-1 model is clearly more compliant with the characteristics of the slowly varying high-speed mobile communication channel.

2.4 OFDM channel estimation based on improved SOMP algorithm

Through the analysis of time correlation of slow time-varying high-speed mobile communication channel, it can be seen that the JSM-1 model in DCS theory is more consistent with the characteristics of high-speed mobile communication channel than the JSM-2 model. In the JSM-1 model, each signal is composed of the common sparse part and the separate sparse part, so we can divide the channel path delay into two different parts, namely, the common channel tap and the dynamic channel tap. On the basis of this separation, an improved SOMP algorithm is proposed in this paper to reconstruct a time-varying sparse channel, the key idea of the proposed algorithm is to first simultaneously detect the common channel tap of the sparse channel in all OFDM data symbols, and then to use the path of the common channel tap as the initialization set in the dynamic channel tap tracking procedure, in order to track the dynamic channel tap and eliminate the wrong tap of the initialization set.

Combined with the symbol-by-symbol OFDM channel estimation model, the OFDM channel estimation process based on the improved SOMP algorithm is presented. The steps are as follows:

1. Input: receive pilot signal $\mathbf{Y}_P = [\mathbf{Y}_P^{(1)}, \mathbf{Y}_P{}^{(2)}, \cdots, \mathbf{Y}_P{}^{(T)}]$; the perception matrix $\boldsymbol{A}$; maximum sparsity K of high-speed mobile communication channel.
2. Part 1: detection of common channel tap

$$\Lambda_0 = SOMP(\mathbf{Y}_P, \mathbf{A}, K) \tag{10}$$

3. Part 2: detection of dynamic channel tap
   Initialization: the index set $\Gamma_0^{(i)} = \Lambda_0$; reconstructed atomic set $\Phi_0^{(i)} = \mathbf{A}_{\Lambda_0}$; channel impulse response $\hat{\mathbf{h}}_0^{(i)} = \mathbf{A}_{\Lambda_0}^{\dagger}\mathbf{Y}_P^{(i)}$; residual $\mathbf{r}_0^{(i)} = \mathbf{Y}_P^{(i)} - \Phi_0^{(i)}\hat{\mathbf{h}}_0^{(i)}$; number of iterations $t = 1$. In all the above symbols, the superscript represents the ith OFDM symbol, and the subscript represents the number of iterations.
   Iteration process: the tth iteration.
   Step 1: Find out the column index λ_t corresponding to the atom with the greatest correlation between residual error $\mathbf{r}_{t-1}^{(i)}$ and the perception matrix A, and update the index set and reconstructed atomic set.;

$$\begin{aligned} \lambda_t^{(i)} &= \arg\max_j \left| \left\langle \mathbf{A}_j, \mathbf{r}_{t-1}^{(i)} \right\rangle \right| \\ \Gamma_t^{(i)} &= \Gamma_{t-1}^{(i)} \cup \lambda_t^{(i)} \\ \Phi_t^{(i)} &= \Phi_{t-1}^{(i)} \cup \mathbf{A}_{\lambda_t^{(i)}} \end{aligned} \tag{11}$$

   Step 2: Using the LS algorithm to find the channel response in each symbol;

$$\hat{\mathbf{h}}_t^{(i)} = \arg\min \left\| \mathbf{Y}_P^{(i)} - \Phi_t^{(i)} \mathbf{h}^{(i)} \right\|_2 \tag{12}$$

   Step 3: Retain only the largest K coefficients in $\hat{\mathbf{h}}_t^{(i)}$, and update the index set $\Gamma_t^{(i)}$ and reconstructed atomic set $\Phi_t^{(i)}$, use the LS algorithm to recalculate $\hat{\mathbf{h}}_t^{(i)}$;
   Step 4: Update the residual:

$$\mathbf{r}_t^{(i)} = Y_P^{(i)} - \Phi_t^{(i)} \hat{\mathbf{h}}_t^{(i)} \tag{13}$$

   Step 5: Judging whether the condition $\|\mathbf{r}_t^{(i)}\|_2 < \|\mathbf{r}_{t-1}^{(i)}\|_2$ is satisfied. If so, let $t = t + 1$ then return to step 1. Otherwise $\Gamma_t^{(i)} = \Gamma_{t-1}^{(i)}$, $\hat{\mathbf{h}}_t^{(i)} = \arg\min \|\mathbf{Y}_P^{(i)} - \Phi_t^{(i)}\mathbf{h}^{(i)}\|_2$, the iteration process ends.
4. Output: channel impulse response $\hat{\mathbf{h}} = [\hat{\mathbf{h}}^{(1)}, \hat{\mathbf{h}}^{(2)}, \cdots, \hat{\mathbf{h}}^{(T)}]$

The core steps of improved SOMP algorithm include two parts: common channel tap detection and dynamic channel tap tracking. The first part of the common channel tap detection is intended to estimate the common path delay set Λ_0 of each symbol, which is considered to contain the most real common channel taps. In this algorithm, this step is completed by the classical SOMP algorithm. After this, the time delay set Λ_0 is used as the initial support set for dynamic channel tap tracking within each symbol.

In the second part of the algorithm, the dynamic channel tap tracking is designed to track the different time-varying channel taps within different OFDM symbols, namely, the separate sparse part in the JSM-1 model. Within each symbol, the common path delay set Λ_0 obtained in the first part of the algorithm is used to initiate the dynamic tracking process. Through each iteration, the size K of the path delay set remains unchanged and only the reconstruction effect is enhanced. In addition, reliable or incorrect channel taps can be added or removed from the path delay set during each iteration. Benefit from the detection of the common channel tap, the dynamic channel tap tracking process can rapidly approach the real path delay set, and when the residual error signal $\mathbf{r}_t^{(i)}$ is no longer reduced, namely, $\|\mathbf{r}_t^{(i)}\|_2 < \|\mathbf{r}_{t-1}^{(i)}\|_2$, the process will be terminated.

3 Simulation results and discussion

3.1 Parameter settings

In order to verify the effectiveness of the algorithm, the proposed channel estimation based on the improved SOMP algorithm is simulated and analyzed, and the proposed algorithm is compared with other similar algorithms in this section. In the simulation system, the total number of subcarriers of OFDM system is set to 256, the protection interval length is 64 cyclic prefixes, and the number of pilot frequencies inserted by a single OFDM symbol is 32 pilot frequencies according to the data. The pilot insertion mode is determined according to the channel estimation method. The conventional LS channel estimation adopts comb pilot insertion with equal interval, but the channel estimation based on CS adopts random pilot insertion. All modulation modes adopt QPSK modulation, and there is no channel coding, the receiver assumes full synchronization.

Because the proposed algorithm is based on the JSM-1 model, the sparse multipath setting in channel simulation includes the common sparse multipath and dynamic sparse multipath. Where the common sparse multipath exists within each symbol, the delay position remains unchanged, only the amplitude changes with time, and the dynamic sparse multipath within each symbol is generated randomly. All sparse multipath are independent of each other, and the gain follows the complex Gaussian distribution of zero mean and decays exponentially. Assuming that the channel length $L = 60$, each multipath time delay is an integer multiple of the system sampling interval without energy leakage. Under the condition of high-speed mobile communication channel, the channel state remains unchanged within an OFDM symbol period, and each OFDM symbol is independent of each other.

3.2 Results discussion

Based on the above system simulation parameters, the following sets of the simulation are made in this paper:

(a) The effect of the number of common channel tap on the performance of the channel estimation algorithm
Since the proposed algorithm is based on the JSM-1 model, the proportion of common and separate sparse paths in the channel will influence the channel estimation algorithm. When the number of common sparse multipath increases, the correlation of the channel multipath sparse structure within the adjacent OFDM symbol is enhanced, and vice versa. In this simulation, the maximum sparsity K of the channel is 10, and the number of the public channel taps is represented by the time correlation L. The specific value varies from 2 to 8. The simulation results are shown in Fig. 1. In addition to the algorithm mentioned in this paper, LS, OMP, SOMP, and Oracle-LS algorithms are also used as comparison algorithms. It can be seen from Fig. 1 that SOMP algorithm is the one most affected by time correlation L of all algorithms, and the other algorithms have no significant influence. Because the proposed algorithm includes dynamic channel tap detection, the change of L only affects the computational complexity of the algorithm, but not the performance of the algorithm. As for the SOMP algorithm, it shares the same path delay set by default, so when L increases, the same path delay number increases, the algorithm performance improves. On the contrary, the performance of the algorithm reduces. As can be seen from the figure, when $L = K = 10$, the channel degenerates into jsm-2 model. At this point, the algorithm proposed in this paper also degenerates into SOMP algorithm, so the estimated performance of the two algorithms is equivalent.

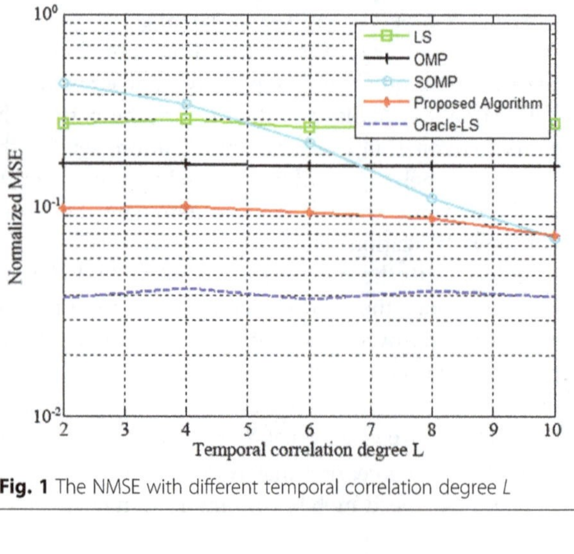

Fig. 1 The NMSE with different temporal correlation degree L

(b) The effect of the number of joint symbol on detection of common channel tap.
As an improved algorithm of SOMP algorithm, the proposed algorithm is also based on joint sparse model. As can be seen from the description of the algorithm above, detection of common channel tap in the first part of the algorithm is to use SOMP algorithm to combine multiple OFDM symbols. Because the delay set of the common channel tap will be the initial support set of the dynamic channel tap detection in the second part, the success rate of the common channel tap detection directly affects the performance and speed of the second part of the algorithm. Figure 2 below simulates the effect of four different joint symbol numbers on the success rate of common channel tap detection. It is not hard to find that the bigger the number of joint symbols, the higher the detection success rate of common channel tap. For this phenomenon, it can be explained that the public sparse multipath exists at all times. When combined with SOMP algorithm recovery, the energy of the weak path will be enhanced by superimposing the same path in multiple symbols, so that it can be detected more easily. If the number of symbols involved in the superposition increases, the detected probability will also increase.
(c) NMSE and BER curves of various algorithms with different SNR.
The channel estimation performances of the proposed algorithm in this paper are compared in this group of simulation. The normalized mean square error and error rate are adopted for the specific evaluation indexes. In this paper, normalized mean square error (NMSE) is defined as follows:

$$\mathrm{NMSE} = \frac{1}{T}\sum_{t=1}^{T}\frac{\left\|\hat{\mathbf{h}}^{(t)}-\mathbf{h}^{(t)}\right\|_2^2}{\left\|\mathbf{h}^{(t)}\right\|_2^2} \tag{14}$$

In Eq. (14), $\mathbf{h}^{(t)}$ represents the impulse response of the ideal channel, and $\hat{\mathbf{h}}^{(t)}$ represents the channel estimated value obtained by various algorithms. The simulation results of NMSE and BER are shown in Fig. 3 and Fig. 4 respectively. In these two simulations, the number of common channel taps and the maximum sparsity of the channel are set as $L = 8$ and $K = 10$ respectively. The channel estimation algorithms involved in the comparison include LS, OMP, SOMP, and Oracle-LS algorithms. In order to ensure the performance of LS algorithm, the pilot insertion mode is comb pilot with uniform spacing of 8, and the rest of CS algorithms adopt random pilot. It is not hard to find from Fig. 3 and Fig. 4 that the performance curves of NMSE and BER tend to be basically the same. Among them, LS algorithm, as a traditional algorithm, has the worst estimation performance because the pilot frequency is smaller than the channel length and the algorithm itself is greatly affected by noise when estimating the sparse path. For CS-based channel estimation, SOMP algorithm has a great advantage over OMP algorithm when SNR is low. The reason is that when SNR is low, SOMP algorithm can combine multiple symbols for channel reconstruction, and its detection ability of weak path is better than that of OMP algorithm which considers only one symbol. When SNR increases, due to the inherent defects of SOMP

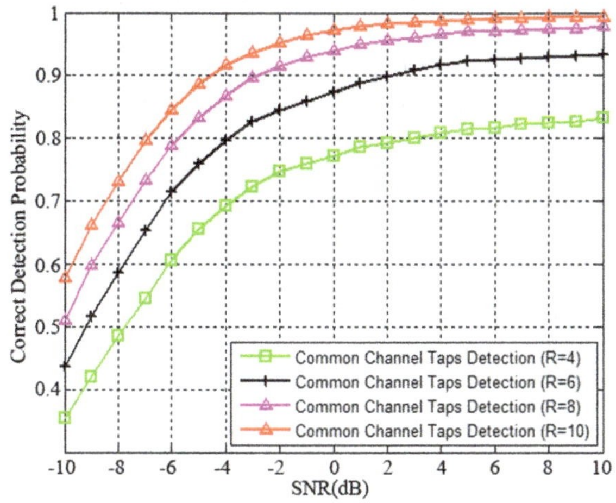

Fig. 2 Correct detection probability of common channel taps with different SNR

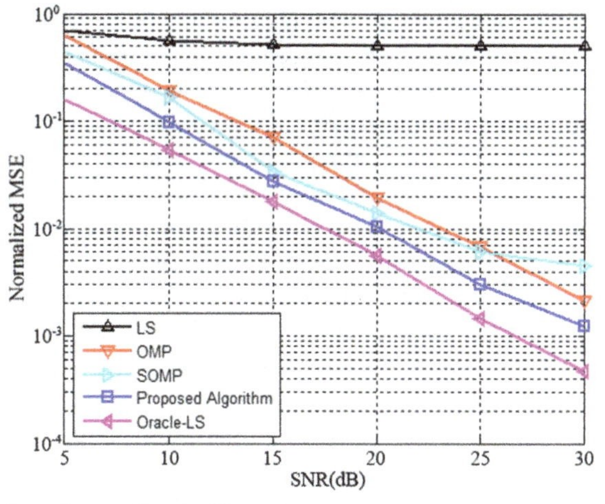

Fig. 3 The NMSE with different SNR

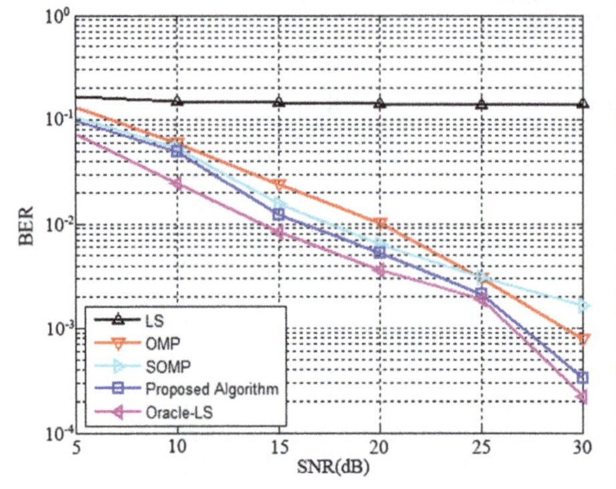

Fig. 4 The BER with different SNR

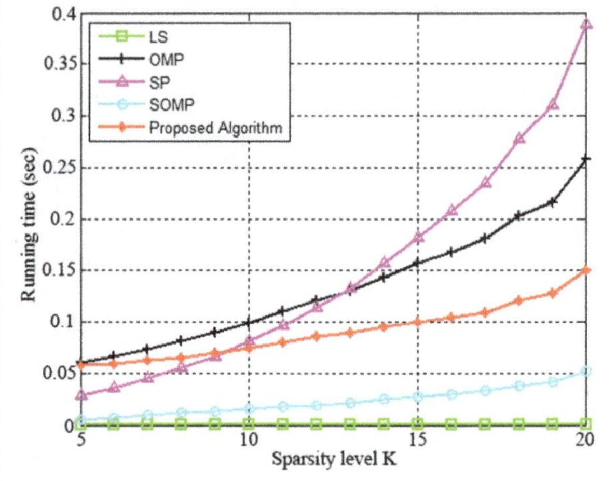

Fig. 5 CPU running time with different *K* sparse channels

algorithm model in time-varying channel, the performance improvement effect is limited. OMP algorithm is less affected by noise in high SNR environment, every time the reconstructed channel can be completed K iterations, so the performance is better under high SNR environment. As for the improved SOMP algorithm proposed in this paper, it is better than OMP and SOMP algorithm because it considers the detection of dynamic channel tap, which is consistent with the results of the previous mathematical analysis. In addition, the comparison algorithm Oracle-LS in the simulation of this group is the LS estimation when the multipath delay set of time-varying sparse channel is known, and this result is the theoretical limit value under the LS criterion.

(d) Analysis of algorithm complexity under different channel sparsity.

For greedy algorithm, the number of iterations is generally signal sparsity *K*. The higher the *K* value of signal sparsity is, the higher the complexity of greedy algorithm. The proposed algorithm is also a greedy algorithm, so the complexity of the algorithm is analyzed by calculating the CPU running time of channels estimation with different *K* sparsity. In the simulation, the estimated channel time of 10 consecutive OFDM symbols is taken as CPU running time, a total of 100 simulations were performed. The comparison algorithm of this group of the simulation consists of LS, OMP, SP, and SOMP algorithms. The simulation results are shown in Fig. 5. As it can be seen from Fig. 5, except the LS algorithm, the CPU running time of the other algorithms increases exponentially with the increase of *K* value. Among the four greedy algorithms, the OMP channel estimation and the SP channel estimation are performed symbol by symbol, so it is much larger in time than the channel estimation algorithm based on the joint sparse model. Particularly, the SP algorithm, because each iteration takes *K* column atoms, so when *K* increases, the algorithm complexity also has a rapidly increase, which is most obvious in all the algorithms. Among the SOMP algorithm and the proposed algorithm in this paper, due to the improved SOMP algorithm increases the dynamic channel tap detection, so the CPU running time is longer, which is equivalent to sacrificing speed for higher estimation accuracy.

4 Conclusion

This paper mainly studies the joint estimation method of high-speed mobile communication channel based on improved SOMP algorithm. Firstly, the distributed compressed sensing theory and joint sparse model are introduced. Secondly, the existing time domain joint channel estimation method based on SOMP algorithm is introduced. On this basis, this paper analyzes the defects of existing algorithms that do not take into account channel time-varying and proposes an improved SOMP algorithm based on JSM1 model for recovery of joint channel estimation. Simulation results show that the proposed algorithm in this paper has greater advantages of estimation performance and lower algorithm complexity compared with the existing symbol-by-symbol CS channel estimation algorithm in the time-varying channel environment.

Abbreviations

BER: Bit error rate; CS: Compressive sensing; DCS: Distributed compressed sensing; SOMP: Simultaneous orthogonal matching pursuit

Acknowledgements

The research presented in this paper was supported by National Natural Science Foundation of China, Natural Science Foundation of Jiangsu Province of China.

Funding
The authors acknowledge the National Natural Science Foundation of China (grant 11574120, U1636117), the Natural Science Foundation of Jiangsu Province of China (grant BK20161359), The Science and Technology on Underwater Acoustic Antagonizing Laboratory, Systems Engineering Research Institute of CSSC (grant MB80038).

Authors' contributions
YG and BW are the main writers of this paper. They proposed the main idea, deduced the algorithm theory, completed the simulation, and analyzed the result. CH and YW give some work for simulation. ZZ gave some important suggestions for improving the SOMP. All authors read and approved the final manuscript.

Competing interests
The authors declare that they have no competing interests.

References
1. D.L. Donoho, Compressed sensing [J]. IEEE Trans. Inf. Theory **52**(4), 1289–1306 (2006)
2. E.J. Candès, J. Romberg, T. Tao, Robust uncertainty principles: Exact signal reconstruction from highly incomplete frequency information [J]. IEEE Trans. Inf. Theory **52**(2), 489–509 (2006)
3. E.J. Candès, M.B. Wakin, An introduction to compressive sampling [J]. IEEE Signal Process. Mag. **25**(2), 21–30 (2008)
4. S.G. Mallat, Z. Zhang, Matching pursuits with time-frequency dictionaries [J]. IEEE Trans. Signal Process. **41**(12), 3397–3415 (1993)
5. S.F. Cotter, B.D. Rao, Matching pursuit based decision-feedback equalizers [C]// IEEE international conference on acoustics, speech. Signal Process., 2713–2716 (2000)
6. S.F. Cotter, B.D. Rao, Sparse channel estimation via matching pursuit with application to equalization [J]. IEEE Trans. Wirel. Commun. **50**(3), 374–377 (2002)
7. Wang D, Han B, Zhao J, et al. Channel estimation algorithms for broadband MIMO-OFDM sparse channel [C]// IEEE 2003 international symposium on personal, Indoor and Mobile Radio Communications. 2003: 1929–1933
8. W. Li, J.C. Preisig, Estimation of rapidly time-varying sparse channels [J]. IEEE J. Ocean. Eng. **32**(4), 927–939 (2008)
9. Mason S, Berger C, Zhou S, et al. An OFDM design for underwater acoustic channels with Doppler spread [C]// Digital Signal Processing Workshop and, IEEE Signal Processing Education Workshop. 2009: 138–143
10. C.R. Berger, Z. Wang, J. Huang, et al., Application of compressive sensing to sparse channel estimation [J]. IEEE Commun. Mag. **48**(11), 164–174 (2010)
11. J. Huang, S. Zhou, J. Huang, et al., Progressive inter-carrier interference equalization for OFDM transmission over time-varying underwater acoustic channels [J]. IEEE Journal of Selected Topics in Signal Processing **5**(8), 1524–1536 (2011)
12. J.A. Tropp; A.C. Gilbert; M.J. Strauss. Simultaneous sparse approximation via greedy pursuit [C].// proceedings. (ICASSP '05). IEEE international conference on acoustics, speech, and Signal Process., 2005:721–724

5

Secure data sharing scheme for VANETs based on edge computing

Jingwen Pan, Jie Cui[*], Lu Wei, Yan Xu and Hong Zhong

Abstract

The development of information technology and the abundance of problems related to vehicular traffic have led to extensive studies on vehicular ad hoc networks (VANETs) to meet various aspects of vehicles, including safety, efficiency, management, and entertainment. In addition to the security applications provided by VANETs, vehicles can take advantage of other services and users who have subscribed to multiple services can migrate between different wireless network areas. Traditionally, roadside units (RSUs) have been used by vehicles to enjoy cross-domain services. This results in significant delays and large loads on the RSUs. To solve these problems, this paper introduces a scheme to share data among different domains. First, a few vehicles called edge computing vehicles (ECVs) are selected to act as edge computing nodes in accordance with the concept of edge computing. Next, the data to be shared are forwarded by the ECVs to the vehicle that has requested the service. This method results in low latency and load on the RSUs. Meanwhile, ciphertext-policy attribute-based encryption and elliptic curve cryptography are used to ensure the confidentiality of the information.

Keywords: VANETs, edge computing, data sharing, ECC, ABE, cross-domain service

1 Introduction

Vehicular ad hoc networks (VANETs) is an application of the mobile ad hoc network in the transportation field and is a multi-hop mobile wireless communication network that was first mentioned in 2001 [1]. The core idea of VANETs is that vehicles are automatically connected to the mobile network within a specific communication range and these connected vehicles are able to exchange information, such as speed, location, and data sensed by on-board sensors. VANETs communication modes include inter-vehicle communication (V2V, vehicle-to-vehicle) and communication with public infrastructure (V2I, vehicle-to-infrastructure) that realize real-time information exchange and serve people's transportation [2].

VANETs typically consist of three parts, namely, a trusted authority (TA), a roadside unit (RSU), and an on-board unit (OBU). The TA, which is a trusted management center with high computing capacity and storage, is responsible for registering and issuing secret key materials. The RSU, located on both sides of the roads, interacts with vehicles through wireless channels and serves as a bridge between the vehicles and the TA. OBU, which is a computing device equipped in a vehicle, is in charge of V2V and V2I communications, dealing with the release and reception of traffic messages and improving the user's driving experience [3].

Through RSUs, VANETs can provide convenient services such as notifying the nearest restaurant and gasoline stations and acting as a gateway by connecting with the Internet and mobile communication network to provide electronic toll collection service and in-vehicle entertainment services, such as downloading movies. Users can effectively monitor the driving condition of vehicles with respect to different functional requirements and provide comprehensive services that can greatly facilitate passengers' travels and enrich their travel journeys.

Development of information technology has given rise to the demand for information gathering and information sharing among different networks. Data collected within one domain may not satisfy users because the required data may be present in another management domain. For example, a Twitter user may want to share data with another user who has an account on Instagram but not on Twitter. Therefore, a secure way to share data between different domains is needed. The main

* Correspondence: cuijie@mail.ustc.edu.cn
School of Computer Science and Technology, Anhui University, Hefei 230039, Anhui, China

problems encountered during cross-domain sharing are data security and authentication among different domains [4]. The dynamic movement of vehicle nodes in VANETs has given rise to the authentication problem during cross-domain access. The traditional method for vehicles to enjoy cross-domain services relies on RSUs and causes significant delays and large loads on the RSUs.

We have introduced the concept of edge computing to overcome the cross-domain sharing problem. Edge computing is a technology that allows computing to be performed at the edge of a network so that computing occurs near data sources [5]. Traditionally, transmission, storage, and computing have been designed separately for the convenience of management; however, these separate resources cannot satisfy the latency and quality requirements of the service. By contrast, edge computing allows deep integration of transmission, storage, and computing. For example, because edge computing allocates a large amount of computing power near a mobile device, such as a vehicle, the majority of the data are processed and stored at the edge, thereby decreasing the delay in providing information and computing resources to the user. In addition, at present, the shared pools for configuring various resources (e.g., computing networks, servers, storage, applications and services) are centrally located to facilitate management, coherence, and economy. However, a fully centralized approach impedes large-scale connections, large-scale transmissions, and latency. Therefore, edge computing is used to realize the decentralization of shared resources that are distributed along the continuum from the cloud to things. Centralization improves efficiency and flexibility, while decentralization decreases latency and improves capacity and scalability [6, 7].

The resources of computing, communication, storage, and control are distributed among all the nodes. The structure of nodes along with their capabilities of storage, computing, and networking are different. It is imperative that the various communication modes as well as the nodes cooperate with each other to optimize the use of resources and improve the performance of data sharing. Furthermore, a few edge computing nodes are needed to function as data sources. An edge computing node can be perceived as the administrator of one domain and is responsible for sending and processing data [5]. Edge computing nodes can reduce delays by processing and analyzing simple data generated by edge devices, passing on only the necessary results and complex data to the remote cloud. By aggregating information from edge computing nodes, a cloud service provider (CSP) can obtain real-time traffic and data requests and, subsequently, schedule data caching and coordinate resources among different domains. In our research, we select several vehicles to act as edge computing nodes. The points to be considered when selecting an edge computing node are as follows: (1) availability of adequate computing resources, (2) high social centrality implying a greater possibility to contact other nodes and a high data sharing efficiency, and (3) availability and selection of vehicles that offer a wider coverage of routes to share data among a large number of nodes. All OBUs communicate with edge computing nodes and provide information that includes the list of their current neighbors, the channel capacity of each neighbor's link, and the identifiers of the cached and un-cached data items. Subsequently, edge computing nodes inform the published scheduling decisions to all the relevant OBUs. Next, each node obtains the shared data that they requested from their neighbors based on the scheduling decisions [3, 6].

In this paper, we propose a secure scheme based on edge computing for data sharing among different domains. As shown in Fig. 1, in accordance with the concept of edge computing, we select some vehicles, called edge computing vehicles (ECVs), to act as edge computing nodes. ECVs integrate the information obtained from the OBUs and the RSUs and schedule the data conforming to the requests made by vehicles. Each ECV manages the domain it resides in. Requests and responses for data sharing between two domains are transmitted through the CSP and their respective ECVs. This shared data is encrypted by elliptic curve cryptography (ECC). The message that needs to be shared among users in two domains uses the resource pool as its carrier for access and storage and is encrypted by ciphertext-policy attribute-based encryption (CP-ABE) to ensure confidentiality. Different from the traditional public key encryption, the attribute-based encryption (ABE) algorithm embeds the attribute set and policy into the ciphertext and user's private key so that the decryption process is actually matching the set of attributes with the strategy. If the matching is successful, the algorithm will complete the decryption operation and the user will recover the plaintext data. CP-ABE, with policy being embedded in ciphertext, which means that the data owner can determine those who have access to the ciphertext by setting policies, makes an encryption access control for the data that can refine the granularity to the attribute level. Therefore, from the perspective of encryption calculation overhead and storage overhead, CP-ABE has an advantage in performance compared to the traditional public key encryption algorithm in the data encryption sharing scenario. Throughout this process, we use system parameters that are set by TA and stored in a tamper-proof device (TPD) to perform pseudo-identity-based message signing and authentication that guarantees the anonymity of the message owner and security of the message transmission.

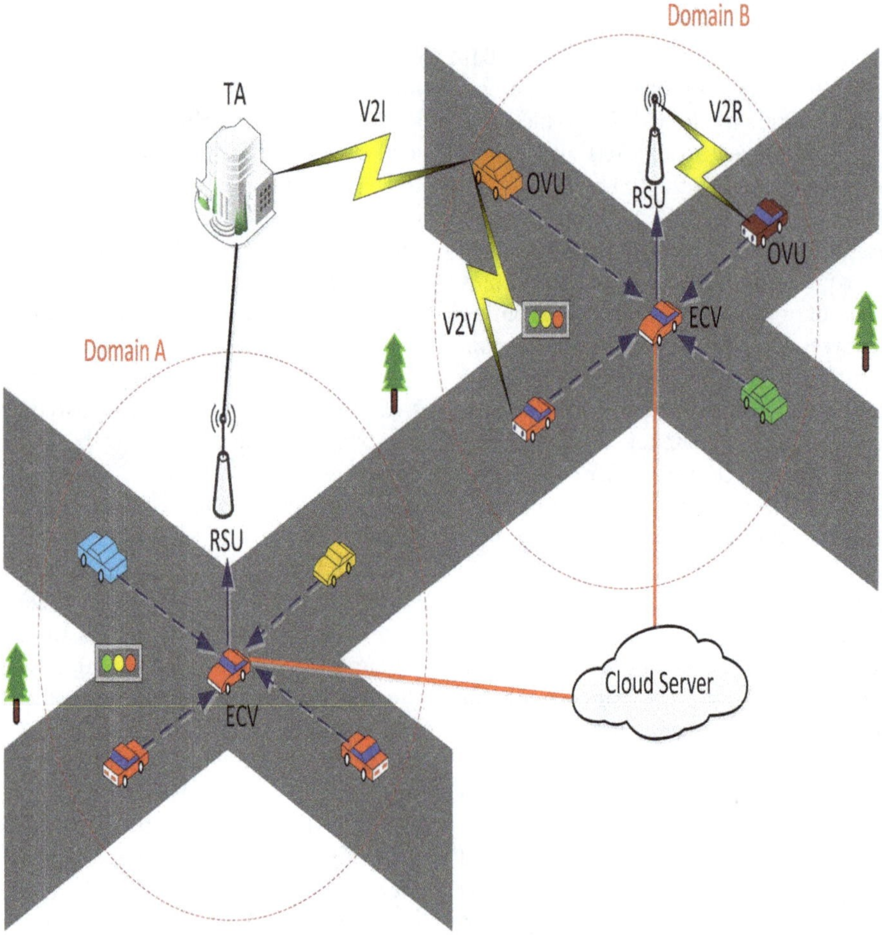

Fig. 1 System model

Our solution enables efficient and dynamic data sharing between vehicles in different domains, which will greatly facilitate vehicle users in practical applications. For example, there is a moving vehicle with a lower oil level that needs to know the location of nearby gas stations, then by using our scheme, it can quickly obtain real-time information through the interaction of ECVs. As far as we know, there is no research on combining ECC with edge computing for cross-domain data sharing, so we explored and proved its feasibility and significance.

The main contributions of this paper can be summarized as follows:

(1) We propose a scheme that uses edge computing to achieve cross-domain sharing, which can improve the efficiency of vehicles to obtain the data they need for service and greatly reduce the load on RSUs.
(2) We apply a method that combines ECC, CP-ABE, and a message signature authentication mechanism and provide a security analysis that proves the security of our scheme.

The rest of this paper is organized as follows: Section 2 briefly mentions the related work. In Section 3, we introduce the fundamental background of ECC and CP-ABE. Thereafter, our proposed scheme is described in Section 4. Proof and analysis are provided in Section 5. Finally, we conclude our work and discuss future research directions in Section 6.

As for the experiment in our paper, we conduct an experiment to evaluate the performance of the proposed solution. Our experiments are tested on Ubuntu 14.04 platform with two cryptography libraries including MIRACL [8] and Crypto++ [9]. By measuring the computation time of some basic cryptography operations, we learn about the computation performance of our proposed scheme and compare it with other related schemes. The experimental results show that compared with the other two existing data sharing schemes, our scheme can reduce the computation overhead in the process of transmitting messages which means that our proposed scheme can suit the real VANETs scenario better. Besides, the cryptography and correctness analysis are used to guarantee that our proposed can resist common attacks of VANETs.

2 Related work

2.1 Privacy protection of VANETs

In 2006, Zeng [10] proposed a pseudonym public key infrastructure (PKI) solution based on public key infrastructure, in which vehicles can generate pseudonyms themselves so as to reduce the overhead of communication with the certification authority (CA). In 2007, Lin et al. [11] presented a privacy protection protocol based on the combination of a group signature and an identity-based signature (IBS). Anonymity and traceability can be guaranteed by using a short group signature to sign messages, while bandwidth can be saved by using the identity-based signature scheme. However, the group signature schemes have such problems as maintenance of the revocation list. In 2008, Zhang et al. [12] introduced an efficient batch signature verification scheme, intending to solve the problem that it is difficult for RSUs to simultaneously verify multiple received signatures in V2I communication mode. The scheme can greatly reduce the overall time and transmission overhead, but it is vulnerable to replays and non-repudiation attacks. In 2015, Horng et al. [13] proposed a scheme based on the certificateless signature, which solved the complex certificate management problem of the traditional PKI-based schemes and the key escrow problem in IBS. Conditional privacy protection will be achieved by mapping the message broadcast by vehicles to different pseudo-identities. And authorities can retrieve real identity from any controversial pseudo-identity. However, this scheme only considers V2I communication and lacks support for malicious vehicles revocation. In 2016, Vijayakumar et al. [14] proposed a dual authentication scheme and a dual group key management scheme, which have high computational efficiency and can safely distribute group keys to vehicle groups. However, it is vulnerable to replay attacks when reusing previously acquired messages. In 2017, Azees et al. [15] presented an efficient anonymous authentication scheme which has an efficient tracking method to avoid malicious vehicles entering VENETs. But the scheme leaves out of considering non-framework, which can guarantee that members' signatures are not forged by others. A new efficient certificateless short signature (CLSS) scheme is designed by Tsai [16] using bilinear pairing, which takes a group element as the signature length and takes on the lower computational cost of signature generation and signature verification. After a formal security analysis, the solution proposed proved to be safe for both super I and super II opponents. In 2018, Pournaghi et al. [17] proposed a scheme based on a combination of RSUs and TPD. Storing the system key and main parameters in the TPD of RSUs ensures that the entire network will not be affected too much when a single OBU hazard or attack occurs. Asaar et al. [18] found that the authentication scheme proposed by Liu et al. [19] using proxy vehicles to reduce the computational overhead of RSUs does not guarantee the authenticity of the message, nor can it resist the modification of the attack and the invalid signature of the batch. So, they designed a new identity-based message authentication scheme using proxy vehicles and demonstrated the security of the scheme on the elliptic curve discrete logarithm problem. Islam et al. [20] introduced a password-based conditional privacy protection authentication and group key generation (PW-CPPA-GKA) protocol for VANETs, which can provide some functions like group-key generation, user departing, user joining, and password modification. Because PW-CPPA-GKA is bilinear-pairing-free, it is lightweight in terms of computation and communication. Cui et al. [21] proposed an authentication scheme using the Cuckoo filter. In their scheme, the ideal TPD is no more necessary and the computation overhead is very low.

In recent years, researchers have made great progress in the privacy protection of vehicle network [22–24].

2.2 Data downloading or sharing

In 2007, Sago et al. [25] grouped vehicles according to locations and estimated future routes. Then, they made predictions about the data items that might be transmitted between different groups in the near future in order to improve the availability of data shared between vehicles. In 2010, Zhang et al. [26] intended to improve the performance of content sharing in VANETs and proposed Roadcast, a popularity-based P2P sharing scheme. On the one hand, Roadcast relaxes the query requirements of users and makes it faster for users to query the content they want. On the other hand, Roadcast returns the most popular content related to queries under the influence of two components (popularity aware content retrieval and popularity aware data replacement) and increases opportunities for spread and share of popular data. Therefore, the overall query delay is reduced. However, data transmitted between any two parties may get compromised, and several attack method has been proposed such as [27]. Hence, data privacy and security issue should be paid attention to. Some efficient schemes that focus on solving these issues have been proposed such as [28–30]. In 2013, Hao et al. [31] proposed a secure co-downloading framework for paid services of VANETs. Data downloading takes place when the vehicles enter the range of RSUs and data sharing takes place after the vehicles leaving it. The application layer data sharing protocol they proposed coordinates the vehicles based on location to transfer the data to be shared. This cooperative sharing can effectively avoid conflicts in the media access control (MAC) layer and hidden terminal problems in multi-hop transmission and can ensure that each vehicle near the RSUs can receive the requested data. In 2014, Wu et al. [32] used evolutionary games (EG) to implement multimedia services and data sharing among VANETs vehicles. This scheme presents a repeated game "More Pay for More Work (RGMPMW)" incentive mechanism based on service evaluation information. In 2017, Lai et al. [33] proposed an effective cloud-assisted scheme for data storage and query in VANETs. The cloud calculates the transfer strategy of the data query result by solving the linear programming problem. This scheme integrates the cloud, in-vehicle network, and 4G technology, and processes and transfers queries to corresponding communication channels based on the cost and time of the query, which greatly improves efficiency.

2.3 Edge computing

Cloud computing service which mainly contains SaaS (software as a service), LaaS (infrastructure as a service), PaaS (platform as a service) [34] is very popular in recent years because it can decrease the terminal running costs. However, cloud computing cannot process data timely. Given the recent proliferation in the number of smart devices connected to the Internet, the era of Internet of Things (IoT) is challenged with massive amounts of data generation. Edge computing or fog computing is gaining popularity and is being increasingly deployed in various latency-sensitive application domains including industrial IoT [35].

In 2016, Shi et al. [36, 37] described the application prospects of edge computing, pointing out that edge computing will play an important role in solving delays, limited battery life, bandwidth cost, data security, and privacy issues. In 2017, Mao et al. [38] conducted a comprehensive survey of MEC from the perspective of communication, discussed some challenges and directions of the research, including MEC system deployment, mobility management, and privacy awareness, and introduced some typical application scenarios of edge computing. Ren et al. [39] studied the application of edge computing in the field of Internet of Things. They implemented an extensible Internet of Things platform based on transparent computing using edge computing, which proves that edge computing can enhance the scalability of lightweight Internet of Things devices. In 2018, Roman et al. [40] introduced several of the most important edge examples, which indicated the challenges and potential synergies of mobile edge computing (MEC). Yuan et al. [41] studied how to meet the need of real-time access services in the autonomous driving process using MEC technology and proposed a two-level edge computing architecture to coordinate vehicular content sharing by making full use of base stations on wireless edge. The simulation results show that the proposed solution can significantly reduce the backhaul and wireless bottleneck of the cellular network.

Fan et al. [5] first linked edge computing to cross-domain access. The proposed edge computing model effectively solves the authentication problem between different domains through edge computing nodes and cloud links. Meanwhile, the RSA algorithm and CP-ABE guarantee scheme security to achieve cross-domain sharing. Luo et al. [6] studied the distribution problem of vehicular content in 5G-VANETs. In order to allocate large amounts of data, a two-layer hierarchical structure based on edge computing was designed. The upper layer coordinates base station resources and handles unbalanced traffic, while the macro base station (MBS) of the lower layer supports cooperation among different communications and coordinates content requests among vehicles. After data prefetching into RSUs and vehicles, the RSUs and the vehicles act as data sources to provide content download services for the neighboring vehicles, and the data is scheduled by the MBS and propagated between RSUs and the vehicles.

3 Background

3.1 Elliptic curve cryptography

ECC is an algorithm based on elliptic curve mathematics for public key cryptography, which is first proposed by Miller [42] and Koblitz [43]. Under the same security conditions, ECC has a key with a shorter length compared to other public key cryptographic algorithms. An elliptic curve is a collection of points that satisfy a particular equation, while a definite elliptic curve cryptosystem can be determined by a maximum prime number, an elliptic curve equation, and a common point on the curve.

For an elliptic curve equation E, there are an infinity point O and an operator + called addition in the mathematical principle. It has the following properties:

(1) Unit element: $P + O = O + P = P$, for all $P \in E$.
(2) Reversibility: $P + (-P) = O$, for all $P \in E$.
(3) Associative law: $(P + Q) + R = P + (Q + R)$, for all P, Q, $R \in E$.
(4) Commutative law: $P + Q = Q + P$, for all P, $Q \in E$.
(5) Specific calculation: Given two points P_1 and P_2 on E, there must be a third point $P_3 = P_1 + P_2$ on E, and it can be determined by the connection between P_1 and P_2.
(6) Multiplication: Ellipse scale multiplication is an extension of elliptical addition. Given the point P on E, then $kP = P + P + \ldots + P$(k times).

In ECC, given the elliptic curve E, the base point G, and the point xG, then we take xG as the public key and take x as the private key. According to the natures of the elliptic curve, we can know that it is very simple to obtain the public key when the private key is known, but it is quite hard to find the private key when the public key is known. This is the elliptic curve discrete logarithm problem (ECDLP), whose difficulty guarantees the security of the elliptic curve cryptography.

3.2 Ciphertext-policy attribute-based encryption

In 2017, Bethencourt et al. [44] proposed the CP-ABE scheme. In CP-ABE, the ciphertext corresponds to the access structure while the key corresponds to the set of attributes. The encryption part encrypts data using public parameters and the user decrypts the ciphertext using the attribute-based private key. Since the policy is embedded in the ciphertext, the data owner can define access control policy to determine users who can access the ciphertext. The process of CP-ABE is as follows:

(1) Setup: The setup algorithm outputs the public parameter PK and a master key MK.
(2) Encryption: The encryption algorithm inputs a message m and an access structure A and PK, then outputs the ciphertext C.
(3) Generate key: The algorithm inputs a set of attributes S, MK, and PK, and outputs a decryption key D.
(4) Decryption: The decryption algorithm inputs the key D, the public parameter PK, and the ciphertext C encrypted based on the access structure A. If the number of attributes that can satisfy A in all the attributes corresponding to the user D reaches a certain threshold, then the user can decrypt and gain the message m.

4 Our scheme

4.1 System model

As shown in Fig. 1, our model consists of five types of entities: a trusted authority (TA), a roadside unit (RSU), a cloud service provider (CSP), edge computing vehicle (ECV), and ordinary vehicle user (OVU).

(1) TA: As a trusted management center with high computing capacity and storage, TA generates and publishes common system parameters about the secret key to all vehicles.
(2) RSU: Besides serving as a bridge between the vehicles and the TA, RSUs also provide the information obtained from vehicles to the ECV and participate in the transferring of data under the control of ECV.
(3) CSP: It links all ECVs so that those domains managed by ECVs can have contact with each other and the sharing of data can be implemented between those different domains.
(4) ECV: An ECV, which is responsible for transmission and storage of data as well as users' registration and revocation, manages a domain. After receiving the request of data sharing from another domain, the ECV encrypts the list of attributes of its domain with ECC.
(5) OVU: OVUs can either encrypt data according to the policy and send it to the resource pool as data requesters or can access and decrypt the data of the resource pool as other users.

4.2 System initialization phase

We need some necessary system parameters which are generated by the TA and preloaded into the tamper-proof device (TPD) of all vehicles.

TA randomly selects two large prime numbers p and q, a non-singular elliptic curve$E: y^2 = x^3 + ax + b \bmod q$, and a generator element G randomly selected in the group.

TA randomly selects $k_s \in Z_q^*$ as the system private key and calculates $K_s = k_sG$ as the system public key.

TA randomly selects $k_R \in Z_q^*$ as the private key of RSUs, calculates $K_R = k_RG$ as the public key of RSUs, and sends k_R to RSUs.

TA chooses a secure hash function: $h: \{0, 1\}^* \to Z_q$.

TA assigns a real identity RID and password PWD to each vehicle and preloads $\{\text{RID}, \text{PWD}, k_s\}$ into the TPD of the vehicle.

TA randomly selects two numbers α, $\beta \in Z_q$ as public parameters for encryption and decryption.

TA publishes common system parameters $\{p, q, a, b, G, K_s, K_R, h, \alpha, \beta\}$ to all vehicles.

4.3 ECV election method

In this paper, we consider two criteria when selecting ECVs: closer distance from RSUs and enough available computing resources. In our proposed scheme, we adopt the same election method as that of the scheme of Cui et al. [3].

4.4 Process of constructions

Assuming that the OVU in domain A denoted by OVU_A wants to share data with users in domains B, first, OVU_A sends data sharing request and public key K_A to the edge computing vehicle ECV_B in domain B through ECV_A and CSP. Then, ECV_B derives the symmetric encryption key S from K_A and returns the attributes set of domain B which encrypted by S to OVU_A through CSP and ECV_B. Finally, OVU_A defines policy according to the set of attributes and encrypts the message as well as sends it to the resource pool where users in domain B can access to. Here, we describe the specific process of the proposed scheme in detail, as shown in Fig. 2.

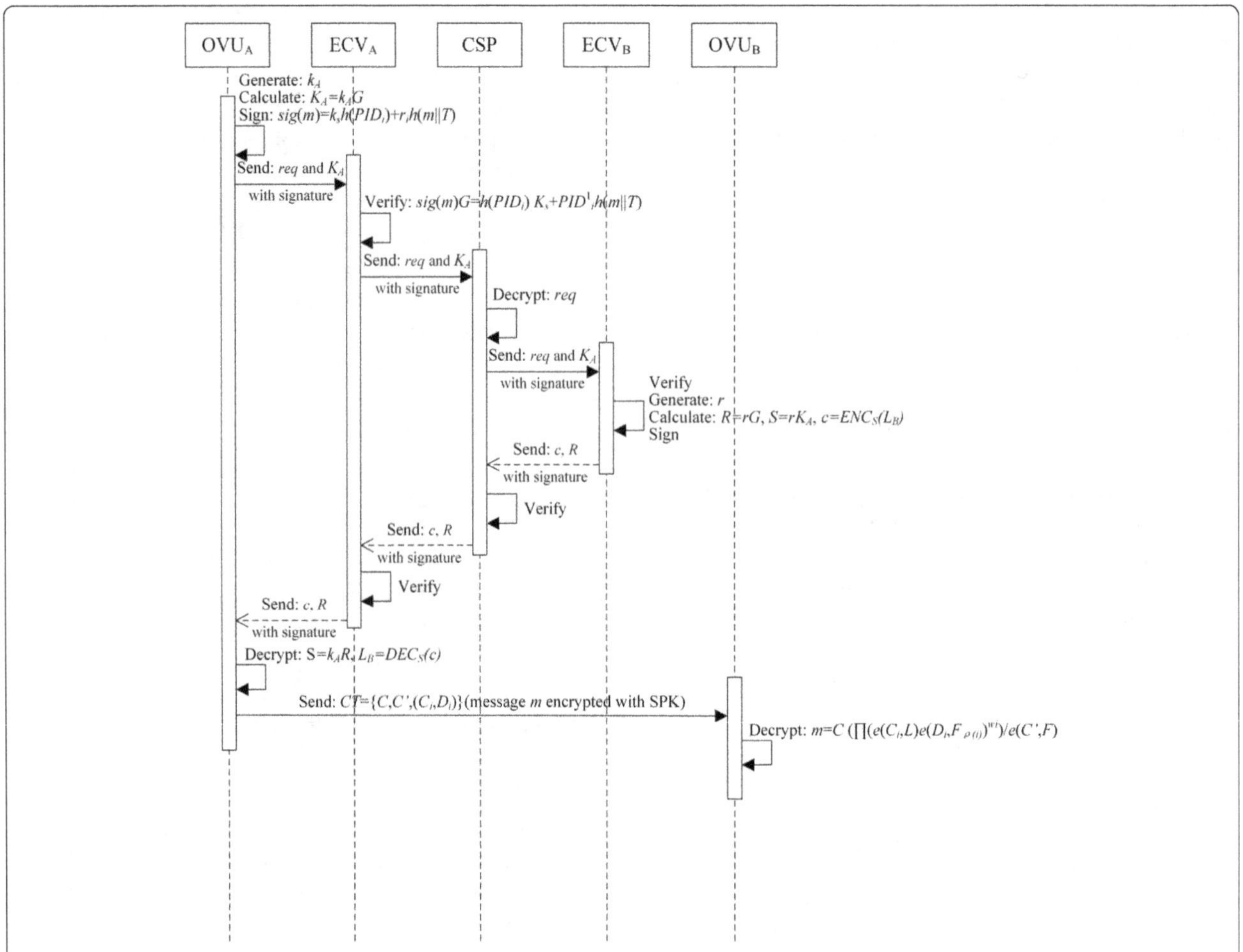

Fig. 2 Operations and interaction process of proposed scheme

(1) Requests by $\mathrm{OVU_A}$: $\mathrm{OVU_A}$ randomly generates $k_A \in [1, n-1]$ as the private key and then calculates $K_A = k_A G$ as the public key. And req denotes the request of sharing data with users who are in domain B. Before sending a message, $\mathrm{OVU_A}$ must complete the following work so that K_A and req can be sent securely to $\mathrm{ECV_A}$.

First, $\mathrm{OVU_A}$ needs to send its real identity RID and password PWD to TPD for authentication. If these two values are inconsistent with the pre-stored values in the TPD, the authentication will fail and the next service will be rejected. After the identity is successfully verified, TPD will calculate the pseudo-identity $\mathrm{PID}_i = \{\mathrm{PID}_i^1, \mathrm{PID}_i^2\}$, where i denotes the number of the vehicle, r_i is a random number generated by TPD, $\mathrm{PID}_i^1 = r_i \cdot G$, $\mathrm{PID}_i^2 = \mathrm{RID} \oplus h(r_i \cdot K_s)$. Besides, in order to prevent messages from being tampered with during transmission, $\mathrm{OVU_A}$ must provide signatures for all messages to be sent. To send the message m, the signature function is defined as $\mathrm{sig}(m) = k_s h(\mathrm{PID}_i) + r_i h(m \| T)$, where T is the current timestamp. So, the content sent by the vehicle is such a message signature pair $\{\mathrm{PID}_i, T, m, \mathrm{sig}(m)\}$.

Therefore, $\mathrm{OVU_A}$ sends $\{\mathrm{PID_{OVU_A}}, T_{\mathrm{OVU_A}}, K_A, \mathrm{req}, \mathrm{sig}(K_A), \mathrm{sig}(\mathrm{req})\}$ to $\mathrm{ECV_A}$ finally.

(2) $\mathrm{ECV_A}$ and CSP process: After receiving the message, $\mathrm{ECV_A}$ verifies the integrity of the message and the legality of the message signature. According to some formulas above, we can know that:

$$\begin{aligned}\mathrm{sig}(m) \cdot G &= G \cdot k_s h(\mathrm{PID}_i) + G \cdot r_i h(m \| T) \\ &= K_s \cdot h(\mathrm{PID}_i) + \mathrm{PID}_i^1 \cdot h(m \| T)\end{aligned}$$

Thus the equation $\mathrm{sig}(m) \cdot G = K_s \cdot h(\mathrm{PID}_i) + \mathrm{PID}_i^1 \cdot h(m \| T)$ can be used to verify messages. If the calculation results on the left are equal to the one on the right, the verification is successful. Otherwise, the verification fails.

$\mathrm{ECV_A}$ sends the message to CSP. After receiving the message, CSP decrypts the request to determine that the domain OVU_A wants to share with is domain B, and then forwards the message to ECV_B.

(3) ECV_B return: After verifying the public key K_A and the request for data sharing *req* from domain A, ECV_B encrypts the attribute list of domain B with symmetrical encryption to return to domain A.

ECV_B generates a random number $r \in [1, n-1]$ and calculates the intermediate parameter $R = rG$ and symmetric key $S = rK_A$.

After that, ECV_B uses S and symmetric encryption scheme to encrypt the attribute list of domain B denoted by L_B, and the encrypted ciphertext is $c = ENC_S(L_B)$.

Finally, ECV_B sends the message signature pair $\{PID_{ECV_B}, T_{ECV_B}, R, c, sig(R), sig(c)\}$ signed for R and c to OVU_A through CSP and ECV_A.

(4) Decryption by OVU_A: After OVU_A successfully verifies the message received, decryption is needed to obtain L_B.

First, according to the known conditions, the following equation exists: $S = rK_A = rk_AG = k_AR$. Then, we can calculate $S = k_AR$ to get the same key S as generated by ECV_B and take it as the session key. Finally, the attributes list L_B can be gained by decryption with the symmetric encryption scheme: $L_B = DEC_S(c)$.

(5) Encryption by OVU_A: Based on the attribute list, OVU_A can use CP-ABE scheme to encrypt data and define a policy to determine users who can access the ciphertext.

We define system public key as $SPK = g, e(g,g)^{\alpha}, g^{\beta}, f_1, \ldots, f_X$, where g is a generator, random numbers $\alpha, \beta \in Z_q$, and random numbers $f_1, \ldots, f_X$ correspond to the x attributes of L_B.

OVU_A defines access control policy (M, ρ). M is a share-generating matrix with x rows and y columns. For $i = 1, \ldots, x$, function $\rho(i)$ associates the ith row of matrix M to an attribute of list L_B. OVU_A selects a random vector $\overrightarrow{v} = (\gamma, t_2, \ldots, t_y) \in Z_q^y$, where $t_2, \ldots, t_y$ are randomly chosen to share γ, then calculates $\lambda_i = \overrightarrow{v} \cdot M_i$, where M_i is the vector corresponding to the ith row of M. $\{\lambda_i = (M\overrightarrow{v})_i\}_{i\in\{1,\ldots,x\}}$ are valid shares only when there is a set of constants $\{w_i \in Z_q\}_{i\in\{1,\ldots,x\}}$ such that the equation $\sum_{i\in\{1,\ldots,x\}} w_i\lambda_i = \gamma$ holds. In this case, the user can decrypt the ciphertext.

Assuming that OVU_A wants to share the message m, the system public key SPK is used to encrypt m. First, calculate $C = me(g,g)^{\alpha\gamma}$ and $C' = g^{\gamma}$. Meanwhile, for all rows of the matrix M, i.e., for $i = 1, \ldots, x$, calculate $C_i = g^{\beta\lambda_i} f_{\rho(i)}^{-r_i}$ and $D_i = g^{r_i}$, where $r_1, \ldots, r_x \in Z_q$ are chosen randomly by OVU_A.

Therefore, the ciphertext that published by OVU_A is $CT = \{C, C', (C_i, D_i)\}_{i\in\{1,\ldots,x\}}$.

Finally, OVU_A sends CT to the public resource pool to which the users in domain B can access.

(6) Users decryption: Users use attribute-based privacy key to decrypt the ciphertext. The private key of a user with attributes A in domain B is defined as:

$$F = g^{\alpha} g^{\beta\varepsilon}, L = g^{\varepsilon}, \forall i \in A : F_i = f_i^{\varepsilon}$$

Users whose attributes satisfy the access structure can gain the message m by calculating the following formula:

$$m = C \cdot \left(\Pi_{i\in\{1,\ldots,x\}} \left(e(C_i, L) e(D_i, F_{\rho(i)})\right)^{w_i}\right) / e(C', F)$$

5 Analysis of our scheme

In this section, the correctness proof and security analysis and efficiency analysis of our scheme are given.

5.1 Correctness of the CP-ABE scheme

A user who is qualified to access and decrypt the ciphertext has attributes that satisfy the access structure, which means his $\{\lambda_i = (M\overrightarrow{v})_i\}_{i\in\{1,\ldots,x\}}$ are valid. Thus, there exist constants $\{w_i \in Z_q\}_{i\in\{1,\ldots,x\}}$ to make the equation $\sum_{i\in\{1,\ldots,x\}} w_i\lambda_i = \gamma$ set up. The correctness of the decryption algorithm is proved as follows:

$$\frac{C \cdot \left(\Pi_{i\in\{1,\ldots,x\}} \left(e(C_i, L) e(D_i, F_{\rho(i)})\right)^{w_i}\right)}{e(C', F)} = \frac{C \cdot \left(\Pi_{i\in\{1,\ldots,x\}} e(g,g)^{\varepsilon\beta\lambda_i w_i}\right)}{e(g,g)^{\alpha\gamma} e(g,g)^{\beta\gamma\varepsilon}} = \frac{m \cdot \left(e(g,g)^{\sum_{i\in\{1,\ldots,x\}} \varepsilon\beta\lambda_i w_i}\right)}{e(g,g)^{\beta\gamma\varepsilon}} = m$$

5.2 Security analysis

(1) Anonymity: Vehicles use pseudo-identities instead of their real identities during the communication process, and the real identities of vehicles are stored in the non-attackable TPD, which effectively protects the privacy of their identities. Additionally, for a malicious vehicle, TA can obtain its real

identity according to its pseudo-identity so as to investigate the responsibility of this vehicle. The calculation formula is:

$$\mathrm{PID}_i^2 \oplus h\left(k_s \cdot \mathrm{PID}_i^1\right) = \mathrm{RID} \oplus h(r_i \cdot K_s) \oplus h(k_s \cdot r_i \cdot g) = \mathrm{RID}$$

(2) Message authentication: In our scheme, signing message ensures that the message will not be tampered with during transmission, so the integrity of the message and the legitimacy of the message owner are guaranteed.

(3) Data confidentiality: We use ECC to transmit the list of attributes L_B. The session key S for decrypting L_B can be calculated only when the private key k_A is known. However, according to ECDLP, we can know that it is difficult for other vehicles to get the private key. Therefore, the confidentiality of the data can be ensured.

(4) Unlinkability: Unlinkability is an effective complement to anonymity, which makes it impossible for a receiver to link one user who is interacting with it currently with another who was previously authenticated by it. Every time the sender sends a message, it needs to select a random number which will be used in the signature function. And some system parameters are safely stored in the TPD. Therefore, the malicious attacker cannot judge whether he has authenticated the same vehicle twice according to pseudo-identity.

5.3 Efficiency analysis

Our experiment was run on an Intel Core i3 2.4-GHz processor with MIRACL library [8] and Crypto++ library [9]. We compared our scheme with other two scheme [5, 45]. Some operations about execution time are defined as follows.

(1). T_{ab}: The execution time of a multiplication operation $ab \bmod n$, where $a, b \in Z_q^*$.

(2). T_{xP}: The execution time of a scale multiplication operation $x \cdot P$, where $x \in Z_q^*$ and $P \in E$.

(3). T_{g^x}: The execution time of a modular exponentiation $g^x \bmod n$, where $x \in Z_q^*$.

(4). T_{pair}: The execution time of a bilinear pairing operation $e(aP, bP)$, where $a, b \in Z_q^*$ and $P \in E$.

(5). T_{hash}: The execution time of SHA256 hash function operation.

(6). T_{AES}: The execution time of the encryption or decryption operation of AES-CCM algorithm.

(7). $T_{\mathrm{RSA-ED}}$: The execution time of the encryption or decryption operation of RSA1024 algorithm.

(8). $T_{\mathrm{RSA-SV}}$: The execution time of the signature or verification operation of RSA1024 algorithm.

(9). T_{ECIES}: The execution time of the encryption operation of elliptic curve integrate encrypt scheme.

(10)T_{ECDSA}: The execution time of the signature or verification operation of elliptic curve digital signature algorithm.

As we all know, it was difficult to measure accurately due to the short single-step execution time in the experiment. So, we choose more steps in the program and choose a longer input on the data to improve the accuracy of the measurement results. For the four operations of hashing, signing, encryption, and decryption, we set the number of for loops to 1000 and select the random bit string with the maximum length as the input. Then, the average value, dividing the time spent by 1000, is taken as the execution time of the operation. For the AES encryption/decryption algorithm, we use the counter with CBC-MAC mode. For the RSA encryption/decryption and sign/verification algorithm, we use the 1024-bit key and RSA encryption with PKCS v1.5 padding. For the ECIES and ECDSA algorithm, we use the secp256r1 as the initial parameter of the elliptic curve and SHA256 as the hash function.

All the parameters in the above operations including a, b, x, P are selected randomly from their domains of definition. Finally, we got the time cost of above operations from the experiment and listed them in Table 1.

Throughout the interaction process in our scheme, the whole time cost includes the time to encrypt, sign, verify, and decrypt. The time needed to perform the calculation operation $K_A = k_A G$ in step (1) of Section 4.4 is $T_{xP} = 1.258ms$. Then, the time to sign a message by the function $\mathrm{sig}(m) = k_s h(\mathrm{PID}_i) + r_i h(m \| T)$ is $T_{\mathrm{sign}} = 2T_{ab} = 0.0692ms$. Similarly, the time to verify the message that the vehicle receives by calculating the equation $\mathrm{sig}(m) \cdot G = K_s \cdot h(\mathrm{PID}_i) + \mathrm{PID}_i^1 \cdot h(m \| T)$ is $T_{\mathrm{ver}} = 3T_{xP} = 3.774ms$. In step (3) and step (4), the respective execution time of $R = rG$, $S = rK_A$, and $S = k_A R$ is also equal to $T_{xP} = 1.258ms$, and

Table 1 Execution time cost of different cryptographic operations

Operations	Times (ms)
T_{ab}	0.0346
T_{xP}	1.258
T_g^x	3.3421
T_{pair}	23.625
T_{hash}	0.005
T_{AES}	0.022
$T_{\mathrm{RSA\text{-}ED}}$	0.13/1.51
$T_{\mathrm{RSA\text{-}SV}}$	1.49/0.13
T_{ECIES}	4.35
T_{ECDSA}	3.01/8.89

the encryption process $c = \text{ENC}_S(L_B)$ and decryption process $L_B = \text{DEC}_S(c)$ take approximately $2T_{\text{AES}} = 0.044ms$ in total. It should be noted that since the three comparison schemes all use the ABE algorithm identically, we have not taken into account the time overhead of this part. In a similar way, the total time overhead for the encryption and decryption operations of [45] is $2T_{\text{pair}}$ and that of [5] is $2T_{\text{pair}}5T_{ab} + 2T_{g^x} + 2T_{xP} + T_{\text{RSA-ED}} + T_{\text{RSA-SV}}$. As for the signature and verification operations, [45] needs to calculate the time overhead of signature generation, the certificate verification, and the message signature verification. And for [5], we use the same calculation method as ours to maintain consistency since its author does not specify the specific signature and verification method. The result is showed in Fig. 3.

5.4 Result and discussion

From Fig. 3, we can see that our scheme, requiring less execution time when transmitting the same number of messages, has better performance than the other two schemes.

The reason for such a result is that the elliptic curve encryption algorithm we use is more efficient than the other two papers' algorithms. As shown in Table 1, the time of the bilinear pair encryption algorithm used in [45] T_{pair} is much larger than T_{xP} and T_{AES} of our scheme. The execution time of a modular exponentiation T_{g^x} and that of the encryption or decryption operation of RSA used in [5] are also greater than our T_{xP} and T_{AES}. Therefore, our method has the best performance.

However, the main time overhead for our scheme is spent on the message signing and authentication operations, so its limitations will be clearly reflected when the number of vehicles participated in data sharing is greatly large.

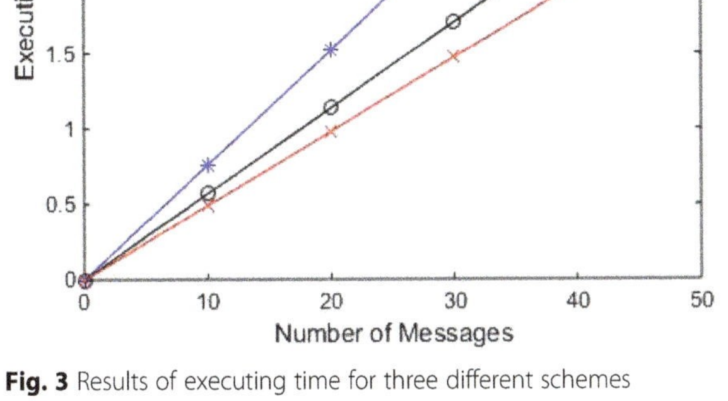

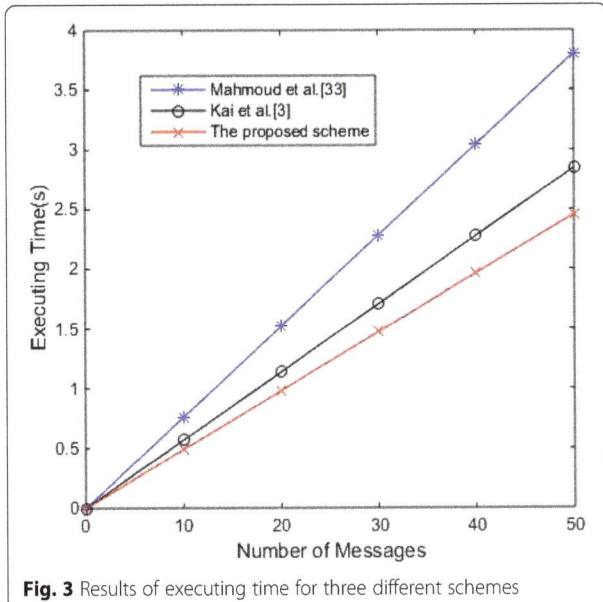

Fig. 3 Results of executing time for three different schemes

6 Conclusions

We propose a secure scheme based on edge computing to achieve data sharing among different domains. Next, we use ECC, CP-ABE, and the message authentication mechanism during the phase of data encryption and transmission. Finally, an analysis of our scheme demonstrates its security and efficiency.

In our scheme, the method used for selecting ECVs takes into consideration the number of available computing resources and their distances to the RSUs. In the future, we plan to include the social centrality of vehicles and select vehicles that can contact and interact with more vehicle nodes as edge computing nodes.

Abbreviations

AES: Advanced Encryption Standard; AES-CCM: Advanced Encryption Standard-Counter with CBC MAC; CA: Certification authority; CBC-MAC: Cipher Block Chaining Message Authentication Code; CLSS: Certificateless short signature; CPABE: Ciphertext Policy Attribute Based Encryption; CSP: Cloud service provider; CT: Cipher Text; ECC: Elliptic curve cryptography; ECDLP: Elliptic curve discrete logarithm problem; ECDSA: Elliptic curve digital signature algorithm; ECIES: Elliptic curve integrated encryption scheme; ECV: Edge computing vehicle; EG: Evolutionary games; IBS: Identity-based signature; IoT: Internet of Things; MAC: Media access control; MBS: Macro base station; MEC: Mobile edge computing; MK: Master key; OBU: On-board unit; OVU: Ordinary vehicle user; PID: Pseudo-identity; PK: Public key; PKCS: Public key cryptography standards; PKI: Public key infrastructure; PW-CPPA-GKA: Password-based conditional privacy protection authentication and group key generation; RID: Real identity; RSU: Roadside unit; SHA256: Secure Hash Algorithm 256; SPK: System public key; TA: Trusted authority; T_{ab}: The execution time of a multiplication operation; T_{AES}: The execution time of the encryption or decryption operation of AES-CCM algorithm; T_e: The execution time of a bilinear pairing operation; T_{ECDSA}: The execution time of the signature or verification operation of elliptic curve digital signature algorithm; T_{ECIES}: The execution time of the encryption operation of elliptic curve integrate encrypt scheme; T_g: The execution time of a modular exponentiation; T_h: The execution time of a hash function operation; TPD: Tamper-proof device; $T_{\text{RSA-ED}}$: The execution time of the encryption or decryption operation of RSA1024 algorithm; $T_{\text{RSA-SV}}$: The execution time of the signature or verification operation of RSA1024 algorithm; T_{xP}: The execution time of a scale multiplication operation; V2I: Vehicle-to-infrastructure; V2V: Vehicle-to-vehicle; VANETs: Vehicular ad hoc networks

Acknowledgements

The authors would like to thank Jing Zhang for her comments and suggestions.

Authors' contributions

JP carried out the study and drafted the manuscript. LW conceived the idea and participated in the design of the algorithm. JP and LW performed the experiment and analyzed the result. JC, YX, and HZ participated in the technical discussion and helped to perform the data analysis. All authors read and approved the final manuscript.

Funding

The work was supported by the National Natural Science Foundation of China (No. 61872001, No. 61572001, No. 61702005), the Open Fund of Key Laboratory of Embedded System and Service Computing (Tongji University), Ministry of Education (No. ESSCKF2018-03), the Open Fund for Discipline Construction, Institute of Physical Science and Information Technology, Anhui University, and the Excellent Talent Project of Anhui University.

Competing interests

The authors declare that they have no competing interests.

References

1. C.K. Toh, Ad hoc mobile wireless networks: protocols and systems. Pearson Education (2001)
2. J.J. Cheng, J.L. Cheng, M.C. Zhou, et al., Routing in internet of vehicles: A review[J]. IEEE Transactions on Intelligent Transportation Systems **16**(5), 2339–2352 (2015)
3. J. Cui, L. Wei, J. Zhang, et al., An efficient message-authentication scheme based on edge computing for vehicular ad hoc networks. IEEE Transactions on Intelligent Transportation Systems, 1–12 (2018)
4. J. Cui, H. Zhong, W. Luo, et al., Area-based mobile multicast group key management scheme for secure mobile cooperative sensing[J]. Science China(Information Sciences), 286–292 (2017)
5. K. Fan, Q. Pan, J. Wang, et al., *Cross-domain based data sharing scheme in cooperative edge computing* (2018 IEEE International Conference on Edge Computing, 2018), pp. 87–92
6. G. Luo, Q. Yuan, H. Zhou, et al., Cooperative vehicular content distribution in edge computing assisted 5G-VANET. China Communications **15**(7), 1–17 (2018)
7. X. Liu, R. Zhu, B. Jalaian, et al., Dynamic spectrum access algorithm based on game theory in cognitive radio networks. Mobile Networks and Applications **20**(6), 817–827 (2015)
8. Scott M, Multiprecision integer and rational arithmetic C/C++ library (MIRACL), (2003). https://www3.cs.stonybrook.edu/~algorith/implement/shamus/implement.shtml.
9. Dai W, Crypto++ library 5.1-a free C++ class library of cryptographic schemes, (2004). https://www.cryptopp.com/.
10. K. Zeng, Pseudonymous PKI for ubiquitous computing. European Public Key Infrastructure Workshop, 207–222 (2006)
11. X. Lin, X. Sun, P.H. Ho, et al., GSIS: a secure and privacypreserving protocol for vehicular communications. IEEE Transactions on Vehicular Technology **56**(6), 3442–3456 (2007)
12. C. Zhang, R. Lu, X. Lin, et al., in *IEEE INFOCOM 2008-The 27th Conference on Computer Communications*. An efficient identity-based batch verification scheme for vehicular sensor networks (2008), pp. 246–250
13. S.J. Horng, S.F. Tzeng, P.H. Huang, et al., An efficient certificateless aggregate signature with conditional privacy-preserving for vehicular sensor networks. Information Sciences **317**, 48–66 (2015)
14. P. Vijayakumar, M. Azees, A. Kannan, et al., Dual authentication and key management techniques for secure data transmission in vehicular ad hoc networks. IEEE Transactions on Intelligent Transportation Systems **17**(4), 1015–1028 (2016)
15. M. Azees, P. Vijayakumar, L.J. Deboarh, EAAP: efficient anonymous authentication with conditional privacy-preserving scheme for vehicular ad hoc networks. IEEE Transactions on Intelligent Transportation Systems **18**(9), 2467–2476 (2017)
16. J.L. Tsai, A new efficient certificateless short signature scheme using bilinear pairings. IEEE Systems Journal **11**(4), 2395–2402 (2017)
17. S.M. Pournaghi, B. Zahednejad, M. Bayat, et al., NECPPA: a novel and efficient conditional privacy-preserving authentication scheme for VANET. Computer Networks **134**, 78–92 (2018)
18. M.R. Asaar, M. Salmasizadeh, W. Susilo, et al., A secure and efficient authentication technique for vehicular ad-hoc networks. IEEE Transactions on Vehicular Technology **67**(6), 5409–5423 (2018)
19. Y. Liu, L. Wang, H.H. Chen, Message authentication using proxy vehicles in vehicular ad hoc networks. IEEE Transactions on Vehicular Technology **64**(8), 3697–3710 (2015)
20. S.K.H. Islam, M.S. Obaidat, P. Vijayakumar, et al., A robust and efficient password-based conditional privacy preserving authentication and group-key agreement protocol for VANETs. Future Generation Computer Systems **84**, 216–227 (2018)
21. J. Cui, J. Zhang, H. Zhong, et al., SPACF: A secure privacy-preserving authentication scheme for VANET with cuckoo filter. IEEE Transactions on Vehicular Technology **66**(11), 10283–10295 (2017)
22. J. Cui, J. Wen, S. Han, et al., Efficient privacy-preserving scheme for real-time location data in vehicular ad-hoc network. IEEE Internet of Things Journal **5**(5), 3491–3498 (2018)
23. H. Zhong, B. Huang, J. Cui, et al., *Efficient conditional privacy-preserving authentication scheme using revocation messages for VANET. 2018 27th International Conference on Computer Communication and Networks* (2018), pp. 1–8
24. T. Jing, Y. Pei, B. Zhang, et al., An efficient anonymous batch authentication scheme based on priority and cooperation for VANETs. EURASIP Journal on Wireless Communications and Networking **277** (2018)
25. H. Sago, M. Shinohara, T. Hara, et al., in *21st International Conference on Advanced Information Networking and Applications Workshops*. A data dissemination method for information sharing based on inter-vehicle communication, vol 2 (2007), pp. 743–748
26. Y. Zhang, J. Zhao, G. Cao, Roadcast: a popularity aware content sharing scheme in VANETs. ACM SIGMOBILE Mobile Computing and Communications Review **13**(4), 1–14 (2010)
27. Y. Zhu, Y. Zhang, X. Li, et al., Improved collusion-resisting secure nearest neighbor query over encrypted data in cloud. Concurrency and Computation: Practice and Experience, e4681 (2018)
28. X. Li, Y. Zhu, J. Wang, et al., On the soundness and security of privacy-preserving SVM for outsourcing data classification. IEEE Transactions on Dependable and Secure Computing, 1-1 (2017)
29. J. Xu, D. Zhang, L. Liu, et al., Dynamic authentication for cross-realm SOA-based business processes. IEEE Transactions on services computing **5**(1), 20–32 (2012)
30. J. Wang, R. Zhu, S. Liu, A differentially private unscented Kalman filter for streaming data in IoT. IEEE Access **6**, 6487–6495 (2018)
31. Y. Hao, J. Tang, Y. Cheng, Secure cooperative data downloading in vehicular ad hoc networks. IEEE Journal on Selected Areas in Communications **31**(9), 523–537 (2013)
32. D. Wu, H. Liu, Y. Bi, et al., Evolutionary game theoretic modeling and repetition of media distributed shared in P2P-based VANET. International Journal of Distributed Sensor Networks **10**(6), 718639 (2014)
33. Y. Lai, L. Zheng, T. Wang, et al., in *International Conference on Security, Privacy, and Anonymity in Computation, Communication, and Storage*. Cloud-assisted data storage and query processing at vehicular ad-hoc sensor networks (2017), pp. 692–702
34. J. Li, Y. Jia, L. Liu, et al., CyberLiveApp: A secure sharing and migration approach for live virtual desktop applications in a cloud environment. Future Generation Computer Systems **29**(1), 330–340 (2013)
35. D. Miao, L. Liu, R. Xu, et al., An efficient indexing model for the fog layer of industrial internet of things. IEEE Transactions on Industrial Informatics **14**(10), 4487–4496 (2018)
36. W. Shi, S. Dustdar, The promise of edge computing. Computer **49**(5), 78–81 (2016)
37. W. Shi, J. Cao, Q. Zhang, et al., Edge computing: vision and challenges. IEEE Internet of Things Journal **3**(5), 637–646 (2016)
38. Y. Mao, C. You, J. Zhang, et al., A survey on mobile edge computing: the communication perspective. IEEE Communications Surveys & Tutorials **19**(4), 2322–2358 (2017)
39. J. Ren, H. Guo, C. Xu, et al., Serving at the edge: a scalable IoT architecture based on transparent computing. IEEE Network **31**(5), 96–105 (2017)
40. R. Roman, J. Lopez, M. Mambo, Mobile edge computing, Fog et al.: a survey and analysis of security threats and challenges. Future Generation Computer Systems **78**, 680–698 (2018)
41. Q. Yuan, H. Zhou, J. Li, et al., Toward efficient content delivery for automated driving services: an edge computing solution. IEEE Network **32**(1), 80–86 (2018)
42. V.S. Miller, Use of elliptic curves in cryptography. Conference on the theory and application of cryptographic techniques, 417–426 (1985)
43. N. Koblitz, Elliptic curve cryptosystems. Mathematics of computation **48**(177), 203–209 (1987)
44. J. Bethencourt, A. Sahai, B. Waters, Ciphertext-policy attribute-based encryption. IEEE Symposium on Security and Privacy. IEEE Computer Society, 321–334 (2007)
45. M.H. Eiza, Q. Ni, Q. Shi, Secure and privacy-aware cloud-assisted video reporting service in 5G-enabled vehicular networks. IEEE Transactions on Vehicular Technology **65**(10), 7868–7881 (2016)

6

Performance analysis of wireless-powered cognitive radio networks with ambient backscatter

Daniyal Munir[1], Syed Tariq Shah[1,2], Kae Won Choi[1], Tae-Jin Lee[1] and Min Young Chung[1*]

Abstract

Ambient backscatter is a promising wireless communication technique where low-power users communicate with each other without any dedicated power source. These communicating users transmit their information by reflecting ambient radio-frequency (RF) signals. In this paper, we propose an ambient backscatter communications-assisted wireless-powered underlay cognitive radio network (CRN). The proposed CRN consists of a single primary transmitter (PT) and multiple primary receivers (PRs), secondary transmitters (STs), and secondary receivers (SRs). For efficient utilization of radio resources, the STs in the proposed scheme dynamically adopt either harvest-then-transmit mode or backscatter mode. Furthermore, PRs cooperate with STs to select an appropriate mode for their communication with SRs. To evaluate the performance of our proposed scheme, we conduct system-level simulations. Numerical results show that the performance of the secondary system can be improved in terms of throughput with minimum effect on the communication of primary users.

Keywords: Ambient backscatter, Underlay cognitive radio, Harvest-then-transmit

1 Introduction

For the realization of *Internet of things (IoT)*, low-power wireless sensor networks (WSNs) have been adopted in numerous applications, such as health care, traffic control, surveillance, and so on. In general, WSNs operate in the unlicensed industrial, scientific, and medical (ISM) spectrum [1]. However, with the extensive use of Wi-Fi, ZigBee, and Bluetooth in modern day applications, the mutual interference problem has become more challenging [2]. To alleviate this problem for a large number of sensors, cognitive radio technology was incorporated in *WSNs*, which yields cognitive radio sensor networks (CRSNs) [3]. Sensors, in a CRSN, can exploit more channel access opportunities in an underutilized licensed spectrum.

In CRSNs, secondary users can share the spectrum with primary users by adopting one of the two known modes: overlay and underlay modes. In overlay mode, secondary users can access radio resources (i.e., channels) owned by the primary network only if the primary channels are idle. However, for some wireless broadcast services such as TV and FM radio, the primary channels may be busy most of the time and secondary users may be unable to access these channels opportunistically. On the other hand, in underlay mode, secondary users can concurrently access the channels owned by the primary network, provided that the resulted interference at primary users remains below a predefined threshold value [4]. Operating on this mode may become challenging for secondary users in densely deployed area since these users will cause more interference to closely located primary users. Therefore, we need to find substitute techniques to solve this problem and enhance the performance of the secondary system.

In order to enable opportunistic communication between secondary transmitters (STs) and secondary receivers (SRs), ambient backscatter communication was proposed recently [5]. Ambient backscatter users operate on the principle of reflecting received radio frequency (RF) signals from an ambient source such as TV tower, WiFi access point (AP), and cellular base station [6].

*Correspondence: mychung@skku.edu
[1]Department of Electrical and Computer Engineering, Sungkyunkwan University, 2066 Seobu-Ro, Jangan-Gu, 16419 Suwon, Gyeonggi-Do, South Korea
Full list of author information is available at the end of the article

Note that instantaneous excitation signals provide enough power to a backscatter user to carry out its operations [7]. Hence, a dedicated time for energy harvesting is not required.

From the perspective of primary network, the backscattered (reflected) signals received at the primary receiver (PR) are treated as multipath signals from the primary transmitter (PT). This is because of the fact that the modulation rate of backscatter is very low as compared to the modulation rate of the PT's transmission. Hence, the backscattered signals are treated as constant signals from the viewpoint of a PR. Moreover, PR can prevent the effect of multipath distortion from backscattering users by implementing existing techniques used in modern wireless networks, such as cyclic prefix in OFDMA networks [5]. Therefore, ambient backscatter communications with its ultra low-power and low-complexity characteristics can be conveniently employed in a CRSN.

Recently, RF energy harvesting has been introduced and implemented in CRSNs [8]. This leads to a new type of networks, called wireless-powered CRSNs. In this network, the STs are able to harvest energy from primary signals, and use the harvested energy to transfer data to their SRs through a primary channel. Therefore, the transmission used in wireless-powered CRSN is known as harvest-then-transmit mode [9]. A wireless-powered CRSN can be integrated with ambient backscatter communication to improve the performance of the secondary system. Two possible communication modes can be adopted in such networks: (i) harvest-then-transmit mode and (ii) ambient backscatter mode [10]. However, integrating ambient backscatter communications with underlay CRSN may raise different issues as compared to overlay CRSN model considered in [10]. Some of the major challenges faced in underlay CRSN are as follows: (a) how can a secondary user choose between backscatter mode or harvest-then-transmit mode; (b) while using harvest-then-transmit mode, how much time is used for harvesting energy and information transmission; and (c) how to estimate the interference at the PRs.

In order to address the above mentioned issues, we use ambient backscatter-assisted wireless-powered network proposed by [10] in an underlay environment. The proposed CRN consists of a PT and multiple PRs, STs, and SRs. All PRs, STs, and SRs have the capability of performing RF communication and/or ambient backscatter communication. STs considered in our proposed network are energy-constraint devices and harvest energy from ambient RF signals. Thus, STs transmit their information using either harvest-then-transmit mode or backscatter mode. STs select one of the two modes based on the estimated interference they may cause to nearby PRs. STs estimate this interference with the help of channel state information (CSI) between PT and PRs, which is reported by PRs using backscatter communication. The main contributions of this paper are as follows:

- We propose a transmission mode selection mechanism for secondary users, where STs can select either harvest-then-transmit mode or ambient backscatter mode based on the estimated interference at PRs.
- In order to estimate the interference at the PR, we propose an ambient backscattering-based CSI reporting mechanism. PRs report their CSI to nearby STs using backscatter communication.
- To evaluate the performance of the proposed scheme, we conduct system-level simulations. The numerical analysis shows that the proposed scheme improves the performance of the secondary system in terms of average data rates with a minimal effect on the primary communications average data rates.

A preliminary version of this work has been presented at a conference, which briefly describes the integration of ambient backscatter with underlay CRSN [11].

The rest of this paper is organized as follows. In the next section, we present related works. System model and problem formulation are introduced in Section 3. Section 4 explains the proposed scheme in details. Performance of the proposed solution is evaluated in Section 5 and finally, we conclude this paper in the last section.

2 Related work

In CRSN, the battery life of a sensor user is very important, because it highly affects the network topology once the battery runs out. To improve the network lifetime and energy efficiency, a lot of effort has been put together by the research society [12–16]. Nonetheless, battery replacement is still required to keep the CRSN active. However, changing the batteries usually costs a lot and can even be dangerous where the sensors are used for monitoring radioactive and toxic materials. Instead of replacing the batteries, wireless power transfer with simultaneous information transfer (SWIPT) was first proposed by Varshney [17]. RF energy transfer was further studied for WSNs in [18–20].

In CRSN, despite reliable channel state between PT and secondary users, there would be no gain if the secondary users experience lack of energy. This motivated [21] to introduce the RF energy harvesting to the cooperative relaying CRSN. The users can cooperate at both information and energy levels, through which the energy-limited secondary users can operate continuously. In a similar environment, the authors in [22] proposed an optimal resource allocation scheme for secondary users to maximize the sum throughput. However, for wireless-powered CRSNs, when the secondary users have less opportunity to access the primary channels, backscatter communications can be used to improve the low overall transmission rate for secondary networks [10].

Recently, backscatter communications powered by RF energy harvesting has received a lot of attention. Parks et al. in [23] proposed a coding mechanism for backscatter users that enables long range communication between them as well as improves the transmission rate as high as 1 Mbps. To provide Internet connectivity for these low power devices, authors in [24] reuse existing Wi-Fi infrastructure to demonstrate its feasibility. Authors in [6] have proposed a network architecture where two users can communicate with each other by integrating energy harvesting and backscatter communications. They consider a scenario where power beacons are deployed for wirelessly powering these users. The users transmit data to their potential receiver by modulating and reflecting a portion of received signals from the power beacons. However, deploying power beacons to recharge an enormous amount of low power users may prove to be costly.

Backscatter communications was also addressed in some recent works [7, 25, 26]. To enhance the throughput of backscatter communications, authors in [25] proposed a reader design with multi-antenna wireless energy beamforming for multiple backscatter users. Lyu et al. [7] have proposed resource allocation policies for multi-user backscatter communication systems. They have considered a dedicated reader antenna, which excites the backscatter user for their concurrent communication with the reader. Authors in [26] have proposed a wireless powered communications network assisted by backscatter communications. Their model consists of a power station, an information receiver and multiple users that can work in either backscatter mode or harvest-then-transmit mode. These works provide useful insight into the backscatter communications and, however, require special infrastructure to generate the excitation RF signals that backscatter users can reflect. Therefore, ambient backscatter communications should be given more attention to exploring cost-effective possibilities for low-power devices.

Recent research works such as [10] have reduced the need for specialized infrastructure. In [10], an ambient backscatter-assisted CRSN was proposed, where two types of users are present in the network, which can operate in backscatter and harvest-then-transmit modes. For flexible network deployment, the authors assumed that a user could operate in one of the two modes. A single-user CRSN was considered, and a tradeoff between the backscatter and harvest-then-transmit modes was analyzed. Based on [10], the authors in [27] extended the work for a multi-user case and studied the optimal time allocation policy. In [10] and [27], authors have studied an overlay CRSN to maximize the secondary system throughput. However, in practical systems, the possibility of finding idle channels is very low for overlay-based CRSN, which motivates us to study this scenario in an underlay CRSN.

3 System model

As shown in Fig. 1, we consider an underlay CRSN scenario, where energy-constrained secondary users simultaneously access the licensed spectrum to transmit their own data. The network consists of primary users, primary

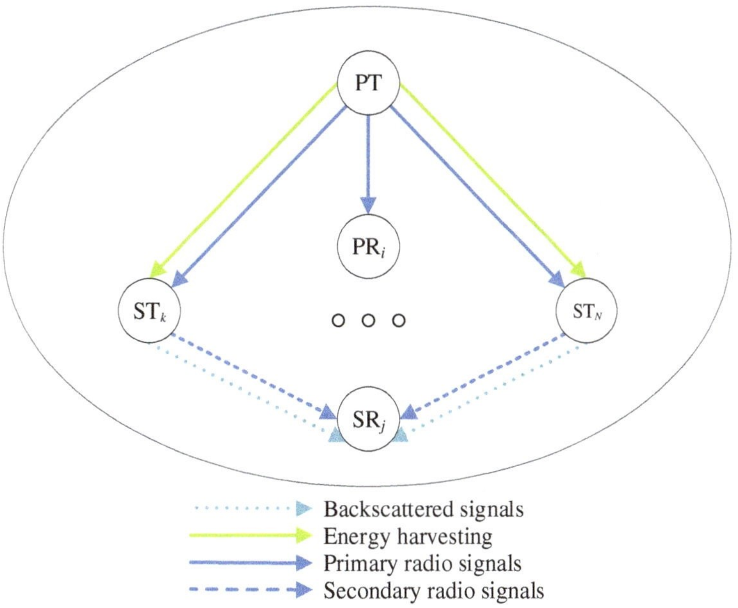

Fig. 1 System model for the proposed scheme

transmitter (PT) and primary receiver (PR), and secondary users, secondary transmitter (ST) and secondary receiver (SR). We denote set of PRs, set of SRs, and set of STs as $\mathcal{I}$, $\mathcal{J}$ and $\mathcal{K}$, respectively, where $\mathcal{I} = \{1, 2, 3, \cdots, L\}$, $\mathcal{J} = \{1, 2, 3, \cdots, M\}$ and $\mathcal{K} = \{1, 2, 3, \cdots, N\}$. The PRs and SRs are randomly deployed within the disc radius of PT using Poisson point process (PPP) with densities λ_{pr} and λ_{sr}, respectively. Then, STs are deployed randomly using PPP within the radius of each SR with density λ_{st}.

The PT continuously broadcasts radio signals within its coverage area to transfer information to PRs. PRs receive these broadcast signals using their RF communication interface. Furthermore, PRs concurrently report CSI between PT and PR to STs located nearby, using their backscatter interface. Those STs that will receive the CSI from a PR are considered as its nearby STs. Based on the information received from the PR, nearby STs estimate the interference that may be induced to the PR. This estimated interference is then used by the STs to select either harvest-then-transmit mode or backscatter mode. It should be noted that STs cannot operate on harvest-then-transmit mode and backscatter mode simultaneously [5]. Therefore, STs switch between one of the two modes to transfer their information to SRs. To receive useful information from STs, SRs may tune to either RF communication or backscatter communication, depending upon the communication mode used by STs.

For the harvest-then-transmit mode, STs rely only on harvested energy from RF signals transmitted by PT and store the harvested energy in their energy storage units for further data transmission. On the other hand, when the backscatter mode is activated, the STs can backscatter the modulated PT signals to the SRs instantaneously. It is assumed that SRs have the perfect knowledge of each ST's operating mode and are able to select the corresponding demodulators for effective communication [10]. Moreover, the channel gains between the users are modeled by the quasi-static block-fading process. In other words, the channel remains constant over the block time T, but may vary independently from one block to another following an identical distribution. In addition, the small-scale fading experienced by these channels is represented by Rayleigh fading model.

The transmission block structure of the proposed scheme is shown in Fig. 2. A time block in which PT transmits signals to PRs can be utilized by STs in two main phases: CSI reporting phase and communication phase. For βT duration, i.e., CSI reporting phase, PRs report their CSI to nearby STs by using backscatter communication. Based on the received CSI, each ST estimates the interference it may cause to its nearby PRs. Depending on the estimated interference in the first phase, ST decides to select one of the two modes for $(1 - \beta)T$ duration in communication phase.

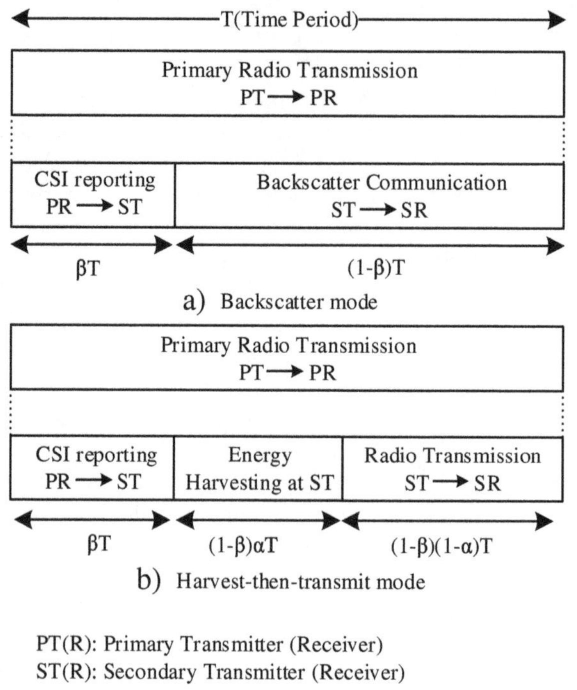

Fig. 2 Transmission block structure for the proposed scheme

In backscatter mode, since STs do not use conventional radio components such as oscillator, mixer, and power amplifiers, they activate their backscatter module from received RF signal power. In addition, STs use the same signals to modulate their data and reflect these signals to communicate with SRs. The SRs may use either passive or active components to demodulate the backscattered signals received from STs. Figure 2a illustrates the operation for backscatter mode. STs communicate with SRs by using backscatter communications for the whole $(1 - \beta)T$ duration.

In order to harvest energy and decode information simultaneously, we adopt time switching-based receiver architecture at STs [28]. In other words, STs use a portion of time for energy harvesting and use that harvested energy for radio transmission for the rest of the time. We assume that STs consume all the harvested energy to transmit the signals to SR. Figure 2b shows a time switching scheme for the harvest-then-transmit mode of the proposed scheme. During this mode, the communication phase $(1-\beta)T$ is further divided into α and $(1-\alpha)$, denoting the time ratio between energy harvesting and radio transmission, respectively. ST harvests energy from the received signals of PT for α $(1 - \beta)T$ time and uses this harvested energy to transfer its information to SR through radio transmission for $(1 - \alpha)$ $(1 - \beta)T$ time. The details of the proposed procedure are described in the following sections.

4 Proposed scheme

In this section, we explain the proposed scheme for underlay CRSN assisted by backscatter communications. Figure 3 presents a graphical overview of the proposed scheme. In step 1, PT broadcasts signals to all the users located within its coverage area. PRs backscatter the received signals to report their CSI to nearby STs in step 2. Based on the received CSI, an ST estimates interference it may cause to the nearby PRs in step 3. ST takes a decision to choose one of the two communication modes based on the estimated interference. If the interference is less than a certain threshold, in step 4a, ST chooses to harvest energy from PT signal power and transmits its information to SR using the harvested energy. On the other hand, if the estimated interference is greater than a threshold, ST chooses to communicate with the SR using backscatter communications in step 4b. The detailed procedure is explained in the following subsections.

4.1 CSI reporting phase

In conventional underlay CRSNs, PRs do not assist STs to estimate the channel conditions. This problem is more intractable for broadcast networks, where receivers do not send any acknowledgment of the reception of their data. However, authors in [29] have considered that primary users cooperate with secondary users to enhance mutual performance. In order to maintain high-performance gains for STs and minimize interference at PRs, STs should continuously monitor channel gains between PT and PRs [30]. If an ST can determine the information about the received signal strength at a PR, it will be able to estimate the interference it may cause to that PR. These channel gains can be estimated by using pilot-aided approaches or by employing sensors near all the PRs [30]. Therefore, for reliable estimation of interference at PRs, we assume that each PR is equipped with a backscatter interface, instead of a separate sensor, which is only used for reporting the channel conditions between PR and PT to its nearby STs. This channel condition information can then be used by the nearby STs to estimate interference which may be induced to the PR. Due to low power consumption and simple communications mechanism of backscatter communications, it is rational to assume that PRs use backscatter communications for reporting channel condition to nearby STs.

In broadcast networks such as TV transmissions, the transmissions periodically encode specific symbols for synchronization [31]. These special symbols are used by the receivers to synchronize the timing and compute different channel characteristics such as multipath. Once a PR is synchronized, it reports CSI between PR and PT to its nearby STs using its backscatter interface for βT time. In case more than one PRs are located near a ST, there will be collision if both the PRs report CSI simultaneous. To avoid this collision, βT time slot is further divided into X sub-slots. In each slot, independent of others, a

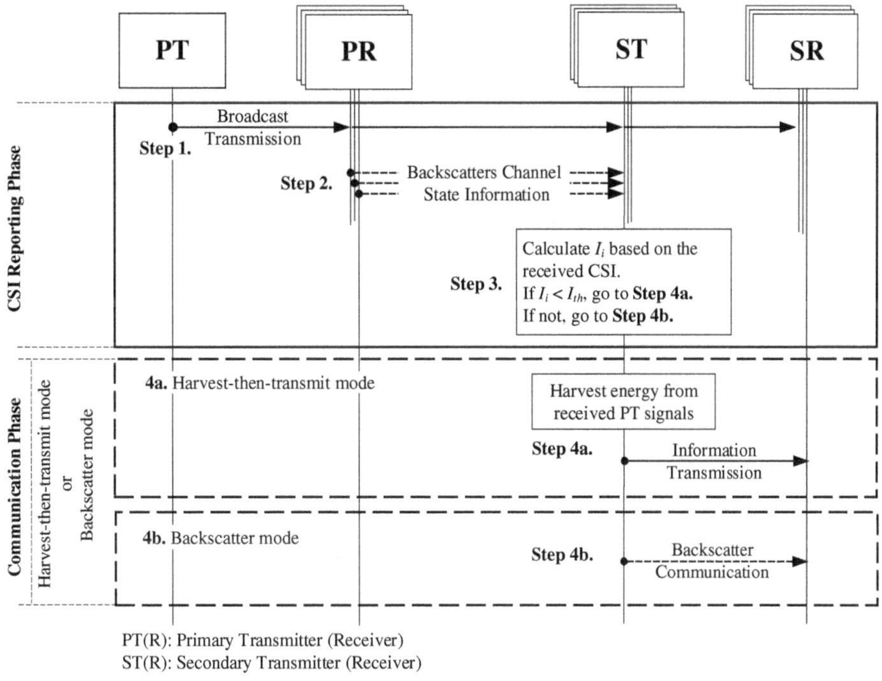

Fig. 3 Operational procedure for the proposed scheme

backscattering PR randomly selects a single sub-slot to transmit its signal. This divides each slot into a backscatter phase and a waiting phase of durations $1/X$ and $(1 - 1/X)$, respectively. If two or more PRs select the same sub-slot, then there will be collision, which will be reported back by the receiving ST. Those users will then back-off for a random time and then try to send their information again by randomly selecting another sub-slot in the next time frame. We adopt this simple anti-collision method for our system, however, one can further reduce the number of collisions by adopting more complex anti-collision mechanisms reported in [32].

4.2 Communication phase

Based on the interference estimated after the CSI phase, an ST will be allowed to operate on either harvest-then-transmit mode or backscatter mode to communicate with the associated SR. Since there are N number of STs present in the network, some may cause more interference to the nearby PRs and select backscatter mode. On the other hand, some STs may select harvest-then-transmit mode if the interference caused by this mode is less then a specific threshold. The average capacity of SR is computed by averaging the individual capacity of each ST. The main objective of the proposed scheme is to improve the overall data rates of the secondary system in the communication phase, which can be mathematically represented as:

$$C_{sr} = \begin{cases} C_b & , \text{Backscatter mode} \\ C_{rt} & , \text{Harvest-then-transmit mode} \end{cases} \tag{1}$$

where C_b and C_{rt} are the data rates of backscattering and harvest-then-transmit, respectively. The procedure for calculating the data rates of ST using two different modes is explained in the following.

4.2.1 *Backscatter mode*

The transmission rate of backscatter communication depends on the RC circuit of the backscatter module [5]. The rate may vary according to the different settings of the circuit elements (such as resistance and capacitance of the modulator). In addition, reported data rates of ambient backscatter can range from 1 Kbps [5] to 1 Mbps [24]. Therefore, data rates of the backscatter communications with a fixed transmission rate can be calculated as:

$$C_b = (1 - \beta)TB_b, \tag{2}$$

where B_b is the transmission rate of the backscatter mode considered for a unit time block. Note that instantaneous excitation signals received from PT provide enough power to the ST for carrying out backscatter operations [10]. Therefore, in Eq. (2), there is no need to consider the circuit energy consumption for the backscatter mode.

4.2.2 *Harvest-then-transmit mode*

The signals transmitted by PT with transmission power P_p are received at PR i and can be given by:

$$y_{i,p} = \frac{1}{\sqrt{d_{p,i}^m}}\sqrt{P_p}h_{p,i}x_p + n_i, \tag{3}$$

where $h_{p,i}$ is the *channel coefficient* between PT and PR, $d_{p,i}$ is the distance between the two users, and n_i is the additive white Gaussian noise (AWGN) at PR. Similarly, the signals received from PT at ST k can be written as:

$$y_{k,p} = \frac{1}{\sqrt{d_{p,k}^m}}\sqrt{P_p}h_{p,k}x_p + n_k, \tag{4}$$

where $h_{p,k}$ and $d_{p,k}$ are the channel coefficient and distance between PT and ST, respectively, and n_k is the AWGN at ST. For energy harvesting duration $(1 - \beta)\alpha T$, the ST k harvests energy from the received signal power (P_0), which can be given as:

$$E_{H,k} = (1 - \beta)\alpha TP_0 = (1 - \beta)\alpha T\eta\frac{P_p|h_{p,k}|^2}{d_{p,k}^m}, \tag{5}$$

where η is the energy harvesting efficiency.

For radio transmission, ST k utilizes harvested energy $E_{H,k}$ as a source of transmission power. Thus, for a successful transmission, $E_{H,k}$ should be greater than circuit energy consumption E_c, i.e., $E_{H,k} > E_c$. Substituting the value of $E_{H,k}$ from (5), the value of energy harvesting time α can be obtained as:

$$\alpha > \frac{E_c d_{p,k}^m}{\eta(1 - \beta)TP_p|h_{p,k}|^2}. \tag{6}$$

The minimum energy harvesting time $\left(\alpha' = \frac{E_c d_{p,k}^m}{\eta(1-\beta)TP_p|h_{p,k}|^2}\right)$ is required such that sufficient energy is acquired to activate the circuit of the ST for radio transmission. After accumulating enough energy for radio transmission, the ST transmits the signal x_s for $(1 - \beta)(1 - \alpha)T$ duration and the transmitted power P_k can be given as:

$$P_k = \frac{E_{H,k}}{(1 - \alpha)(1 - \beta)T} = \frac{\eta\alpha P_p|h_{p,k}|^2}{(1 - \alpha)d_{p,k}^m}, \tag{7}$$

where P_k must be non-negative if $\alpha \geq \alpha'$.

On the other hand, the PR i reports its CSI to nearby STs by reflecting the received signals from PT by using its backscatter interface. All the STs that will receive this CSI are considered as nearby STs. When CSI is received at STs, they will then estimate the interference that may be caused to the nearby PRs. If nearby STs choose to transmit their information through radio transmission, signal-to-noise-plus-interference-ratio (SINR) at ith PR can be estimated from (3) as:

$$\gamma_i = \frac{P_p|h_{p,i}|^2}{\underbrace{\sum_{k=1}^{N'} \eta\alpha P_p|h_{p,k}|^2|h_{k,i}|^2 + (1-\alpha)d_{p,k}^m d_{k,i}^m d_{p,i}^m \sigma_i^2}_{\text{Interference } I_i}}. \quad (8)$$

Nearby ST k, where $k \in \{1, 2, \cdots, N'\} \subseteq \mathcal{K}$, induce interference to PR i, as indicated in (8). ST k should limit this interference to a certain level such that the SINR (γ_i) at PR i does not drop below a certain threshold (γ_{th}). If γ_i at PR i is greater than γ_{th} (i.e. $\gamma_i > \gamma_{th}$) after experiencing interference from ST k, ST k will use harvest-then-transmit mode. Otherwise, if $\gamma_i < \gamma_{th}$, then ST k will use backscatter mode for its data transfer. The data rates of PR i can be calculated as:

$$C_{pr} = \mathbb{E}(\log_2(1+\gamma_i)),$$
$$= \mathbb{E}\left(\log_2\left(1 + \frac{P_p|h_{p,i}|^2}{\sum_{k=1}^{N'} \eta\alpha P_p|h_{p,k}|^2|h_{k,i}|^2 + (1-\alpha)d_{p,k}^m d_{k,i}^m d_{p,i}^m \sigma_i^2}\right)\right). \quad (9)$$

The main focus of the proposed scheme is to improve the performance of the secondary system without compromising a certain level of quality of service for the primary system. In the communication process of primary system, when PT transmits signals to PR, SR also receives those signals, which can be given as:

$$y_{j,p} = \frac{1}{\sqrt{d_{p,j}^m}}\sqrt{P_p}h_{p,j}x_p + n_j, \quad (10)$$

where $h_{p,j}$ and $d_{p,j}$ are the channel coefficient and distance between PT and SR j, respectively, and n_j is the AWGN at the SR j. These signals serve as interference for SR j. The signals received at SR j from ST k can be given as:

$$y_{j,k} = \frac{1}{\sqrt{d_{k,j}^m}}\sqrt{P_k}h_{k,j}x_s + n_j, \quad (11)$$

where $h_{k,j}$ represents the channel coefficient between ST k and SR j, $d_{k,j}$ is the distance between them, and n_j is the AWGN at SR. Substituting the value of P_k from (7) into (11), the received signal $y_{j,k}$ can be rewritten as:

$$y_{j,k} = \frac{h_{k,j}}{\sqrt{d_{k,j}^m}}\sqrt{\frac{\eta\alpha P_p|h_{p,k}|^2}{(1-\alpha)d_{p,k}^m}}x_s + n_j. \quad (12)$$

SR j receives useful information from ST k and interference from PT. The subsequent SINR at SR j can be written as:

$$\gamma_j = \frac{\eta\alpha P_p|h_{p,k}|^2|h_{k,j}|^2 d_{p,j}^m}{(1-\alpha)P_p|h_{p,j}|^2 d_{p,k}^m d_{k,j}^m + (1-\alpha)d_{p,j}^m d_{p,k}^m d_{k,j}^m \sigma_j^2}, \quad (13)$$

As mentioned before, ST k can successfully use radio transmission only if the harvested energy is greater than the circuit energy consumption. Only then the ergodic capacity of the SR can be non-negative and can be given as:

$$C_{rt} = \mathbb{E}((1-\beta)(1-\alpha)T\log_2(1+\gamma_j)),$$
$$= \mathbb{E}\big((1-\beta)(1-\alpha)T\log_2(1+M\alpha)\big) \quad (14)$$

where

$$M = \frac{\eta P_p|h_{p,k}|^2|h_{k,j}|^2 d_{p,j}^m}{(1-\alpha)P_p|h_{p,j}|^2 d_{p,k}^m d_{k,j}^m + (1-\alpha)d_{p,j}^m d_{p,k}^m d_{k,j}^m \sigma_j^2}. \quad (15)$$

5 Performance evaluation

5.1 Simulation setup

We consider a CRSN where PT is a TV broadcast tower operating on 539 MHz frequency and a bandwidth of 6 MHz [5]. We consider that the PT is a source with unlimited power supply and a transmission power P_p = 10 kW [10]. All STs within the coverage area of PT, which is set as a disc of radius 1000 m, harvest energy from the signal power of PT. We set energy harvesting efficiency, $\eta = 0.6$ and path loss exponent $m = 2.7$ (for urban wireless network environment). The system parameters are listed in Table 1. The mean values of randomly generated channel gains $|h_{x,y}|^2$, $x \in \{p, k\}$, $y \in \{k, i, j\}$, are set to 1.

For simplicity, the noise variance at PR, SR, and ST are assumed to be the same, i.e., $\sigma_i^2 = \sigma_j^2 = \sigma_k^2 = 0.01$. To ensure minimum effect on the communication of primary system, SINR threshold (γ) at PR is set as -5 dB. The circuit power consumption is set to -35 dBm [10]. Unless otherwise stated, PRs and SRs are deployed in the network using PPP with densities λ_{pr} and λ_{sr} set to 0.2/km^2 and 0.1/km^2, respectively. STs are also distributed using PPP with a density λ_{st} = 0.002/m^2 within a radius of 50 m centered at each SR j. Therefore, in terms of distance, nearby PRs are those which are located within 50m radius of STs. An example of our considered network deployment is shown in Fig. 4.

Each ST present in the simulation environment operates using the proposed scheme presented in Section 4. The capacity for both the modes is obtained from the

Table 1 Simulation parameters

Parameters	Studied value
Energy harvesting ratio (α)	0 - 1
CSI reporting ratio (β)	0.2
Conversion efficiency (η)	0.6
Coverage area (radius) of PT	1000 meters
Max distance between ST and SR	50 meters
Transmission power PT (P_p)	10 kW
Circuit power consumption	− 35 dBm
Backscatter transmission rate (B_b)	33 Kbps

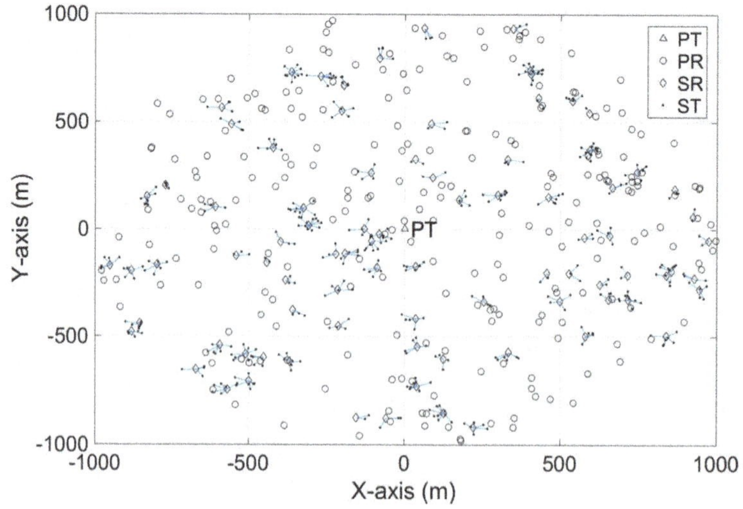

Fig. 4 The spatial distribution of an underlay CRSN modeled using PPP

equations derived in Subsection 4.2. We obtain the average capacity of the secondary system by averaging the individual capacity of each ST, operating on either backscatter mode or harvest-then-transmit mode. Other results in our simulations are also obtained from the mathematical equations presented in Section 4. All simulation results are achieved by averaging 10,000 Monte Carlo runs.

5.2 Simulation results

The performance of the proposed scheme is evaluated through results shown in this subsection. Different performance metrics are shown in this subsection to provide an insight to the wireless-powered underlay CRSN with backscatter communications. In the proposed scheme, when the harvest-then-transmit mode is selected, we observe that there is a tradeoff between time ratio for energy harvesting and average data rates of ST. Figure 5 shows the variation of the average data rates of ST for different values of α. The figure plots average data rates of ST for different transmit powers of PT. As shown in the figure, the average data rate of the secondary system is low for lower values of α and increases as the value of α increases. When $\alpha > \alpha'$, the average data rate C_{rt} is concave, and it exhibits the highest value at $\alpha = 0.56$. We use this value of α to evaluate the affect of other system

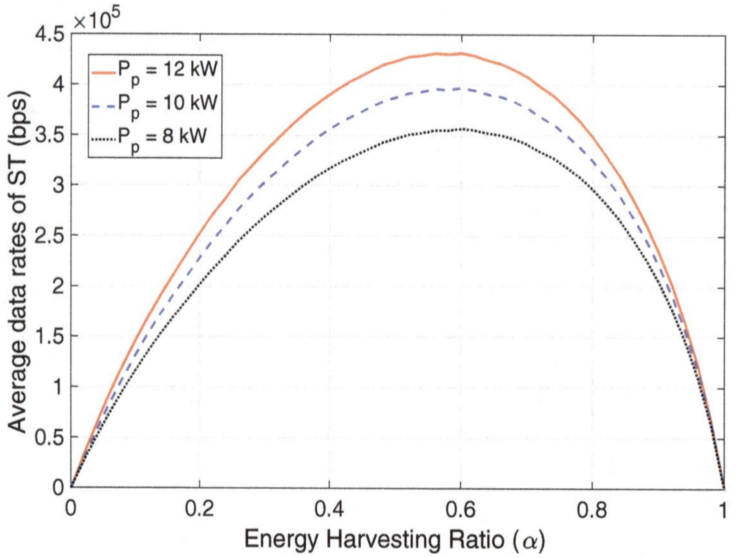

Fig. 5 Performance of the secondary system under the variation of α, for the harvest-then-transmit mode. Other parameters: $\eta = 0.6$ and $\beta = 0.2$

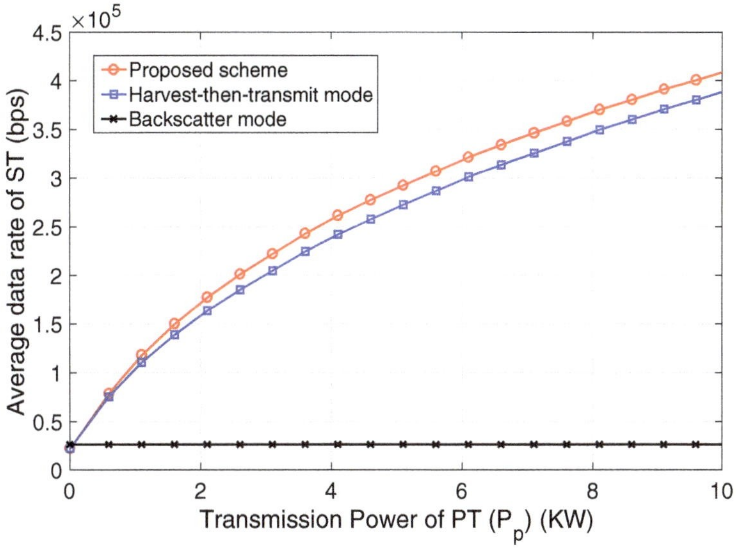

Fig. 6 Performance of the secondary system under the variation of the transmission power of PT (P_p). Other parameters: $\alpha = 0.56$, $\eta = 0.6$ and $\beta = 0.2$

parameters on the system performance. α represents the time spent for harvesting energy from the received signals of PT; if more energy is harvested, more power will be available for the transmission of data through radio transmission; as a result, the data rates will be improved. On the other hand, a higher value of α means there will be less time for radio transmission and ST will not be able to send more data which results in lower data rates.

To provide insight of the effect of the transmission power of PT (P_p) on the proposed scheme, Fig. 6 shows the average data rate of ST for different values of P_p. For the value of $\alpha = 0.56$, the figure shows that as the values of P_p increases, the average data rate of ST increases. This is because of the fact that more energy can be harvested for higher values of P_p. The harvested energy is then used as the transmission power of ST, which results in higher data rates. It can be seen from the figure that the proposed scheme has the best performance in terms of the average data rate of ST as compared to backscatter mode and harvest-then-transmit mode.

The effect of PR density (λ_{pr}) on the SINR of STs is shown in Fig. 7 for the proposed scheme. The figure plots the CDF of average SINR at STs for different values of λ_{pr}. Increase in the density of the PR degrades the

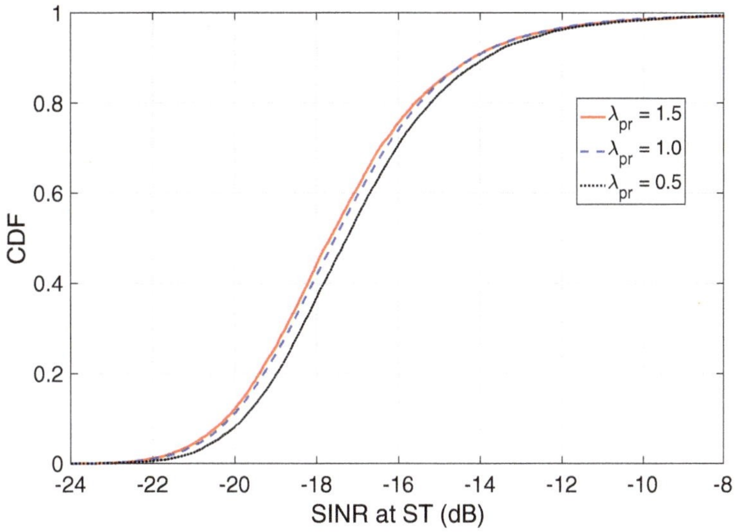

Fig. 7 Performance of the secondary system under the variation of PR density. Other parameters: $\alpha = 0.56$, $\eta = 0.6$ and $\beta = 0.2$

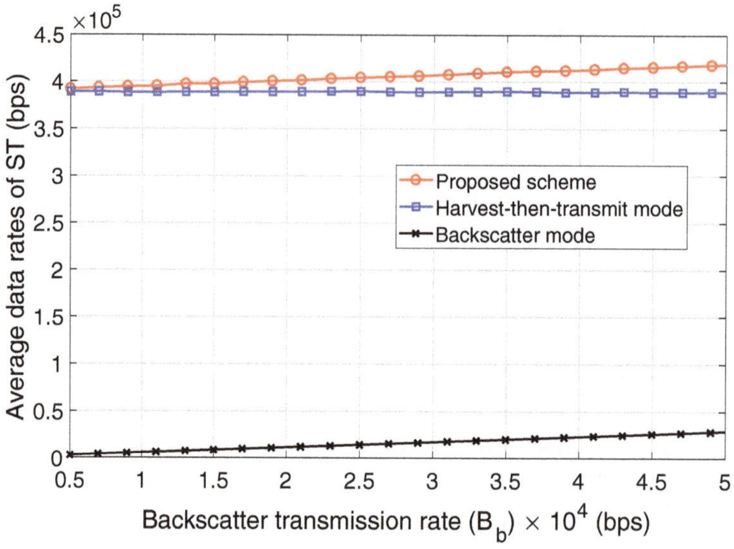

Fig. 8 Performance of the secondary system under the variation of backscatter transmission rate (B_b). Other parameters: $\alpha = 0.56$, $\eta = 0.6$ and $\beta = 0.2$

performance of STs. This is aligned with the basic concept of the underlay CRSN model because less number of STs will be able to use harvest-then-transmit mode as the density of PRs increases. Note that backscatter mode will be still used for the transmission of information from ST to SR. However, the transmission rate of backscatter mode is much less than that of the harvest-then-transmit mode, which results in the degradation of average SINR at STs.

Furthermore, we present the results to provide insight on the performance of the proposed scheme by varying different system parameters. To evaluate the performance of the secondary system, Fig. 8 shows the effect on the average data rate of ST with the variation of backscatter transmission rate. For this performance metric, we use the value of $\alpha = 0.56$ because it achieves the highest data rates for the harvest-then-transmit mode. It can be seen from the figure that the average data rate of ST achieves the best performance for the proposed scheme as compared to backscatter mode and harvest-then-transmit mode separately. This is because when the ST is unable to operate at harvest-then-transmit mode due to interference at the PR, it can use backscatter mode to transfer its information to the SR.

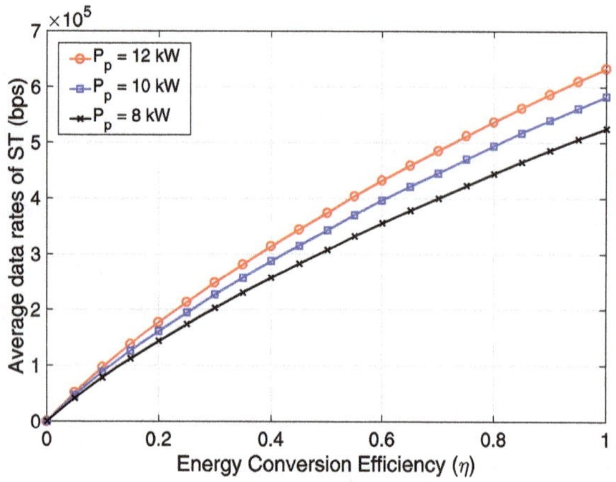

Fig. 9 Performance of the secondary system under the variation of energy harvesting efficiency (η). Other parameters: $\alpha = 0.56$, and $\beta = 0.2$

We then vary the energy harvesting efficiency η shown in Fig. 9, for different values of the transmission power of PT. Average data rates of ST increases as the value of η increases. Energy harvesting efficiency is the capability of energy harvesting circuit to convert the RF signal power into DC current. This harvested energy is then used by the ST to transmit information towards SR using active radio component. As the value of η increases average data rate of ST also increases because ST can more efficiently harvest energy and use that energy as its transmission power. Furthermore, for higher values of P_p the data rates increase which can be confirmed by previous figures.

For the secondary system, we finally evaluate the performance of the proposed scheme by varying the CSI reporting ratio (β). Figure 10 shows the effect on the average data rate of ST with the variation of CSI reporting ratio (β). It is intuitive that the average data rates of the secondary system will decrease with an increase in the value of β. The reason for this decrease is the direct relation of

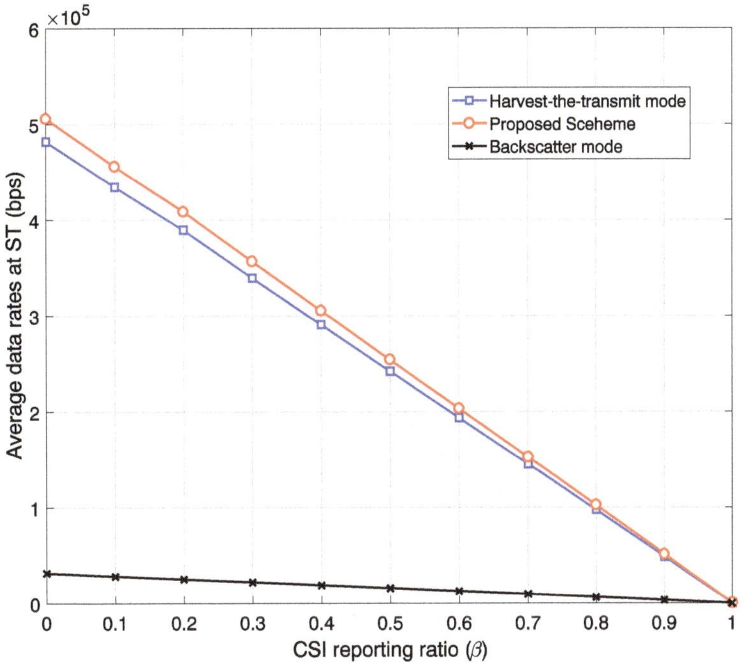

Fig. 10 Performance of the secondary system under the variation of CSI reporting ratio (β). Other parameters: $\alpha = 0.56$, and $\eta = 0.6$

β with both data rates of backscatter mode and harvest-then-transmit mode. Increasing values of β imply that more time will be used for CSI reporting and less time will be available for communication phase.

To see the effect of energy harvesting ratio (α), Fig. 11 plots the average data rate of PR in the presence of STs. Since for smaller values of α, the amount of harvested energy is less, consequently, the transmission power of STs is also less and the interference ST may cause will be lower. On the other hand, although less time is available for transmission, STs may cause a significant interference to the nearby PRs. This is because of the fact that large number of STs are located near the PRs and the transmission power of STs is higher for higher values of α. As a result, we choose an optimal value of α, which maximizes average data rate of secondary

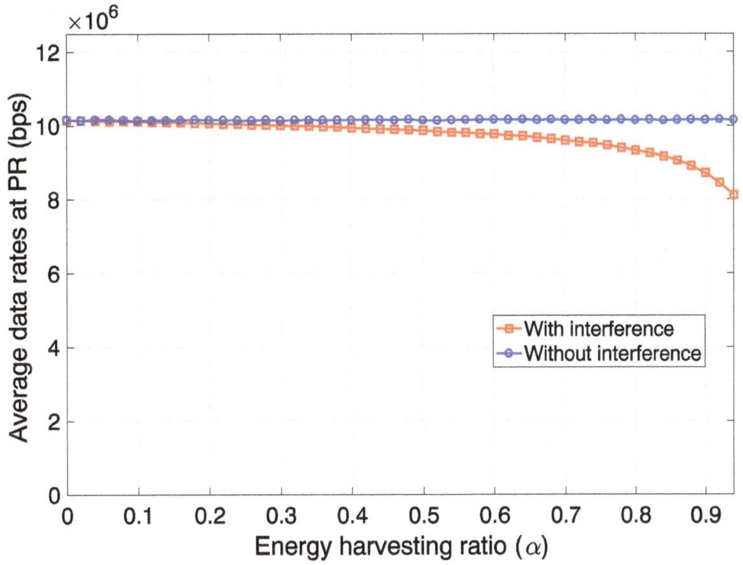

Fig. 11 Average data rate of PRs under the variation of energy harvesting ratio (α). Other parameters: $\gamma = -5dBm$, $\eta = 0.6$, and $\beta = 0.2$

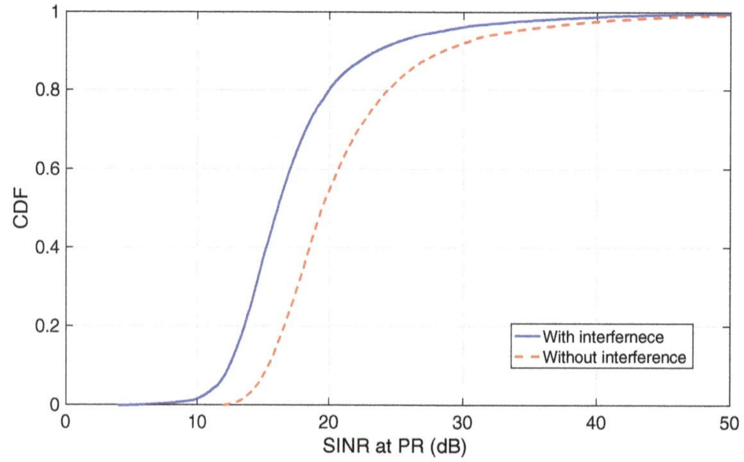

Fig. 12 Performance of primary receivers with and without secondary interference. Other parameters: $\gamma = -5dBm, \alpha = 0.56, \eta = 0.6$ and $\beta = 0.2$

system and minimizes the effect on the communication of primary system.

Communication of the primary system is of prime concern in CRSN and should be guaranteed a certain level of quality of service. Figure 12 plots the CDF of the average SINR at PR. The figure also shows the effect of the proposed scheme on the SINR of PR. It can be observed from the figure that the performance of PR experience a slight degradation due to the presence of STs operating in underlay model. This minor degradation to the performance of the primary system is a limited cost to pay for providing communication opportunity to a large number of users in a CRSN. However, SINR of all the PRs is above a certain threshold level which shows that the proposed scheme minimally affect primary communications.

6 Conclusion

In this paper, we proposed a new concept of integrating ambient backscatter communications with the cognitive radio network, where STs are wirelessly powered by radio signals transmitted by the PT. Considering an underlay model for the secondary system, we proposed a scheme where STs choose between backscatter mode or harvest-then-transmit mode, based on the estimated interference that will be induced to PRs. The secondary system minimally affects the primary users' performance while coexisting with the primary network. Under the constraints of interference on PRs, the proposed scheme improves the average data rate of the secondary system in underlay CRSN. Furthermore, we evaluate the performance of the proposed scheme through system level simulations. Finally, the results are presented for the proposed scheme to show the improvement in the performance of the secondary system. We also highlight the slight degradation of the primary system which is a tradeoff for providing communication services to a large number of low power secondary users.

Acknowledgements

This work was supported by the National Research Foundation of Korea (NRF) grant funded by the Korean government (MSIP) (2014R1A5A1011478).

Authors' contributions

DM and MYC conceived the idea and designed the algorithm. STS helped in the evaluation of the proposed scheme. DM wrote the manuscript. KWC and T-JL helped in improving the algorithm and reviewing the manuscript. All authors read and approved the final manuscript.

Competing interests

The authors declare that they have no competing interests.

Author details

[1]Department of Electrical and Computer Engineering, Sungkyunkwan University, 2066 Seobu-Ro, Jangan-Gu, 16419 Suwon, Gyeonggi-Do, South Korea. [2]Department of Telecommunication Engineering, FICT, Balochistan University of Information Technology, Engineering and Management Sciences, Airport Road, Baleli, 87300 Quetta, Pakistan.

References

1. A. Ahmad, S. Ahmad, M.H. Rehmani, N.U. Hassan, A survey on radio resource allocation in cognitive radio sensor networks. IEEE Commun. Surv. Tutor. **17**(2), 888–917 (2015)
2. G.I. Tsiropoulos, O.A. Dobre, M.H. Ahmed, K.E. Baddour, Radio resource allocation techniques for efficient spectrum access in cognitive radio networks. IEEE Commun. Surv. Tutor. **18**(1), 824–847 (2016)
3. O.B. Akan, O.B. Karli, O. Ergul, Cognitive radio sensor networks. IEEE Netw. (23)4, 34–40 (2009)
4. M.E. Tanab, W. Hamouda, Resource allocation for underlay cognitive radio networks: a survey. IEEE Commun. Surv. Tutor. **19**(2), 1249–1276 (2017)
5. V. Liu, A.N. Parks, V. Talla, S. Gollakota, D. Wetherall, J.R. Smith, *Ambient Backscatter: Wireless Communication out of Thin Air*. (Association for Computing Machinery (ACM), Hong Kong, 2013)
6. K. Han, K. Huang, Wirelessly powered backscatter communication networks: modeling, coverage and capacity. IEEE Trans. Wirel. Commun. **16**(4), 2548–2561 (2017)
7. B. Lyu, Z. Yang, G. Gui, Y. Feng, Optimal resource allocation policies for multi-user backscatter communication systems. Sensors. **16**(12), 2016 (2016)
8. D.T. Hoang, D. Niyato, P. Wang, D.I. Kim, Opportunistic channel access and RF energy harvesting in cognitive radio networks. IEEE J. Sel. Areas Commun. **32**(11), 2039–2052 (2014)

9. S. Bi, Y. Zeng, R. Zhang, Wireless powered communication networks: an overview. IEEE Wirel. Commun. **23**(2), 10–18 (2016)
10. D.T. Hoang, D. Niyato, P. Wang, D.I. Kim, Z. Han, *The Tradeoff Analysis in RF-Powered Backscatter Cognitive Radio Networks. Global Communications Conference (GLOBECOM)*. (Institute of Electrical and Electronics Engineers (IEEE), Washington, 2016)
11. K.H. Park, D. Munir, J.S. Kim, M.Y. Chung, *Integrating RF-Powered Backscatter with Underlay Cognitive Radio Networks*. (Institute of Electrical and Electronics Engineers (IEEE), Da Nang, 2017)
12. K. Hareesh, P. Singh, *An Energy Efficient Hybrid Co-operative Spectrum Sensing Technique for CRSN*. (Institute of Electrical and Electronics Engineers (IEEE), Kottayam, 2013)
13. Y. Lin, C. Wang, J. Wang, Z. Dou, A novel dynamic spectrum access framework based on reinforcement learning for cognitive radio sensor networks. Sensors. **16**(10), 1675 (2016)
14. Y. Zhang, S. He, J. Chen, Y. Sun, X. Shen, Distributed sampling rate control for rechargeable sensor users with limited battery capacity. IEEE Trans. Wirel. Commun. **12**(6), 3096–3106 (2013)
15. S. Gao, L. Qian, D.R. Vaman, Distributed energy efficient spectrum access in cognitive radio wireless ad hoc networks. IEEE Trans. Wirel. Commun. **8**(10), 5202–5213 (2009)
16. J.A. Han, W.S. Jeon, D.G. Jeong, Energy-efficient channel management scheme for cognitive radio sensor networks. IEEE Trans. Veh. Techol. **60**(4), 1905–1910 (2011)
17. L.R. Varshney, *Transporting Information and Energy Simultaneously*. (Institute of Electrical and Electronics Engineers (IEEE), Toronto, 2008)
18. S. He, J. Chen, F. Jiang, D.KY. Yau, G. Xing, Y. Sun, Energy provisioning in wireless rechargeable sensor networks. IEEE Trans. Mob. Comput. **12**(10), 1931–1942 (2013)
19. Y. Zhang, S. He, J. Chen, Y. Sun, X.S. Shen, Distributed sampling rate control for rechargeable sensor users with limited battery capacity. IEEE Trans. Commun. **12**(6), 3096–3106 (2013)
20. I. Yoon, M.Y. Jun, J. Semi, K.N. Dong, Adaptive sensing and compression rate selection scheme for energy harvesting wireless sensor networks. Int. J. Distrib. Sens. Netw. **13**(6), 1–10 (2017)
21. Y. Wang, W. Lin, R. Sun, Y. Huo, Optimization of relay selection and ergodic capacity in cognitive radio sensor networks with wireless energy harvesting. Pervasive Mob. Comput. **22**, 33–45 (2015)
22. S. Kalamkar, J. Jeyaraj, A. Banerjee, K. Rajawat, Resource allocation and fairness in wireless powered cooperative cognitive radio networks. IEEE Trans. Commun. **64**(8), 3246–3261 (2016)
23. A.N. Parks, A. Liu, S. Gollakota, J.R. Smith, *Turbocharging Ambient Backscatter Communication*. (ACM SIGCOMM, Chicago, 2014)
24. B. Kellogg, V. Talla, S. Gollakota, J.R. Smith, *Passive Wi-Fi: Bringing Low Power to Wi-Fi Transmissions*. (The Advanced Computing Systems Association (USENIX), Santa Clara, 2016)
25. C. Boyer, S. Roy, Backscatter communication and RFID: coding, energy, and MIMO analysis. IEEE Trans. Commun. **62**(3), 770–785 (2014)
26. B. Lyu, Z. Yang, G. Gui, H. Sari, Optimal time allocation in backscatter assisted wireless powered communication networks. Sensors. **17**(6), 1258 (2017)
27. D.T. Hoang, D. Niyato, P. Wang, D.I. Kim, *Optimal Time Sharing in RF-Powered Backscatter Cognitive Radio Networks*. (Institute of Electrical and Electronics Engineers (IEEE), Paris, 2017)
28. R. Zhang, C.K. Ho, MIMO broadcasting for simultaneous wireless information and power transfer. IEEE Trans. Wirel. Commun. **12**(5), 1989–2001 (2013)
29. W. Su, J.D. Matyjas, N.B. Stella, Active cooperation between primary users and cognitive radio users in heterogeneous ad-hoc networks. IEEE Trans. Signal Process. **60**(4), 1796–1805 (2012)
30. L.B. Le, E. Hossain, Resource allocation for spectrum underlay in cognitive radio networks. IEEE Trans. Wirel. Commun. **7**(12), 5306–5315 (2008)
31. ATSC digital television standard (1995). Advance Television System Committee (ATSC) Digital Television System Standards, ATSC Standard A/53, Online available at: https://www.atsc.org/standard/a53-atsc-digital-television-standard/
32. D.K. Klair, K.-W. Chin, R. Raad, Asurvey survey and tutorial of RFID anti-collision protocols. IEEE Commun. Surv. Tutor. **12**(3), 400–421 (2010)

7

User rate and power optimization for HetNets under Poisson cluster process

Xinqi Jiang and Fu-Chun Zheng*

Abstract

Heterogeneous cellular networks (HetNets) consist of different tiers of base stations (BSs) to meet the ever-increasing mobile traffic demand. Due to the random deployment of BSs, Poisson point process (PPP) is often used to model the BS distribution. However, low power small cells are usually clustered around the popular areas, and PPP can not reflect such a feature. To this end, we in this paper consider base station (BS) cooperation and analyze user rate and energy efficiency of HetNets based on a Poisson cluster process (PCP). A calculable formula for the average data rate (or spectral efficiency) and its approximated closed form are derived. Based on this closed form, a power minimization solution with certain spectral efficiency constraint is proposed, and the optimal cooperation radii are derived. Furthermore, we analyze spectral efficiency under a limited number of cooperative BSs in a two-tier network. Finally, we propose a range expansion (RE) scheme and examine the impact of this scheme. The theoretical analyses are verified by simulation results.

Keywords: Heterogeneous networks, Poisson cluster process, Energy efficiency, User rate, Stochastic geometry, Range expansion

1 Introduction

Comprising macro base stations (BSs) overlaid with dense low power BSs (hence small cells) [1], heterogeneous networks (HetNets) are being deployed to meet the rapid growth in data demand from wireless users, especially those users at the edge of macro cells. Due to the nature of dense deployment for HetNets and the power consumption of all the BSs, energy efficiency (EE) has now been considered as another key performance indicator (KPI) for future wireless networks (e.g., 5G) on top of the traditional KPI of data rate or spectrum efficiency. As a result, not only the data rate but also the EE has recently received enormous attention from the communications community [2–4].

BS locations in dense HetNets tend to be irregular and randomly placed compared with traditional cellular networks, which makes the analysis of the HetNets' performances much more intricate. Fortunately, for the user rate and energy efficiency analysis, some tools from stochastic geometry have proved to be powerful, such as the probability generating functional (PGFL) of Poisson cluster process (PCP), the Palm characterization of PCP, and the Campbell-Mecke theorem. Indeed, these tools have made possible the theoretical analysis of many network metrics, such as average rate, energy consumption, as well as transmission delay.

One major challenge for HetNets is interference management. Coordinated multi-point (CoMP) transmission [5] is an effective scheme to reduce interference by transforming interfering signals to useful signals or to increase space diversity, hence enhancing the performances of HetNets. As such, CoMP is expected to play a key role in future cellular networks as an effective means of meeting higher data rates as well as expanding the indoor and cell-edge coverage. In [6, 7], the authors apply CoMP transmission in HetNets to enhance performances of network, and simulation results illustrate that CoMP can significantly improve the average SE and EE compared to non-CoMP. Thus, a single-stream CoMP transmission scheme (also termed macro-diversity CoMP) is exploited in this paper.

*Correspondence: fzheng@ieee.org
School of Electronic and Information Engineering, Harbin Institute of Technology, Shenzhen, China

1.1 Related work and motivations

As mentioned above, there has been much effort on applying stochastic geometry to modeling and analyzing HetNets. In most works so far, cellular networks have been modeled as a tractable distribution such as Poisson point process (PPP) [8–10]. In [10], the authors derived the spectral efficiency based on the PPP model and determined the optimal received signal strength (RSS) thresholds under certain approximation. However, most small cells are usually clustered around densely populated areas, and PPP is not always suitable. Investigating a more accurate model therefore becomes imperative. To this end, the Laplace transform of the total interference in HetNets using a PCP was given in [11]. In [12], the authors compared PPP, aggregative point process, and repulsive point process to choose the optimal model for BS deployment in urban areas. In [13], it was stated that the PPP models in some case is more accurate than the hexagonal grid model, but the PCP is able to even more accurately model BS deployment. Assessing KPIs of HetNets based on the PCP model therefore becomes necessary. However, an integrated characterization for aggregate interference as well as the PGFL, based on the PCP network model, has not been reported, although such a characterization is crucial to the theoretical analysis of network performances, such as SE and EE.

Indeed, as both EE and SE are now viewed as crucial performance metrics in 5G networks, analysis of EE and SE has gained much attention recently. Reference [14] proposed a new cooperative multiple-input multiple-output (MIMO) scheme to improve the spectrum efficiency while maintaining the same spatial multiplexing and diversity gains as traditional MIMO schemes. Reference [15] studied the EE and area SE with respect to the number and size of microcells. References [16, 17] studied the performances of cross-tier cooperation scheme based on the PPP network model, and the numerical results illustrate that such a cooperation strategy can significantly boost ergodic capacity and reduce outage probability. Cross-tier cooperation scheme is expected to be a major technique to provide higher data rates. However, none of the above works has applied this cooperation schemes to a PCP network. As a result, Reference [18] derived new analytical expressions for the coverage probability and area spectral efficiency on clustered device-to-device millimeter wave networks. In [19], the authors investigated the outage probability, the coverage probability, and the average achievable rate based on a tractable lower bound of the PGFL when the nodes follow a PCP. Reference [20] found that the coverage probability in [19] is not always accurate for analyzing PCP network performances. However, the EE and SE of PCP modeled HetNets, based on cross-tier cooperation, have not been analyzed so far.

Over the above situation, this paper targets at the single-stream CoMP over a PCP HetNet, where cross-tier cooperation is based on the RSS threshold. From the perspectives of EE and SE, we derived a tractable result for SE based on PCP and then proposed an energy optimization solution. Moreover, we also analyzed spectral efficiency in case of the limited number of cooperative BSs in a two-tier HetNet and proposed a range expansion (RE) scheme based on CoMP, which balances the load of macro BS.

1.2 Contributions

The main contributions of this paper are therefore as follows: (1) a computationally efficient formula for spectral efficiency is derived under the PCP model, and then approximated as a closed form; (2) a comprehensive comparison with the corresponding PPP results is carried out; (3) a power optimization scheme is proposed by optimizing the cooperative radius of each tier; and (4) a scheme based on a limited number of cooperative BSs and a range expansion (RE) strategy are proposed, which can ease the load of macro BS and enhance the network EE.

1.3 Paper organization

The remainder of this paper is structured as follows. In Section 3, we give a brief description on a particular PCP model and present the downlink system model. The formula of spectral efficiency is derived based on PCP and an optimal solution for each tier's cooperative radius is determined in Section 4. In Section 5, we propose an effective BS cooperation and RE strategy based on CoMP. Section 6 presents the numerical results and related illustrations. A list of major symbols and their meanings are shown in Table 1.

2 Methods/experimental

The methods and analysis in this paper are based on stochastic geometry and are described in Section 3, while the verification is based on Monte Carlo simulations as presented in Section 6. This paper contains no experiments.

3 PCP and downlink system model

3.1 Poisson cluster process (PCP)

A Poisson cluster process (PCP) consists of two kinds of point processes: the parent points in each tier following homogeneous Poisson point process (PPP) with the density $\lambda_{p,k}$ form the centers of the clusters, while the daughter nodes are scattered around cluster centers. Matern cluster process (MCP) is a special case of PCP, in which the daughter nodes are of uniform distribution within a circle of radius R_a with the corresponding parent point being the cluster center. The parent points indicate the centers of hotspots, and the daughter nodes represent the locations of BSs. In this paper, we mean MCP by PCP. This special PCP model and the PPP model are illustrated in Fig. 1. In each cluster of the kth tier, the number of daughter nodes is fixed to be c_k, the mean density of the kth tier is therefore $c_k\lambda_{p,k}$. The probability density function of such PCP daughter nodes is given by

$$f(d) = \begin{cases} \frac{1}{\pi R_a{}^2} & ||d|| \le R_a \\ 0 & \text{otherwise,} \end{cases} \tag{1}$$

where d is a daughter node's position relative to the cluster center, and $||d||$ is the distance between the daughter node and the center of the relevant cluster.

3.2 System model

We consider a general model of downlink HetNets which includes K network tiers, and the BS positions of each tier form a PCP denoted by Φ on $\mathbb{R}^2$. BSs across tiers are distinguished by their deployment mean density $c_k\lambda_{p,k}$, transmit power p_k, and path loss exponents α_k. User density is denoted by λ_u. Both BSs and users are equipped with a single antenna. Without any loss of generality, the general path loss is given by $PL(x) = \|x\|^{-\alpha_k}$, where $\|x\|$ indicates the distance between a typical user at the origin and BS at x. Inspired by [10], we also consider a user-centric cross-tier BS collaboration scenario. Figure 2 illustrates a two-tier HetNet involving a mix of macro and pico BSs, where the typical user can be served by cooperative BSs from both tiers.

Table 1 Major symbols used in this paper

Variable	Meaning	Unit
λ_u	User density	m^{-2}
$\lambda_{p,k}$	Parent points density of the kth tier	m^{-2}
c_k	Number of daughter points within each cluster of the kth tier	None
$R_{a,k}$	Cluster radii of the kth tier	m
R_k	Cooperative radii of the kth tier	m
$R_{o,k}$	Optimal cooperative radii of the kth tier	m
P_u	Average power consumption on serving one user	W
p_k	Transmit power of the kth tier	W
$P_{bh,k}$	Backhaul power dissipation of the kth tier on serving one user	W
$P_{sp,k}$	Signal processing power consumption of the kth tier on serving one user	W
θ_k	RSS threshold of the kth tier	W
a_k	Slope of the kth tier power consumption	None
U_k	Number of cooperative BSs in the kth tier	None
α_k	Path loss exponents of the kth tier	None
h_k	Power fading coefficient of the kth tier	None

We employ the Palm measure [21] to characterize the total signal and the aggregate interference at the typical user located at the origin and served by cross-tier cooperative BSs. For convenience, the set of cooperative BSs is denoted by C_C. BSs in the C_C transmit the same data to the typical user (hence a non-coherent macro-diversity CoMP strategy) and the RSS at the user exceeds some RSS threshold. Namely, for the kth tier, the BS located at x_k belongs to the cooperative group C_C if $p_k h_k \|x_k\|^{-\alpha_k} \ge \theta_k$, where θ_k is the RSS threshold for the kth tier, and h_k denotes the power fading coefficient of the kth tier.

Assuming that the cooperative region in each tier is circular, the cooperative radius R_k can then be given by

$$R_k = \left(\frac{p_k h_k}{\theta_k}\right)^{1/\alpha_k}. \tag{2}$$

Denoting by $C_I = \Phi \backslash C_C$ the set of the interfering BSs, the signal-to-interference-plus-noise ratio (SINR) at the typical user is given by

$$\mathrm{SINR} = \frac{\sum\limits_{x_k \in C_C} p_k h_k \|x_k\|^{-\alpha_k}}{\sum\limits_{x_k \in C_I} p_k h_k \|x_k\|^{-\alpha_k} + \sigma^2}, \tag{3}$$

where σ^2 is the noise power.

4 Spectral efficiency and power optimization

In this section, we derive and optimize two key metrics, i.e., the spectral efficiency (SE) and energy efficiency (EE) based on the probability generating functional (PGFL) of the Poisson cluster process in a K-tier HetNet. Both metrics depend on the SINR in [3]. As in [10], the Laplace transform is used to characterize the aggregate interference (the denominator) and the total signal power (the numerator plus the denominator). For comparison, the same denotations have also been employed as in [10].

4.1 Laplace transform of the total signal power

For the proposed PCP model, the number of daughter nodes in each cluster is assumed to be fixed, and the parent points are not included. All cooperative BSs are assumed to transmit the same signal to the "typical user" at the origin. The total signal power (the signal plus the interference) received at this user is given by

$$P = \sum_{x \in \Phi} p_k h_k \|x\|^{-\alpha_k} = \sum_{k=1}^{K} \sum_{x_k \in \phi_k} p_k h_k \|x_k\|^{-\alpha_k}, \tag{4}$$

where ϕ_k denotes the set of BS at the kth tier.

From the above formula, the Laplace transform of total signal power P is

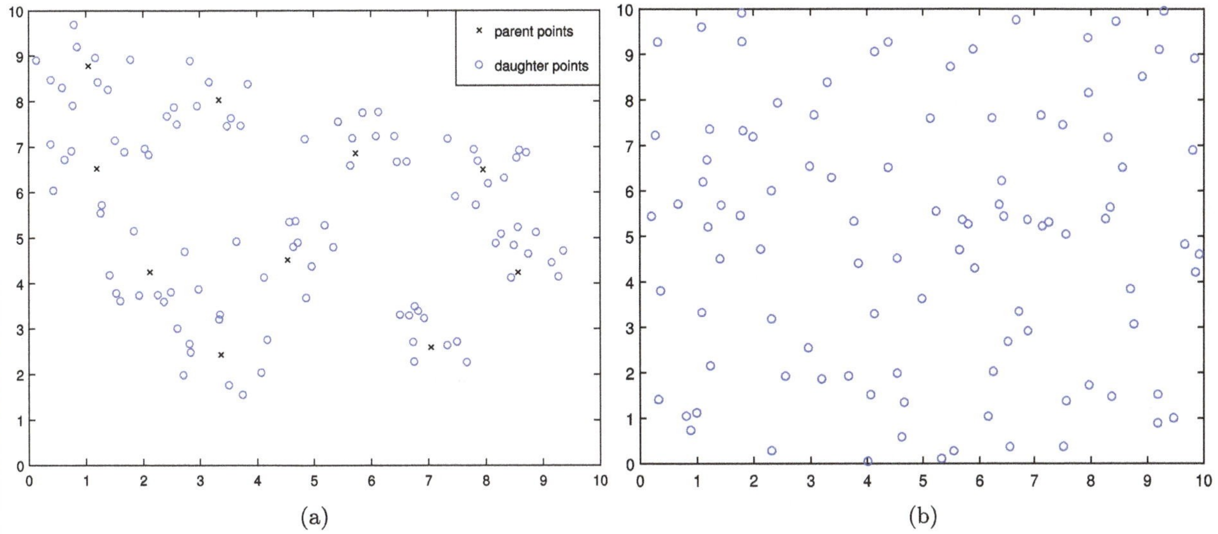

Fig. 1 The PCP model and PPP model with the same average intensity $\bar{\lambda} = 1$ for comparison. **a** Poisson cluster process ($\lambda_p = 0.1, \bar{c} = 10$). **b** Poisson point process ($\bar{\lambda} = 1$)

$$\begin{aligned}\mathcal{L}_P(s) &= E_P[\exp(-sP)]\\ &= \prod_{k=1}^{K} E_{h_k,x_k}\left[\exp\left(-s\sum_{x_k\in\phi_k} p_k h_k \|x_k\|^{-\alpha_k}\right)\right]\\ &= \prod_{k=1}^{K} E\left[\prod_{x_k\in\phi_k} \mathcal{L}_h(sp_k\|x_k\|^{-\alpha_k})\right]\\ &\overset{(a)}{=} \prod_{k=1}^{K} E\left[\prod_{x_k\in\phi_k} \frac{1}{1+sp_k\|x_k\|^{-\alpha_k}}\right],\end{aligned} \tag{5}$$

where $\mathcal{L}_h(sp_k\|x_k\|^{-\alpha_k})$ denotes the Laplace transform of received signal power at x_k, and (a) is due to the fact that h_k is exponentially distributed with mean one.

On the other hand, the above expression is directly related to the PGFL of PCP as given in [22]

$$\tilde{G}(v) = E\left[\prod_{x\in\phi} v(x)\right] = \exp\left\{-\lambda_p \int_{\mathbb{R}^2}\left[1-\beta(y)^c\right]dy\right\}, \tag{6}$$

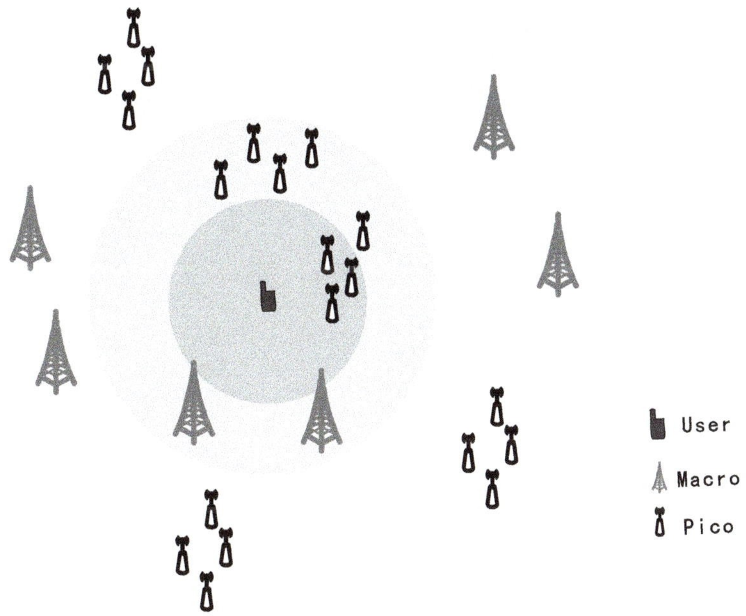

Fig. 2 A typical user is served by macro and pico BSs, where a small circle and a large circle denote pico BSs and macro BSs cooperative areas, respectively

where

$$\beta(y) = \int_{\mathbb{R}^2} \nu(x-y) f(x)\, dx,$$

and c denotes the number of daughter nodes in each cluster.

The lower bounded PGFL and the upper bounded conditional PGFL of the PCP are given by [19]. In order to evaluate accurately the spectral efficiency, we derive below the precise expression by algebraic operations.

Lemma 1 *The Laplace transform (i.e., (5)) of total signal power P is estimated as*

$$\mathcal{L}_P(s) = \prod_{k=1}^{K} \exp\left\{-\lambda_{p,k} 2\pi \int_0^{\infty} [1 - B_k(r)^{c_k}] r dr\right\}, \quad (7)$$

where

$$B_k(r) = \frac{1}{1 + s p_k \|r\|^{-\alpha_k}}.$$

Proof Let $\nu(x) = \frac{1}{1+sp_k\|x\|^{-\alpha_k}}$. By applying (6) to (5), we have □

$$\begin{aligned}
&\int_{\mathbb{R}^2} [1 - \beta_k(y)^{c_k}]\, dy \\
&= \int_{\mathbb{R}^2} \left(1 - \left(\int_{\mathbb{R}^2} \frac{f(x)}{1 + sp_k\|x-y\|^{-\alpha_k}} dx\right)^{c_k}\right) dy \\
&\overset{(a)}{=} \int_{\mathbb{R}^2} \left(1 - \left(\int_{\mathbb{R}^2} \frac{f(x)}{1 + sp_k\|r\|^{-\alpha_k}} dx\right)^{c_k}\right) dr \\
&= \int_{\mathbb{R}^2} \left(1 - \left(\frac{1}{1 + sp_k\|r\|^{-\alpha_k}}\right)^{c_k}\right) dr \\
&\overset{(b)}{=} 2\pi \int_0^{\infty} [1 - B_k(r)^{c_k}] r dr,
\end{aligned} \quad (8)$$

where (a) uses the change of variables $\|r\| = \|x - y\|$, and (b) follows from the polar representation.

Then by substituting (8) into (6) and then (5), we obtain (7) in Lemma 1. Computing the integral in (6) numerically is a very arduous task, but (7) in Lemma 1 is now much more computationally efficient.

4.2 Spectral efficiency

From the information theory, we can achieve Shannon bound $ln(1 + \text{SINR})$ for an instantaneous SINR. Thus, the spectral efficiency in units of nats/s/Hz is calculated as

$$\begin{aligned}
\tau &= E_{\text{SINR}}[\ln(1 + \text{SINR})] \\
&= E_{h_k, x_k}\left[\ln\left(1 + \frac{\sum_{x_k \in C_C} p_k h_k \|x_k\|^{-\alpha_k}}{\sum_{x_k \in C_I} p_k h_k \|x_k\|^{-\alpha_k} + \sigma^2}\right)\right] \\
&= E_{P,I}\left[\ln\left(\frac{P + \sigma^2}{I + \sigma^2}\right)\right],
\end{aligned} \quad (9)$$

where P and I denote the total signal power and aggregate interference power, respectively.

Theorem 1 *In a K-tier HetNet, the spectral efficiency of a typical user at the origin under the proposed model can be evaluated as*

$$\tau = \int_0^{\infty} \left[\prod_{k=1}^{K} \exp\left\{\lambda_{p,k} 2\pi \int_{R_k}^{\infty} [B_k(y)^{c_k} - 1] y dy\right\} - \prod_{k=1}^{K} \exp\left\{\lambda_{p,k} 2\pi \int_0^{\infty} [B_k(y)^{c_k} - 1] y dy\right\}\right] \frac{e^{-s\sigma^2}}{s} ds. \quad (10)$$

Proof Please see Appendix 1. □

Although it is not a closed form, the above integral can be analysed numerically. As $B_k(y)^{c_k} - 1 < 0$, the spectral efficiency is an increasing function of the cooperative radii R_k. As such, we can design a larger cooperative region of BS in HetNets to meet higher data rates (subject to the overhead incurred).

In order to compare the performances (i.e., SE and EE) under PPP and PCP, we adopt the same setup for a typical user as in a PPP model. Assuming fading coefficient $h \sim \exp(1)$, the spectral efficiency under PPP [10] with the density $\lambda_{ppp,k} = c_k \lambda_{p,k} = \lambda_k$ can be express as

$$\tau_{ppp} = \int_0^{\infty} \frac{e^{-s\sigma^2}}{s} \left\{\exp\left[-\sum_{k=1}^{K} \pi \lambda_k B(R_k, sp_k)\right] - \exp\left[-\sum_{k=1}^{K} \pi \lambda_k B(0, sp_k)\right]\right\} ds, \quad (11)$$

where

$$B(R_k, sp_k) = \int_{R_k^2}^{\infty} \frac{sp_k}{u^{\frac{\alpha_k}{2}} + sp_k} du,$$

$$B(0, sp_k) = (sp_k)^{\frac{2}{\alpha_k}} \Gamma\left(1 + \frac{2}{\alpha_k}\right) \Gamma\left(1 - \frac{2}{\alpha_k}\right),$$

and $\Gamma(\cdot)$ denotes the gamma function.

Due to the complexity of the spectral efficiency expression in (10), we now derive a closed form by ignoring the noise (interference is the dominated issue in a dense HetNet).

Lemma 2 *Ignoring the noise in dense HetNets, Eq.* (10) *can be simplified as follows*

$$\tau \overset{(a)}{\approx} q - \ln f, \tag{12}$$

where

$$q = \int_0^\infty \left[e^{-s} - \prod_{k=1}^{K} \exp\left(2\pi\lambda_{p,k} \int_0^\infty [B_k(y)^{c_k} - 1]\, y dy \right) \right] \frac{ds}{s},$$

$$f = \sum_{k=1}^{K} \pi c_k \lambda_{p,k} p_k \frac{2}{\alpha_k - 2} R_k^{2-\alpha_k}.$$

Proof Please see Appendix 2. □

Note that q is not related to the cooperative radii, and it is a constant for a given deployment density and transmit power. As shown in Appendix 2, $q - ln(f)$ is in fact the lower bound of τ in (10). When R_k becomes larger, however, the gap between (10) and $q - ln(f)$ rapidly tapers off, hence the close approximation of (12) for τ. Equation (12) is a closed formula, which makes the following optimization problem easier to solve.

4.3 Minimizing energy consumption via optimizing cooperative radii

For the proposed model, the BSs inside the cooperative clusters can communicate with the typical user located at the origin, and the average power consumption when serving such a user is given by [10]

$$P_u = \sum_{k=1}^{K} [\pi R_k^2 c_k \lambda_{p,k} (P_{bh,k} + P_{sp,k} + a_k \cdot p_k) + \frac{c_k \lambda_{p,k}}{\lambda_u} P_{0,k}], \tag{13}$$

where $P_{bh,k}$ is the backhaul power dissipation when serving one user (i.e., the typical user) in the kth tier, $P_{sp,k}$ denotes the corresponding signal processing power consumption, and a_k denotes the slope of the kth tier power consumption with respect to transmit power p_k. The second term of P_u in (13) denotes the average static power for serving one user, and it is independent of the load of BSs.

In the following, we will provide a solution for optimizing the average power consumption of the HetNet for the typical user in (13). In reality, it is very important to determine the network parameters which can minimize the network power consumption. Since the average energy consumption P_u per user and the average consumed energy P_{av} are related with $P_{av} = \lambda_u \times P_u$, minimizing P_u is critical to designing an energy-saving network.

For a given cooperative radius R_1 of the first tier, we want to find out the optimal cooperative radius R_2 of the second tier in a two-tier HetNet, which minimizes the energy consumption while ensuring a certain rate to the typical user. Note that the energy consumption in (13) increases with the cooperative radii, which suggests that we should determine the minimum cooperative radii while guaranteeing the minimum user rate. Intuitively, the user rate should increase with cooperative areas, and the rate in (10) is indeed an increasing function of the cooperative radii. Assuming that the first-tier cooperative radius R_1 is known, we can determine the optimal value of the second-tier cooperative radius R_2 through dichotomy. The problem will be transformed into a simple problem with a single variable (with the constraint from (10)):

$$\begin{aligned} &\min R_2 \\ &s.t. \int_0^\infty \left[\prod_{k=1}^{2} \exp\left\{ \lambda_{p,k} 2\pi \int_{R_k}^\infty [B_k(y)^{c_k} - 1] y dy \right\} - \right. \\ &\left. \prod_{k=1}^{2} \exp\left\{ \lambda_{p,k} 2\pi \int_0^\infty [B_k(y)^{c_k} - 1] y dy \right\} \right] \frac{e^{-s\sigma^2}}{s} ds \geq \tau_0 \end{aligned} \tag{14}$$

The above problem formulation can be solved by the well-known dichotomy algorithm[1].

Now, we want to form the general problem to determine the optimal values for $R_1, R_2, \cdots, R_K$ in a K-tier network, which minimize the power consumption in (13) while satisfying the minimum spectral efficiency requirement. Note that the second term of P_u in (13) is independent of $R_1, R_2, \cdots, R_K$, and can be ignored. Hence, by applying (12) and (13), the optimization problem can be formulated as follows:

$$\begin{aligned} &\min_{\{R_1, R_2, \cdots, R_K\}} \quad P_{u1} = \sum_{k=1}^{K} \pi R_k^2 \lambda_{p,k} c_k (P_{bh,k} + P_{sp,k} + a_k p_k) \\ &s.t. \quad q - \ln f = \tau_0 \end{aligned} \tag{15}$$

Clearly, both the constraint condition and the optimization objective function are of closed form. The problem can then be solved easily by a linear programming method as follows.

Theorem 2 *The optimal cooperative radii approximately satisfy*

[1]The dichotomy algorithm (also commonly known as binary search algorithm) means, in our case, the following operation: select a very large R_2 value at the beginning and then test the constraint in (14) at the middle point of R_2 (i.e., 0.5 R_2). If (14) is satisfied, take the middle point of the lower half of R_2 and test the constraint in (14) again; if not, take the middle point of the upper half of R_2 and test (14). Repeat the same steps until the process converges to a certain value (which is the optimal vale for R_2).

$$R_{o,k} = \left(\frac{p_k P_{a,i}}{p_i P_{a,k}} R_{o,i}^{\alpha_i}\right)^{\frac{1}{\alpha_k}} \quad (16)$$

$$f = \sum_{k=1}^{K} \pi c_k \lambda_{p,k} p_k \frac{2}{\alpha_k - 2} R_{o,k}^{2-\alpha_k} = \exp(q - \tau_0) \quad (17)$$

where $R_{o,k}$ denotes the optimal value of the kth tier cooperative radius, and $P_{a,i} = P_{bh,i} + P_{sp,i} + a_i p_i$ denotes the total power dissipation when serving a typical user in the ith tier.

Proof Please refer to Appendix 3. □

Combining (16) and (17), the optimal value of each tier can be determined by solving the resultant system of nonlinear equations for a given data rate τ_0, which optimizes the PCP networks' energy consumption.

Once the power P_u has been minimized, the network energy efficiency η_{ee}, defined as the ratio of the average spectral efficiency to the average power consumption, can be calculated as

$$\eta_{ee} = \frac{\tau}{P_u}. \quad (18)$$

5 Limited BS number and range expansion

In this section, we examine two implementation variations of the above collaboration strategy: (a) limited number of cooperative BSs and (b) range expansion (RE).

5.1 Limited number of cooperative BSs

In order to accommodate the ever-increasing traffic demand, low power BSs are densely deployed in HetNets. The number of cross-tier cooperative BSs for each user may be too high in popular areas, due to the overhead incurred and the marginal rate increase.

As a result, the number of associated BSs at a typical user should be limited to an acceptable level. Therefore, we now propose a cooperation tactic where the number of cooperative BSs is restricted to N ($N \leq 4$). In other words, the typical user at the origin communicates with N BSs based on maximum RSS, and all other BSs become noncooperative. Of course, the SE always increases when a new BS is added to the cooperative BS set, but the SE gain becomes more and more marginal and the overhead still increases linearly. Hence, there exists a tradeoff between SE and the overhead (including designing costs of BS and energy consumption). It is therefore meaningful to put a limit on the number of cooperative BSs, which can optimize EE and the overhead cost efficiency. As an example, we consider a two-tier HetNet, and the cooperative BSs number in this case is given by

$$\sum_{k=1}^{2} U_k = N \quad (N \leq 4) \quad (19)$$

where U_k is the number of cooperative BSs in the kth tier.

The effect of this BS constraint will be shown via simulation.

5.2 Range expansion

In HetNets, deployment of macro BSs are for large coverage ranges, which may lead to a heavy load for macro BSs in crowded areas, and the energy consumption also heavily depends on the loading of macro BSs due to macro BSs' high power consumption. Therefore, we now propose a range expansion (RE) scheme based on CoMP to alleviate the loading of macro BSs and improve the EE.

Generally, the transmission power of macro BSs is much larger than that of pico BSs, and as a result, many users will still receive the strongest signal from the macro BS even in hotspots. The RE scheme is to favor the selection of low power BSs by introducing a bias to the association threshold for low power BSs [23]. Due to the smaller coverage, lower power BSs can provide more resource blocks to each user than macro BSs. References [23–25] have shown that RE can achieve load balancing and enhance performances of networks without a loss of average user throughput. For example, in a two-tier HetNet, due to the bias, users are more likely to connect to the pico BSs than the macro BSs even if $p_1 h_1 r_1^{-\alpha_1} > p_2 h_2 r_2^{-\alpha_2}$. Figure 3 illustrates how the typical user is served based on the RE scheme in such an example. Here, we consider a two-tier HetNet comprising macro BSs and pico BSs, which follow a PCP. Consequently, the macro BSs can collaborate to serve the typical user at the origin only when $p_1 h_1 \|x\|^{-\alpha_1} \geq \varepsilon_1 \theta^*$ and the pico BSs can serve this user if $\varepsilon_2 p_2 h_2 \|x\|^{-\alpha_2} \geq \theta^*$, where θ^* denotes normalized RSS threshold (each tier then still has the same RSS threshold), ε_1 and ε_2 are the first power bias and the second power bias, respectively. Therefore, the cooperative radii after biasing are given by

$$R_1^* = \left(\frac{p_1 E[h_1]}{\varepsilon_1 \theta^*}\right)^{1/\alpha_1} \quad (20)$$

$$R_2^* = \left(\frac{\varepsilon_2 p_2 E[h_2]}{\theta^*}\right)^{1/\alpha_2} \quad (21)$$

where R_1^* and R_2^* denote the modified macro-tier radius and modified pico-tier radius after utilizing a power bias, respectively.

From above expressions, the macro BSs cooperation region shrinks but the pico BSs cooperation region expands when the power bias coefficient is greater than 1. The energy efficiency under various bias values will be examined via simulation in the next section.

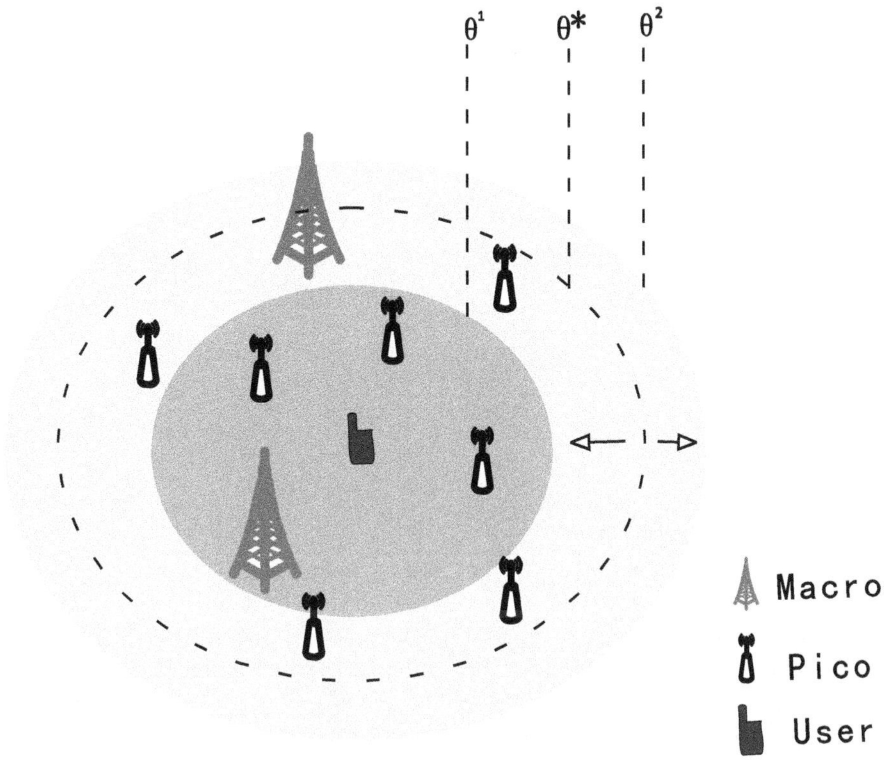

Fig. 3 The typical user is served by macro and pico BSs, where the dashed circle denotes the normalized RSS threshold θ^*, while θ^1 and θ^2 denote the macro-tier and pico-tier RRS thresholds, respectively

6 Numerical results and discussions

In this section, we present some numerical results of the user rate and energy efficiency for a two-tier HetNet consisting of macro and pico BSs. The PCP cluster radii of this two-tier HetNet are $R_{a,1} = 200m$ and $R_{a,2} = 100m$, respectively. Unless otherwise specified, the parameters are those from [26]: the transmit power $p_1 = 20W, p_2 = 0.13W$, and the slope of the power consumption: $a_1 = 4.7, a_2 = 4$. We now compare the results under the same parameters for two different point processes: PCP and PPP.

6.1 Validation of Laplace transform on PCP model

Figure 4 shows simulation and numerical results of the Laplace transform for a two-tier HetNet computed (i.e., (7)) under the PCP model. The PCP parameters are $\lambda_{p,1} = 1/(\pi \times 300^2), \lambda_{p,2} = 1/(\pi \times 200^2), p_1 = 40W, p_2 = 0.13W, c_1 = 1, c_2 = 10$, and the cooperative radii $R_1 = 400m, R_2 = 100m$. The simulation results match well with the analytical integrations in the whole range of s and different path loss exponents (assuming $\alpha_k = \alpha$ here), which verifies the validity of the derived Laplace transform of the aggregate interference and total signal powers. The reason for the differences among the α curves is that for larger a α value, the total interference attenuates more, while the Laplace transform is an exponentially decaying function.

6.2 Comparison between PPP and PCP

Figure 5 compares the spectral efficiency (SE) versus the first-tier and the second-tier cooperative radii for PPP and PCP models. In order to compare PCP and PPP, we have chosen some parameters as in [10] : $\lambda_1 = 1/(\pi \times 250^2), \lambda_2 = 1/(\pi \times 50^2), \alpha_1 = 4.3, \alpha_2 = 3.8$, and set $c_1 = 1, c_2 = 4$. As can be seen from Fig. 5, the average user rate of PCP is worse than that of PPP, and the reason is that the main interferers are more likely to be closer to the typical user for PCP network (hence the higher interference level) than in the case of PPP. Furthermore, the spectral efficiency increases with the macro-tier cooperative radius and the pico cooperative radius, and this is due to the number of cooperative BSs increasing for larger cooperative regions. We note that the pico-tier BSs after some sufficiently large distance contribute very marginally to the user rate. Therefore, the RSS threshold of the second tier should not be too low (to limit the number of the second-tier cooperative BSs), considering the corresponding collaboration overhead.

Figure 6 shows the energy efficiency (EE) versus the mean network cooperative radii R_1 and R_2 for PCP and PPP. The network model parameters are $\lambda_u = 1/(\pi \times 50^2), c_1 = 1, c_2 = 4, P_{bh,1} = P_{bh,2} = 11W, P_{sp,1} = 55W, P_{sp,2} = 2.5W, P_{0,1} = 75W$, and $P_{0,2} = 4.3W$. We can see that the pico-tier cooperative radius R_2

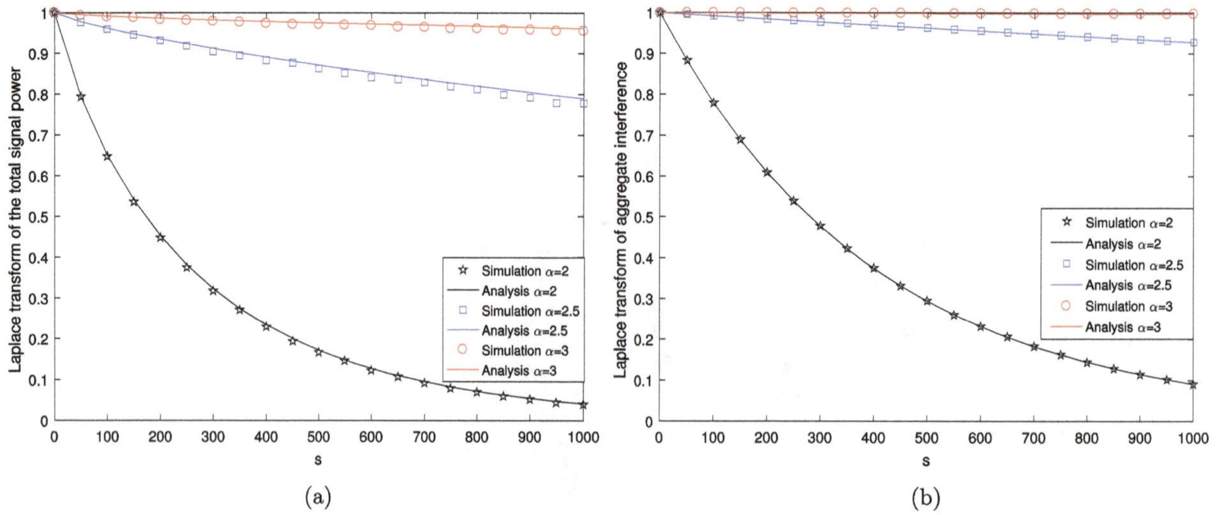

Fig. 4 The Laplace transform for PCP with different α. **a** The Laplace transform of the total signal power. **b** The Laplace transform of the aggregate interference

has an optimal value for a given R_1 which leads to a maximum EE. The case is similar to that of PPP, but the EE of PCP is poorer compared with that of PPP, since the average rate of PPP is better than that of PCP, and energy consumption of the two cases are the same due to the same mean network density. Note that the EE is a decreasing function of the macro-tier cooperative radius, and this is because a large cooperative area of a macro-tier results in an increase in the number of the cooperative macro BSs, while the main energy consumption is heavily dependent on the number of cooperative macro BSs.

6.3 Validity of Lemma 2

Figure 7 shows that the optimal and the upper bound cooperative radius values of the second tier versus the minimum user rate for two different cooperative radii of the first tier . By utilizing (12), we can acquire the upper

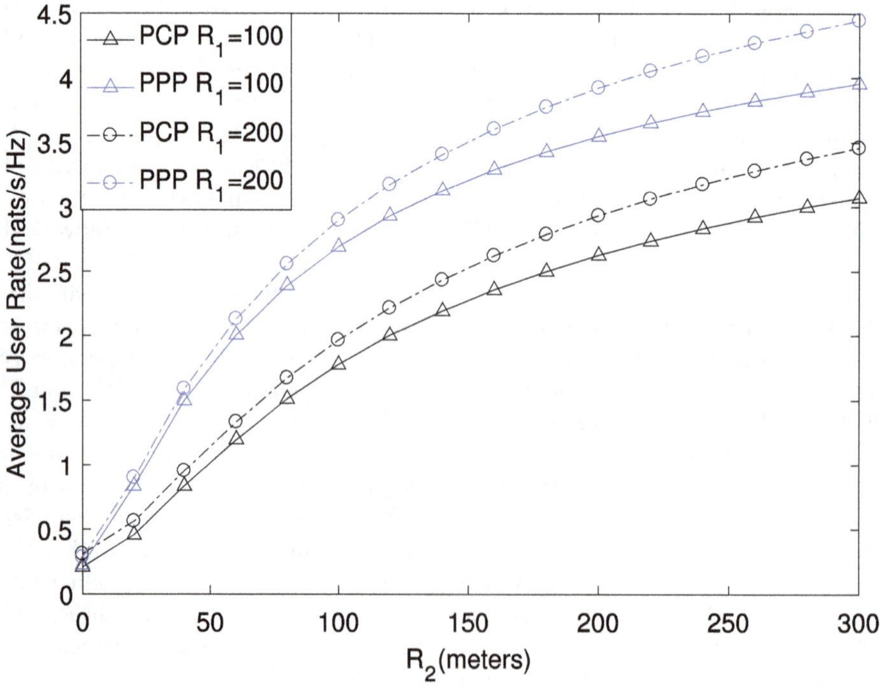

Fig. 5 The average user rates of PPP model and PCP model as a function of the second-tier cooperative radius R_2 with two different cooperative radii of the first tier

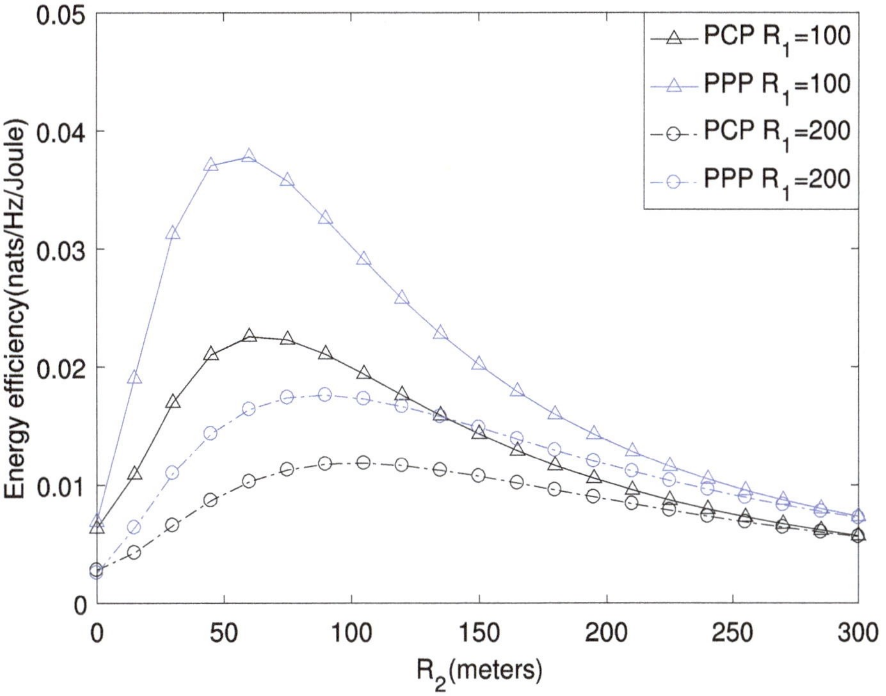

Fig. 6 The energy efficiency of a two-tier network as a function of the first-tier and the second-tier cooperative radius for PPP and PCP

bound curves, and the optimal radius can be calculated approximately by applying the binary search algorithm to (14). It can be seen that the optimal radius increases with the minimum rate τ_0, since the increase of the minimum rate τ_0 means that more BSs need to be connected to the user, which results in an increase of cooperative radius. Furthermore, the gap between the optimal radius in the case of $R_1 = 300m$ and that in the case of $R_1 = 600m$ increases with the minimum rate τ_0. This is because the macro BSs in a specific cooperative area contribute limited impact on user rate, and more pico BSs are therefore needed to satisfy the increment of the τ_0. As τ_0 increases, the upper bound increasingly approaches the optimal value, which testifies that the derived closed form is effective, especially for high data rates.

6.4 Limited number of cooperative BSs and range extension

Figure 8 shows the average achievable rate versus the second-tier cooperative radius for a different number of cooperative BS based on PPP and PCP. We can see that the achievable rate of PCP is still worse than that of PPP for a fixed number of cooperative BSs, and the reason is the same as for Fig. 5. It can be seen that the achievable rate does not change when R_2 is greater than a specific value, which reveals that the achieved rate has an upper bound for a given number of cooperative BSs, under either PPP or PCP. The reason is that the number of cooperative BS reaches the set limit N in the cooperative areas. Moreover, the increase in the number of associated BSs can improve the achievable rate, but the rate gain increment is a decreasing function of N. In the meantime, adding one additional BS means more energy consumption. As a result, BSs located at the edge of cooperative areas is likely to decrease the EE, which also justifies limiting the number of cooperative BSs.

Figure 9 presents the simulation results of the EE based on RE with respect to power bias. Here, we assume that the two-tiers' power biases are both equal to ε. It can be seen that the EE is a concave function of the bias. No matter PCP or PPP, there exists an optimal bias (approximately $\varepsilon = 7$ dB) and it seems to be nearly independent of the normalized RSS threshold. This figure reveals the RE can improve EE as a consequence of decreasing the macro BS loading in comparison with the non-RE scheme (i.e., $\varepsilon = 0$ dB). Equivalently, it is obvious that the EE decreases with the normalized RSS threshold θ^* decreasing because of more cluster edge BSs. It is interesting to observe that for $\theta^* = -65$ dBm, the bias ε has a relatively weaker effect on the EE (due to the large pre-expansion collaboration range). These curves also reveal that the EE of PCP is always lower than that of PPP, as the major interfering BSs in a PCP network are more likely to be closer to the user, which worsens the performances.

7 Conclusions

In this paper, we have derived computationally efficient new formulas for Laplace transforms of the aggregate

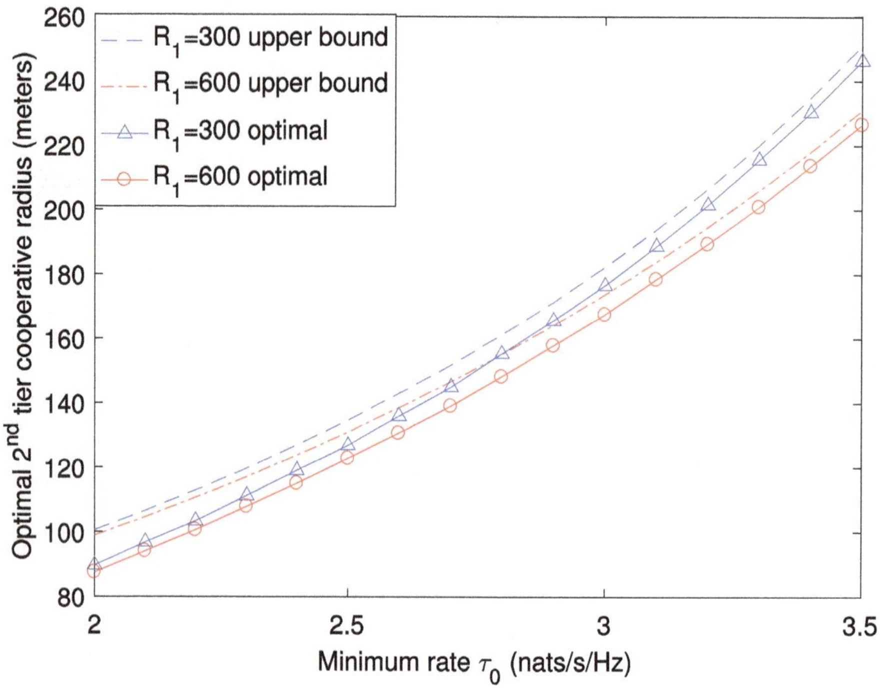

Fig. 7 The optimal cooperative radius as a function of the minimum rate with two different cooperative radii of the first tier

interference and total signal power for PCP HetNets. By using these formulas, we have derived the spectral efficiency and the energy efficiency expressions for a typical user. We have demonstrated that the performances of PCP networks are worse than those of PPP networks. The optimal cooperative radii are also determined under the condition of a minimum rate, which minimizes the energy consumption. Additionally, we examined the case of limited number of cooperative BSs and the RE scheme (to balance the load between macro BSs and pico BSs).

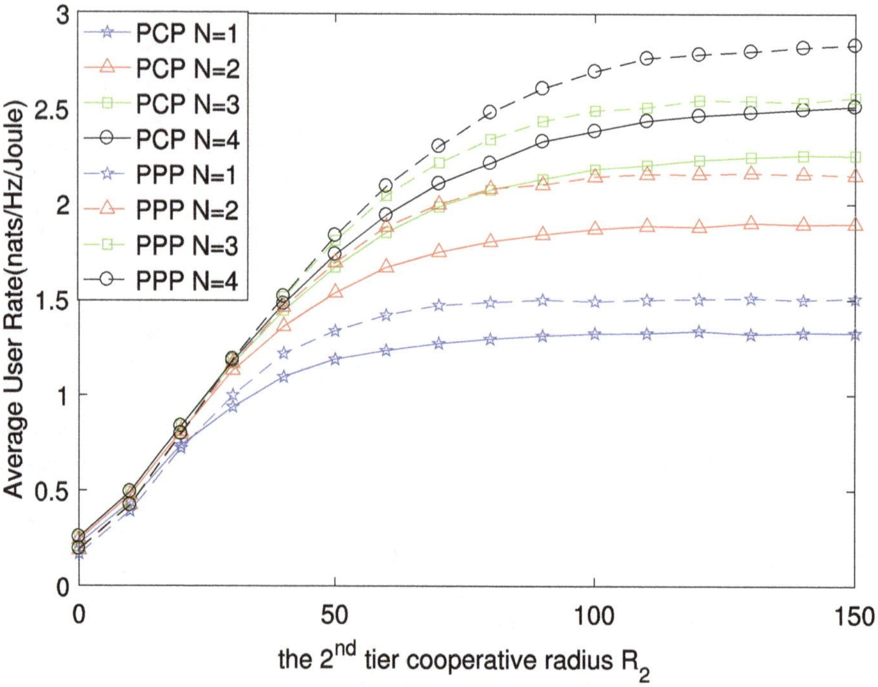

Fig. 8 The average achievable rates as a function of the number of cooperative BSs ($N = 1, 2, 3, 4$)

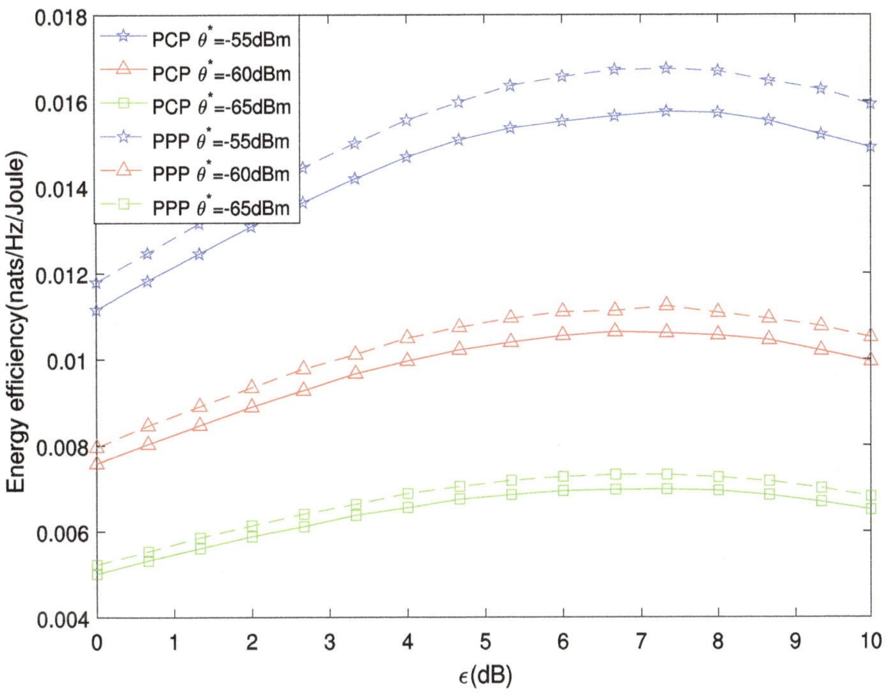

Fig. 9 The simulation results of the energy efficiency based on RE with the cooperative bias factors ε

In terms of energy efficiency, we observed that there exists an optimal solution for the bias, which maximizes the energy efficiency.

Appendix 1. Proof of Theorem 1

From [27], we have the following useful lemma

$$\ln(1+x) = \int_0^\infty \frac{1}{t}(1-e^{-xt})e^{-t}dt = \int_0^\infty \frac{1}{t}\left(e^{-t} - e^{-(x+1)t}\right) dt \;\; \forall x > 0. \tag{22}$$

By applying (22) to (9), the rate is given by

$$\begin{aligned}\tau &= E_{P,I}\left[\int_0^\infty \frac{1}{t}\left[\exp(-t) - \exp\left(-\frac{P+\sigma^2}{I+\sigma^2}t\right)\right] dt\right] \\ &\overset{(a)}{=} E_{P,I}\left[\int_0^\infty \frac{e^{-s\sigma^2}}{s}[\exp(-sI) - \exp(-sP)]\, ds\right] \\ &= \int_0^\infty \frac{e^{-s\sigma^2}}{s}[\mathcal{L}_I(s) - \mathcal{L}_P(s)]\, ds,\end{aligned} \tag{23}$$

where (a) follows by substituting $t = s(I+\sigma^2)$, and $\mathcal{L}_I(s)$ denotes the Laplace transform of the aggregate interference power of HetNets.

The group of interfering BSs is denoted by $C_I = \Phi \backslash C_C$, and the aggregate interference range is outside the circle with radius R_k. Therefore, following a similar process to Lemma 1, we can express approximately the Laplace transform of aggregate interference as

$$\begin{aligned}\mathcal{L}_I(s) &= E_I(\exp(-sI)) \\ &\overset{(a)}{\approx} \prod_{k=1}^{K} \exp\left\{\lambda_{p,k} 2\pi \int_{R_k}^{\infty} [B_k(y)^{c_k} - 1] y dy\right\},\end{aligned} \tag{24}$$

where (a) follows from the fact that BS position can be approximated by the relevant cluster center.

Finally, by substituting (7) and (24) into (23), we can obtain (10) in Theorem 1.

Appendix 2. Proof of Lemma 2

Proof A useful result from [28] is given by □

$$\int_0^\infty \left[\exp\left(-\nu x^p\right) - \exp\left(-\mu x^p\right)\right]\frac{dx}{x} = \frac{1}{p}\ln\frac{\mu}{\nu}, \tag{25}$$

where a real part of μ and ν is greater than zero.

The first integral in (10) is

$$\int_{R_k}^{\infty}\left[1-\left(1+sp_k\|y\|^{-\alpha_k}\right)^{-c_k}\right]ydy$$
$$\overset{(a)}{=}\frac{(sp_k)^{\frac{2}{\alpha_k}}}{2}\int_{(sp_k)^{-\frac{2}{\alpha_k}}R_k{}^2}^{\infty}\left[1-\left(1+u^{-\frac{\alpha_k}{2}}\right)^{-c_k}\right]du$$
$$\overset{(b)}{\leq}\frac{(sp_k)^{\frac{2}{\alpha_k}}}{2}\int_{(sp_k)^{-\frac{2}{\alpha_k}}R_k{}^2}^{\infty}c_k u^{-\frac{\alpha_k}{2}}du$$
$$=\frac{c_k}{\alpha_k-2}p_k R_k^{2-\alpha_k}s, \tag{26}$$

where (a) is from substituting $\|u\| = (sp_k)^{-\frac{2}{\alpha_k}}\|y\|^2$, and (b) follows from the fact that $1-(1+x)^{-n} \leq nx$.

Then, by substituting (26) into (10), we ignore the noise and focus on the interference, and the spectral efficiency is approximately calculated as

$$\tau \geq \int_0^{\infty}\left[\exp\left\{-\sum_{k=1}^{K}\lambda_{p,k}\pi\frac{2c_k}{\alpha_k-2}p_k R_k^{2-\alpha_k}s\right\}-\right.$$
$$\left.\prod_{k=1}^{K}\exp\left\{-\lambda_{p,k}2\pi\int_0^{\infty}[1-B_k(y)^{c_k}]ydy\right\}\right]\frac{1}{s}ds$$
$$=\int_0^{\infty}\frac{1}{s}\left[\exp(-fs)-\exp(-s)\right]ds+q$$
$$\overset{(a)}{=}\ln\frac{1}{f}+q=q-\ln f, \tag{27}$$

where (a) follows from fact that $\nu = f$, $\mu = 1$ and $p = 1$ in (25).

Appendix 3. Proof of Theorem 2

By applying the Lagrangian multiplier method [29], we can establish the objective function as follows:

$$\min_{\{R_1,R_2,\cdots,R_K\}} \quad F = P_{u1} + \zeta\left(q-\ln f-\tau_0\right) \tag{28}$$

Differentiating (28) with respect to $R_1, R_2, \cdots, R_K$, and ζ, respectively, we can obtain

$$\frac{\partial F}{\partial R_1}=\frac{P_{a,1}}{p_1}L_1-\frac{\zeta L_1}{\exp(q-\tau_0)}R_1^{-\alpha_1}=0$$
$$\frac{\partial F}{\partial R_2}=\frac{P_{a,2}}{p_2}L_2-\frac{\zeta L_2}{\exp(q-\tau_0)}R_2^{-\alpha_2}=0$$
$$\cdots\cdots$$
$$\frac{\partial F}{\partial R_i}=\frac{P_{a,i}}{p_i}L_i-\frac{\zeta L_i}{\exp(q-\tau_0)}R_i^{-\alpha_i}=0 \tag{29}$$
$$\cdots\cdots$$
$$\frac{\partial F}{\partial \zeta}=q-\ln f-\tau_0=0,$$

where $L_i = 2\pi\lambda_{p,i}c_i p_i$.

Obviously, we can select one equation arbitrarily from (29) to determine the coefficient ζ:

$$\zeta = f\frac{P_{a,i}}{p_i}R_i^{\alpha_i} \tag{30}$$

$R_{o,1}, R_{o,2}, \cdots, R_{o,k}, \cdots, R_{o,K}$ satisfy these equations. Then, we can obtain (16) and (17) in Theorem 2.

Abbreviations

BS: Base station; CoMP: Coordinated multi-point; RSS: Received signal strength; EE: Energy efficiency; HetNets: Heterogeneous cellular networks; KPI: Key performance indicator; MCP: Matern cluster process; MIMO: Multiple-input multiple-output; PCP: Poisson cluster process; PGFL: Probability generating functional; PPP: Poisson point process; RE: Range expansion; SE: Spectral efficiency; SINR: Signal-to-interference-plus-noise ratio

Authors' contributions

F-CZ proposed the main ideas and directions, supervised the research, and modified and rewrote some sections of the paper. XJ performed simulations and wrote a draft of the paper. Both authors read and approved the final manuscript.

Funding

This work was supported in part by a Shenzhen city/HITSZ start-up grant entitled "Energy-Efficient Low-Latency Wireless Networks," the Shenzhen Science and Technology Innovation Commission under Grant JCYJ20180306171815699, the National Basic Research Program of China (973 Program) under grant 2012CB316004, and the UK Engineering and Physical Sciences Research Council under Grant EP/K040685/2.

Competing interests

The authors declare that they have no competing interests.

References

1. M. Kamel, W. Hamouda, A. Youssef, *Ultra-Dense Networks: A Survey*, vol. 18, (2016), pp. 2522–2545
2. Y. Jiang, Q. Liu, F. Zheng, X. Gao, X. You, Energy-efficient joint resource allocation and power control for D2D communications. IEEE Trans. Veh. Technol. **65**(8), 6119–6127 (2016)
3. T. Zhang, J. Zhao, L. An, D. Liu, Energy efficiency of base station deployment in ultra dense HetNets: A stochastic geometry analysis. IEEE Wirel. Commun. Lett. **5**(2), 184–187 (2016). https://doi.org/10.1109/LWC.2016.2516010
4. S. Khan, S. A. Mahmud, in *IEEE International Conference on Computer and Information Technology; Pervasive Intelligence and Computing, Liverpool*. Power Optimization Technique in interference-limited femtocells in LTE and LTE advanced based femtocell networks, (2015), pp. 749–754. https://doi.org/10.1109/cit/iucc/dasc/picom.2015.110
5. S. Geirhofer, P. Gaal, in *2012 IEEE Globecom Workshops*. Coordinated multi point transmission in 3GPP LTE heterogeneous networks, (Anaheim, 2012), pp. 608–612. https://doi.org/10.1109/glocomw.2012.6477643
6. R. Irmer, et al., Coordinated multipoint: Concepts, performance, and field trial results. IEEE Commun. Mag. **49**(2), 102–111 (2011)
7. A. He, D. Liu, Y. Chen, T. Zhang, in *Annual International Symposium on Personal, Indoor, and Mobile Radio Communication (PIMRC)*. Stochastic geometry analysis of energy efficiency in HetNets with combined CoMP and BS sleeping, (Washington DC, 2014), pp. 1798–1802. https://doi.org/10.1109/pimrc.2014.7136461
8. J. G. Andrews, F. Baccelli, R. K. Ganti, A tractable approach to coverage and rate in cellular networks. IEEE Trans. Commun. **59**(11), 3122–3134 (2011)
9. R. W. Heath, M. Kountouris, T. Bai, Modeling heterogeneous network interference using Poisson point processes. IEEE Trans. Signal Proc. **61**(16), 4114–4126 (2013)
10. W. Nie, F. C. Zheng, X. Wang, W. Zhang, S. Jin, User-centric cross-tier base station clustering and cooperation in heterogeneous networks: Rate improvement and energy saving. IEEE J. Sel. Areas Commun. **34**(5), 1192–1206 (2016)
11. V. Suryaprakash, J. Møller, G. Fettweis, On the modeling and analysis of heterogeneous radio access networks using a Poisson cluster process. IEEE Trans. Wirel. Commun. **14**(2), 1035–1047 (2015)
12. Q. Ying, Z. Zhao, Y. Zhou, R. Li, X. Zhou, H. Zhang, *Characterizing spatial patterns of base stations in cellular networks*, (Shanghai, 2014), pp. 490–495. https://doi.org/10.1109/iccchina.2014.7008327

13. C.-H. Lee, C.-Y. Shih, Y.-S. Chen, Stochastic geometry based models for modeling cellular networks in urban areas. Wirel. Netw. **19**(6), 1063–1072 (2013)
14. C. Kong, et al., *VSMC MIMO: A spectral efficient scheme for cooperative relay in cognitive radio networks*, (2015), pp. 2137–2145. https://doi.org/10.1109/infocom.2015.7218599
15. M. Demirtas, A. Soysal, in *IEEE 81st Vehicular Technology Conference (VTC Spring)*. Energy and spectral efficient microcell deployment in heterogeneous cellular networks, (2015), pp. 1–5. https://doi.org/10.1109/vtcspring.2015.7145803
16. Y. Lin, W. Yu, Ergodic capacity analysis of downlink distributed antenna systems using stochastic geometry. Proc. IEEE Int. Conf. Commun. (ICC), 3338–3343 (2013). https://doi.org/10.1109/icc.2013.6655062
17. J. Zhao, Q. Wang, Y. Dong, W. Wei, in *IEEE Wireless Communications and Networking Conference*. Performance analysis for cross-tier cooperation in heterogeneous cellular networks: A stochastic geometry approach, (Doha, 2016), pp. 1–6. https://doi.org/10.1109/wcnc.2016.7564964
18. W. Yi, Y. Liu, A. Nallanathan, Modeling and Analysis of D2D Millimeter-Wave Networks With Poisson Cluster Processes. IEEE Trans. Commun. **65**(12), 5574–5588 (2017). https://doi.org/10.1109/TCOMM.2017.2744644
19. Y. J. Chun, M. O. Hasna, A. Ghrayeb, Modeling heterogeneous cellular networks interference using Poisson cluster processes. IEEE J. Sel. Areas Commun. **33**(10), 2182–2195 (2015)
20. Y. Wang, Q. Zhu, Modeling and analysis of small cells based on clustered stochastic geometry. IEEE Commun. Lett. **21**(3), 576–579 (2017)
21. S. N. Chiu, D. Stoyan, W. S. Kendall, J. Mecke, *Stochastic Geometry and Its Applications*, 3rd ed. (Wiley, Hoboken, 2013)
22. R. K. Ganti, M. Haenggi, Interference and outage in clustered wireless ad hoc networks. IEEE Trans. Inf. Theory. **55**(9), 4067–4086 (2009)
23. L. Zhang, S. Zhao, P. Shang, J. Liu, F. Han, in *IEEE 85th Vehicular Technology Conference (VTC Spring)*. Distributed adaptive range extension setting for small cells in heterogeneous cellular network, (Sydney, NSW, 2017), pp. 1–7. https://doi.org/10.1109/vtcspring.2017.8108474
24. Q. Ye, B. Rong, Y. Chen, M. Al-Shalash, C. Caramanis, J. G. Andrews, User association for load balancing in heterogeneous cellular networks. IEEE Trans. Wirel. Commun. **12**(6), 2706–2716 (2013)
25. H. S. Jo, Y. J. Sang, P. Xia, J. G. Andrews, Heterogeneous cellular networks with flexible cell association: A comprehensive downlink SINR analysis. IEEE Trans. Wirel. Commun. **11**(10), 3484–3495 (2012)
26. G. Auer, et al., How much energy is needed to run a wireless network? IEEE Wirel. Commun. Mag. **18**(5), 40–49 (2011)
27. K. A. Hamdi, Capacity of MRC on correlated Rician fading channels. IEEE Trans. Commun. **56**, 708–711 (2008)
28. A. Jeffrey, D. Zwillinger, *Table of Integrals, Series, and Products*. (Academic, New York, 2007)
29. S. Boyd, L. Vandenberghe, *Convex Optimization*. (Cambridge Univ Press, Cambridge, 2004)

8

MMIR: A microscopic mechanism for street selection based on intersection records in urban VANET routing

Zhen Cai[1], Mangui Liang[1*] and Qiushi Sun[2]

Abstract

In urban vehicular ad hoc networks (VANETs), the intersection-based routing scheme has represented its greater applicability and better efficiency to adapt to high and constrained mobility. How to make an accurate decision for street selection is a challenging issue due to the rapid topology changes in VANETs. In this paper, we propose a microscopic mechanism based on intersection records (MMIR) in which the intersection vehicle nodes maintain and update a records table with every passing vehicle's individual information. By analyzing and processing these entries, we evaluate these vehicles' current positions so as to compute the connectivity probability or estimated delivery delay for all candidate streets to support street selection. In contrast to the statistical and macroscopic information for the common condition, we firstly make use of the individual and microscopic data to enhance the accuracy of estimated results. Furthermore, according to the quantity and the running interval, we classify vehicles into two categories: individual and queue vehicles, in order to effectively decrease the complexity of position estimation. Lastly, since there are no dedicated control packets generated in MMIR, the network overhead is low. The simulation results show that the proposed MMIR outperforms existing approaches of street selection in terms of the accuracy of computed connectivity probability and estimated delay.

Keywords: VANETs, Intersection-based routing, Connectivity probability, Delivery delay

1 Introduction

With the advance in wireless network technology in recent years, each vehicle running in the urban streets can exchange data with the nearby vehicles through vehicle-to-vehicle (V2V) communications [1, 2] or with the roadside units (RSU) via vehicle-to-infrastructure communications (V2I) [3]. Vehicular ad hoc networks (VANETs) have attracted extensive attention from both academic and commercial communities. VANETs play an important role in safety-related (collision avoidance, cooperative driving, etc.), information services (real-time traffic, weather information, etc.), and infotainment (multiplayer games, multimedia sharing, etc.) for drivers and passengers. As a particular type of mobile ad hoc network (MANETs), especially in the urban scenario, VANETs have some unique features. First, due to the high mobility of the vehicles running in the street, the topology of the vehicular networks changes rapidly and thus the end-to-end connection is frequently broken. Second, the trajectories of vehicle nodes which only move along with the existing streets and change them one by one when they pass the intersections result in that the routes of multi-hop delivery between the vehicle nodes have to follow the urban traffic map restrictively. Third, the network connectivity in the street between two adjacent intersections depends on not only the vehicle nodes' density which is mainly related to the location of the street and the time of day, but also the vehicle nodes' evenness which is frequently affected by the traffic lights, vehicle accidents, and the difference of various vehicles' speeds, etc. Due to these characteristics of urban VANETs, the classical topology-based routing protocols [4–6] and the traditional position-based routing protocols [7–9] for MANETs are not suitable.

To adapt to the urban VANETs with high but constrained mobility, the intersection-based routing scheme

* Correspondence: mgliang@m.bjtu.edu.cn
[1]Institute of Information Science, Beijing Jiaotong University, Beijing 100044, People's Republic of China
Full list of author information is available at the end of the article

[10] has represented its greater applicability and better efficiency. Its working mechanism is that geographical greedy forwarding strategy or its improved versions (e.g., [11–15]) are still used for packet transmission in the intra-streets, but when packets reach the intersection, a selection of the next street (i.e., direction) for forwarding is decided based on which one can provide a higher delivery rate and lower network delay in the entire multi-hop routing. The routing path from source to destination is separated into a series of streets connected one by one so that with the divide-rule policy it can better deal with the rapidly changing topology in VANETs. And as intersections are the only places where routing decisions are made, it is adaptable to the constrained mobility of the vehicle nodes and effective to avoid the local optimum problem caused by the street layouts and some towering obstacles in the urban environment.

How to make an accurate decision for street selection is one of the key issues in the intersection-based VANET routing. According to different application scenarios, there are different metrics for the street selection in VANET routing such as distance to destination, connectivity probability, delivery delay, and delivery ratio. No matter which metric, its calculation needs to be based on some information, e.g., the street length, the vehicle density, and the number of neighboring vehicles. The data acquisition should have characteristics of accuracy and real time in order to support the right decision. However, there are some interference factors to street selection. For instance, owing to traffic lights at intersections, the network in each street may be partitioned into several segments to impact network connectivity [16]; with the drivers' diverse customs, the vehicle nodes have different not only driving velocities but also velocity variations to a certain condition (e.g., high density). These concrete factors make it more difficult to capture the real and current network state by applying some macroscopic statistical data such as average vehicle velocity, traffic flow, and vehicle density in the past period of time.

Most of the existing mechanisms and models of connectivity estimation were designed at the macroscopic level. In this way, it is not suitable for the network environment with high mobility and rapid topology changing. In this paper, we focus on designing a more accurate strategy of real-time street selection with low overhead from the micro point of view. To this end, we propose a *microscopic mechanism based on intersection records* (MMIR) in which the intersection vehicle nodes maintain and update a records table with every passing vehicle's information. By analyzing and processing these entries, we can evaluate these vehicles' current positions so as to compute the connectivity probability or estimate the delivery delay for all candidate streets which connected with this intersection. In contrast to the statistical and macroscopic data or unreliable topology information, in MMIR, we firstly make use of the concrete and microscopic information recorded at the intersection when each vehicle passed so as to enhance the accuracy of estimated results. In this way, MMIR can detect the actual status of network connectivity, in which it may be disconnected even with high density. Furthermore, according to the vehicle quantity and intervals during them, we classify the recorded vehicles into two categories to handle respectively, i.e., individual nodes and queue nodes. The nodes in a queue are regarded as a whole in order to effectively decrease the complexity of position estimation caused by the different velocity variance of every vehicle to the relatively high density and respective location in the queue. Lastly, since there is no dedicated control packet generated (i.e., the information for intersection recording is contained in periodical beacon messages for sending), there is barely extra network overhead. MMIR is suitable to the intersection-based VANET routing in urban environments especially the signalized arterial street networks. Simulation results show that the proposed MMIR outperforms existing approaches of street selection in terms of the accuracy of computed connectivity probability and estimated delay.

The remainder of this paper is organized as follows: Section 2 summarizes the related work. Section 3 describes the detail of MMIR including free velocity, intersection recording, connectivity calculation, and delivery delay estimation. Section 4 determines the key parameters of MMIR. Section 5 evaluates the performance of MMIR by simulations. Finally, Section 6 concludes the paper.

2 Related work

Selection of an optimal street for delivering data packets is the critical issue in designing an intersection-based VANET routing protocol. Generally speaking, the performance of a street selection strategy greatly depends on what information it adopted. The *greedy perimeter coordinator routing* (GPCR) [17] is a classical intersection-based VANETs protocol which was proposed to solve the local optimum problem in the *greedy perimeter stateless routing* (GPSR). In GPCR, with supports of GPS and static street maps, the street through which there is the shortest path for packet delivery to destination is selected by using Dijkstra's algorithm. Yang et al. [18] proposed an *adaptive connectivity aware routing* (ACAR) and indicated that the width of a street can be used to assess the candidate street: a wider street implies a higher probability of vehicle density and consequent network connectivity. But only introducing the static information is obviously insufficient. The delay model in the *vehicle-assisted data delivery* (VADD) protocol proposed by Zhao and Cao [19] combines static data (Euclidean distance of street) in digital maps and statistical data (average velocity and vehicle density) from third-party services to estimate the packet delivery delay for every adjacent outgoing street and select one with the shortest delay towards the destination at the current intersection. Jeong et al. [20] proposed another link delay model with one-way vehicular traffic given the vehicle arrival rate, the vehicle average speed, and the length of the street. The model separates the street length into two parts: forwarding distance and carrying distance. By ignoring the small communication delay for forwarding packets, the delivery delay along a street is the corresponding carry delay with carrying distance which is calculated by the analytical method of probability function. However, there might be some broken links with length greater than transmission range due to low traffic density, splitting the network into multiple clusters. To improve the connectivity probability, Panichpapiboon et al. [21] took advantage of the opposing vehicles on a two-way street and proposed a connectivity model by applying the bidirectional statistics of the street. Furthermore, it is well known that speeds of vehicles are not constant and normally distributed in the free-flow traffic state [22, 23]. Yousefi et al. [24] proposed an analytical model for connectivity probability and connectivity distance by considering not the constant but the normal distribution (mean and variance) of the vehicles' speeds. Al-Mayouf et al. and Ding et al. [25, 26] also make use of the average speed of vehicles to calculate the connectivity for the next street selection. Apart from the above macroscopic models which adopted static data and traffic statistics, some other studies focused on applying the real-time control information exchanged with neighboring vehicles to estimate network connectivity or delivery delay in the street. In the *landmark overlays for urban vehicular routing environments* (LOURE) [27] and the *virtual vertex routing* (VVR) [28], similarly, a node obtains the number of its current neighbors by received beacon messages and adds this new information into its next beacon to broadcast. Thus, all vehicle nodes including that located at the intersection can collect the density and topology information in the street to calculate the network connectivity for routing selection in real time. Zhang et al. [29] considered the phenomenon of the link correlation which represented the influence of different link combinations in network topology to transmit a packet and deigned an opportunistic routing metric called the *expected transmission cost over a multi-hop path* (ETCoP) for the selection guidance of the relaying node in intra-streets and the next street at an intersection. Likewise, the topology information used to calculate ETCoP is obtained via beacon packets. The *link-delay update* (LDU) module in the *static-node-assisted adaptive data dissemination protocol for vehicular networks* (SADV) proposed by Ding [30] measures the transfer delay for each street in real time and propagates the up-to-date estimation among the static nodes which were deployed at intersections, so that each static node can get a more complete delay matrix and contribute to making an accurate decision of street selection. Nzouonta et al. [31] proposed two *road-based* (the same as intersection-based) *using vehicular traffic* (RBVT) routing protocols: a reactive protocol RBVT-R and a proactive protocol RBVT-P. Especially in the RBVT-P, the *periodical connectivity packets* (CPs) are generated to visit connected streets and store the graph that they form. By dissemination, all nodes in the network can maintain the information of entire topology and calculate the shortest connected paths to the destination. In consideration of network overhead and freshness of information, another routing protocol, *diagonal-intersection-based routing* (DIR) [32], only gathers topology information within the range of the successive three streets; moreover, it takes into account the probability of the green light at intersections for delay estimation. With the affection by traffic lights, various vehicles' different speeds, etc., the network connectivity in the street depends on not only the average density but also the vehicles' distribution in real time. In the *improved greedy traffic-aware routing* (GyTAR) *protocol* proposed by Jerbi et al. [33], each street is dissected into small fixed-area cells in advance depending on the transmission range of vehicles. By acquiring the number of vehicles within every cell of the street in real time, the intersection vehicle nodes consider traffic density information included in the *cell data packet* (CDP) and the curve metric distance to the destination extracted from digital maps, then calculate a score for every candidate street and select the one with the highest score for forwarding packets. Furthermore, with the development of sensor technology and intelligent transportation system, the real-time status of traffic lights was also considered a deciding factor for selecting streets, e.g., in [12] and [34].

3 Method

Apart from the distance to destination, the connectivity of network in the candidate street is also the crucial element for street selection in VANET routing. As well known, it mainly depends on the density and distribution of the vehicle nodes in the street. In this paper, we study the connectivity from the microscopic point of view, describing the traffic flow by tracking individual vehicles rather than on an aggregated basis [35]. MMIR which we proposed aims to give an accurate estimation in real time and with low overhead. It is organized into three parts: (1) free-velocity analysis (the definition of free velocity in this paper), (2) recording at the intersection, and (3) connectivity calculation and delivery delay estimation.

MMIR considers that each vehicle is equipped with a Global Position System (GPS) and a street-level digital map, and then it can easily acquire the information about its own position, velocity, moving direction, etc. The information can be also obtained by their neighboring vehicles with the aid of the periodical beacon messages exchanged with each other. Furthermore, a source node knows the current geographical position of the destination which can be achieved by the location service. It would draw support from a low power wide area (LPWA) network [36] such as LoRa [37], Narrowband Internet of Things (NB-IoT) [38], etc. In addition, all the vehicles are assumed to be synchronized by GPS.

3.1 Free-velocity analysis

MMIR's main approach to the calculation of the street connectivity is using the data of individual vehicles to estimate their positions at a certain time. To achieve this objective, each running vehicle needs to gather and calculate its accurate and effective driving data to support the connectivity calculation in which the free velocity is the crucial one.

In this paper, we refer to free velocity as the driver's desired velocity in the free-flow traffic state in which there is little influence from other vehicles and no occurrence of traffic incidents nearby. In terms of the definition, to gather the real-time data, the free velocity needs to meet the following conditions and principles:

- Within a certain range in the front, the number of vehicles which are moving in the same direction is not sufficient to affect the driver to make a reaction on velocity. For instance, the threshold value of the number can be determined as $N_{\text{ln}} - 1$, where N_{ln} is the number of lanes in one direction. It means that the vehicle still has a free lane to move at its desired velocity without the influence of the slow vehicles in front of it.
- The free-velocity collection cannot be executed in the vicinity of intersections in consideration of the forced decelerating, waiting, accelerating processes of vehicles due to the traffic lights and security considerations.
- The free-velocity collection can be executed only when the condition mentioned above has been active for a certain time. It ensures that there is enough time for the driver to convert to his desired velocity from the previous state.
- According to different conditions such as the number of lanes, lane width, and the value of speed limit, we classify the streets into several classes in advance. Thus, the individual vehicle needs to gather and calculate its free velocity for each class respectively. Such is helpful to the accuracy of information collection.

The free velocity may also be affected by some other factors which are hardly recognized quantitatively, such as the weather, driver's mental status, and even the mood. However, generally speaking, the samples of free velocity gathered in the above conditions follow Gaussian distributions [23, 39]. In this way, for the later works, it needs to calculate the average (1) and the variance (2) of free velocity by each vehicle in real time using the following two formulas:

$$\mu_{\text{Vfree}} = \frac{1}{n} \cdot \sum_{i=1}^{n} X_i, \tag{1}$$

$$\sigma^2_{\text{Vfree}} = \frac{1}{n-1} \left(\sum_{i=1}^{n} X_i^2 - n \cdot \mu^2_{\text{Vfree}} \right). \tag{2}$$

3.2 Recording at the intersection

The street intersection plays an important role in urban VANETs as it is the junction between different streets. At the intersection, vehicles leave their last street and enter a new one by going straight or taking a turn. Correspondingly, the forwarding direction of a packet in VANETs may also be changed depending on the destination location and the network connectivity in the candidate street. Vehicle nodes at the intersection always act as the decision makers of street selection in most of the intersection-based routing protocols. In MMIR, among these vehicle nodes, one or some are considered the intersection-server node (ISN) according to their current locations and other features. They are in charge of receiving and storing the records for all vehicles that passed the intersection in recent time. In general, the closest vehicle to the intersection center (optimal position), and with the slowest velocity (longest duration), is the optimum one for ISN election. With respect to the mechanism of the server's selection and replacement, it could draw lessons from related ideas of the location server, e.g., in [40] which is a quorum-based location service protocol. As it is not the main aspect for study in this paper, the details are not given here.

In practical situations, every vehicle, when it passed the center zone of the intersection and entered a new street, packets its information and attaches them to the beacon message in the next time for sending. After (usually less than) a beacon time interval, once the ISN received the modified beacon message, it extracts the information and generates a new entry in its intersection records table for the vehicle that the message comes from. In other words, only the ISNs maintain the records table. Additionally, when an ISN left the intersection, it removes its status as a server, also generates a new entry for itself, and then sends the whole records table included in the beacon message to the other ISNs or the optimal vehicle-node which will be elected as a new ISN. Note that due to all the information of the passing vehicles which are contained in the periodical and mandatory beacon messages, MMIR does not introduce much additional network overhead from the recording process at the intersection.

As illustrated in Table 1, the intersection record includes vehicle ID, last street ID, new street ID, transmission time (t_{rec}), current position in the new street (pos_{rec}), current velocity (v_{rec}), free velocity (average and variance), normal acceleration (acc), transmission range (R), TTL (time-to-live), by which we can estimate the position of the vehicle at a later time.

Note that, in order to reduce the data volume and calculation quantity at ISN, the term of TTL is added. It means the estimated time which will be taken to pass through the new street by the vehicle. However, sometimes a vehicle cannot run at its free velocity throughout the whole street due to the vehicles around it. In practice, we should set the valid time bigger than the estimated time. The detail will be discussed in the following section.

Privacy protection is a critical issue for the drivers and passengers in the vehicles [41]. To make sure that the vehicle's trajectory cannot easily be traced by the others, the vehicle ID in an intersection record is denominated as a temporal and unique character string which is not its real ID in the network. From the prospective of intersection recording, it only needs this unique string to avoid the occurrence of duplicated records for the same vehicle in its records table, rather than to know which vehicle it is in the whole network for other uses such as location service, etc.

3.3 Connectivity calculation and delay estimation for street selection

After study of intersection recoding, we introduce connectivity calculation and delay estimation in detail below.

3.3.1 Connectivity probability in light traffic

In the urban environment especially in the arterial streets, for vehicle driving, there is a more comfortable condition relatively which commonly includes three or four lanes for each direction, a greater width of the lane, a smaller ratio of the curved section, no crosswalks, few parking points, etc. Vehicles in such situation, and in free-flow traffic state (i.e., the density is sufficiently low), could be running at the free velocity all the way. Thus, their positions (pos_{est}, i.e., the distance from the street entrance), at a certain time (t_{cur}), can be calculated based on its record at last intersection as

$$\mu_{pos} = \boldsymbol{E}(pos_{est}) = \begin{cases} \mu_{v_{free}} \cdot (t_{int} - t_{acc}) + \frac{1}{2} \cdot acc \cdot t_{acc}^2, & t_{int} \geq t_{acc}, \\ \frac{1}{2} \cdot acc \cdot t_{int}^2, & t_{int} < t_{acc}, \end{cases} \quad (3)$$

$$t_{int} = t_{cur} - t_{rec}, \quad (4)$$

$$t_{acc} = \frac{\mu_{v_{free}} - \min\left(\mu_{v_{free}}, v_{rec}\right)}{acc}. \quad (5)$$

where t_{int} is the interval between the current time and the transmission time in the record, t_{acc} is the time which the vehicle needs to accelerate to its free velocity. At the intersection with traffic lights, vehicles which begin to move when the light turns green from red, and vehicles which have slowed down for passing the intersection, need an accelerating process to attain their free velocity when they have entered the new street. In most cases, t_{acc} is less than t_{int}, because the transmission time t_{rec} is the last beacon message to ISN when the vehicle is in the communication range (around 250 m) and for this distance the vehicle's acceleration has been normally completed.

Since the free velocity is not a constant value and is distributed following Gaussian distribution, the estimated position follows Gaussian distribution correspondingly when $t_{int} \geq t_{acc}$, as

$$pos_{est} \sim N\left(\mu_{pos}, \sigma_{pos}^2\right), \quad (6)$$

$$\sigma_{pos}^2 = \left(\sigma_{v_{free}} \cdot (t_{int} - t_{acc})\right)^2. \quad (7)$$

Let N be the total number of vehicles of which the estimated positions are still in the current street or in the next intersection area. Hence, we can calculate and sort the vehicles' estimated positions as $\mu_{pos1} < \mu_{pos2} < \cdots < \mu_{posN}$, where μ_{posi} is the ith vehicle.

As the free-velocities of vehicles are independent and identically distributed (i.i.d.), according to the property of Gaussian distribution, the distance (dis_i) between

Table 1 Intersection record format

Intersection Record			
Vehicle ID	$Street_{last}$ ID	$Street_{new}$ ID	*TTL*
v_{rec}	pos_{rec}	t_{rec}	
$\mu_{v_{free}}$	$\sigma_{v_{free}}$	*acc*	*R*

$vehicle_i$ and $vehicle_{i+1}$ follows Gaussian distribution as well (8). The probability of connectivity (P_i) between them can be calculated as (9).

$$\mathrm{dis}_i \sim N\left(\mu_{\mathrm{pos}(i+1)} - \mu_{\mathrm{pos}i}, \sigma^2_{\mathrm{pos}(i+1)} + \sigma^2_{\mathrm{pos}i}\right) \quad (8)$$

$$P_i = P(|\mathrm{dis}_i| \leq R_i) = \int_{-R_i}^{R_i} \frac{1}{\sqrt{2\pi\left(\sigma^2_{\mathrm{pos}(i+1)} + \sigma^2_{\mathrm{pos}i}\right)}} e^{\frac{-\left(x-\left(\mu_{\mathrm{pos}(i+1)} - \mu_{\mathrm{pos}i}\right)\right)^2}{2\left(\sigma^2_{\mathrm{pos}(i+1)} + \sigma^2_{\mathrm{pos}i}\right)}} dx \quad (9)$$

The distance between any two consecutive vehicles must be smaller than the transmission range R to ensure that the network from the first to the last is connected. Thus, it is required that $\mathrm{dis}_i \leq R$ for $i = 1, 2, \ldots, N - 1$. Note that even in the multi-lane streets, the connections between vehicles mainly depend on the distance along the street (in the parallel direction) and the distance in the transverse direction can be negligible relatively. In other words, VANETs in the urban streets are considered a one-dimensional network.

Furthermore, the street connectivity considered in MMIR is that there is an end-to-end connection from the last intersection area to the next one. Accordingly, it needs at least one vehicle in the next intersection. We can calculate the probability of each vehicle to be in the next intersection area (P_{next}) in sequence as

$$P_{\mathrm{next}} = P(\text{street length} - R_{\mathrm{int}} \leq \mathrm{pos}_{\mathrm{est}} \leq \text{street length} + R_{\mathrm{int}}), \quad (10)$$

where R_{int} is the range of intersection area which is the half transmission range so as to ensure any two vehicles in the area can communicate with each other by one-hop link. If the P_{next} of the nth vehicle is greater than the threshold value (such as 0.8) that we set and for $i = 1, 2, \ldots, n - 1$, the P_{next} of the ith vehicle is less than it, we can calculate the connectivity probability in the street (P_c) as

$$P_c = \prod_{i=1}^{n-1} P_i. \quad (11)$$

3.3.2 Queues and individuals

Until now, we have described the calculation of connectivity probability in the street based on free-flow traffic state. However, in many cases, a vehicle cannot be driven

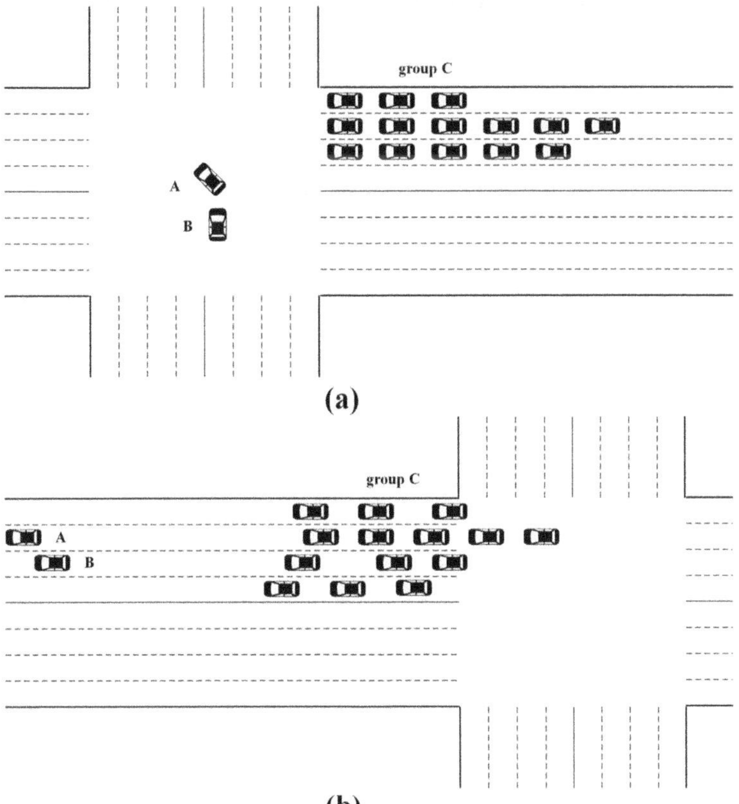

Fig. 1 Formation of individual and queue vehicles by traffic lights

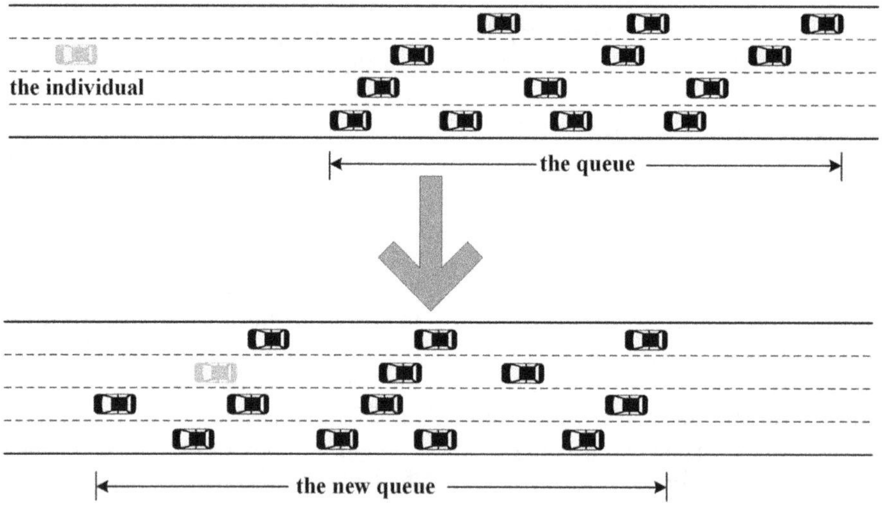

Fig. 2 a, **b** An example of new queue formation (a queue caught up with an individual)

at its free velocity all the way due to the interaction with a crowd of vehicles around it. It needs to make some actions such as acceleration, deceleration, and frequent lane changing, and these events may interrupt network connection. To deal with these disturbances to our calculation of connectivity probability, in MMIR, we classify the vehicles into two categories to handle respectively, i.e., individual vehicle and queue vehicle.

As the most significant source of fixed interruptions, the traffic lights at intersections periodically halt vehicle flow for each movement which on a given set of lanes is possible only on the green light, and then partition the flow in the street into several clusters which are called as queues in this paper due to the consideration of one dimensional network. As illustrated in Fig. 1(a), it is a typical intersection in the urban environment, of which in each entrance direction there are two dedicated straight lanes, one straight lane sharing with right turn and one dedicated left turn. Vehicle A and B have arrived at the intersection and want to turn left and enter street-L. On the other side, some vehicles (group C) of which the number is enough to let them affect each other in velocity when they are starting to move together, have stopped opposite to street-L and are waiting for the straight-moving signal to turn green. At the next moment in Fig. 1b, vehicle A and B have entered street-L and moved a distance; behind them, group C has obtained the permit (green signal) and also entered street-L. As we mentioned above, the individual vehicles, A and B, could run at their respective free velocity. On the contrary, the vehicles in group C formed a queue and their velocities might be affected by each other.

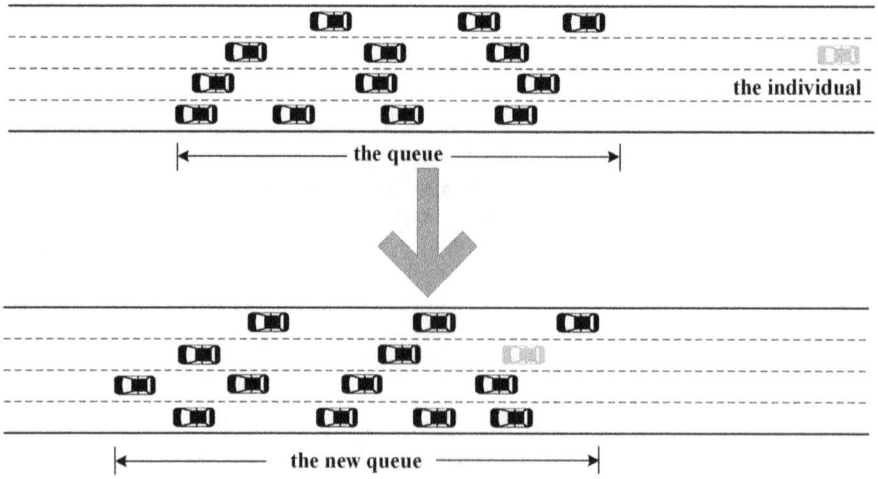

Fig. 3 An example of new queue formation (an individual caught up with a queue)

Furthermore, after group C passed the intersection, there is a queue-discharging process representing that all the vehicles can be back to their free-velocities until the queue fully dissipates. Let us analyze the discharging queue in terms of the following points:

- Connectivity in the queue. The influence of queue in the urban street is very likely to be negligible when the length exceeds 2 mi (3.21 km) [42]. However, in an urban environment, the length of the street between adjacent intersections is generally less than such 2 mi. In other words, the queue generated at last traffic lights will not be dispersed in the current street. And in view of the transmission range of about 250 m, we consider that the connection in the queue is linked from the head vehicle to the last one in the whole street which they entered.
- Head vehicle and tail vehicle. In MMIR, we refer to the head vehicle in the queue as the headmost vehicle at the time of executing the connectivity calculation rather than the time when the queue formed. On the contrary, the tail vehicle is also the meaning. Common sense says that with fewer disturbances from other vehicles, the one with the fastest free velocity in the front of the queue accelerates and more likely runs at its free velocity without loss. On the other side, from starting to move to the last communication for intersection records, the rear vehicles in the queue have more time and practicable distance (is about queue length plus transmission distance) to accelerate than others. Furthermore, the one with the slowest free velocity in the rear can get its free velocity more quickly and then run without disturbance (the vehicles behind have overtaken it almost). Therefore, from respective recording time in the intersection records, we consider both the processes of head vehicle and tail vehicle as acceleration (it is not needed if the vehicle has reached its free velocity) and then running at the free velocity without loss until reaching the range of next intersection or catching the queue ahead.
- Integration and overlap. Once a queue is formed and enters the new street, there are three occurrences we need to notice: the queue catches up with an individual (Fig. 2), an individual catches up with the queue (Fig. 3), and the queue catches up with another queue (Fig. 4). In the first case, the head vehicle overtakes the individual vehicle which means the individual is integrated into the queue, and then we no longer consider it independently. The second case is similar to the first, after the individual vehicle overtakes the tail vehicle in the queue, and then we no longer consider it. Note that the individual vehicle can hardly overtake or be overtaken by all the vehicles in the queue within the distance of usual urban street length, and moreover, there is little probability that its velocity is faster or slower than all the vehicles. In the last case, two queues overlap with each other and are integrated into a new queue. Then, we consider the head vehicle in the queue in front as the new head vehicle and the tail vehicle in the queue behind as new tail vehicle.

As discussed above, in a queue which is generated due to traffic lights at the intersection, only the head vehicle and the tail vehicle can be considered, and between them there is still a connection link under common circumstance. Furthermore at a given time, if the estimated

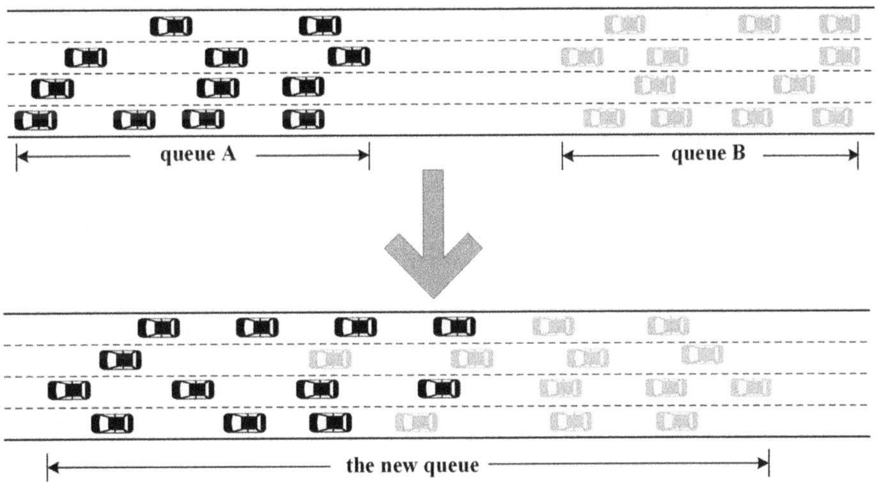

Fig. 4 An example of new queue formation (a queue caught up with another queue)

position of an individual vehicle is in the range between the head and tail vehicles in the queue or even strides over the queue, the individual can be ignored. The pseudo code of the calculation of connectivity probability is shown below.

Algorithm 1 Calculation of Connectivity Probability

$qint$: maximum time interval enabled of consecutive vehicles in a queue
$qnum$: minimum number enabled of vehicles in a queue
n_{head} / n_{tail} : optional range of head/tail vehicle selection
$threshold$: used to decide whether the vehicle is in the next intersection area
m : the number of the vehicles which are not in the next intersection area
Start calculation of connectivity probability
input: intersection records in ISN, current time t
analyze records in ISN, differentiate between individuals or queues according to $qint$ and $qnum$
tag individual and queue vehicles respectively
for every queue
{ **for** i in n_{head} vehicles (in the front of the queue), tag the one with maximum free-velocity as head vehicle
for i in n_{tail} vehicles (in the rear of the queue), tag the one with minimum free-velocity as tail vehicle
omit other vehicles in the queue
}
calculate estimated position pos_{est} at t for all vehicles
for every individual
{**if** pos_{est} fall in the range of a queue **then** omit it
if pos_{est} passes through a queue backwards or forwards **then** omit it
}
resort vehicles according to pos_{est} as v_i, $i = 1, \ldots, n$
$m = 0$
for $i = n$ to 1
{**if** P_{next} (for V_i) > $threshold$, **then** $m = j$}
if $m = 0$ **then** return 0
$P_c = 1$
for $i = 1$ to m
{**if** (V_i is individual and V_{i+1} is individual) or (V_i is individual and V_{i+1} is tail vehicle) or (V_i is head vehicle and V_{i+1} is individual) **then** $P_c = P_c * P_i$} /* if V_i is tail vehicle and V_{i+1} is head vehicle, then $P_i = 1$ */
return P_c
End

3.3.3 Delivery delay

In a sparse traffic circumstance, sometimes there is probably not an existing connection link in the street. However for delay-tolerant applications, the *carry-and-forward* approach can be adopted, where the vehicle carries the packet when connection does not exist, and forwards the packet when there is an appropriate receiver that appears. The delivery delay which is taken to deliver the packet through the street is commonly constituted by transmission delay and carrying delay. By ignoring the transmission delay which is very small relatively, we consider delivery delay mainly as the corresponding carrying delay. In MMIR, by means of the position estimation of vehicles, we can estimate the connection status over time at equal intervals and then calculate a score of delivery delay for every candidate street. Note that we just give a score to compare for street selection on our original purpose, not to precisely model the delivery delay of packet forwarding in a street, which is a complicated work especially from the microscopic point of view due to many uncertainties. Algorithm 2 describes the process of score calculation: at a certain time if the connectivity probability P_i between two consecutive vehicles is greater than 0.5, it is considered the packet is sent to the front vehicle without increasing carrying delay; on the contrary, the packet is left at the current vehicle and will be judged again at the next moment (e.g., next second); when the packet arrived in the range of the next intersection, the ratio of the time spent to the expiration time is the score of delivery delay. If the expiration time ran out and the packet cannot arrive at the next intersection, the score is set as 0. In MMIR, the street which has a higher $\mathrm{score}_{\mathrm{delay}}$ is regarded as that with lower delivery delay relatively for forwarding packets.

Algorithm 2 Calculation of Delivery Delay Score

t_{exp} : the expiration time of the packet
v_{pkt} : the vehicle which is carrying the packet
Start calculation of $score_{delay}$
$t = 0$
while ($t < t_{exp}$ and $pos_{v_{pkt}}$ is not in the range of next intersection area)
{calculate the estimated positions of vehicles at time t
m = the order of v_{pkt}
for $i = m$ to n
{**if** $P_i > 0.5$, **then** $v_{pkt} = v_{i+1}$
else break
}
if $pos_{v_{pkt}}$ is in the range of next intersection, **then break**
$t = t + 1$
}
return $1 - t / t_{exp}$
End

In Section 3.3, we discussed the estimation of connectivity and delay for a street in MMIR. Thus, like [33], combining the distance to the destination, we conclude the calculations of the total score for street selection, (12) and (13), which focus on connectivity probability and delivery delay respectively:

$$\mathrm{socre}_{\mathrm{con}}(\mathrm{street}_i) = \alpha_1 \cdot \left(1 - \frac{d_i}{d_{\mathrm{cur}}}\right) + \beta_1 \cdot P_c, \tag{12}$$

$$\mathrm{socre}_{\mathrm{del}}(\mathrm{street}_i) = \alpha_2 \cdot \left(1 - \frac{d_i}{d_{\mathrm{cur}}}\right) + \beta_2 \cdot \mathrm{score}_{\mathrm{delay}}, \tag{13}$$

where d_i is the curve metric distance from the next intersection of the candidate street to the destination and it should be less than d_{cur} which is the distance from the current intersection to the destination. α_1, α_2, β_1, and β_2 are weighting factors for the distance and connectivity or delay respectively with $\alpha_1 + \beta_1 = 1$ and $\alpha_2 + \beta_2 = 1$. The candidate street with higher score is preferred here. Note that if the destination of packet delivery is in a candidate street, this direction will be chosen without calculation.

3.3.4 Improvement and adjustment for connectivity calculation

Forwarding packets during the vehicles with the same running direction can enhance the stability of network connection. However, in a two-way street, sometimes packets can be relayed by the opposing vehicles to improve connectivity, which may also be considered in MMIR. In advance, we can get the average value of spatial density λ_{opp} in opposing directions over a recent period of time from the intersection records in the other side of the candidate two-way street. In the connectivity calculation between two consecutive vehicles, if the distance ($d = \mu_{\text{pos}_{i+1}} - \mu_{\text{pos}_i}$) is greater than the transmission range R, and then we can make use of the opposing vehicles to fix the connectivity. We get the number which is needed at least: $n = d/R$, and then recalculate the connectivity probability adding opposing vehicles as

$$P_i^{'} = 1-\left((1-P_i)\cdot\sum_{k=0}^{n-1}\frac{(d\cdot\lambda_{\text{opp}})^k}{k!}e^{-d\cdot\lambda_{\text{opp}}}\right). \quad (14)$$

So far, for most street selection strategies based on monitoring traffic density, there are some social disturbance factors existing. For instance, there is a large residential community on the side of the street. In the morning of working days, many vehicle nodes *appear* in the street and start their trips to work, and after work they pass the intersection connecting the street, come back to the community, and *disappear*. If the motoring point is at the upstream intersection (such as in MMIR), the *disappeared* vehicles will decrease the actual connectivity in the street. On the contrary, if the connectivity evaluation is based on the information gathered at the downstream intersection, the connectivity probability calculated will be larger than the real value due to the *appeared* vehicles. In MMIR, we introduce two solutions to this problem. In consideration of the limitation of length, the details of this study are not given here. First is to utilize the vehicles' trajectory information like [20]: each vehicle's destination position will be acquired at the intersection if its destination is in the candidate street, and according to it we can correct our calculation. Second is to set the *variation* factor by means of the statistics information: the ratios of *appeared* (n_{app}) and *disappeared* (n_{dis}) vehicles to the recorded vehicles (n_{rec}) at the intersection in terms of the street location and the time of day will be used, and the *variation* factor can be set in the form of $(\text{pos}/l) \times (1 + \alpha' \cdot n_{\text{app}}/n_{\text{rec}} - \beta' \cdot n_{\text{dis}}/n_{\text{rec}})$, where pos is the position (relative to the current intersection) of the disturbance point such as a large community and factory and l is the length of the street.

4 Parameter setting

In this section, we determine the threshold values: arriving time interval and the number of vehicles which satisfy the requirement of a queue in MMIR. As interpreted in Section 3.3.2, the generation of a queue in MMIR needs to satisfy two conditions: the arriving time interval between every two consecutive vehicles in the queue is short enough to avoid a communication break occurring (i.e., the distance between any two consecutive vehicles is larger than the transmission range after driving a while), the number of queue vehicles is large enough to differ from the individual vehicles. Therefore, to determine the threshold value of them, we simulate a scenario using SUMO (Simulation of Urban Mobility) [43] which is an open source program to generate realistic vehicular mobility, and the parameters are shown in Table 2. There is a street with 1-km length and 4 lanes in a single direction which is the common environment in urban arterial streets. The free-velocities of vehicles are not the same and follow a normal distribution N (70, 10.5) KPH (kilometers per hour). The value of sigma in SUMO which describes the random influence on velocity from the driver imperfection (i.e., uncertain factors) is set as 0.5 to achieve realistic vehicle behavior. The transmission range of each vehicle is set as 250 m. In the simulation, the vehicles are injected from one side and travel through the street with different combinations of the average arrival time interval and total number. For every combination, we repeated sufficient times to compute the mean broken rate which is the proportion of disconnecting time (the multi-hop connection from the head vehicle to the last vehicle is broken) to the queue existence time (between the time when the last vehicle entered and the time when the head vehicle left the street), and the average lost time due to driving slower than desired velocities of all vehicles.

As expected in Fig. 5, the broken rate in the queue increases as the arrival time interval increases. When it is set at 2 s or less, whether the number of vehicles is 12, 30, or 50, the broken rate can be controlled less than 1%. However, when it is set at 2.5 s or above it, the

Table 2 Simulation parameters

Parameters	Value
Street length	1 km
Transmission range	250 m
Vehicle free-velocity	$v \sim N$ (70, 10.5) KPH
Sigma (driver imperfection)	0.5
Number of vehicles	12/30/50
Average time interval	0.5/1/1.5/2/2.5/3 s

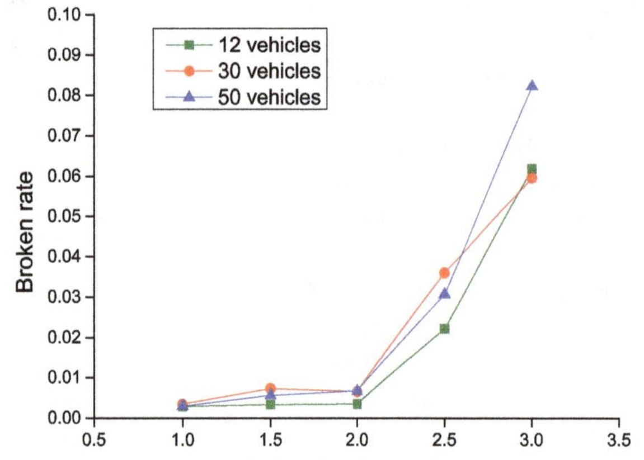

Fig. 5 Broken rate vs. arrival time interval

broken rate increases roughly. And Fig. 6 shows that the average lost time is inversely proportional to the number of queue vehicles. As we know, the more vehicles there are in a queue, the more interactions to the velocity with each other the vehicles have and then the slower they drive. When the number is 8, the average lost time is less than 1 s which means 8 vehicles do not need to be considered a queue but individuals. In order to classify the vehicles (individual and the queue) appropriately and avoid excessive queues, combining of the actual feature of the street (e.g., the urban arterial street), we set qint to 2 s which is the maximum time interval enabled of consecutive vehicles in a queue and set qnum to 12 which is minimum number allowed of the vehicles in a queue. And we use these parameters in the following simulations. Note that with different street conditions (e.g., the number of street lanes, velocity limit.), there are different values of qint and qnum we should adopt.

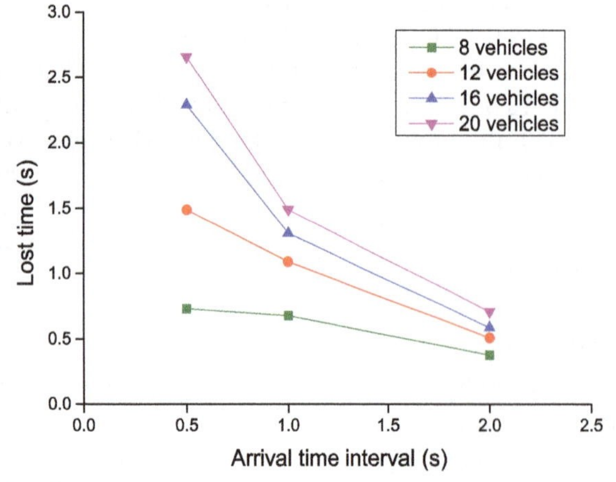

Fig. 6 Average lost time vs. arrival time interval

5 Results and discussion

In this section, we evaluate the performance of MMIR. In terms of connectivity probability, we run the simulation to verify its accuracy. And in an intuitive simulation test about street selection for routing, the estimated delivery delay in MMIR is evaluated and compared with two classical methods based on traffic statistics and GyTAR, respectively. The traffic simulations are conducted with SUMO and the trace files are injected into OMNet++ tools [44] to analyze.

5.1 Accuracy of connectivity probability

To evaluate our algorithm of connectivity probability, in the simulation, we set two intersections and a street connected them. We adopt some of the same parameters in Table 2 here (e.g., street length, vehicle velocity, and sigma), and the vehicles through the street are deployed as different degrees of traffic flow (50/100/150/200/250/300 per lane per hour) to test and verify the accuracy of the connectivity probability. The test packet is generated per 10 s and sent to another intersection relayed by the vehicle nodes in a single direction. In the meantime, we calculate the street connectivity probability P_c for the street. And if the packet reaches the destination without carrying delay which means every distance between two consecutive vehicles is smaller than the transmission range, the multi-hop network between two intersections is connected at the moment, and we set the value of real connectivity as 1, otherwise 0. As the evaluation indicator, we adopt probability deviation which equals $|\text{real connectivity} - P_c|$. Obviously, the smaller value of probability deviation means a more accurate prediction of the real-time street connectivity, vice versa. The total simulation time is 3000 s, and we gather the data from 500 s to ensure the traffic state has reached stability. For MMIR, in advance, to obtain the free-velocity data of vehicles, we performed the *free-flow traffic* test to simulate the light traffic environment to help us to collect the data for free-velocity calculation. Furthermore, to investigate the impact of traffic lights on the connectivity calculation, the simulations are performed in two scenarios, respectively: no traffic lights and existing traffic lights, of which the period is 150 s. Furthermore, the threshold in Algorithm 1 is set to 0.8, the n_{head} and n_{tail} are set to 8.

In Fig. 7, there are no traffic lights at the intersection. We can learn that the probability deviation is proportional to the traffic flow. It ranges from 0 to 0.13, while the traffic flow increases from 50 to 300. The reason is that the more vehicles there are in the street, the more interactions they have, and then they add more uncertainty to the connectivity. Moreover, when the traffic flow gets its critical value (about 150) to just support the end-to-end connection through the street statistically (i.e., the real connectivity value is swung between 0 and

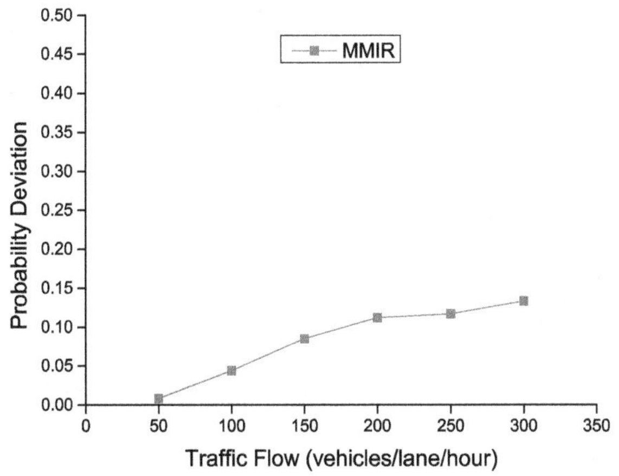

Fig. 7 Connectivity probability deviation vs. traffic flow (no traffic lights)

1 frequently), the probability deviation is about 0.09 and does not have fluctuation. It shows its accuracy and real-time performance. Figure 8 illustrates the variation under the condition of the traffic lights existing at the intersection, which means the traffic flow is separated by the lights, and even though there are more vehicles in the street to support the network connectivity, it will still have the breaks. However, MMIR displays similar results like in the scenario of no traffic lights. Its probability deviation ranges from 0 to 0.12. It is due to the reason that MMIR records every vehicle's entrance time directly (i.e., when they passed the intersection and traffic lights), instead of only the statistical traffic flow. In consideration of many uncertain influential factors (we set sigma in SUMO as 0.5) to the real-time connectivity, the result of MMIR is accurate enough.

5.2 Analysis of estimated delay

In order to evaluate the performance of estimated delivery delay in MMIR directly, we built a simple and intuitive scenario. It includes three intersections: I1, I2, and I3. I2 and I3 connected I1 by street A and B which have the same condition (e.g., the number of lanes, street length). The traffic lights are set at these intersections. There is a static sender node at the center of I1, and two static receiver nodes are deployed at I2 and I3, respectively. At regular intervals, the sender generates the test packet including its id and transmission time to the receivers. The TTL (time-to-live) of the packet is set to 100 s. When the packet arrives at the receiver node at I2 and I3, the time will be recorded to check which street has a shorter delivery delay and then is the better routing choice. Before the transmission of the test packet, we use the estimated delivery delay of MMIR to choose the better street and mark the corresponding packet. If the marked packet arrives at its receiver node earlier than another one, the selection is correct. Note that, due to the same condition, the optimal selection from street A and B depends mainly on their respective connectivity in the whole packet transmission time (i.e., $\alpha_2 = 0$ and $\beta_2 = 1$ in Eq. 13). The pair of average traffic flows in street A and B are deployed as the same and different degrees to evaluate the performance respectively. The detail values of the simulation parameters are shown in Table 3.

For comparison, we introduce two classical street selection methods from various intersection-based routings. The first is a classical connectivity model [21] which uses the statistic traffic information like traffic flow and average velocity to calculate the network connectivity in the candidate street and then chooses the best one to forward the packet. For every calculation, we use the statistical data in the past 300 s before that moment in our simulation. And we called this method as statistical model in this paper. The similar method for street selection is also adopted in VADD. The second is GyTAR in which the forwarding node at the intersection assigns a score to each candidate street considering the

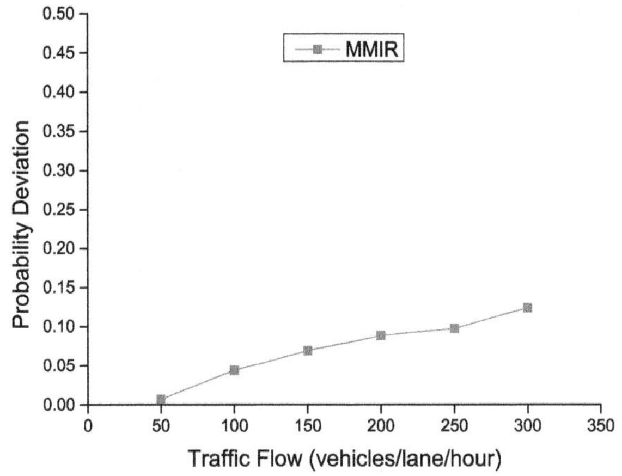

Fig. 8 Connectivity probability deviation vs. traffic flow (with traffic lights)

Table 3 Simulation parameters

Parameters	Value
Streets length	1 km
Transmission range	250 m
Vehicle free-velocity	$v \sim N$ (70, 10.5) KPH
Sigma (driver imperfection)	0.5
Traffic flow	50/150/200/250/300 vehicles/lane/h
Simulation time	3000 s
Beacon interval	1 s
Test packet sending rate	10 s
Test packet's TTL	100 s
Traffic lights' period	160 s

traffic density and curve metric distance to destination. The street with the highest score is selected to forward the packet. The information about traffic density in the street is gathered by its dedicated control packets—CDP (cell density packets) which are generated by the dynamic vehicles in the next intersection regularly and traverse the street to the current intersection. Note that as well as MMIR, due to the same street condition as we set, the optimal selection from street A and B depends mainly on their respective connectivity (i.e., traffic density and vehicles' distribution) in GyTAR. Then, we can compare their accuracy of the street selection with MMIR by connectivity-related metric directly.

As shown in Fig. 9, we deploy five combinations of average traffic flows to street A&B as 200&150, 200&100, 250&100, 250&50, and 300&50 (the differences between A&B are 50, 100, 150, 200, and 250 respectively) in order to observe the performances of three street selection methods. The accuracy of the statistical model increases incrementally when the difference increases. Since it selects the street with higher flow actually based on the traffic information of the past period. And the higher average traffic flow means the lower delivery delay probabilistically. On the contrary, GyTAR and MMIR do not display the similar change. As we know, they use the practical measured information to verify the current status as possible, and then their decisions are rarely influenced by the non-real-time statistics. MMIR achieves a higher accuracy that all the results are greater than 80%, thanks to its capability to estimate the end-to-end delay directly making use of the effective individual information in the intersection records. GyTAR does not perform as well as MMIR due to the relatively low updating rate of real-time traffic state which is about 250 m (transmission range)/average velocity (for details, please refer to [33]). Sometimes, it cannot follow the rapid change of vehicles' distribution caused by the differences in vehicles' velocities and the alteration of traffic lights.

In the next set of tests, the traffic flows with the same degree are deployed to street A&B as 50, 100, 150, 200, 250, and 300. Besides the selection accuracy in Fig. 10, we provide the delivery delay in Fig. 11 to show the performance difference among three methods precisely. The statistical model has the longest delivery delay and does not perform as well as it in the last set. Street A and B have the same average traffic flow, so that the macroscopic statistical data cannot further support it to make the correct choice between street A and B. GyTAR displays the similar results when the traffic flow is large relatively. But when it is 50 or 100, the accuracy is lower and the delivery delay is relatively longer. Since, in those cases, there are fewer vehicles which are not enough to form an end-to-end connection through the whole street in many times. Moreover, the existing traffic lights increase the probability of broken time further. Inevitably, the CDPs do not reach the current intersection regularly and cannot update the traffic information timely. Based on the unfresh data, GyTAR is difficult to make an accurate decision well. As can be seen, MMIR still achieves a high accuracy (greater than 80%) and outperforms others in delivery delay at every level of traffic flow.

6 Conclusions

In this paper, we study street selection in the intersection-based routing for urban vehicular ad hoc networks. We show that existing methods and models which utilize macroscopic information are not suitable for VANETs with high mobility and rapid topology changing. In summary, macroscopic data (e.g., traffic density, average velocity) can be used to make a good

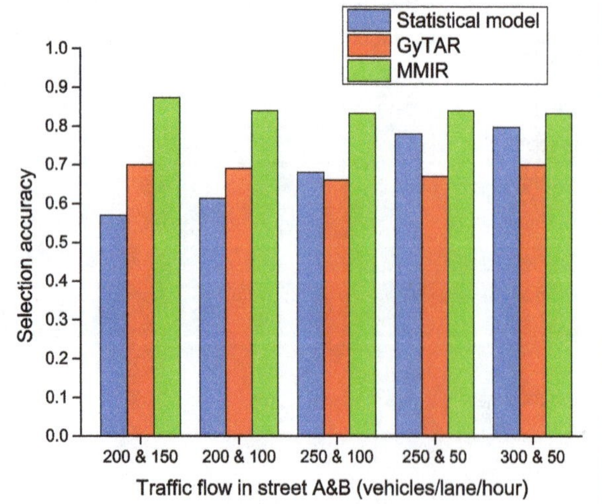

Fig. 9 Selection accuracy vs. different degrees of traffic flow in street A&B

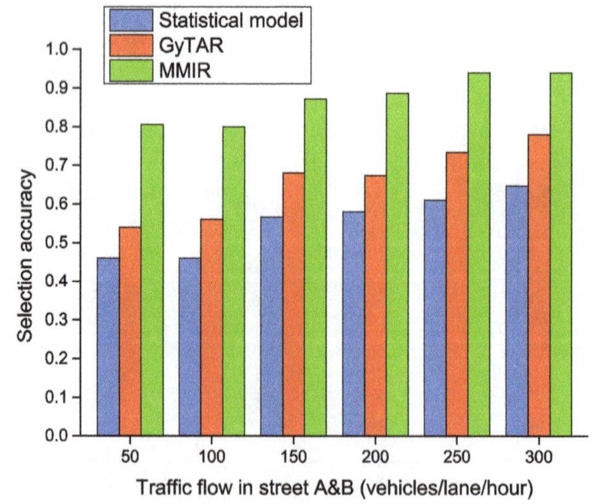

Fig. 10 Selection accuracy vs. same degrees of traffic flow in street A&B

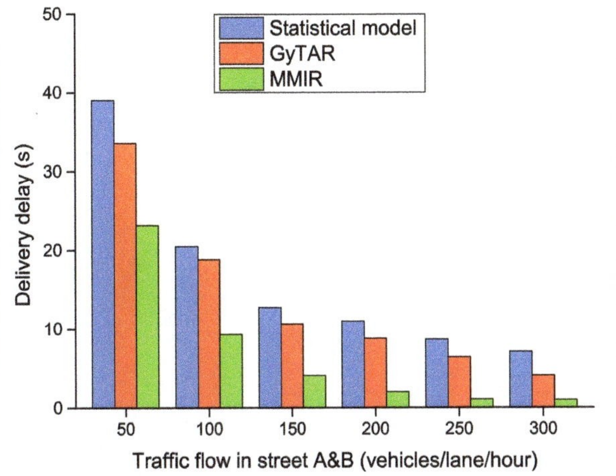

Fig. 11 Delivery delay vs. same degrees of traffic flow in street A&B

decision only for the general condition not every concrete condition and moment actually. To address this problem, we proposed a microscopic mechanism based on intersection records (MMIR), which makes use of vehicle' individual information recorded at the intersection to estimate their current positions and calculate the connectivity probability or estimated delay for candidate streets. The simulation results show that in terms of connectivity probability and delivery delay, MMIR provides an accurate estimation and outperforms existing schemes. In the future, based on the microscopic mechanism, we will improve our method (e.g., take more microscopic individual factors into consideration) and support more metrics satisfying the quality of service in urban VANET routings.

Abbreviations

CDP: Cell density packets; ISN: Intersection-server node; KPH: Kilometers per hour; LoRa: Long range; LPWA: Low power wide area; MANETs: Mobile ad hoc networks; NB-IoT: Narrowband Internet of Things; TTL: Time-to-live; VANETs: Vehicular ad hoc networks

Authors' contributions

ZC is the main writer of this paper. He proposed the research ideas, designed the scheme, and completed the writing. ML contributed to the conception and design of the study. ZC and QS performed the simulations and data analysis. All authors read and approved the final manuscript.

Authors' information

Zhen Cai received a BS degree in Computer Science and Technology, and a MS degree in Computer Application Technology from North China Electric Power University, in 2003 and 2010, respectively. He is currently a Ph.D. candidate at the Institute of Information Science, Beijing Jiaotong University. His research interests include vehicular ad hoc networks, wireless sensor networks, and network behavior.
Mangui Liang received his Ph.D. in 1998. Currently, he is a Professor and Ph.D. supervisor at the Institute of Information Science, Beijing Jiaotong University, Beijing, China. His research interests include vector network, next-generation network, and speech processing.
Qiushi Sun received a M.S. degree in Computer Science from Beijing Jiaotong University, Beijing, China, in 2015. He is currently a Ph.D. student at the Department of Computer Science, Beijing Jiaotong University, Beijing, China. His research interests include traffic flow theory, machine learning, and biological image analysis.

Funding

This research was supported by the NSFC-Joint Funds for Basic Research of Generic Technology (U1636109) and the National High Technology Research and Development Program 863 (2007AA01Z2035).

Competing interests

The authors declare that they have no competing interests.

Author details

[1]Institute of Information Science, Beijing Jiaotong University, Beijing 100044, People's Republic of China. [2]Beijing Key Lab of Traffic Data Analysis and Mining, Beijing Jiaotong University, Beijing 100044, People's Republic of China.

References

1. A.T. Giang, A. Busson, A. Lambert, et al., Spatial capacity of IEEE 802.11p based VANET: Models, simulations and experimentations. IEEE Trans. Veh. Technol. **65**(8), 6454–6467 (2016)
2. X. Cheng, Q. Yao, M. Wen, et al., Wideband channel modeling and intercarrier interference cancellation for vehicle-to-vehicle communication systems. IEEE J Sel Areas in Commun **31**(9), 434–448 (2013)
3. I Guler, MENENDEZ, Monica, et al., Using connected vehicle technology to improve the efficiency of intersections. Transp. Res. C 46(46), 121–131(2014)
4. D.B. Johnson, D.A. Maltz, Dynamic source routing protocol for mobile ad hoc networks. Mob Comput **353**(1), 153–181 (1996)
5. C.E. Perkins, E.M. Royer, in Proceedings of the Second IEEE Workshop on Mobile Computing Systems and Applications (WMCSA'99). Ad-hoc on-demand distance vector routing (IEEE, New Orleans, 1999), pp. 90–100
6. P. Jacquet, P. Muhlethaler, T. Clausen, et al., in *Proceedings of IEEE International Multi Topic Conference*, 2001. Optimized link state routing protocol for ad hoc networks (Lahore, Pakistan, 2001), pp. 62-68
7. B. Karp, H.T. Kung, in Proceedings of Mobile Computing and Networking. GPSR: Greedy perimeter stateless routing for wireless networks (ACM, Boston, 2000), pp. 243–254
8. P.M. Ruiz, V. Cabrera, J.A. Martinez, F.J. Ros, in Proceedings of the Second IEEE International Workshop on Intelligent Vehicular Networks (InVeNET'10). BRAVE: Beacon-less routing algorithm for vehicular environments (IEEE, San Francisco, 2010), pp. 709–714
9. S. Basagni, I. Chlamtac, V.R. Syrotiuk, B.A. Woodward, in *Proceedings of the ACM/IEEE international conference on Mobile computing and networking*. A distance routing effect algorithm for mobility (DREAM) (DBLP, Dallas, 1998), pp. 76–84
10. H. Saleet, R. Langar, K. Naik, R. Boutaba, Intersection-based geographical routing protocol for VANETs: A proposal and analysis. IEEE Trans. Veh. Technol. **60**(9), 4560–4574 (2011)
11. A.P. Mazumdar, A.S. Sairam, TOAR: Transmission-aware opportunistic ad hoc routing protocol. EURASIP J. Wirel. Commun. Netw. **2013**(1), 1–19 (2013)
12. X. Cai, Y. He, C. Zhao, L. Zhu, C. Li, LSGO: Link state aware geographic opportunistic routing protocol for VANETs. EURASIP J. Wirel. Commun. Netw. **2014**(1), 96 (2014)
13. V. Namboodiri, L. Gao, Prediction-based routing for vehicular ad hoc networks. IEEE Trans. Veh. Technol. **56**(4), 2332–2345 (2007)
14. D. Lin, J. Kang, A. Squicciarini, et al., MoZo: A moving zone based routing protocol using pure V2V communication in VANETs. IEEE Trans. Mob. Comput. **16**(5), 1357–1370 (2017)
15. X. Zhang, K. Chen, X. Cao, et al., A street-centric routing protocol based on micro topology in vehicular ad hoc networks. IEEE Trans. Veh. Technol. **65**(7), 5680–5694 (2016)
16. J.J. Chang, Y.H. Li, W. Liao, I.C. Chang, Intersection-based routing for urban vehicular communications with traffic-light considerations. IEEE Wirel. Commun. **19**(1), 82–88 (2012)
17. C Lochert, M Mauve, H Füssler, H Hartenstein, Geographic routing in city scenarios, in Proceedings of the SIGMOBILE Mobile Computing and Communications Review, vol. 9(1), 2005, pp. 69–72
18. Q. Yang, A. Lim, S. Li, J. Fang, P. Agrawal, ACAR: Adaptive connectivity aware routing for vehicular ad hoc networks in city scenarios. Mob Netw Appl **15**(1), 36–60 (2010)
19. J. Zhao, G. Cao, in Proceedings of the 25th IEEE International Conference on

Computer Communications (IN-FOCOM 2006). VADD: Vehicle-assisted data delivery in vehicular ad hoc networks (IEEE, Barcelona, 2006), pp. 1–12
20. J. Jeong, S. Guo, Y. Gu, T. He, D.H.C. Du, Trajectory-based statistical forwarding for multihop infrastructure-to-vehicle data delivery. IEEE Trans. Mob. Comput. **11**(10), 1523–1537 (2012)
21. S. Panichpapiboon, W. Pattara-Atikom, Connectivity requirements for self-organizing traffic information systems. IEEE Trans. Veh. Technol. **57**(6), 3333–3340 (2008)
22. R.P. Roess, E.S. Prassas, W.R. Mcshane, Traffic Engineering, 4th edn. (Pearson/Prentice Hall, New Jersey, 2011)
23. Z. Li, Y. Song, J. Bi, CADD: Connectivity-aware data dissemination using node forwarding capability estimation in partially connected VANETs. Wirel. Netw **25**(1), 379–398 (2019)
24. S. Yousefi, E. Altman, R. El-Azouzi, M. Fathy, Analytical model for connectivity in vehicular ad hoc networks. IEEE Trans. Veh. Technol. **57**(6), 3341–3356 (2008)
25. Y. Al-Mayouf, N. Abdullah, O. Mahdi, et al., Real-time intersection-based segment aware routing algorithm for urban vehicular networks. IEEE Trans. Intell. Transp. Syst. **19**(7), 2125–2141 (2018)
26. Q. Ding, B. Sun, X. Zhang, A traffic-light-aware routing protocol based on street connectivity for urban vehicular ad hoc networks. IEEE Commun. Lett. **20**(8), 1635–1638 (2016)
27. K.C. Lee, M. Le, J. Harri, M. Gerla, in Proceedings of the Vehicular Technology Conference. LOUVRE: Landmark overlays for urban vehicular routing environments (IEEE, Calgary, 2008), pp. 1–5
28. H. Lee, Y. Lee, T. Kwon, Y. Choi, in Proceedings of the Wireless Communications and Networking Conference. Virtual vertex routing (VVR) for course-based vehicular ad hoc networks (IEEE, Kowloon, 2007), pp. 4405–4410
29. X. Zhang, X. Cao, L. Yan, K.S. Dan, A street-centric opportunistic routing protocol based on link correlation for urban VANETs. IEEE Trans. Mob. Comput. **15**(7), 1586–1599 (2016)
30. Y. Ding, L. Xiao, SADV: Static-node-assisted adaptive data dissemination in vehicular networks. IEEE Trans. Veh. Technol. **59**(5), 2445–2455 (2010)
31. J. Nzouonta, N. Rajgure, G. Wang, C. Borcea, VANET routing on city roads using real-time vehicular traffic information. IEEE Trans. Veh. Technol. **58**(7), 3609–3626 (2009)
32. Y.S. Chen, Y.W. Lin, C.Y. Pan, DIR: Diagonal-intersection-based routing protocol for vehicular ad hoc networks. Telecommun. Syst. **46**(4), 299–316 (2011)
33. M. Jerbi, S.M. Senouci, Y. Ghamri-Doudane, T. Rasheed, Towards efficient geographic routing in urban vehicular networks. IEEE Trans. Veh. Technol. **58**(9), 5048–5059 (2009)
34. C. Guo, D. Li, G. Zhang, Z. Cui, Data delivery delay reduction for VANETs on bi-directional roadway. IEEE Access **4**, 8514–8524 (2016)
35. D. Yin, T. Qiu, Compatibility analysis of macroscopic and microscopic traffic simulation modeling. Can. J. Civ. Eng. **40**(7), 613–622 (2013)
36. U. Raza, P. Kulkarni, M. Sooriyabandara, Low power wide area networks: An overview. IEEE Commun Surv Tutorials **19**(2), 855–873 (2017)
37. O. Georgiou, U. Raza, Low power wide area network analysis: Can LoRa scale? IEEE Wireless Commun Lett **6**(2), 162–165 (2017)
38. V. Petrov, A. Samuylov, V. Begishev, et al., Vehicle-based relay assistance for opportunistic crowdsensing over narrowband IoT (NB-IoT). IEEE Internet of Things Journal 5(5), 3710-3723 (2018)
39. C. Wu, Y. Ji, F. Liu, et al., Toward practical and intelligent routing in vehicular ad hoc networks. IEEE Trans. Veh. Technol. **64**(12), 5503–5519 (2015)
40. S.M. Zaki, M.A. Ngadi, S.A. Razak, Location service protocol for highly mobile ad hoc network. Arab. J. Sci. Eng. **39**(2), 861–873 (2014)
41. H.J. Lim, T.M. Chung, in Proceedings of the 11th International Conference on Algorithms and Architectures for Parallel Processing. A survey on privacy problems and solutions for VANET based on network model (Springer-Verlag, Heidelberg, 2011), pp. 74–88
42. Transportation Research Board (ed.), *HCM2010 : Highway Capacity Manual*, 5th edn. (2010)
43. D Krajzewicz, Traffic simulation with SUMO - simulation of urban mobility. in Fundamentals of Traffic Simulation, Ed. by Jaume Barcelópp. International Series in Operations Research & Management Science, 145 (Springer, New York, 2010), p. 269–293
44. C. Sommer, I. Dietrich, F. Dressler, Simulation of ad hoc routing protocols using OMNeT++. Mob Netw Appl **15**(6), 786–801 (2010)

9

Selfish node detection based on hierarchical game theory in IoT

Solmaz Nobahary[1], Hossein Gharaee Garakani[2*], Ahmad Khademzadeh[2] and Amir Masoud Rahmani[1]

Abstract

Cooperation between nodes is an effective technology for network throughput in the Internet of Things. The nodes that do not cooperate with other nodes in the network are called selfish and malicious nodes. Selfish nodes use the facilities of other nodes of the network for raising their interests. But malicious nodes tend to damage the facilities of the network and abuse it. According to reviews of the previous studies, in this paper, a mechanism is proposed for detecting the selfish and malicious nodes based on reputation and game theory. The proposed method includes three phases of setup and clustering, sending data and playing the multi-person game, and update and detecting the selfish and malicious nodes. The process of setup and clustering algorithm are run in the first phase. In the second phase, the nodes of each cluster cooperate with each other in order to execute an infinite repeated game while forwarding their own or neighbor nodes' data packets. In the third phase, each node monitors the operation of its neighbor nodes for sending the data packets, and the process of cooperation is analyzed for determining the selfish or malicious nodes which forwarded the data packets with delay or even not sent them. The other nodes reduce the reputation of the nodes which does not cooperate with them, and they do not cooperate with the selfish and malicious nodes, as punishment. So, selfish and malicious nodes are stimulated to cooperate. The results of simulation suggest that the detection accuracy of the selfish and malicious nodes has been increased by an average of 12% compared with the existing methods, and the false-positive rate has been decreased by 8%.

Keywords: Internet of Things (IoT), Selfish node, Malicious node, Game theory

1 Introduction

Nowadays, the Internet of Things (IoT) is introduced as a global infrastructure to establish communication between the physical and virtual worlds via the available technologies. The IoT is an intelligent network, and things attempt to transfer information through the network equipment. Its applications impress all human life aspects, including smart cities, smart environment, smart water control, security and emergency, smart transportation, smart agriculture, industrial control, and health. Its purposes are to facilitate the works and increase the quality of life. The IoT is popular for its capability to connect different kinds of things to the virtual world, and sensing data from different detectors are sent to the center [1–4]. Due to the improvement in wireless communications, it becomes a way to send and receive the data packets in IoT. However, the low range of wireless communications makes it possible for multi-hop communications, and the life of these communications depends on the cooperation of each node [5, 6].

One of the most important challenges is the lack of cooperation in some nodes due to the connection between the things and the Internet in the IoT network when sending data in multi-hop communications [7, 8]. Such nodes are called selfish nodes. The selfish nodes use the network facilities for personal purposes and only send their own data packets but do not help to forward the other neighbor nodes' data packets to save their energy power. The other group of nodes is called malicious nodes which tend to harm and exploit the network facilities. By increasing the number of such nodes, the network throughput and lifetime will reduce, and the energy consumption, average end-to-end delay, and network traffic will increase. As a result, it will disturb the network operation [7–9].

* Correspondence: gharaee@itrc.ac.ir
[2]Iran Telecom Research Center (ITRC), Tehran, Iran
Full list of author information is available at the end of the article

To overcome the side effects of the selfish and malicious nodes, it is necessary to detect and identify them. To this purpose, different strategies have been proposed, and the reputation-based method is one of the most popular methods in which each node receives a specific reputation according to the nodes' feedback. The nodes with higher reputations are recognized by the network as more reliable nodes, and the nodes with lower reputation are known as selfish nodes. The throughput of these methods is low, the energy consumption is high, the selfish and malicious nodes can collude, there are no punishments or incentives in the selfish nodes, and no second chances are given to the selfish or malicious nodes to cooperate with other nodes [10–15]. The other group of strategies discover the selfish and malicious nodes in the credit-based methods in which the nodes should pay the cost to send the data packets and/or the nodes trade the data packet and sell it at a higher price when they have purchased a packet. Collision attack, lack of punishment, and incentives are the disadvantages of these methods [16–22] (Table 1). The acknowledgment-based methods guarantee to send a packet of a node by using an acknowledgment message. In these methods, the throughput is low; it suffers from the collusion of the selfish and malicious nodes; it has high overhead (communicative, data packet, etc.) and end-to-end delay increases in the network due to the high traffic generated by the acknowledgment messages [23–26]. Another method to detect selfish nodes is game theory-based methods. The game theory is an applied mathematical theory that models and analyzes the systems where each individual attempts to find the best strategy selected by others to reach success [27]. Game theory-based approaches take advantage of the incentive mechanisms through payoffs and have lower false-positive rate and overhead (time, supervisory, hardware) compared to the other aforementioned methods in the network to detect the selfish and malicious nodes. However, the prior game theory-based methods have less throughput and more end-to-end delay, and the proposed methods attempt to increase these parameters [28–31].

The proposed mechanism is a multi-step method based on reputation and game theory for the stimulation of selfish and malicious nodes in Internet of Things, and the mechanism has been designed in three steps including setup and clustering, sending data and playing the multi-person game, and update and detecting selfish and malicious nodes. In the first step, a set of things are placed in a cluster due to being communicated with the destination base station, and they choose the base station as a cluster head. In the second step, the nodes cooperate with each other to forward the data packets to the cluster head, and for this purpose, during sending and forwarding their own or the neighbor nodes' data packets, they run a multi-person game and the results of this game is sent to the next step and determine the reputation of each of the nodes. In the third phase, the reputation of each node is assigned a value in the network and updated by its neighbor node. Each of the nodes whose reputation value is less than the predetermined threshold is detected as a selfish or malicious node. In the following, we highlight the novel contributions of our paper:

- The proposed method is a hybrid method which takes advantage of reputation-based method and game theory-based method to detect and stimulate the non-cooperation nodes. The strategy assigns a reputation value earned by playing the game to all nodes. The nodes with a low reputation cannot be active and send the data packet so the node tries to earn more reputation. The nodes that want to earn reputation should cooperate with other nodes, and it means stimulation of the nodes.
- If the reputation of a node is not less than the threshold, the mentioned node is provided the opportunity of cooperating with other nodes.

Table 1 Advantages and disadvantages of systems

Systems	Advantages	Disadvantages
Reputation-based	High throughput Less end-to-end delay High detection rate Less channel traffic	High energy consumption High overhead and complex systems No robustness against collusions The high false-positive rate
Credit-based	Less channel traffic High throughput Less end-to-end delay	No robustness against collusions No second chance Less detection rate Less energy consumption
Acknowledgment-based	The less false-positive rate High detection rate	High channel traffic Less throughput High end-to-end delay
Game theory-based	Less end-to-end delay High detection rate Less channel traffic	No second chance High energy consumption The high false-positive rate Less throughput

- We propose a multi-phase method to detect non-cooperation nodes based on the multi-person game in each round and prevent resending the data packets to such nodes. Then, it does not have high energy consumption. The network throughput is increased due to the source node that does not need to resend the data packets. Also, less traffic is caused by not resending the data packets, and it leads to decreased average of end-to-end delay of the data packets in the network. We have a lower false-positive rate and false-negative rate and a higher detection accuracy of selfish nodes. It causes the throughput of the networks to increase, and resending the data packets is prevented which might be needed due to failure of the data packets to reach the destination.
- We carry out several simulations to survey the different metrics of the theoretical results.

The remaining parts of the paper are as follows: Section 2 addresses the related works. In Section 3, the system model is described, and the multi-person repeated game is formalized. The strategy of detecting the co-operation stimulation node is presented in Section 4. The performance evaluation is addressed in Section 5. The conclusion is presented in Section 6.

2 Related works

Several approaches have been developed to detect non-cooperative nodes and stimulate them to cooperate with other nodes in the network. These approaches, based on their nature, are classified reputation-based approaches, credit-based approaches, acknowledgment-based approaches, and game theory-based approaches. In reputation-based methods, network nodes cooperate with each other to provide feedback for a set of particular nodes. Each node updates the assigned reputation value with respect to its feedback. The nodes that have more reputation value are recognized as cooperative nodes, and the nodes have a lower reputation are recognized as non-cooperative nodes. The most popular approach is the watchdog method based on the reputation mechanism [10]. The approach is proposed for fair cooperation nodes in MANET [11]. A node is used as a reputation manager called CONFIDANT, and it is responsible for maintaining the credibility of watchdog nodes and pathrater. The Observation-based Cooperation, Enforcement in Ad hoc Networks (OCEAN) approach is the DSR protocol developed based on the reputation-based approach and monitoring methods [12]. An intelligent central organization approach called the Separation of Detection Authority (SDA) is designed to consider the reliability of the network [13]. A Payment Punishment Scheme (PPS) is proposed to send messages, monitor, and report the neighbor nodes by using three watchdogs and stimulate them to cooperate with the network [14]. The method has clustered the nodes, and the cluster head applies the modified Extended Dempster-Shafer model by using watchdogs to detect the selfish node. A trust model approach is proposed based on the Dempster-Shafer evidence theory [15]. The proposed method is a clustered approach based on the direct trust of nodes to each other and the indirect trust of the neighboring nodes based on the Dempster-Shafer evidence theory. The cluster's nodes send their trust values about other neighboring nodes to their cluster head. The cluster head uses Dempster-Shafer theory to compute the indirect trust, and if a node trust is less than the threshold, it will be known as a non-cooperative node and isolated in the network.

In credit-based, if the nodes have a data packet to send, they pay for it, or the nodes trade their data packets between themselves and sell it at a higher price after buying a packet. The approach is proposed to detect the selfish nodes in the network using Nuglets [16]. It is the combination of a packet purse model (PPM) and a packet trade model (PTM) by the credit-based approaches [17, 18]. A credit-based approach called SPRIT (s simple, cheat-proof, credit-based system) is presented to stimulate the nodes to cooperate with other nodes [19]. I each node after receiving a message, the receipt of the message will store in the node's memory reporting to the credit clearance service (CCS) by transferring receipts for each sent/received message. The SPRITE approach has been improved, called MODSPRITE [20]. In this approach, if a node receives a message, it will store the receipt of the message and then communicates with the cluster head, which is responsible for providing credit and charging it for the other cluster member nodes. The new Nuglet approach is a combination of PPM and PTM approaches to identify selfish nodes in the network [21, 22]. In this approach, some virtual currency is generated by the source node using the PPM approach, and it is traded between nodes by the PTM approach until the content reaches zero.

In acknowledgment-based methods, it ensures sending a packet to a node using an acknowledgment message. Balakrishnan et al. have developed TWOACK to detect selfish nodes in the network [23]. Each intermediate node is sent a TWOACK message with a specific packet identifier to the previous node, and it continues until the packet is received by the destination node. The S-TWOACK scheme is actually an improved TWOACK method [24]. It sent an acknowledgment packet after receiving a certain number of the data packets. The EAACK method consists of three main partitions: ACK, S-ACK, and MRA malicious authentication [25]. It is basically an end-to-end acknowledgment model. For all three consecutive nodes on the path, the third node must send an acknowledgment S-ACK packet to the first node, but the source node does not immediately rely on a misbehaving report and needs to change its MRA state and approve misbehavior reporting. Each of the three EAACK sections uses a digitally signed digital signature and retrieves the message. Bounouni et al. proposed an approach consists of four models [26]. The monitoring model is responsible for controlling the sending of routing packets and data packets by using the acknowledgment packet. The reputation model evaluates each nodes' neighbors. Stimulator model manages and updates nodes' credit accounts, and malicious and selfish nodes are punished by an isolator model.

Game theory is an applied mathematical theory, it models and analyzes systems in which each person tries to find the best strategy that has been chosen by others to find success [27]. It is primarily used in economics to model competition between companies. The game consists of a principle and the finite set of players as $N = \{1, 2 \ldots n\}$. Each of them chooses a $s_i \in S_i$ strategy aimed at improving the utility function $U_i\ (s)$: S → R denotes the sensitivity of each player to everyone's actions. The game theory has been classified into cooperation/non-cooperation games, dynamic/static games, repeated/one-interaction games, finite/infinite games, and n-person/two-person games. An approach based on a dynamic auction framework, non-cooperative, and finite-repetition game theory is presented based on the second-lowest price [28]. In the approach, the source node is trying to find a route with the lowest cost to send packets and at auction uses the second-lowest payment bid. A dynamic and self-learning repeated game was proposed to improve transmission efficiency by considering the non-cooperative network nodes [29]. In this approach, each node has two stages of decision-making, which is the first decision to send its packet, and the next decision is to forward the packets of other nodes. A game theory-based approach has been developed to associate users of wireless stations to prevent heterogeneous and poor performance based on joint resource allocation and association of wireless stations [30]. The payoff of wireless stations is based on the individual power of each station. Vijayakumaran et al. proposed a novel detection of the selfish node, which consists of two phases [31]. The generation phase includes routing task confirmation phase and the routing-report generation phase, and coordination-confirmation report generation phase. In the confirmation phase, the supervising agent will send the request for approval to the middle relay nodes.

The proposed scheme is a selfish node detection and prevention method called SENDER [32]. The scheme consists of two phases: the detection and prevention phases. In the detection phase, an adaptive threshold algorithm has been used to identify all nodes. In the prevention phase, selfish behavior is avoided based on the repeated game. The number of forwarded packets should be compared between current behavior and normal behavior to identify selfish behavior, which consists of three phases. Initially, the threshold value is set to the previous values. Next, the packet forward ratio (PFR) is calculated. Finally, the comparative threshold algorithm is used to compare with a threshold value to determine whether the current node shows selfish behavior or not. If the PFR is lower than the threshold value, the node is selfish, and the alarm is raised. Otherwise, the threshold value will be updated in accordance with the current PFR and the new threshold value for the next interval. In the prevention phase, the proposed method uses repeated games to prevent selfish behavior, and the game with payments are designed so that nodes gain lower payoffs if the nodes choose the selfish strategy; hence, they are unwilling to choose this strategy, and if some nodes sometimes choose the selfish strategy, they will tend to choose a normal strategy after a certain period of time due to reduced payoffs. Therefore, in the prevention phase, selfish behavior can be prevented by choosing a normal strategy.

3 System model

Each of the nodes is aware of the set of its neighbor nodes located at its domain. Each node is either normal or selfish. The nodes are not aware of the selfish or normal nature of their neighbor nodes. Each node uses all of its information and expectations from the behavior of other nodes to find the best strategy for itself. In order to be able to forward the data packets to the destination, the nodes do this by cooperating with each other due to the limitation of the sent frequency range. During the network operation, nodes gain some information about their neighbor nodes' cooperation. Each node tries to achieve the best result and the most payoff in the network. For this purpose, it tries to interact with normal nodes to forward its data packet to the destination. All notation and its description are used in the system model collected in Table 2.

The game is defined as $G = \{N, A^k, u^k\}$. In this definition, N is the number of nodes existing in the network, A^k is the set of the actions of the nodes, and u^k is the set of utility function which is the outcome of players in one round. The round is named as round k. For further explanation, $A^k = [A_1^k, A_2^k, \ldots, A_N^k]$ is the set of the node actions in round k. For example, A_i^k is the action of node i in round k. $A_i^k \in [0,\ \max(\text{pow}_i))$, pow_i is the highest sending energy power required by node i for sending a packet.

A_i^k is the total actions of each node i in the repeated game in one round, and it can be forwarded or not be

Table 2 Notations and its description of systems

Notation	Description
N	The number of nodes
A^k	The set of the actions of the nodes in round k
A_i^k	The action of node i in round k
u^k	Set of the utility function
u_i^k	The utility function of node i in round k
pow_i	The highest sending power required by node i
r_0	The constant value considered as a reward for node i
$n_{iCH_i}^k$	The total number of the received packets from cluster head CH_i by node i
n_{Rij}^k	The total number of the forwarded packets of node j
n_i^k	The number of node i packet in round k
c_i	The total energy required for sending the data packets
$d(i,j)$	The distance between node i and node j
p_i	The probability of node i to run one strategy
P_i	The set of probabilities for node i to run all the strategies
s_i	One strategy of node i
$S_i = \{F, NF\}$	The set of all strategies (forwarding and not forwarding)
$p_i(s_i)$	The probability for every pure strategy of s_i
$\Delta(S_i)$	Strategies by node i in a mixed game
p_i^*	New probability for node i
Δp	The probability changes rate in each round
$u_i(p_1^*, \ldots, p_n^*)$	Payoffs function of node i in new probability
$\pi_1(F, NF)$	The probability distribution for node 1
$\pi_i(F, NF)$	The probability distribution for node i
$\pi_i^k(F, NF)$	The probability distribution for node i in round k
$E_{n_1}(F \mid p_2^*, p_3^*, \ldots, p_n^*)$	The expectation value for node 1
$E_{n_i}(F \mid p_{-i})$	The expectation value for node i for forwarding the packets
$E_{n_i}(NF \mid p_{-i})$	The expectation value for node i for not forwarding the packets

forwarded to the data packet of node j (F, NF). Each player (nodes of each cluster) should choose one of these strategies in each round. After adopting the strategies in the games, the player gains a profit, the value of which is calculated by utility function and $u^k = \{u_1^k, u_2^k, \ldots, u_N^k\}$ is the efficiency value and utility function expected by the nodes in round k, and u_i^k is the utility function of node i in round k according to Eq. (1).

$$u_i^k = r_0\left(n_{i\mathrm{CH}_i}^k + n_{Rij}^k\right) - c_i n_i^k \quad (1)$$

In Eq. (1), r_0 is the constant value considered as a reward for node i. $n_{i\mathrm{CH}_i}^k$ is the total number of the received packets from cluster head chi (cluster head of cluster CH_i) by node i, n_{Rij}^k is the total number of the forwarded packets of node j (member of cluster CH_i) by node i in round k, and n_i^k is the number of node i's packet in round k that has been sent to the destination and c_i is according to Eq. (2).

$$c_i = \sum_{j \in \mathrm{CH}_i} \frac{\mathrm{pow}_i}{d(i,j)} \quad (2)$$

If a node forwards the data packet of the other node, the node consumes the energy power for it. The energy consumption of forwarding the packets should pay by the owner of the packets. In Eq. (2), c_i is the total energy power required for sending the data packets by node i and have been sent to other nodes that are members of cluster CH_i in round k. The notation $d(i,j)$ is the distance between node i and node j. It is clear that c_i is the cost of the data packets which node i should pay it as the owner of the packet.

The proposed method models the system as a mixed or random strategy. The method allocates probabilities for different strategies of each player in Fig. 1.

When node N_1 wants to send the data packet to the destination N_m, it sends the packet to node N_2 but N_2 may forward it or not. According to our definition of mixed strategy, N_2 is independent to select the A_i^k as the strategy of s_i with the probability of p_i then it selects node N_3 to forward the data packet to the destination this procedure is continued up to the data packet forwarded to the destination or it is dropped by nodes. According to the mixed strategy definition, P_i that is $P_i : S_i \overset{s_i \in S_i}{\rightarrow} [0,1]$ allocates a probability equal to $p_i(s_i)$ for every pure strategy of $s_i \in S_i$, and p_i is the probability of node i to run s_i as one strategy. The set of all strategies are in S_i and one strategy of node i defined s_i, so that $\sum_{s_i \in S_i} p_i(s_i) = 1$. In other words, according to the probability rules, the total probability of all strategies run by each player is equal to 1. With the increase of node number in each cluster, the game is expanded and a multiplayer game is done by the cluster member nodes similar to Fig. 1.

If the game assumes as the mixed strategy in Fig. 1, with n nodes, the game is represented $G_{ME} = \{N, (\Delta(S_i)), (u_i)\}$, in which $\Delta(S_i)$ is defined as:

$$\Delta(S_i) = \left\{ \begin{array}{c} (p_{i1}, \ldots, p_{im}) \in R^m : p_{ij} \geq 0 \\ , \forall j = 1, \ldots, m \\ , \sum_{j=1}^{m} p_{ij} = 1 \end{array} \right\} \quad (3)$$

At first, to define $u_i(p_1, \ldots, p_n)$, it is mentioned that the random variables $p_1, \ldots, p_n$ are pairwise independent;

therefore, the probability of a pure position of $(s_1, \ldots, s_n)$ is:

$$p(s_1, \ldots, s_n) = \prod_{i \in N} p_i(s_i) \tag{4}$$

Regarding the fact that after each round of the game, the probabilities change and depend on node i cooperation to forward the data packets; it has been changed to Eq. (5) and is defined as follows:

$$p_i^* = p_i \mp \Delta p, \forall i = 1, \ldots, n \tag{5}$$

Accordingly, we can define the pay-off functions of $u_i(p_1^*, \ldots, p_n^*)$ as follows:

$$u_i(p_1^*, \ldots, p_n^*) = \sum_{(s_1, \ldots, s_n) \in S} P^*(s_1, \ldots, s_n) u_i(s_1, \ldots, s_n) \tag{6}$$

For the game in Fig. 1, we have a probability distribution for nodes in Fig. 1 when $S_i = \{F, NF\}$ and $u_i(p_1^*, \ldots, p_n^*)$ defined in Eq. (7):

$$\pi_i^k(F, \mathrm{NF}) = \sum_{s_1, s_2, \ldots, s_n \in S} p^*(s_1, s_2, \ldots, s_n) * u_i^k(s_1, s_2, \ldots, s_n),$$

$$S = s_1 \times s_2 \times \ldots \times s_n = \{(F, F, \ldots, F), (F, F, \ldots, \mathrm{NF}), \ldots, (\mathrm{NF}, \mathrm{NF}, \ldots, \mathrm{NF})\} \tag{7}$$

The expectation values for all nodes are calculated for the choice of sending or not sending the data packets with Eqs. (8) and (9).

$$E_{n_i}(F|p_{-i}) = \sum_{s_1, s_2, \ldots, s_n \in S_{-i}} p^*(S_{-i}) \times u_i^k(S|s_i = F) \tag{8}$$

$$E_{n_i}(\mathrm{NF}|p_{-i}) = \sum_{s_1, s_2, \ldots, s_n \in S_{-i}} p^*(S_{-i}) \times u_i^k(S|s_i = \mathrm{NF}) \tag{9}$$

In other nodes, the expectation value is calculated by equivalent to each node in Eqs. 8 and 9; the Nash equilibrium is obtained. By calculating the Nash equilibrium, the probability values of each player are calculated. By getting these values at each round of the game, each player chooses the best action against the opponent. Given the fact that no nodes know when the game is over, a repeated game is infinite which has N players. In a repeated game at the beginning of round k, node i makes the decision based on the nodes' behavior in the past.

4 Proposed method

A multi-step mechanism is designed based on the game theory to stimulate the cooperation between the selfish nodes in the Internet of Things in Fig. 2. A multi-step scenario is considered by setting up the nodes in the IoT networks, and the nodes try to distinguish their neighbor nodes by sending Hello messages. Some base stations (BS) are placed in a different location to collect the data packets. In the first step, nodes grouped in the cluster with the cluster head(s) and a BS to collect the data. The member nodes send the data packets in multi-hop to the cluster head, and the nodes cooperate in forwarding the data to the cluster heads or the destination of the base station. Then, they run a multi-person game in the second step (playing the multi-person game and sending data) while forwarding their own data packets or the neighboring nodes. By running the game, each node has

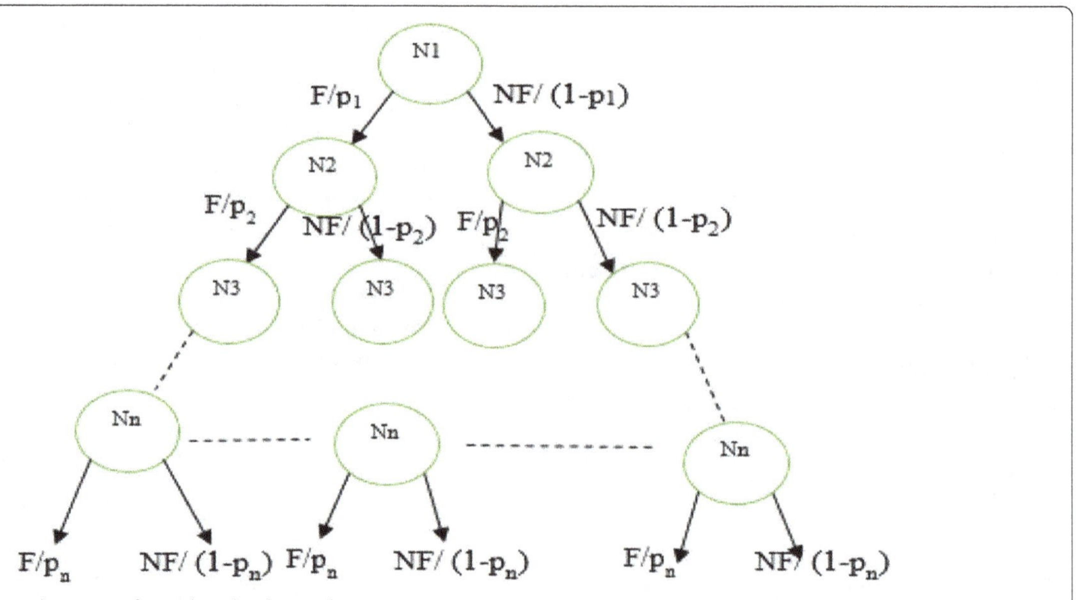

Fig. 1 Chart of forwarding or not forwarding the data packets

the data node to send the destination; it will select one neighbor node to play the game. The node will train about neighbor node status, and each node extracts features of the neighbor nodes. If the neighbor node forwards the data packet and cooperates with others, the node optimizes the probability of the neighbor node in the game (forward the data packets) for the next rounds. In the third step (detecting non-cooperation node and update), the nodes are assumed to be the selfish node and can select their strategy when forwarding the packets. The game theory and the cooperation process analysis are used to determine the selfish nodes and/or malicious nodes that forward the data packets by delay and/or do not send them at all so that the non-cooperative nodes are identified. The nodes classify the other nodes and update the reputation table. The playing multi-person game and sending data and detecting non-cooperation node and update phases are done as a repeated game.

The game theory-based mechanism takes advantage of the punishment-based and incentive mechanism. While the nodes are identified in misbehavior, the power of transferring their data packets (and/or even their co-operation with other nodes) is reduced so that the non-cooperative nodes are stimulated to cooperate via punishment, not cooperating with them reduced their reputation. By detecting non-cooperative nodes, the network throughput and detection accuracy are increased, and average end-to-end delay and energy consumption are decreased. Other metrics to compare the algorithm with other methods are false-positive rate and false-negative rate. If these metrics are low, the network will have a high performance, and the proposed method has a lower false-positive rate and false-negative rate in comparison with other detection methods.

The mechanism is designed based on the game theory to stimulate the cooperation between the selfish nodes in the Internet of Things which Fig. 3 shows the flow of data in the mechanism. Finally, the main focus of the present paper can be summarized as follows.

- The game theory analysis will be presented in the multi-person game. The game is modeled as an infinite repeated game, and the level of the cooperation power is found, which is achieved by the performance of the game by the reputation of each player proportionate with its cooperation with other nodes. It is also shown that the best mode for each thing is the cooperation option with other nodes in the network. The theoretical game approaches are used to stimulate the nodes to cooperate in the internet of things.
- A stimulating strategy based on the game theory and based on punishment is proposed to stimulate the nodes to cooperate in the Internet of Things. A motivational strategy is introduced and applied to overcome the challenge of non-cooperative behaviors in the network. The main idea is that each node monitors the behavior of other nodes and forwarding or not forwarding other nodes' data packets in a specific period of time. It is shown by the theoretical analysis the nodes are stimulated to cooperate by

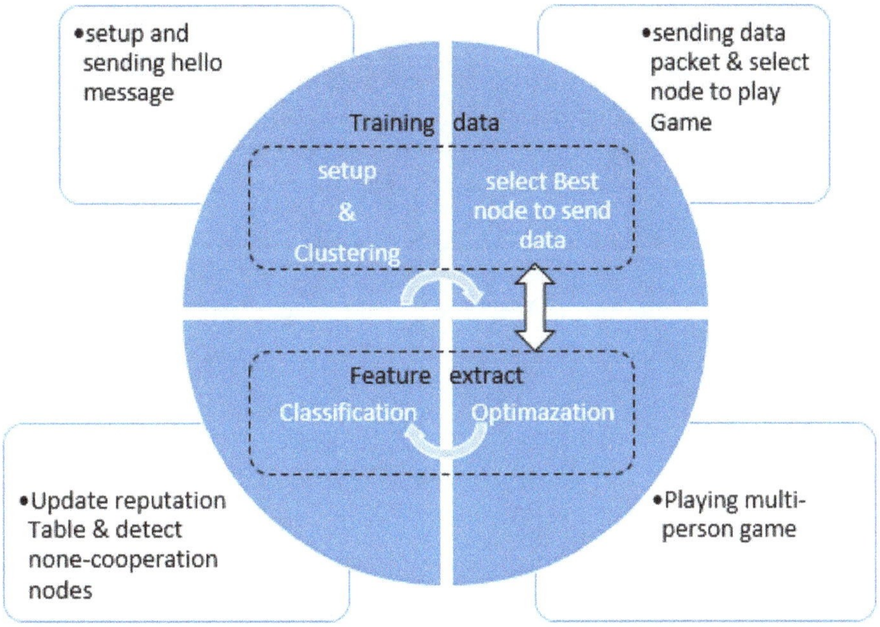

Fig. 2 Multi-step detection mechanism in block diagram

this strategy because each deviation from forwarding the packets leads to less cooperation or even no cooperation, which reduces their reputation.

- The cluster head considers its cluster members for the status of each opponent nodes (neighboring nodes) saved in their reputation table. The nodes with reputations below the threshold are reported as selfish nodes to the cluster head. The cluster head will broadcast to all cluster nodes to know about their neighbors as a selfish node in the games so that they can stimulate the selfish nodes to cooperate with others. The results of the simulation show that the proposed strategy can approximately cooperate effectively to the desirable cooperation condition.

In the following, more details are discussed about each step of the proposed mechanism.

4.1 5-1 Setup and clustering

During this phase, all things are randomly distributed in the area. Then, each node broadcasts a "Hello" message, and the nodes replying to this message are known as a neighbor node. Each node will store information about the neighbor's status in its database as a table consisting of four fields which are shown in Fig. 4

In the following, more details are discussed about each field, as shown in Fig. 4.

- Node's ID: It has 16 bits to save the node's identification.
- Number of hops up to the cluster head: It has 8 bits to save the number of hops between a node and the cluster head
- Node's data: It is an array of n bits and saves the data of each node and its neighbors. The amount of n is per byte.
- Node's status: It is an array of n bits and indicated the status of nodes. Because the nodes can have one of C, S, and LS statues, therefore, the length of each array is considered as 2 bits. The predefined status of this field is C, which is assumed as the cooperation node.

After the setup of the network, the neighbor nodes by using the clustering algorithm proposed by Kumar and Zaveri in 2016 [33], the performance of this method is such that all of the existing things with any features are assumed as a node, and they choose the nodes which are involved in the communications given the cluster heads. Overload decreases communications. Naturally, the nodes are heterogeneous in the Internet of Things, and they are connected to each other from different networks, and this approach has also considered the nodes heterogeneously. The clusters and their cluster head change in regular intervals, and they are dynamic due to the dynamic nature of the Internet of Things. This method promises saving energy by choosing different nodes as a cluster head.

This approach presents a hierarchical model in which the lower layer is known as layer number 2, and it has tools and equipment which have an ID code. The things include sensors, RFID tools, and people. The important assumption is that these things do not have internet protocol due to high energy consumption, and they cannot be directly or indirectly connected to a cloud system. But they are critical points of the networks which are needed by application programs. In this layer, the creation of cluster and choosing the cluster heads are done in a dynamic way. Cluster heads send the accumulated information to the upper layer of the base station.

In the upper layer which is known as layer number 1, the tools and things have an internet protocol, and they are usually less likely to face the problem of an energy shortage, and they have immediate communications and

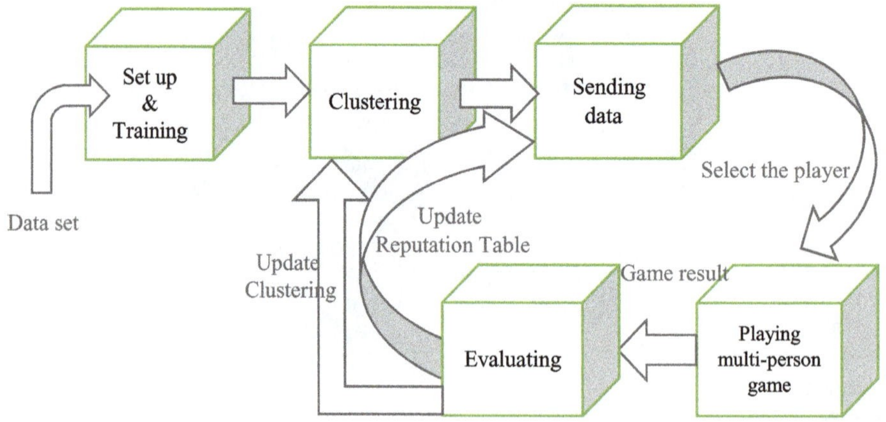

Fig 3 General diagram of the proposed mechanism

Node's ID	# Number of hops to the cluster head	Node's Data	Node's status

Fig. 4 Format of data in cluster heads

processes. These things provide the possibility of communication between all the things existing in the network. The nodes are communicated with their cluster head, and the cluster head is also communicated with the base station. In this way, all the things can communicate with each other, and in the case of necessity, they can exchange information. The proposed algorithm works on the origin of counting the number of neighbors, and the remaining energy of the node and the clusters are formed in the radio range of the node. The proposed algorithm runs three rounds to determine the cluster and the cluster head in layer number 2.

In the first round, a broadcast message is sent, and each of the nodes that receive this message sends a confirmation message in response to the sender node, and finally, all the nodes recognize their neighbors. The ID code of the node or the internet protocol is entered into the neighbor's list. In the second round, a multicast message is sent, and the node declares the number of its neighbors and its remaining energy to all of its neighbors which are located at its own radio range. In the third round, the cluster head selection process is done based on the information of the neighbor nodes, and the node which has the highest amount of remaining energy and the highest number of neighbors is selected as the cluster head.

4.2 5-2 Sending data and playing multi-person game

In the proposed method, sending a packet is considered between two nodes in a cluster from the source node to the destination node as a hierarchical multiplayer game. The players make their decisions independently, and each of them is faced with a minimum probability of losing profit. This game is repeatedly done between all the network nodes, and the number of the repeated game is very large. So, the game is an infinite repeated game. Also, the game is dynamic, because if one of the players cooperate at first, other players choose their actions being aware of the first player's action. Since the nodes do not have any information about the selfish nodes, the game will be done with incomplete information. With the start of network operation and after the setup and clustering phase, all the neighboring nodes in a cluster play with each other many times during the life of the network and exchange their data packets to reach the destination node.

At the beginning of the game, none of the nodes has any information about its neighboring nodes, and they choose one of their neighboring nodes every time of sending data. During sending data packets and receiving an acknowledgment message from the destination, the nodes will be informed of the success or failure of sending the data packet and the status of the neighboring node. However, in order to prevent the data packet from extra forwarding and increased energy consumption in the network, it will even be stuck in the infinite loop; the data packet can pass through all the nodes in the network; the limited lifetime in the data packets are used to solve this problem. In each round and each game, the nodes send the game results to the third phase, and the entry of the related node in the table is updated to take advantage of this information for successfully sending data packets in the next rounds. In other words, as a result of successful or unsuccessful sending of the data packets in each game and per round, each node sends the game results to the third phase, and it is stored in the reputation table by the neighboring nodes.

According to the distance between the nodes calculated using Eq. (1) and the information of the previous games such as the number of unsuccessful and successful sent to the third step then it is sending by the neighboring nodes and the reputation values of each neighboring node, payoff values change in reputation table. In this way, the players play more carefully, and so the selfish nodes cannot reduce the network efficiency by sending unsuccessful packets and unnecessary use of bandwidth.

At each step, node i intends to send the packet to the next step, if it knows the status of all its neighboring nodes, that is, if at least once played with them. At this moment, the mentioned node reviews its reputation table that contains the reputation record of the neighboring node. According to the number of successful and unsuccessful sending in the opponent node, it updates its payoff values and obtains the probability values of choosing each action and their reputation values, which has received the third phase. The node will be able to choose the best strategy and optimize its guess about the nature of the neighbors. But if node i does not know the status of all neighboring nodes, it will continue to play with a neighboring node that does not have any information about it, and this procedure is done for intermediate nodes so that all nodes have the opportunity to play and attend the operation to send the data packet. Further nodes also have the ability to cooperate in network operations. Figure 5 shows the performed operation of sending the data packet by the source node to the destination node.

4.3 5-3 Update and detecting selfish and malicious nodes

The reputation of the network nodes are updated by using the information received from the network nodes from the neighboring nodes for sending data packets, and it becomes possible to identify the selfish nodes in each cluster and consequently in the whole network. The way of detecting the selfish nodes in the network is based on the results of games that directly affect the detecting and updating of the reputation of each node. The game results are analyzed in the third phase, after the nodes play the game in each round. Getting the acknowledgment packet from the node which the game is played, the reputation of the node is increased by a predetermined constant shown by pd_Rep in Fig. 6; in the next round of playing, if the data packet sent to a node to forward them, it will be played according to the previous experience with that node. But if the acknowledgment packet is not received from the node played with it in a game, the reputation of that node will be reduced by a predetermined constant. If the reputation of each node is not lower than the predefined threshold value, this node will have the second opportunity to cooperate with other nodes and participate in the games. So, the selfish and malicious node will increase their reputation with other nodes. And with that single failure, that node is not considered as a selfish or malicious node. Reducing network performance can be prevented by excluding the selfish or malicious nodes. But if with the second opportunity of the nodes the reputation is less than the predefined threshold, the node will be considered as a selfish and malicious node and will be avoided to cooperate in the network.

At certain times, the cluster head controls the cluster members about the situation of each of their opponent nodes (neighboring nodes) that have stored in their reputation table. The nodes with a reputation value of lower than the predefined threshold are reported to the cluster head as selfish nodes. The node that is known as a selfish node based on the most reports of its neighbors in the games will be dismissed from the cluster, and all the cluster nodes will be informed by a message from the cluster head. Detecting the selfish node in the network will increase the network throughput and reduce energy consumption in nodes.

5 Simulation and evaluation

This study is faced with the problem of the selfish and malicious nodes in the Internet of Things. The selfish nodes are the nodes that use the facilities of the network for their own interests. These nodes do not participate in processes of sending and forwarding the packets, so they do not help to save energy and communicating with other nodes. The malicious nodes are considered as those nodes which dropped the data packets or sent them by the delay that the packets are discarded due to the expiration of the packets lifetime [34]. As a result, the throughput of the network is significantly decreased in the presence of such nodes. In order to stimulate the mentioned problem, simulation of such behaviors is done in the network for indicating its effect on the delivery percentage of the data packets, throughput percentage, and end-to-end delay. In this section, the proposed method is evaluated and compared with other similar methods in different metrics. Firstly, the evaluation criteria are introduced. Secondly, the simulation and results are presented.

5.1 Evaluation criteria

The proposed mechanism is evaluated using different criteria. The definition of the evaluation metrics are presented in the following.

5.1.1 Detection accuracy

The detection rate of the selfish node indicates the ratio number of selfish nodes detected to all the selfish nodes in the network as DA, where TP denotes the number of selfish nodes detected, and FN indicates the number of

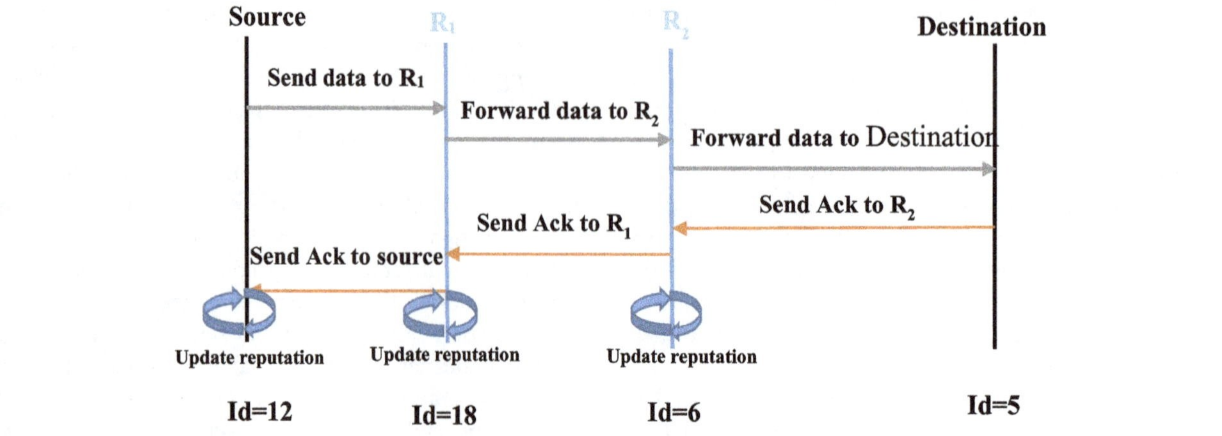

Fig. 5 Example of the packet forwarding with two relay nodes

Algorithm

For (all clusters) **do**
    $n_i$send data packets to source by using n_i^j
    if (Rep ($n^j_{n_i}$) > θ and packet TTL>0) then
        send packet to n_j
    endif
    else do
        For (each neighbor j) **do**
            if (Rep ($n^j_{n_i}$) > θ and packet TTL=0) then
                Drop the packet
            endif
            if (Rep ($n^j_{n_i}$) < θ and packet TTL>0) then
                send packet to another neighbor n_k
            endif
        Endfor
    endelse
    if (packet ack receive) **then**
        Update (Reputation table n_i) , Rep ($n^j_{n_i}$) = Rep ($n^j_{n_i}$) + pd_Rep
    endif
    else do
        Update (Reputation table n_i) , Rep ($n^j_{n_i}$) = Rep ($n^j_{n_i}$) - pd_Rep
    endelse
EndFor
Report Cluster members to Cluster head their neighbor status.

Fig. 6 Simi-code of algorithm

nodes which are selfish nodes but detected as normal nodes in the network. The detection rate of the selfish node is according to Eq. (10).

$$\mathrm{DA} = \frac{\mathrm{TP}}{\mathrm{TP} + \mathrm{FN}} \quad (10)$$

5.1.2 False-positive rate

Another parameter for evaluating selfish node detection is the false-positive rate in the network. The false-positive rate indicates the ratio of the normal node number detected as a selfish node by error to the total number of normal nodes detected by mistake (FP) and the number of normally detected nodes (TN) in the network. Therefore, the false-positive rate (FPR) is calculated using Eq. (11) and is defined in the following:

$$\mathrm{FPR} = \frac{\mathrm{FP}}{\mathrm{FP} + \mathrm{TN}} \quad (11)$$

5.1.3 False-negative rate

The parameter is used to evaluate the efficiency of selfish node detection methods; the false-negative rate is defined in Eq. (12), which is the ratio of the number of the selfish nodes detected in the normal node by error to the total number of selfish nodes detected by normal node (FN) and the number of detected normally nodes (TP) in the network.

$$\mathrm{FNR} = \frac{\mathrm{FN}}{\mathrm{FN} + \mathrm{TP}} \quad (12)$$

5.1.4 Throughput

Throughput is one of the evaluation parameters of the network in most of the fields of IoT. Throughput is actually the number of data packets that are successfully delivered to the destination. Therefore, the average throughput is the ratio of the average number of the data packets delivered to the destination by all nodes to the total number of the packets produced in the network.

5.1.5 End-to-end delay

The average end-to-end delay is the arrival time of a packet from the source node to the destination.

5.1.6 Energy consumption

IoT system nodes consist of sensor nodes, and the nodes have mobility similar MANET nodes. So, each node uses the energy model.

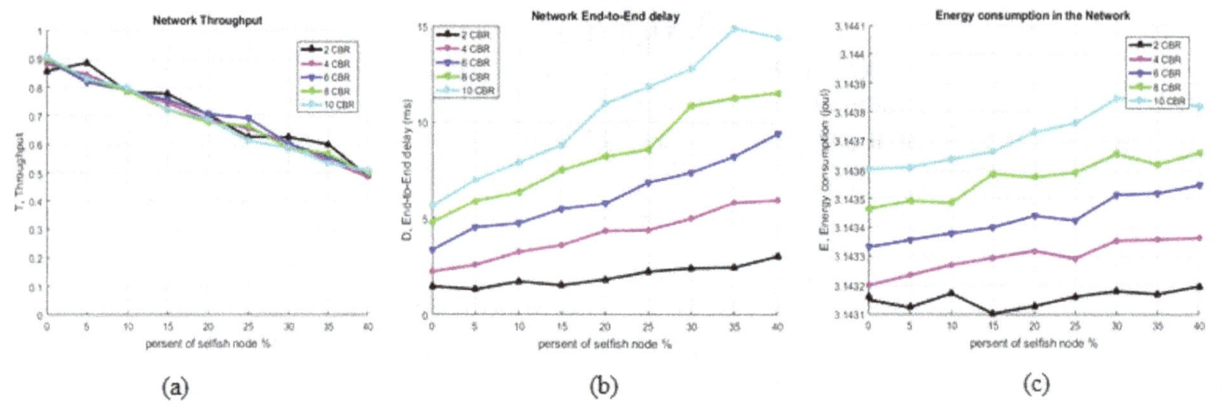

Fig. 7 Impact of the selfish node on the **a** throughput of the network, **b** end-to-end delay, and **c** energy consumption

5.2 Simulation result

A network is consistently distributed in an environment with an area of $1000 \times 1000\,m^2$, and the nodes are randomly dispersed in IoT for 4 different types of IoT nodes with different numbers and parameters. The considered internet network includes immovable things or limited energy source similar to wireless sensor networks with 4 different types of nodes which can be used in agricultural fields as sensor type 1 (controlling water), sensor type 2 (controlling soil), sensor type 3 (controlling weather), and sensor type 4 (controlling temperature). All of these networks have wireless communications. The simulation environment is MATLAB, and in the performed simulation, at the center of all of

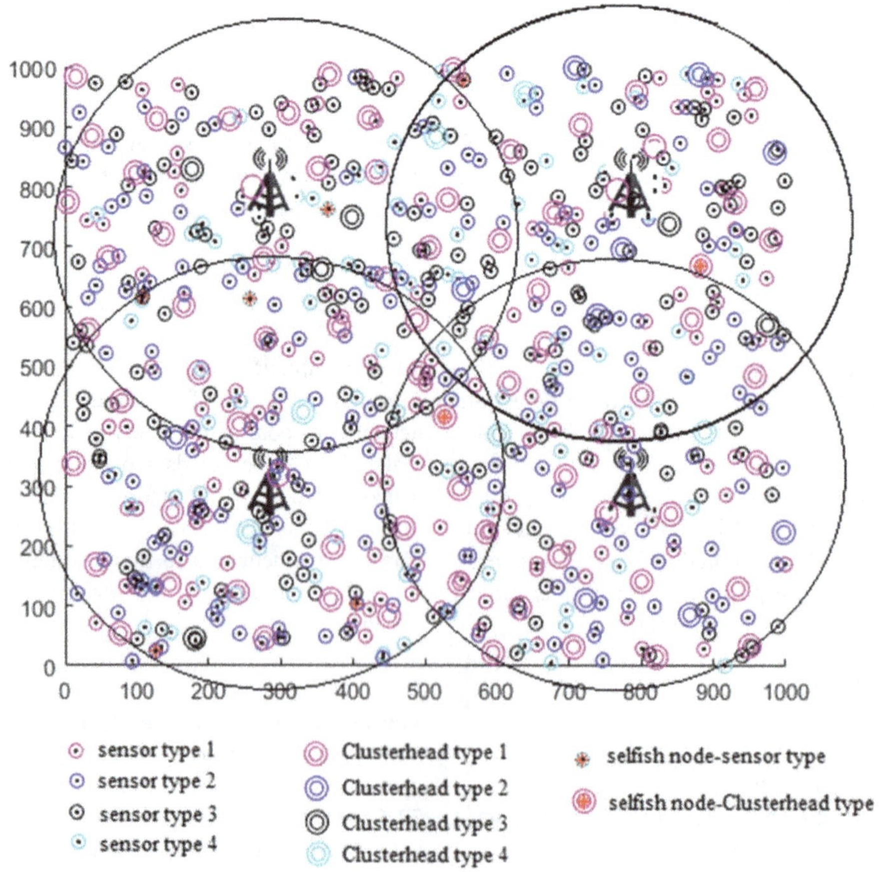

Fig. 8 System models in clustered IoT

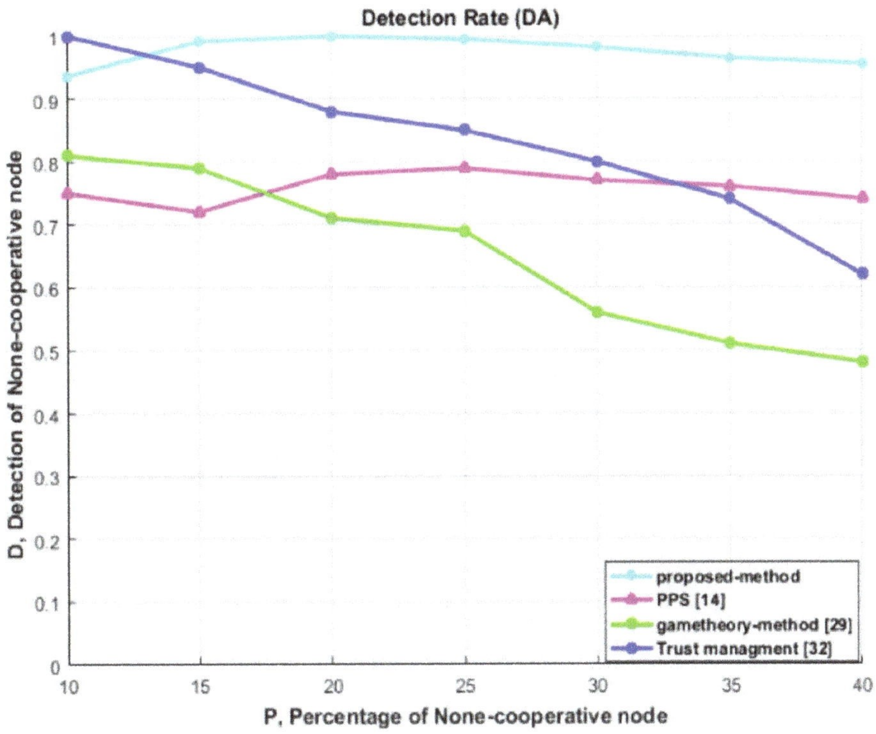

Fig. 9 Comparison of detection accuracy (DA) in IoT

the 4 types of networks; a base station is placed to collect data. Firstly, the network is simulated without any detection mechanism. In Fig. 7, the network service as usual and nodes collect data and send them to the destination, and metrics of throughput, end-to-end delay, and energy consumption are measured in different traffic on the network.

The percentage of the selfish nodes used in the simulation gradually reaches from 0% (without selfish node) to 40% of the total nodes. It means that when some nodes' energy level decreases, they will change to a selfish node. When the nodes forward the data packets, they lose energy and the number of such nodes increases in the network. At first, there is no selfish node, then the percentage of the selfish nodes is 0% of all nodes. The nodes start to collect the data and forward them to the destination, they use energy power, and some of them change their status to the selfish nodes. For example, when we have 100 nodes in the network and 5 nodes change their status to the selfish nodes, we have 5% of total nodes as selfish. In the simulation environment, we assume that the nodes' energy level is lower than the predefined energy level which became a selfish node and numbered them to calculate the percentage of the selfish nodes. In Fig. 7a, network throughput reduces by increasing the selfish and malicious node number. Imposing traffic on networks 2, 4, 6, 8, and 10 (CBR) leads to the increase in the sending of the packets. It is expected that by increasing the transmission of the data packets and the delivery rate of the data packets in the network, the existence of selfish nodes prevents it. In Fig. 7b, the average end-to-end delay for the data packets in the network is represented by the presence of the selfish nodes. Increasing network traffic leads to increase the end-to-end delay in the delivery of the data packets; when the selfish nodes are present in the network, both dropping the data packets and resending them also increases network traffic. Not only have the selfish nodes sent the data packets by delay, but also increasing the network traffic leads to high average end-to-end delay in the data packets. In Fig. 7c, the energy consumption to send the data packets increases by increasing the transmission of the data packets from the source node. When the data packets are dropped by the selfish nodes, they are sent in such a delay that they are not actually destined to be redundant energy consumption, which wastes the energy resources of the nodes. Hence, the detection of selfish nodes in the network and stimulate them to cooperate on the purposes of this article. If we can provide a model to detect the selfish nodes in the IoT, we have provided conditions for increasing throughput, reducing the average end-to-end delay in the delivery rate of the data packets and energy consumption in the network.

Secondly, the proposed mechanism is simulated and evaluated in different metrics. The mechanism clustered the nodes, and the cluster heads have communication

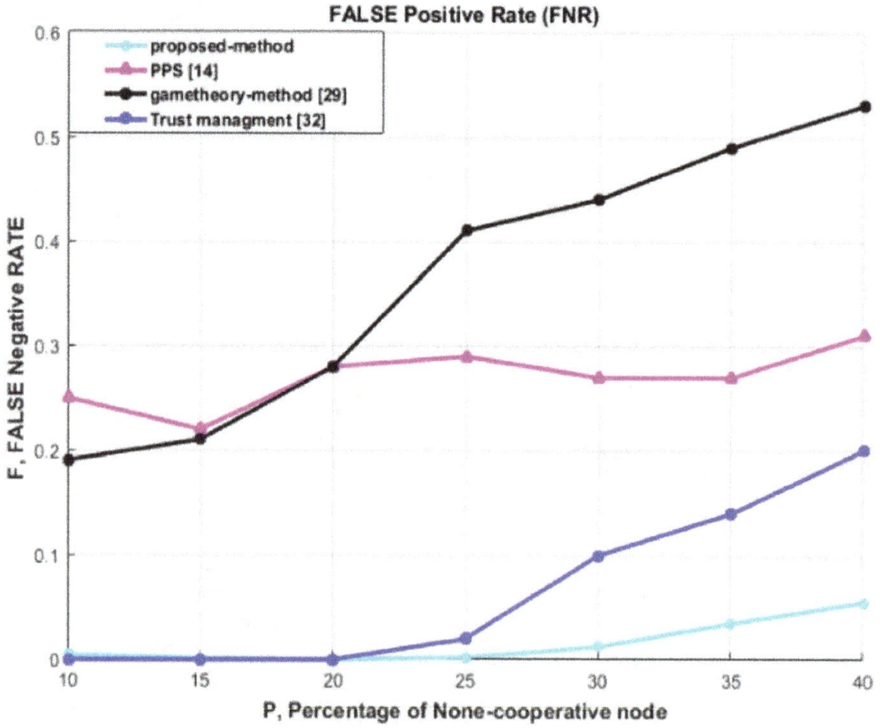

Fig. 10 Comparison of the different algorithm in false-positive rate (FPR) metrics

with cluster members in clusters 1 to 4, as mentioned in Section 4.1. According to the fact that each of these different networks has different simulation models, we will first have a clustering based on the different types of nodes according to Fig. 8. The cluster heads have 4 types: cluster head type 1 for sensor nodes type 1, cluster head type 2 for sensor nodes type 2, cluster head type 3 for sensor nodes type 3, and cluster head type 4 for sensor nodes type 4. In the simulated model, the base stations (cluster heads) are located in the position (250, 250), (750, 750), (250,750), and (750,250), respectively. However, the initial energy of the nodes in clusters 1 to 4 is 0.5, 1.5, 1, and 1.1 J and 200, 100, 200, and 200 number of nodes in clusters with a radio range of 80, 70, 75, and 70 m, respectively. But the energy model and the type of nodes are the same.

The proposed approach has made decisions about both of the cooperation and selfish nodes by cluster heads. To evaluate the proposed scheme, we were simulated in Window 8.1 basic (64-bit), core i7 processors, 370 M processors, 2.40 GHz of speed with a memory of 8 GB, and MATLAB 2015 software. The performance of the proposed method is compared with the PPS [14], game theory-based [29], and trust management [32] protocols for evaluation metrics such as detection accuracy, the percentage of false-positive rate, false-negative rate, throughput, average end-to-end delay, and energy consumption. The simulation result is performed 100 runs, and the average result has shown in different metrics.

In order to examine the important parameter of the detection accuracy (DA) of the selfish node in IoT, at the beginning of the simulation, 10% of the total nodes are selfish nodes in the network; further, the rate of selfish nodes gradually increased by 15%, 20%, and 40%. It is worth mentioning that in the real world, as time goes on, due to lower initial energy levels, nodes tend to maintain their energy resources and they will refuse to send the other packets, and they will be as selfish nodes. According to the diagram in Fig. 9, increased detection accuracy of the selfish node is clearly observed. When the percentage of selfish nodes increases in the network, it leads to a larger number of games running in the proposed method. The reputation of each node is updated during these games and while forwarding the data packets through different nodes. Therefore, when only 10% of the nodes of the network are selfish nodes, lower games have been run for preventing more energy waste, and the reputation of the network nodes is not well known, and only 92% of the selfish nodes have been detected. However, with the increased percentage of the selfish nodes in the network and increased number of games, the reputation of the games is updated, and detection is done accurately, and up to 100% of the selfish nodes will be detected. However, an increased number of selfish nodes in the network needs to run a large number of games, and due to the resulted energy consumption, it is not a rational way. Also, with this

Table 3 Different metrics of the proposed methods in comparison with other methods

Presence of selfish node, algorithms	Metrics	10	15	20	25	30	35	40
Proposed algorithm	Detection accuracy (DA)	0.93	0.99	1	0.99	0.98	0.96	0.95
PPS [14]		0.75	0.72	0.78	0.79	0.77	0.76	0.74
Game theory-based [29]		0.81	0.79	0.71	0.69	0.56	0.51	0.48
Trust management [32]		1	0.95	0.88	0.85	0.8	0.74	0.62
Proposed algorithm	False-positive rate (FPR)	0	0	0	0.002	0.0127	0.0347	0.0549
PPS [14]		0.25	0.22	0.28	0.29	0.27	0.27	0.31
Game theory-based [29]		0.19	0.21	0.28	0.41	0.44	0.49	0.53
Trust management [32]		0	0	0	0.02	0.1	0.14	0.2
Proposed algorithm	Throughput	75.85	73.05	74.14	78.92	86.71	91.07	95
PPS [14]		48	41	43	38	35	32	19
Game theory-based [29]		50	42	41	36	33	30	15
Trust management [32]		85	78	81	84	82	71	73
Proposed algorithm	End-to-end delay (ms)	16.35	17.01	17	12	7.93	4.1	1.47
PPS [14]		48	41	43	38	35	32	19
Game theory-based [29]		50	42	41	36	33	30	15
Trust management [32]		85	78	81	84	82	71	73
Proposed Algorithm	Energy consumption (μJ), 2 CBR	3.14300	3.14302	3.14303	3.14299	3.14301	3.14301	3.14302
PPS [14]		3.1	3.25	4.2	4.8	4.9	5.01	5.2
Game theory-based [29]		2.8	2.9	3.5	3.6	3.66	3.78	3.9
Trust management [32]		2.9	3.1	3.4	3.9	4.12	4.52	4.6

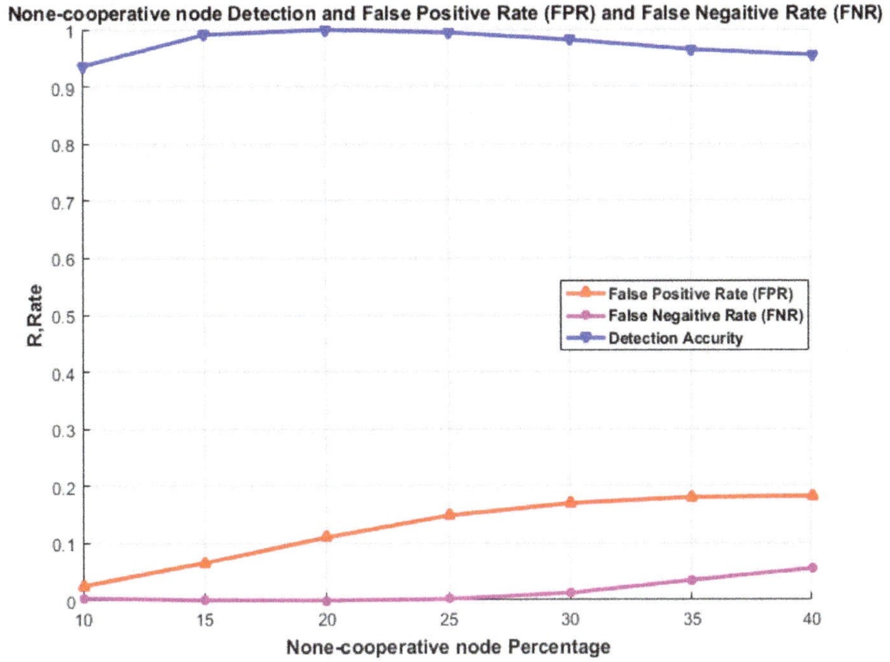

Fig. 11 DA, FPR, and FNR metrics in the proposed method

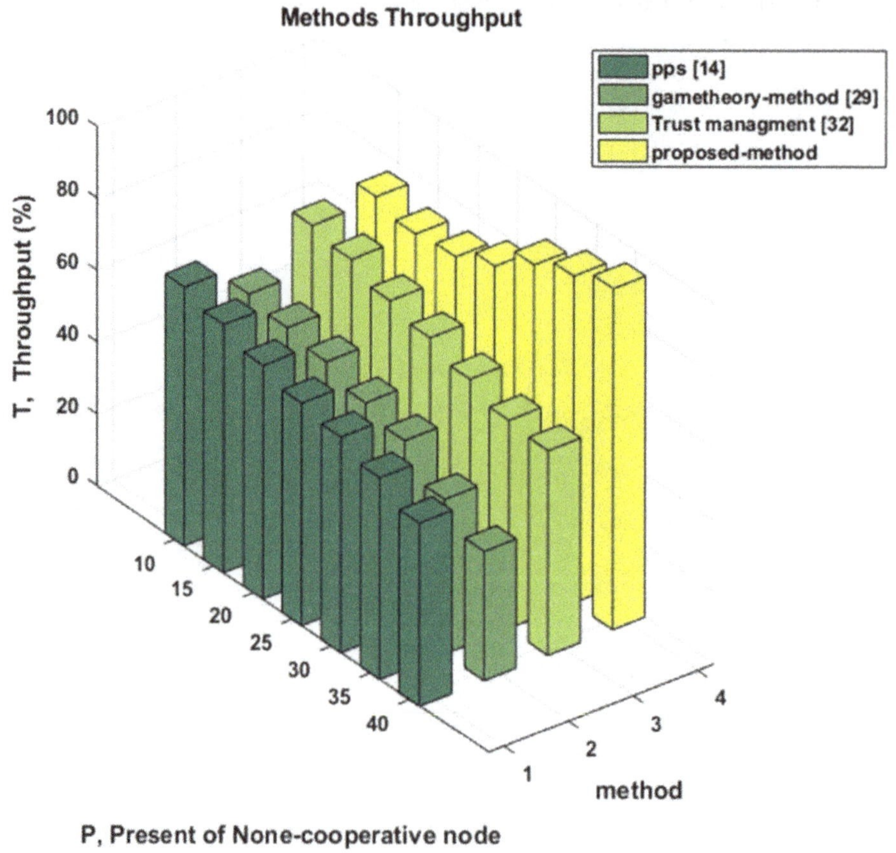

Fig. 12 Comparison of the throughput metrics in a different algorithm

number of games, an acceptable percentage of selfish node detection will be achieved even by a high percentage of the existence of the selfish nodes. The numerical values resulted in comparing the proposed approach showing that with the increased percentage of selfish nodes, the algorithm detection accuracy will be higher than other algorithms and has a slighter slope compared with other similar methods.

The fact that the proposed method has a slighter slope compared to other methods is shown in Table 2. This is obvious that the proposed method uses acknowledgment messages and the reputation of the nodes in each cluster to detect the selfish nodes. While other methods in higher percentages of the selfish nodes are usually unable to detect them in high detection accuracy.

In fact, FPR has an inverse relationship in the network, so its low level shows the accuracy of the proposed approach. That is, the higher the number of cooperation nodes correctly detected, the network performance will also be higher, because the network refuses to cooperate with the node when it detects as a selfish and malicious node. If the detection was mistaken, it would reduce the network efficiency and forward lower the data packets to the destination. Twenty games have been repeated to evaluate the proposed algorithm, and the simulation result is performed 100 runs, then the average result has shown that the higher number of games is the higher selfish node detection accuracy. The algorithm is also implemented in rounds to evaluate the performance of this algorithm in different situations. The numerical comparison has shown that the increase of the selfish nodes in the network, the false-positive rate of the proposed algorithm is lower than the other algorithms. As shown in Fig. 10, the FPR is better than other algorithms when more than 20% of the network nodes are a selfish node and less error than others, but up to 20% is roughly the same as the methods PPS [14], game theory-based [29], and trust management [32] protocols. The fact that the proposed method has a lower false-positive rate compared to other methods is shown in Table 3.

Figure 11 illustrates the proposed algorithm for the detection accuracy (DA), the false-positive rate (FPR), and the false-negative rate (FNR) metrics in the number of selfish nodes from 10 to 40%. The only FNR parameter is the negative point in the proposed method, which increases with the increase in the number of selfish nodes in the network. But it has a low percentage, then it has a disproportionate effect on network performance;

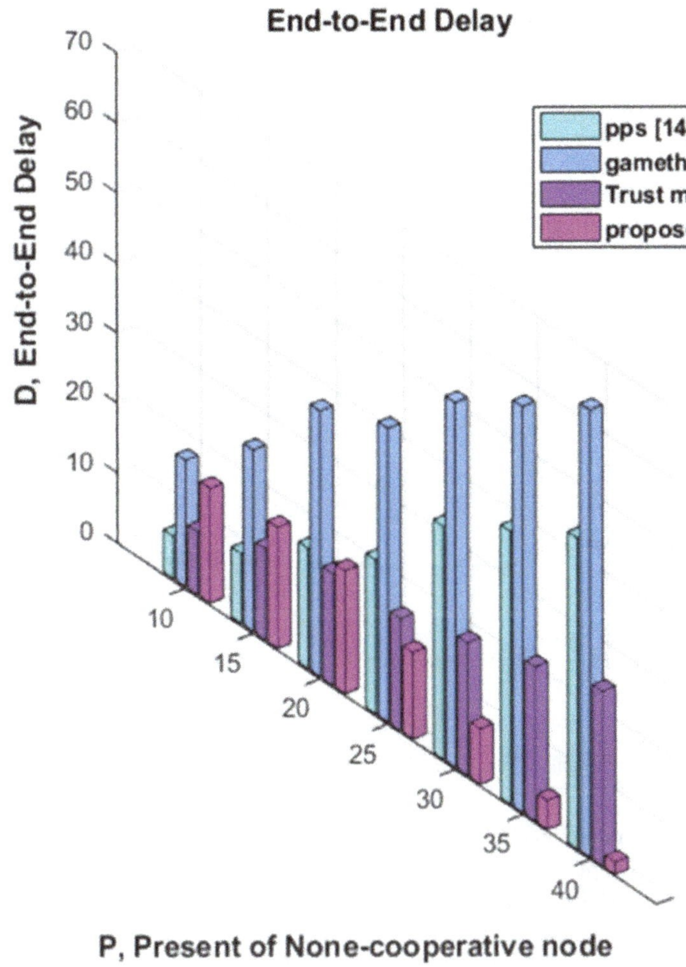

Fig. 13 Comparison of the average end-to-end delay by a different algorithm

considering the diagrams in Fig. 11, this weak point of the proposed approach was negligible, and a further work on this issue will be examined further.

The maximum throughput shows the high efficiency of the proposed method, because when the selfish nodes do not send data packets in the network and the acknowledgment message is not received in the source node, this packet is sent in the network again and it leads to increased traffic and the total number of the produced and sent the packets in the network, which is practically disadvantageous. So, the more throughput in the network led to proper use of the network resources, including bandwidth or limited energy resources in the nodes. According to Fig. 12, high throughput of the proposed approach is observed due to early detection of the selfish nodes in the network which prevents the production of repeated the data packets, increased network traffic, and average delay of the packets in the network. Table 2 shows the network throughput in the proposed method and other methods PPS [14], game theory-based [29], and trust management [32] protocols. Which has a direct relation to the selfish node detection accuracy in which the detection accuracy of the selfish and malicious nodes is higher than other approaches which cannot handle unsuccessful data packets from the nodes, and the data packets are sent from trusted paths to be forwarded to the destination.

Figure 13 depicts the average end-to-end delay in the proposed method compared with different methods. The average end-to-end delay of the proposed method is lower than other algorithms. The increase in the number of selfish nodes increases the average end-to-end delay. As the number of selfish nodes increases, it takes a lot of time to get a packet to the destination. Because of dropping the packets by selfish nodes or delayed, the network has to resend the data packets. Resending the data packets in the network will cause network power loss and decreases network lifetime and increase the average end-to-end delay. So, the proposed method can detect the selfish nodes soon, and it will reduce the end-to-end delay of the data packets. Some selfish nodes on the malicious nature also exist in the network, which increases

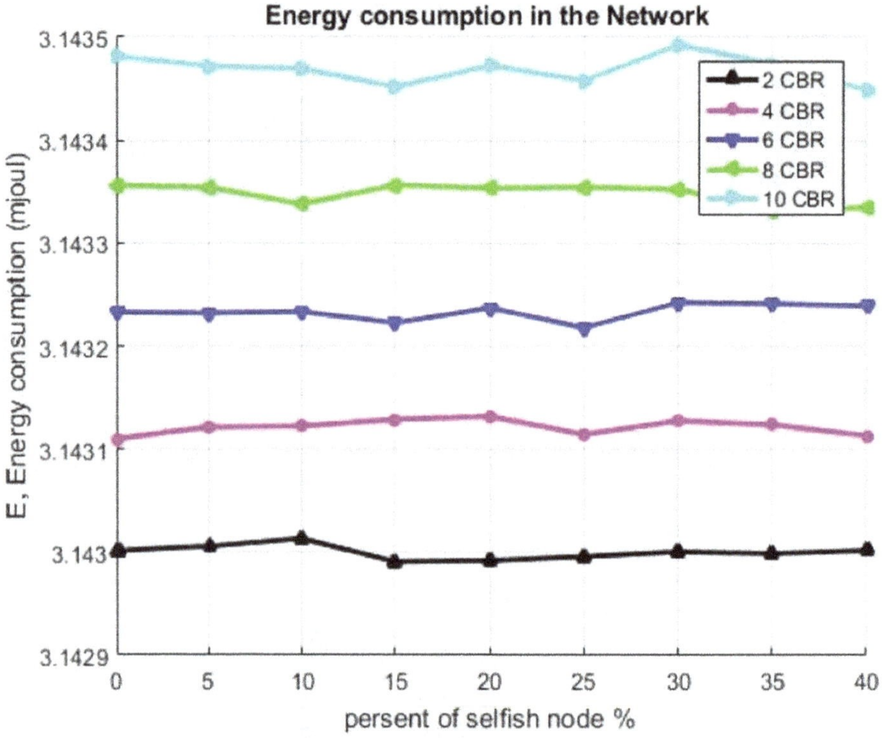

Fig. 14 Comparison of energy consumption in different traffic (2, 4, 6, 8, 10 CBR)

the average end-to-end delay in the network, so that the data packets remain in the buffer of the intermediate node until its survival, then it is sent and the packet to be discarded, because the survival time of the packet is expired, and this increases the average end-to-end delay in the network. Table 2 shows the numerical comparison of this delivery time in milliseconds. The advantages of the proposed method are detecting selfish nodes at high speeds, so the effects on the network are less. The node is detected, and the packets are lost less than other methods at the very beginning of selfish behavior. The proposed method with low end-to-end delay is appropriate for emergency applications, and real-time applications play critical role times and can provide service in that particular application.

The lower the energy consumed in the networks, the higher the efficiency of the proposed method. Due to the random acceleration of mobile nodes in the IoT network, energy consumption varies in a certain range. The average energy consumption in all nodes in the IoT network varies 3.1429~3.1435 μJ. When traffic change in network and the total number of sent packets increase, the power consumption also increases. Figure 14 shows average energy consumption in the network during 20 rounds which are the network continues the normal work, and the proposed method detects the selfish nodes and the network applies different traffic 2 to 10 CBR. Figure 14 illustrates less energy consumption than Fig. 7c. Due to the proposed method, detecting the selfish and malicious nodes prevents further energy consumption and reduces energy consumption.

6 Conclusion

The paper presented a new multi-person game theory based, which is used to detect the selfish and malicious node in the IoT. The proposed method combines advantages from reputation and game theory-based methods. The proposed method is a multi-step method that is performed games between nodes in a clustered network when sending or forwarding the node's data packets. Each player independently chooses their own strategy for forwarding or not forwarding; during the game, each player tries to increase their payoffs in the game. The performance of the method has been tested on the network and compared with PPS [14], game theory-based [29], and trust management [32]. The results have shown that the proposed method can detect selfish and malicious nodes efficiently and prevent increasing of end-to-end delay of the data packets to reach the destination and consumption of node resources (energy, battery, memory, etc.). The average throughput which is the percentage of successful delivery of the data packets to the destination is up to 20%, and simultaneously the average end-to-end delay in the delivery of the data packets is reduced by 12%. Also, the percentage of selfish and malicious nodes increased by 10% compared to similar methods, and the false-positive rate and false-negative rate, indicating the accuracy of the selfish node detection, decreased by 8%. Ultimately, the proposed approach gives the malicious and selfish nodes the second opportunity to cooperate with other nodes, and it does not isolate the nodes immediately from the network, because by increasing the percentage of such nodes, the network actually loses its performance.

Abbreviations

DA: Detection accuracy; FN: Number of nodes which are selfish nodes but detected as normal nodes; FNR: False-negative rate; FP: Normal selfish node detected as normal node; FPR: False-positive rate; TN: Total number of normal nodes detected by mistake; TP: Number of selfish nodes detected

Acknowledgements

Not applicable

Authors' contributions

SN and HGG contributed to the main idea and drafted the manuscript, algorithm design, and performance analysis. AK performed the statistical analysis. AMR conceived the study, participated in its design and coordination, and helped to draft the manuscript. All authors read and approved the final manuscript.

Funding

The research was not funded.

Competing interests

The authors declare that they have no competing interests.

Author details

[1]Department of Computer Engineering, Science and Research Branch, Islamic Azad University, Tehran, Iran. [2]Iran Telecom Research Center (ITRC), Tehran, Iran.

References

1. J. Guth *et al.*, Institute of Architecture of Application Systems A detailed analysis of IoT platform architectures: concepts, similarities, and differences (2018).
2. G. Kotonya, IoT architectural framework: connection and integration framework for IoT systems, pp. 1–17, 2018.
3. Liu, Xin, et al. Multi-modal cooperative spectrum sensing based on Dempster-Shafer fusion in 5G-based cognitive radio. IEEE Access 6. 199-208, 2018.
4. Liu, Xin, et al. A novel multi-channel Internet of Things based on dynamic spectrum sharing in 5G communication. IEEE Internet of Things Journal, 2018.
5. F. Olivier, G. Carlos, and N. Florent, New security architecture for IoT network, *Procedia - Procedia Comput. Sci.*, vol. 52, no. BigD2M, pp. 1028–1033, 2015.
6. T. Revathi, Applied fuzzy heuristics for automation of hygienic drinking water supply system using wireless sensor networks, J. Supercomput., 2018.
7. B. Kim, K. Psannis, H. Bhaskar, Special section on emerging multimedia technology for smart surveillance system with IoT environment. *J. Supercomput.* **73**(3), 923–925 (2017)
8. X. Liu et al., 5G-based green broadband communication system design with simultaneous wireless information and power transfer. *Physical Communication* **28**, 130–137 (2018)
9. J. Choi, Y. In, C. Park, S. Seok, H. Seo, and H. Kim, Erratum. Secure IoT framework and 2D architecture for end-to-end security Erratum J Supercomput, *J. Supercomput.*., vol. 2, no. i, p. 112–127, 2016.
10. S. Marti, T. J. Giuli, K. Lai, and M. Baker, Mitigating routing misbehavior in mobile ad hoc networks, *Proc. 6th Annu. Int. Conf. Mob. Comput. Netw. MobiCom 00*, vol. 1, no. 18, pp. 255–265, (2000).
11. S. Buchegger, Performance analysis of the CONFIDANT protocol (Cooperation Of Nodes: Fairness In Dynamic Ad-hoc NeTworks) background: the DSR protocol, pp. 226–236.
12. S. Bansal and M. Baker, Observation-based Cooperation Enforcement in Ad hoc Networks. http://arxiv.rog/pdf/cs.NI/0307012. (2003)
13. G. B.-M., K.N. B.-C., Z.-K. Chong, S.-W. Tan, Outwitting smart selfish nodes in wireless mesh networks. Int. J. Commun. Syst. **23**(5), 633–652 (2010)
14. A. Jesudoss, S. V. Kasmir Raja, and A. Sulaiman. Stimulating truth-telling and cooperation among nodes in VANETs through payment and punishment scheme, *Ad Hoc Netw.*, vol. 24, no. PA, pp. 250–253, 2015.
15. W. Zhang, S. Zhu, J. Tang, and N. Xiong, A novel trust management scheme based on Dempster–Shafer evidence theory for malicious nodes detection in wireless sensor networks, J. Supercomput., 2017.
16. L. Buttyán and J. Hubaux, Stimulating Cooperation in Self-Organizing Mobile Ad Hoc, pp. 579–592, 2003.
17. Srikanth, B. Detecting selfish nodes in MANETs. Doctoral dissertation, 2014.
18. S. Nobahary, S. Babaie, A credit-based method to selfish node detection in mobile ad-hoc network. *Applied Computer Systems* **23**(2), 118–127 (2018)
19. J. Chen and Y. R. Yang, Sprite: A simple, cheat-proof, credit-based system for mobile ad-hoc networks, vol. 00, no. C, pp. 1987–1997, 2003.
20. R. Kaushik and J. Singhai, MODSPIRITE: a credit based solution to enforce node cooperation in an ad-hoc network, vol. 8, no. 3, pp. 295–302, 2011.
21. J.S.S. Uma, Enhanced intrusion detection & prevention mechanism for selfishness in MANET. *Int. J. Innov. Res. Comput. Commun. Eng.* **3**(10), 10131–10138 (2015)
22. Kumar, Sunil, Kamlesh Dutta, and Girisha Sharma. A detailed survey on selfish node detection techniques for mobile ad hoc networks. Parallel, Distributed and Grid Computing (PDGC), 2016 Fourth International Conference on. IEEE, 2016.
23. K. Balakrishnan, J. D. J. Deng, and V. K. Varshney, TWOACK: preventing selfishness in mobile ad hoc networks, IEEE Wirel. Commun. Netw. Conf. 2005, vol. 4, no. C, pp. 0–5, 2005.
24. Kejun Liu, Jing Deng, P. K. Varshney, and K. Balakrishnan, An acknowledgment-based approach for the detection of routing misbehavior in MANETs, *IEEE Trans. Mob. Comput.*., vol. 6, no. 5, pp. 536–550, 2007.
25. E. M. Shakshuki, S. Member, N. Kang, and T. R. Sheltami, EAACK—a secure intrusion-detection system for MANETs, vol. 60, no. 3, pp. 1089–1098, 2013.
26. M. Bounouni, Acknowledgment-based punishment and stimulation scheme for mobile ad hoc network, J. Supercomput., 2018.
27. Basar, Tamer, and Geert Jan Olsder. Dynamic noncooperative game theory. Vol. 23. Siam, 1999.
28. Z. Ji, W. Yu, K.J.R. Liu, A game theoretical framework for dynamic pricing-based routing in self-organized MANETs. *IEEE J. Sel. Areas Commun.* **26**(7), 1204–1217 (2008)
29. A. H. Networks, Y. Sun, Y. Guo, Y. Ge, S. Lu, and J. Zhou, Improving the transmission efficiency by considering non-cooperation in, vol. 56, no. 8, 2013.
30. M. Touati, R. El-Azouzi, M. Coupechoux, E. Altman, J.M. Kelif, A controlled matching game for WLANs. *IEEE J. Sel. Areas Commun.* **35**(3), 707–720 (2017)
31. C. Vijayakumaran, An integrated game theoretical approach to detect misbehaving nodes in MANETs, pp. 173–180, 2017.
32. W. Zhang et al., A novel trust management scheme based on Dempster-Shafer evidence theory for malicious nodes detection in wireless sensor networks. J. Supercomput. **74**(4), 1779–1801 (2018)
33. K. J. Sathish, M.A. Zaveri, Hierarchical clustering for dynamic and heterogeneous Internet of Things. Procedia Computer Science **93**, 276–282 (2016)
34. T. Sheltami, et al. Video transmission enhancement in presence of misbehaving nodes in MANETs. *Multimedia Systems* **15**(5), 273–282 (2009)

10

Fast compression of OFDM channel state information with constant frequency sinusoidal approximation

Avishek Mukherjee and Zhenghao Zhang*

Abstract

We propose CSIApx, a very fast and lightweight method to compress the channel state information (CSI) of Wi-Fi networks. CSIApx approximates the CSI vector as the linear combination of a small number of base sinusoids on constant frequencies and uses the complex coefficients of the base sinusoids as the compressed CSI. While it is well-known that the CSI vector can be represented as the linear combination of sinusoids, fixing the frequencies of the sinusoids is the key novelty of CSIApx, which is guided by our mathematical finding that almost any sinusoid can be approximated by a set of base sinusoids on constant frequencies. CSIApx enjoys very low computation complexity, because key steps in the compression can be pre-computed due to the constant frequencies of the base sinusoids. We extensively test CSIApx with both experimental and synthesized Wi-Fi channel data, and the results confirm that CSIApx can achieve very good compression ratio with little loss of accuracy.

Keywords: Channel state information, Compression, Approximation

1 Introduction

In Wi-Fi, the channel state information (CSI) for an antenna pair is a vector of complex numbers, representing the channel coefficients of the orthogonal frequency division multiplexing (OFDM) subcarriers. The CSI is needed to calculate the modulation parameters for techniques such as multi-user multiple-input-multiple-output (MU-MIMO). In Wi-Fi, the CSI is typically measured at the receiver and is transmitted back to the sender, which requires significant overhead. For example, on a 20-MHz channel with 64 subcarriers, the full CSI for a single antenna pair has 64 complex numbers, and for 9 antenna pairs, 576. Although Wi-Fi does not use all subcarriers, the feedback for 9 antenna pairs still may exceed 1000 bytes. The Wi-Fi standard [1] defines options to compress the CSI, such as reducing the quantization accuracy or the number of subcarriers in the feedback or using the Given's rotation on the V matrix after the singular value decomposition (SVD) of the CSI matrix. However, these methods either reduce the accuracy of the CSI or only achieve modest compression ratios. For example, a 3 by 3 complex V matrix can only be compressed into 6 real numbers, at a compression ratio of 3.

*Correspondence: zzhang@cs.fsu.edu
Department of Computer Science, Florida State University, Tallahassee, FL, USA

In this paper, we propose CSIApx, a novel CSI compression method for Wi-Fi networks with the following key features: (a) *high compression ratio*, e.g., capable of compressing the CSI of 40 subcarriers in most of our Wi-Fi experiments into just 6 or less complex numbers; (b) *little loss of accuracy*, e.g., the decompressed CSI is very close to the measured CSI; and (c) *very low computation complexity*, suitable for hardware implementation. Compared to the Givens rotation, CSIApx achieves higher compression ratio, e.g., more than 2.5 times higher with our experimental data. Compared to other recent solutions in the literature [2–4], CSIApx has much lower computation complexity, because its main computation is simply the dot products between the CSI and a small number of constant vectors.

CSIApx is based on the well-known fact that the CSI is the linear combination of sinusoids [3, 4], where each sinusoid is the result of a physical path in the channel. Previous approaches solve complex optimization problems to find the parameters of the physical paths, i.e., the frequencies, the phases, and the amplitudes of the sinusoids. Radically different from such solutions, CSIApx does not attempt to find the parameters of the physical paths. Instead, CSIApx uses the linear combination of a set of *base sinusoids*, which are on *fixed frequencies*, to *approximate* the CSI, and our results show that the approximation usually achieves very high accuracy. This approach is guided by our mathematical finding that *the linear combination of sinusoids on constant frequencies can approximate any given sinusoid very well*, which is explained in more details in Section 3. Roughly speaking, as each individual sinusoid can be approximated by just a small number of base sinusoids, the entire CSI, which is the summation of the individual sinusoids, can also be approximated, *regardless of the number and characteristics of actual paths in the channel.* Working with base sinusoids on fixed frequencies has two major benefits. First, it avoids solving complex optimization problems and dramatically reduces the computation complexity. Second, as the frequency values of the base sinusoids are constants, they do not need to be transmitted, further improving the compression ratio. While CSIApx is primarily designed for compressing the CSI of Wi-Fi channels, it will theoretically work on any OFDM-based system that measures the channel state information, as long as the delay spread is not too large.

The rest of the paper is organized as follows. Section 2 discusses related work. Section 3 explains the theoretical foundations. Section 4 describes CSIApx. Section 5 describes the evaluation. Section 6 concludes the paper.

2 Related work

CSI compression has been a major topic of interest due to its practical importance in wireless networks. The differences between CSIApx and the existing methods in Wi-Fi have been discussed in Section 1. Some early approaches [5–7] use quantization and general purpose compression techniques like the Huffman coding, with typical compression ratios around 3:1, lower than that with CSIApx. Another popular approach is to reduce the frequency or the amount of CSI feedback in slow-varying channels similar to those in [8, 9], which complements compression techniques such as CSIApx. AFC [10] chooses from existing compression options for the least degradation of link throughput, which can be complemented by CSIApx as an additional compression option. The problem of consolidating CSI from a small number of settings to predict the CSI under other settings has been studied in [11], which is different from CSI compression.

Sparsity in certain wireless channels has been well-known [3, 4, 12–14] and has been exploited in applications such as CSI compression and CSI estimation. Unlike existing work that usually still attempts to find the actual paths by solving optimization problems, CSIApx focuses on *approximating* the CSI with *constant* frequency sinusoids, which leads to much lower computation complexity than the existing algorithms such as CTDP [3, 4]. CSIApx also has much lower complexity than CSIFit [2], which uses the Levenberg-Marquardt (LM) algorithm to compress the CSI of MIMO channels, because the LM algorithm needs to solve a set of linear equations in each iteration. R2-F2 [15] uses the CSI measured on one direction of the wireless link to predict that of the other, which requires much more computation than CSIApx; in addition, R2-F2 depends on *channel reciprocity*, which may not be true depending on the hardware circuitry, and may also need periodical calibrations [16], which is why Wi-Fi defines explicit CSI feedback and does not solely depend on channel reciprocity to obtain the CSI.

We have presented an initial version of CSIApx in [17] and have obtained a patent [18]. Compared to the conference version, this paper contains significant improvements in the theoretical foundation and the compression ratio in practice.

3 Theoretical foundation

We prove that a *target sinusoid* on frequency g can be approximated as the linear combination of P *base sinusoids* and the approximation error decays exponentially fast as the number of base sinusoids increases, where the frequencies of the base sinusoids are constants. Therefore, the summation of many sinusoids, such as the CSI vector, can still be approximated as the linear combination of only P base sinusoids.

3.1 Approximating real sinusoids

We begin with a theorem on the approximation of real sinusoids.

Theorem 1 *Consider P base sinusoids on evenly spaced frequencies: $\{\cos(0x), \cos(\delta x), ..., \cos[(P-1)\delta x]\}$. Suppose $\cos(gx)$ is to be approximated as the linear combination of the base sinusoids for $x \in [0, X]$, where $0 \leq g \leq (P-1)\delta$. If $X\delta < 1$, there exists an approximation with error decaying exponentially fast as P increases.*

Proof Consider any fixed time instant $x_0 \in [0, X]$ and the function $G(g) = \cos(x_0 g)$ where g is the variable. Let

$$F(g) = \sum_{k=1}^{P} \gamma_k \cos[x_0(k-1)\delta],$$

where

$$\gamma_k = \prod_{h=1,h\neq k}^{P} \frac{(h-1)\delta - g}{(h-k)\delta}.$$

and is called the *coefficient* of base sinusoid k. $F(g)$ is the Lagrange interpolation of function $G()$ at g based on the values of $G()$ at $0, \delta, \ldots, (P-1)\delta$. As function

$G()$ is infinitely differentiable, the Lagrange interpolation guarantees that

$$|F(g) - G(g)| \le \left| \frac{G^{P+1}(\xi) \prod_{k=1}^{P} [\,(k-1)\delta - g]}{(P+1)!} \right|,$$

for some ξ within $[\,0, X]$. Clearly,

$$|G^{P+1}(\xi)| \le X^{P+1},$$

and we claim that

$$\left| \prod_{k=1}^{P} [\,(k-1)\delta - g] \right| \le \delta^P (P-1)!$$

To see this, suppose $g \in [\,(k-1)\delta, k\delta]$ for some k. Clearly,

$$\begin{aligned}\left| \prod_{k=1}^{P} [(k-1)\delta - g] \right| &\le \prod_{h=1}^{k} [k\delta - (h-1)\delta] \prod_{h=k+1}^{P} [(h-1)\delta - (k-1)\delta] \\ &\le \delta^P k!\,(P-k)! \\ &\le \delta^P (P-1)!\end{aligned}$$

Therefore,

$$|F(g) - G(g)| \le \frac{X^{P+1}\delta^P}{P^2 + P}.$$

Clearly, if $X\delta < 1$, the interpolation error decays exponentially fast as P increases. Finally, note that the argument is true for any point in $[\,0, X]$. □

3.2 Extensions

We now discuss a few extensions.

Extension to $\sin(gx)$: With exactly the same arguments, it can be proved that $\sin(gx)$ can be approximated as the linear combination of the base sinusoids. Clearly, for the approximation of $\sin(gx)$, exactly the same coefficients as the coefficients for the approximation of $\cos(gx)$ can be used, because the coefficients are determined only by g and the frequencies of the base sinusoids.

Extension to [−X, X] : The extension to $[-X, X]$ is immediate. That is, if the approximation matches $\cos(gx)$ in $[\,0, X]$, multiplying the base sinusoids by the same coefficients for x in $[-X, 0]$ should also result in a match, because the target sinusoid and the base sinusoids have the same parity.

Extension to e^{igx} : The extension to complex wave e^{igx} is clearly

$$\sum_{k=1}^{P} \gamma_k \cos(f_k x) + i \sum_{k=1}^{P} \gamma_k \sin(f_k x) = \sum_{k=1}^{P} \gamma_k e^{if_k x}.$$

3.3 The summation of many sinusoids

So far, we considered approximating one target sinusoid. The CSI vector, on the other hand, may be the summation of many sinusoids. However, if the base sinusoids are selected correctly, any individual sinusoid in the CSI vector can be approximated, and therefore, the summation can also be approximated with the same set of base sinusoids. The deterministic maximum approximation error will be the summation of all individual approximation errors and may be large. However, in practice, the sinusoids in the CSI have random phases and the errors almost never add up constructively. A probabilistic bound therefore is more suitable; however, it depends on the assumptions of the path delay and power distribution. We instead choose to use both the real-world data and synthesized data to evaluate CSIApx and the results show that CSIApx approximates the CSI very well.

4 CSIApx

CSIApx is a fast compression method based on our theoretical findings. We begin with an overview.

4.1 Overview

According to our theoretical findings, the CSI can be approximated as the linear combination of the base sinusoids. The coefficients of the base sinusoids in the linear combination can be found by minimizing the *fit residual*, defined as the total squared error between the CSI vector and the approximation. The approximation is therefore called an *MSE Fit* and requires very low computation complexity, mainly because the linear system to be solved in the optimization problem is defined by a constant matrix. As any sinusoid can be approximated in this manner, the simplest approach would be to select just one set of base sinusoids to be used for all CSI. However, different types of channels have different delay spreads, which translate to different frequency ranges of the sinusoids in the CSI, the larger the delay spread, the higher the frequency. As sinusoids on lower frequencies can be approximated with fewer base sinusoids, to further improve the compression ratio, a small number of *configurations* are used in CSIApx which have different *orders*, where the order refers to the number of the base sinusoids. CSIApx finds the MSE Fit coefficients for all configurations and selects a configuration, considering both the compression ratio and the fit residual. The computation complexity is also reduced by sharing certain base sinusoids among multiple configurations.

4.2 The MSE Fit

4.2.1 *Finding the MSE Fit*

The core of CSIApx is to find the coefficients of the MSE Fit, denoted as a vector Γ.

To minimize the squared error is to select coefficients to minimize

$$J = \sum_{j=1}^{N} \left| \left(\sum_{k=1}^{P} \gamma_k e^{ijf_k} \right) - y_j \right|^2,$$

where P is the order; γ_k and f_k are the coefficient and frequency of base sinusoid k, respectively; y_j is element j in the CSI vector Y; and N is the length of the CSI vector. By taking the derivatives of J with respect to the coefficients and setting them to 0, Γ that minimizes J is the solution to a linear system $Q\Gamma = S$, where:

- Q is a P by P matrix, in which $q_{k,h} = \sum_{j=1}^{N} e^{i(f_h - f_k)j}$,
- S is a P by 1 vector, in which $s_k = \sum_{j=1}^{N} e^{-if_kj} y_j$.

It can be seen that S is determined by the CSI vector, but is just the dot products of the CSI vector and the conjugate of the base sinusoids. On the other hand, as the frequency values are constants, Q^{-1} is a constant matrix and can be pre-computed. Therefore, after S is obtained, Γ can be found by simply multiplying the constant matrix Q^{-1} with S.

4.3 The compression process of CSIApx

CSIApx has U configurations. Configuration u is defined by the frequencies of its P_u base sinusoids, denoted as $f_{u,k}$ for $k = 1, 2, ..., P_u$, where the frequency is the amount of angular rotation between neighboring points in the CSI vector. For each configuration, CSIApx solves the constant linear system in Section 4 to get the coefficients of the MSE Fit. To evaluate the quality of the fits, for each configuration, CSIApx evaluates the MSE Fit on L evenly spaced sample points, where $L = \frac{N}{4}$, and finds the total fit residual on the sampled points, denoted as η_u for configuration u. The MSE Fit is not evaluated on all points to reduce the computation complexity. CSIApx selects the fit coefficients of configuration u as the compressed CSI, if u is the smallest index satisfying $\eta_u < \zeta \min\{\eta_1, \eta_2, ..., \eta_U\}$, where ζ is a constant, empirically chosen as 1.75 and 4 for CSI with 64 and 40 subcarriers, respectively.

4.4 The configurations and the base sinusoids

The configurations and base sinusoids should be selected considering compression ratio, accuracy, implementation cost, and the range of the fit coefficients. As Wi-Fi has a fixed number of subcarriers and well-studied types of channels [19], we select a configuration for each type of channel.

4.4.1 Finding parameters for a given channel type

For a given type of channel, the parameters to be determined include the number of base sinusoids, denoted here as P, and the frequencies of each base sinusoid, denoted here as $f_1, f_2, ..., f_P$. Note that f_1 is actually always 0 to be able to cover the lowest frequency. Our approach is to first solve the problem of determining the frequencies of the base sinusoids for a given P, then conduct a linear scan on P to find the best P. As the linear scan is straightforward, in the following, we focus on the first problem.

As a channel has multiple paths, while paths of different delays may not have equal power, the selection of the base sinusoids will have to be based on the power profile of the channel. For example, if 90% of the power is concentrated within a certain small delay range, more base sinusoids should be allocated to the corresponding range of frequencies. The power profile is generated using the delay taps and relative power values for the given channel model. It also takes into account possible imperfect symbol-level synchronization, which is assumed to be uniformly within 0 to 50 ns, because such synchronization error will increase the frequencies of the sinusoids.

The problem of finding the base sinusoid frequencies for given P is solved using a standard solver like Levenberg-Marquardt. First, from 0 to the maximum frequency in the power profile, H evenly spaced frequencies are selected, denoted as g_1 to g_H. The objective function passed to the solver basically minimizes the total fit residual of all sinusoid on such frequencies, which are weighted according to the power profile. By doing so, the solver should find a set of base sinusoids that is expected to minimize the fit residual of this type of channel. To be more specific, the objective function is defined as follows with $f_2, f_3, ..., f_P$ as parameters:

$$\sum_{s=1}^{H} F_{f_2, f_3, ..., f_P}(g_s) W(g_s),$$

where $F_{f_2, f_3, ..., f_P}(g_s)$ is the MSE Fit residual of a pure sinusoid of frequency g_s when the base sinusoid frequencies are $0, f_2, f_3, ..., f_P$, and $W(g_s)$ is the power profile weight of frequency g_s. It should be noted that some noise is added to the pure sinusoids such that the signal-to-noise ratio (SNR) is 20 dB, to emulate a real-world scenario.

4.4.2 Results

Figure 1 shows, in log scale, the output of the solver for different numbers of base sinusoids of Wi-Fi channel model B to E, when the number of subcarriers is 64. It can be seen that the fit residual decreases exponentially as the number of bases increases up to a point, after which there is very little change to the final fit residual. In fact in some cases, the fit residual increases slightly with more bases. This is due to over-fitting, i.e., the MSE Fit trying to follow the noise in the data. Based on this result, the selected numbers of bases for each configuration are 5, 7, 11, and 17, respectively.

4.4.3 Manual tuning

It should be noted that two types of channel models, namely model A and model F, are excluded from this study. Channel model A is an ideal case, i.e., modeling a

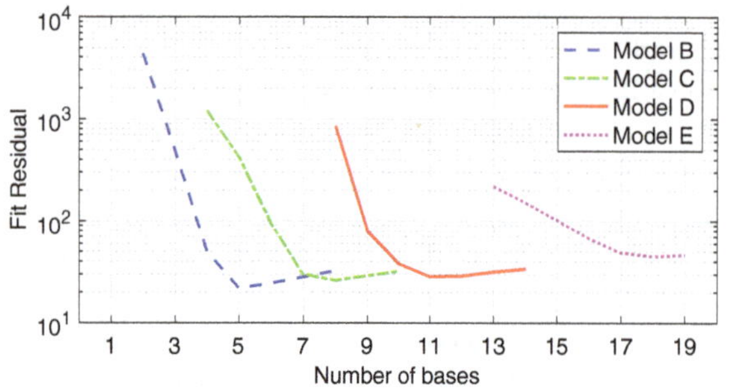

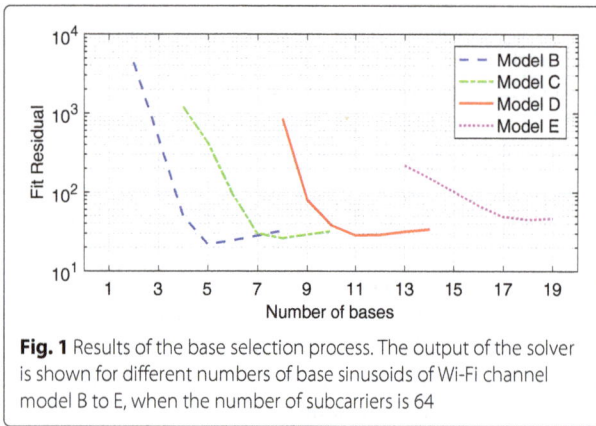

Fig. 1 Results of the base selection process. The output of the solver is shown for different numbers of base sinusoids of Wi-Fi channel model B to E, when the number of subcarriers is 64

channel with a single path, which we have covered with an additional simple configuration with three base sinusoids. Channel model F is excluded because it requires a large number of base sinusoids to approximate and as such does not bring in much in terms of compression. Moreover, based on our experience, model F rarely occurs in indoor Wi-Fi networks.

Figure 2 shows, in log scale, a comparison of the cumulative density function (CDF) of the fit residual per point when running MSE Fit using these optimized base sinusoids as opposed to using uniformly spaced base sinusoids. These tests were performed on 1000 CSI vectors with 20-dB SNR for each channel model. It can be seen that the optimized base sinusoids often resulted in an order of magnitude or greater reduction in the approximation error.

We note that the main computation cost in CSIApx is actually finding the dot products between the base sinusoids and the CSI. If a base sinusoid used in one configuration can be used by another, the computation cost can be reduced.

Therefore, we slightly modified the optimized base sinusoid frequencies, to allow one base sinusoid to be used for more than one configuration when possible. For 64 subcarriers, the selected base frequencies for the five configurations are {0, 0.06, 0.12}, {0, 0.05, 0.1, 0.15, 0.25}, {0, 0.06, 0.12, 0.18, 0.24, 0.3, 0.42}, {0, 0.06, 0.12, 0.18, 0.24, 0.3, 0.36, 0.42, 0.525, 0.6375, 0.75}, and {0, 0.075, 0.15, 0.225, 0.3, 0.375, 0.45, 0.525, 0.6, 0.7, 0.8, 0.9, 1.0, 1.1, 1.2, 1.3}, respectively. In total, there are 27 unique base sinusoids. Figure 1 also shows the CDF of the fit residual when running MSE Fit with this modified set of base sinusoids. It can be seen that there is a small difference under 10% when compared against the optimized base frequencies. However, these modified configurations cut down the complexity of CSIApx considerably. It should also be noted that for channel model E, the number of base sinusoids was reduced from 17 to 16, as there is very little loss in accuracy at 20 dB, but leads to a higher compression ratio.

We repeated the same process also for CSI vectors of 40 subcarriers, which is used in our experiments. The five configurations are {0, 0.05, 0.10}, {0, 0.06, 0.12, 0.2},

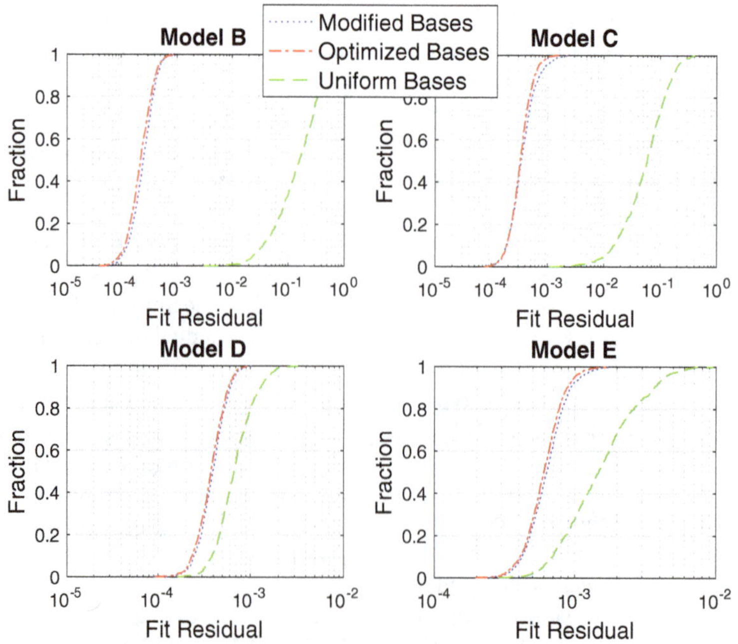

Fig. 2 Fit residual as a function of the number of base sinusoids. A comparison of the cumulative density function (CDF) of the fit residual per point is shown when running MSE Fit using optimized base sinusoids versus uniformly spaced base sinusoids

{0, 0.075, 0.15, 0.225, 0.3, 0.45}, {0, 0.075, 0.15, 0.225, 0.3, 0.375, 0.525, 0.675, 0.825, 0.975}, and {0, 0.09, 0.18, 0.27, 0.36, 0.45, 0.575, 0.7, 0.825, 0.95, 1.075, 1.2, 1.325, 1.45}, respectively. In total, there are 27 unique base sinusoids.

4.5 Complexity of CSIApx

Overall, let W be the total number of unique base sinusoids in all configurations; the complexity of CSIApx, measured by the number of complex multiplications, includes only:

- WN: for computing the dot products between the base sinusoids and the CSI
- $\sum_{u=1}^{U} P_u^2$: for finding the fit coefficients of all configurations
- $\sum_{u=1}^{U} (P_u + 1)L$: for computing the fits at sampled points and the sampled fit residuals

4.6 Coping with shift frequency

In practice, the raw measured CSI often has a *shift frequency*, which is a frequency value added to the frequencies of all sinusoids, caused by the sample timing offset to the OFDM symbol boundary. The shift frequency needs to be removed before running CSIApx, because it may force CSIApx to choose higher configurations to approximate sinusoids on higher frequencies and reduce the compression ratio. This can easily be achieved by multiplying the CSI with a sinusoid on the negative of the shift frequency, a process we call *rotation*. The value of the shift frequency is known to the wireless receiver, because it selects the OFDM symbol boundary. The frequency used in the rotation can also be slightly adjusted to make sure that the sinusoids in the CSI are still on positive frequencies after the rotation.

5 Results and discussion

We report our evaluation of CSIApx on both real-world and synthesized CSI data in this section.

5.1 Evaluation with the experimental CSI data

We first discuss our evaluation of CSIApx with the real-world experimental CSI data.

5.1.1 Data collection and preprocessing

CSI data was collected using the Atheros CSITool [20] installed on two laptops with the Atheros AR9462 wireless card with two antennas on 20-MHz channels. A total of 100 experiments in various location settings were conducted, which include typical environments like office buildings, apartment complexes, and large hallways. The experiments include both line of sight and non-line of sight cases as well as varying channel conditions due to human movements near the machines. Some of the experiment locations are shown in Fig. 3.

The CSITool reports the CSI on 56 selected subcarriers for four antenna pairs. Figure 4 shows the absolute values of some raw CSI vectors, where it can be seen that the data has some level of noise. A few preprocessing steps were taken before passing the data for compression. Firstly, as the signal always seems to be attenuated at both ends of the spectrum, caused most likely by additional filtering in hardware, not representing the characteristics of the actual channel, 8 subcarriers on both ends are removed, with only the middle 40 subcarriers kept. Secondly, as some of the experiments have very weak signals, measurements with RSSI lower than 30 dB were not used. Thirdly, the CSI data of all antenna pairs is normalized by a common factor such that the maximum amplitude is 1. Lastly, as explained before, each CSI vector is rotated to remove the shift frequency. As the shift frequency value is not reported by our current device to the driver level, a simple estimation method is used, which basically keeps rotating the CSI from the same transmitting antenna incrementally, until most energy appears to occupy only a spectrum starting from 0 up to some frequency for both receiving antennas. As it may over-rotate the CSI and lead to negative frequencies, when running CSIApx, the CSI for all

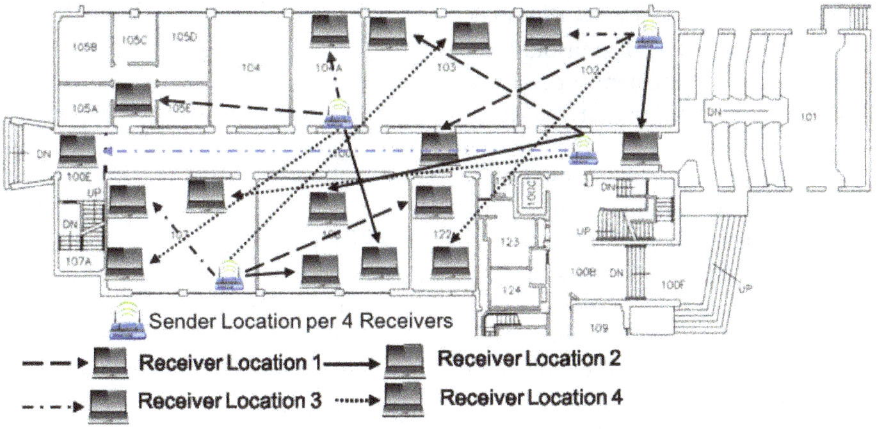

Fig. 3 Some experiment locations. Typical experimental locations are shown

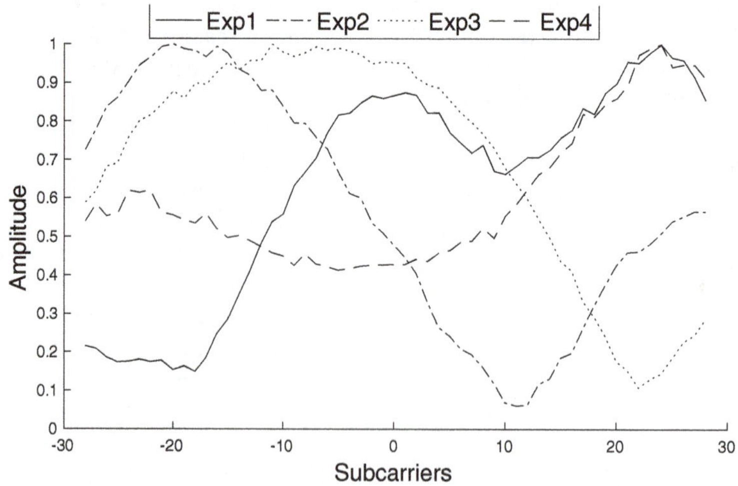

Fig. 4 Absolute values of some measured CSI. The absolute value of the CSI across 64 subcarriers is shown for the strongest antenna pair in four experiments

antenna pairs are multiplied by a sinusoid with a positive frequency of 0.0491 to move most sinusoids in the CSI to positive frequency.

5.1.2 Compared methods

For comparison, we also implement the CTDP extraction algorithm according to [4], referred to as CTDP, which iteratively selects a sinusoid that best matches the current residual signal, until the power of the selected sinusoid is below a threshold. CTDP is chosen because it is one of the more recent methods and has a good performance. As CTDP requires the noise power value, which needs to be estimated with the experimental data, we use the fit residual found by CSIApx as the total noise power, which should be very close. The frequency range of sinusoids scanned in CTDP is $[-0.785, 1.57]$, which should cover all frequencies in the CSI. Another constrained version of CTDP, referred to as cCTDP, is also evaluated, with which the fit residuals of CTDP and CSIApx can be compared when using similar number of sinusoids. That is, with cCTDP, the number of sinusoids used is the smallest upper bound of the average number of sinusoids used by CSIApx for the same CSI measurement, noting that CSIApx may use different configurations for the different antenna pairs. It should also be mentioned that as CTDP has to solve an optimization problem to select the frequency of a sinusoid in each iteration, it has much higher implementation complexity than CSIApx, because CSIApx avoids this problem altogether by using constant frequencies.

Two other methods were also implemented, but the results are not shown in the figures in this section, since their performances are not comparable with CSIApx and CTDP. One of the methods is CSIFit [2], the median of fit residual of which, for example, is more than six times than that with CSIApx for the experimental data, while the compression ratio is about the same. The other maybe obvious approach, i.e., keeping only a small number of significant FFT coefficients, is also not included, because it usually has an order of magnitude higher fit residual than CSIApx even when using twice number of coefficients.

5.1.3 Fit accuracy and compression ratio

As the fit residual and the compression ratio are related, i.e., improving one is often at the cost of the other, they are jointly compared. Figure 5 shows the CDF of the total fit residual of all four antenna pairs in 7923 CSI measurements. Figure 6 shows the compression ratio, which is the number of real numbers in the CSI vector divided by that needed by a compression method to describe the sinusoids, noting that a complex number consists of two real numbers.

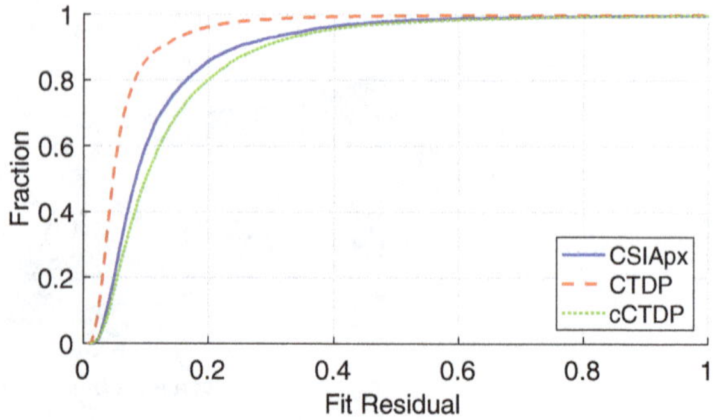

Fig. 5 Fit residual with experimental data. A comparison of the cumulative density function (CDF) of the fit residual per point is shown when running CSIApx versus CTDP. In addition, a comparison is also shown when running a constrained version of CTDP (CCTDP)

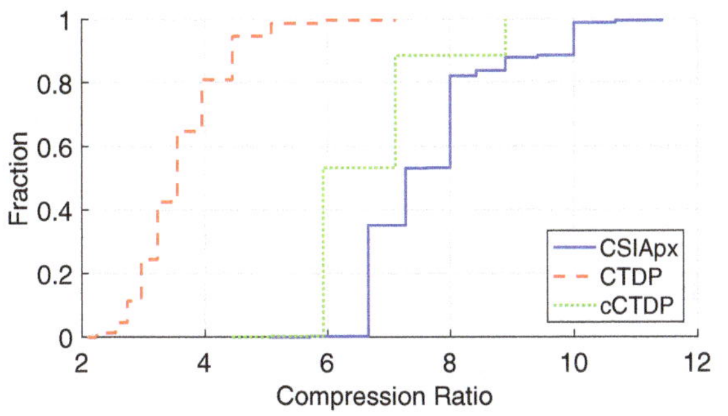

Fig. 6 Compression ratio with experimental data. A comparison of the cumulative density function (CDF) of the compression ratio achieved when running CSIAPx versus CTDP versus CCTDP

It can be seen that the fit residual of CSIApx in most cases are very small with a median of 0.0828 for all antenna pairs, which translates to an error of 0.0005 per data point. The fit residual of CTDP is better with a median of 0.0467, however it is at a cost of a much lower compression ratio, as the average compression ratio of CSIApx against 40 subcarriers is 7.68, much better than CTDP, which is 3.59. By closely examining the fits, we found that CSIApx actually fits the signal very well, and its fit residual is mainly the quantization noise, such as those shown in Fig. 4, which cannot be eliminated unless more sinusoids are introduced to fit the noise. In this sense, CSIApx achieves a better tradeoff between fit residual and compression. The better performance with CSIApx can also be seen from the cCTDP results, as cCTDP actually has higher fit residual, at the same time slightly lower compression ratio.

5.1.4 MU-MIMO rate

The end result of the compression method can be the MU-MIMO data rate of the users. We use the MU-MIMO rate program at [21], which first calculates the modulation parameters with the supplied imperfect CSI, then finds the achievable data rate when the selected parameters are used on the actual channel. We configure the program to use the greedy method for user selection and run at SNR of 20 dB. For each subcarrier, we run the program twice, feeding the compressed and the measured CSI to the program to obtain two values, representing total data rates to all users with imperfect and perfect CSI, respectively. The difference between the two, divided by the latter, is called the *normalized rate difference* and is used as the metric.

A total of 1500 tests are run, where each test has one sender and two receivers. In each test, we use the CSI collected from experiments where the sender was at a fixed location for four receivers, and randomly select two receivers from the four actual receivers. As the link is 2 by 2 but each MU-MIMO receiver has only one antenna, we select the first antenna for each receiver. Figure 7 shows the CDF of the normalized rate difference, where we can see that the rate difference with CSIApx is usually very small, e.g., within − 3% and 3% in over 98.3% of the cases. CTDP performs better reporting 99.0%, but this comes at the cost of its compression ratio. At similar compression ratio, cCTDP performs worse than CSIApx at 95.7%. The rate difference in some very rare cases can also be positive, since the greedy method sometimes selects different sets of users when given the compressed and measured CSI.

5.1.5 Parameter distribution

One of the nice features of CSIApx is that the fit coefficients stay in a small range, making it simple and inexpensive to quantize and transmit the coefficients as the compressed CSI. Figure 8 shows the distribution of the real and imaginary parts of the coefficients found by CSIApx for the strongest antenna pair in each test case, because the distributions for other antenna pairs should just be its scaled versions. We can see that all numbers resides in a small range with smooth density.

5.2 Evaluation with synthesized CSI data

We test CSIApx with synthesized CSI data, which complements the experimental evaluation by challenging CSIApx with more channel types and testing CSIApx under controllable settings like the SNR level.

5.2.1 CSI generation

We use the channel model code at [19] to generate CSI for all 64 subcarriers in Wi-Fi on 20-MHz channels for 3 by 3 links with nine antenna pairs. Four cases, referred to as model B, model C, model D, and model E, are used, which represent typical indoor environments with around 100-ns, 200-ns, 400-ns, and 800-ns delay spread, respectively, and should cover the majority of typical Wi-Fi channels. The maximum amplitude is again normalized to 1. White Gaussian noise is added to the

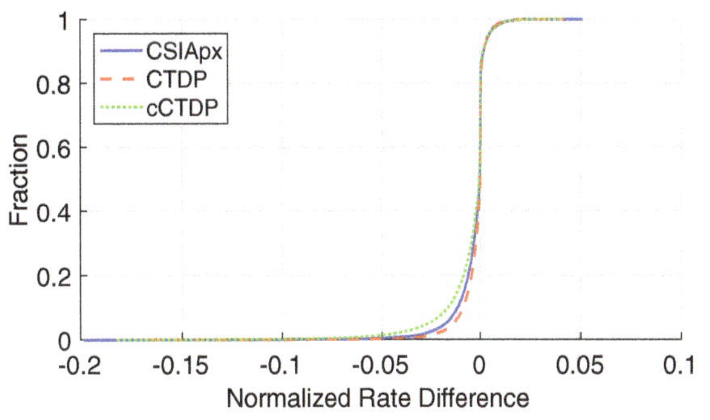

Fig. 7 Normalized rate difference with experimental data. A comparison of the cumulative density function (CDF) of the normalized rate difference is shown when running CSIApx versus CTDP versus CCTDP

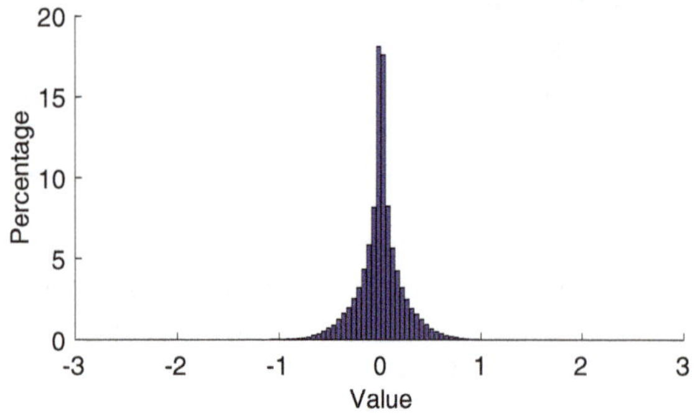

Fig. 8 Coefficient values with experimental data. The distribution of the real and imaginary parts of the coefficients found by CSIApx is shown for the strongest antenna pair in the each experimental test case

CSI vector, and a total of 1000 test cases are performed for each SNR level. To simulate imperfect rotation, the CSI is further multiplied by a sinusoid with random frequency, which translates to within 0 to 50 ns of timing error.

5.2.2 Fit accuracy and compression ratio

As the clean CSI is available, when calculating the final fit residual, the fit is compared with the clean CSI; all prior steps are still based on the noisy CSI. Figure 9 shows the mean of the total fit residual of all antenna pairs in various settings. The fit residual of CSIApx is usually very small, such as about 0.0007 or lower per point at 20 dB or above. In addition, as the noise level reduces by 5 dB, the fit residual in most cases also reduces by roughly 5 dB, suggesting that the fit residual is mostly noise. Figure 10 shows the average compression ratios. It can be seen that CSIApx achieves very high compression ratios in many cases, i.e., above 12.4:1, 7.9:1, 5.5:1, and 4.0:1 against 64 subcarriers for models B, C, D, E, respectively, when the SNR is 20 dB or above. More complicated channel conditions do pose a challenge to CSIApx as it has to use higher configurations. Also, although CSIApx may have slightly larger fit residual, it has much higher compression ratios than CTDP in all cases. In addition, the compression ratio CSIApx is more stable than CTDP for each model when the SNR is 20 dB or higher, suggesting the CSIApx is better at capturing the actual signal and less susceptible to the influence of noise. cCTDP has higher fit residual and lower compression ratios in almost all cases when the SNR is 20 dB or higher.

5.2.3 MU-MIMO rate

MU-MIMO rate is also tested in a similar manner as with the experimental data. Figure 11 shows the percentage of cases that the normalized rate differences are above 3% or lower than -3%, where we can see that the fraction is very low for CSIApx when the SNR is 25 dB or higher, and still reasonably small at 20 dB except for model E which is the most complicated.

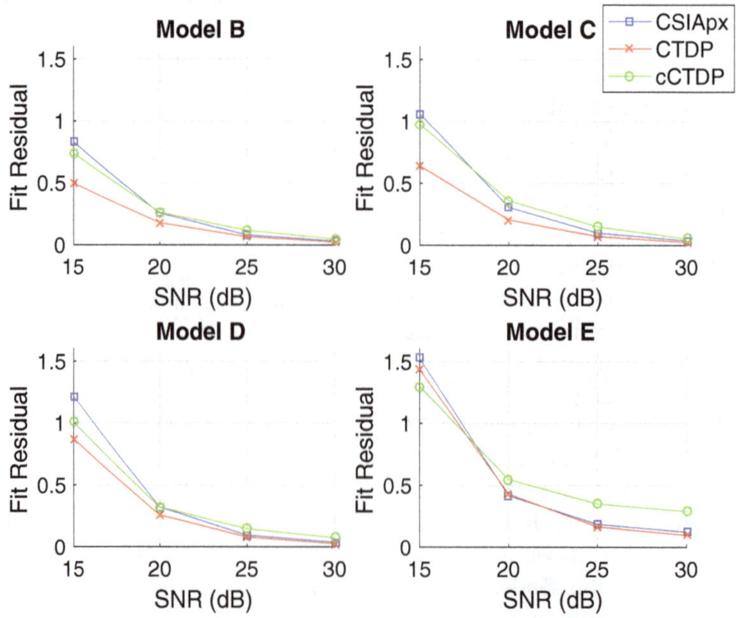

Fig. 9 Fit residual with model data. The mean of the fit residual per point is shown when running CSIApx versus CTDP versus CCTDP on the synthesized CSI data across four channel models B, C, D, and E and four SNR values 15 dB, 20 dB, 25 dB, and 30 dB

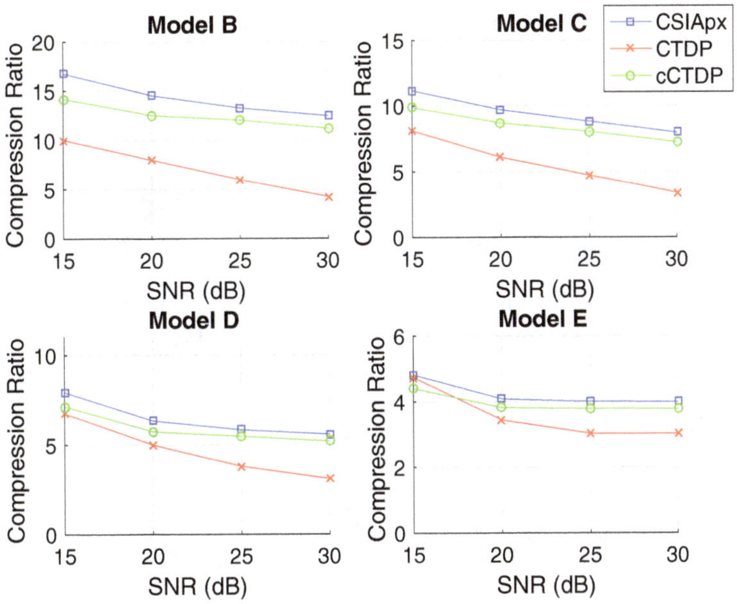

Fig. 10 Compression ratio with model data. The mean of the fit residual per point is shown when running CSIApx versus CTDP versus CCTDP on the synthesized CSI data across four channel models B, C, D, and E and four SNR values 15 dB, 20 dB, 25 dB, and 30 dB

5.2.4 Parameter distribution

Figure 12 shows the distribution of the fit coefficients by CSIApx for the strongest antenna pair, which is similar to that with the experimental data.

5.3 More compression with Huffman coding

Even higher compression ratio can be achieved for CSI-Apx by running Huffman coding on the coefficients, taking advantage of the fact that the distribution of the real and imaginary parts of the coefficients, such as that in Fig. 13, is spiky and has low entropy. The process is explained for the experimental data in the following. We empirically choose 12 bits for quantization in range $[-2.56, 2.56]$, which results in negligible quantization error and includes all coefficients. A training set with 3962 experiments is randomly chosen from the

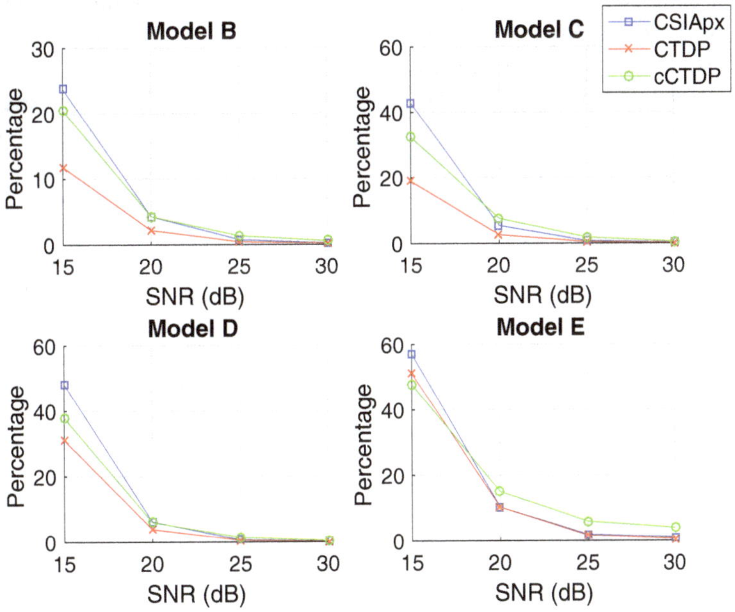

Fig. 11 MU-MIMO rate difference with model data. The percentage of cases where the normalized rate differences are above 3% or lower than −3% is shown when running CSIApx versus CTDP versus CCTDP on the synthesized CSI data

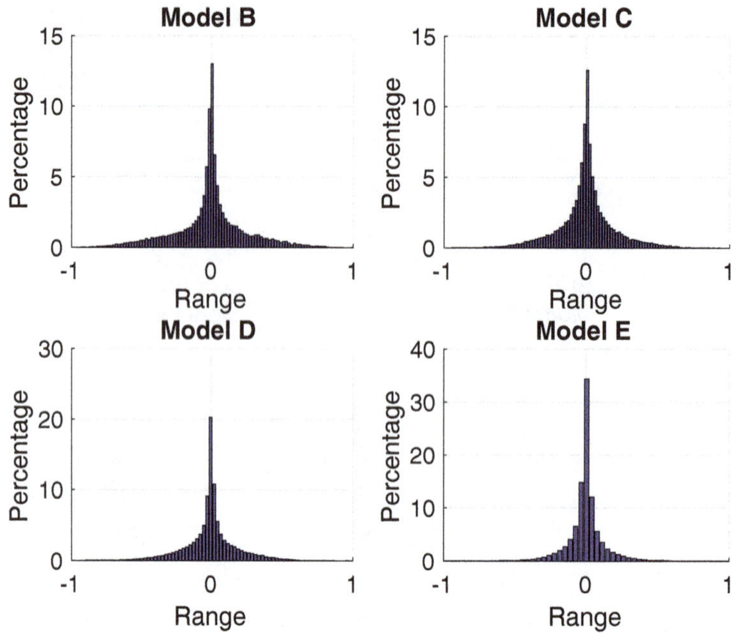

Fig. 12 Coefficient values with model data. The distribution of the fit coefficients found by CSIApx for the strongest antenna pair when evaluated on the synthesized CSI data

data to obtain the dictionary of the Huffman coding, which is then tested on the remaining data. Figure 13 shows the CDF of the compression ratio achieved by Huffman coding, where the ratio is calculated by subtracting the size of the raw coefficients by the size of the compressed coefficients then divided by the former. The average compression ratio is 22.1%, and the ratio is positive for over 98% of the cases. Separate Huffman coding dictionaries can also be built for each configuration; however, our results show that the improvement is marginal.

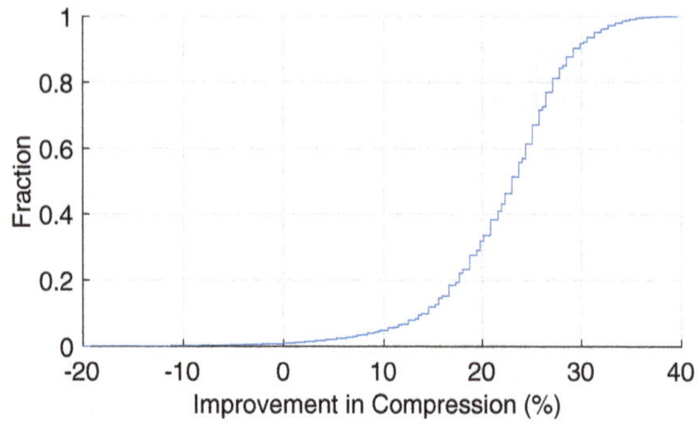

Fig. 13 Improvement with Huffman coding. The improvement in compression ratio when running Huffman coding after CSIApx is shown for the experimental data

5.4 Comparing with Givens rotation

The Wi-Fi standard includes an option to use the Givens rotation to compress CSI. That is, instead of sending the entire CSI, it calculates a compressed feedback matrix by zeroing out some elements, which is later reconstituted to obtain the full CSI. We provide a head-to-head comparison between CSIApx and the Givens rotation method and argue that CSIApx is a better alternative.

5.4.1 Fit accuracy and compression ratio

Givens rotation is lossless in the sense that the other end of the communication link can exactly reproduce the measured CSI. It therefore appears that Givens rotation will have higher accuracy than CSIApx, as CSIApx is based on approximation. However, this is only true when the measured CSI is clean, i.e., without any noise. With measurement noise and quantization noise, we found that CSIApx actually achieves better accuracy than the Givens rotation, i.e., the CSI with CSIApx follows the shape of the actual CSI more closely than the Givens rotation. From a high level, this is because when fitting a CSI curve, CSIApx serves as a filter and filters out most of the noise, whereas Givens rotation will simply preserve the noise.

Figure 14 shows this comparison between CSIApx and Givens rotation on the model data, where we can see that CSIApx indeed achieves lower fit residual. The model data is used because the clean CSI is available. For a more fair

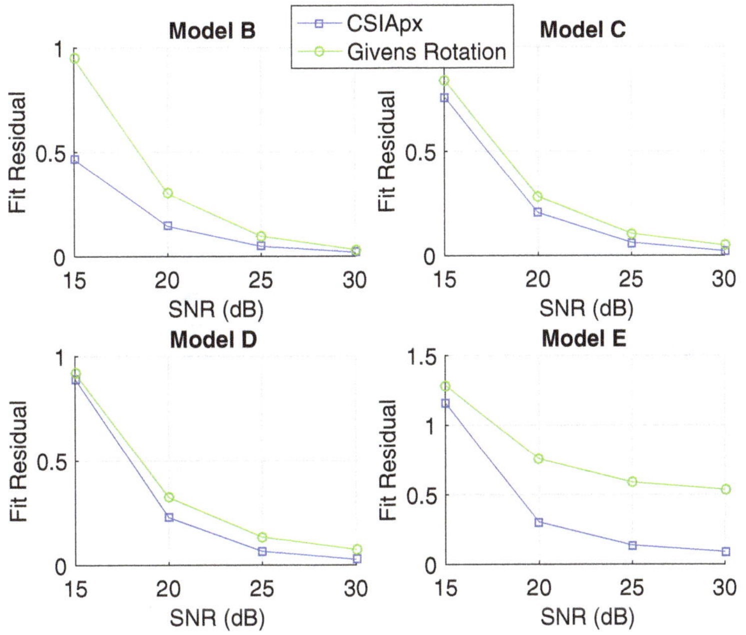

Fig. 14 Fit residual comparison with Givens rotation. The comparison between the fit residual achieved by CSIApx and Givens rotation is shown when running on the synthesized CSI data at four SNR values 15 dB, 20 dB, 25 dB, and 30 dB

comparison, before running the Givens rotation, a low-pass filter is used in an attempt to filter out some noise, as it is expected that such filter will be used in practice. Due to the low-pass filter, only the results of the middle 50 subcarriers are used in this comparison. The performance of CSIApx is better with the middle 50 subcarriers than with all subcarriers, because the subcarriers near the ends have less constraints in the fitting and have larger errors.

CSIApx will also enjoy a higher compression ration than the Givens rotation. Figure 15 shows the compression ratio for the experimental data. As mentioned earlier the average compression ratio achieved by CSIAPX for the 2×2 system in a real-world setting was 7.68. Because CSIApx fits each antenna pair individually, hence, it can remain constant even when the number of pairs increases. The compression ratio achieved by Givens rotation on the other hand will keep decreasing as the antenna order increase is approaching 2.

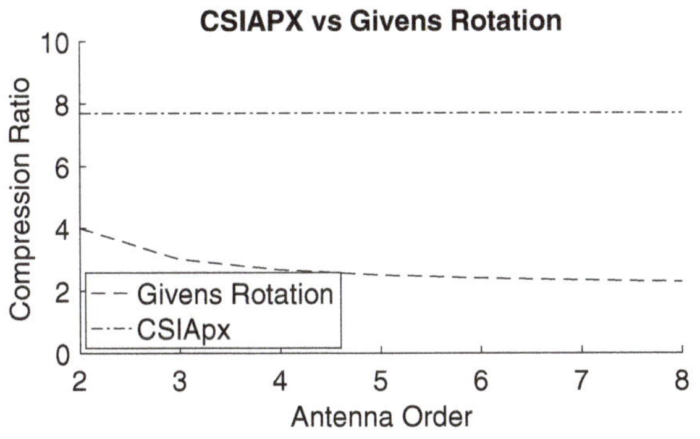

Fig. 15 Compression ratio comparison with Givens rotation. The comparison between compression ratio for CSIAPx versus Givens Rotation is shown for the experimental data

5.4.2 MU-MIMO rate

We further evaluate the fit accuracy by comparing the data rate achieved in a MU-MIMO setting from both CSI-Apx and Givens rotation. Figure 16 shows the percentage of cases where the normalized rate differences are higher than 3% or lower than − 3% when compared against the clean signal. We see that CSIApx outperforms Givens rotation.

6 Conclusion

We propose CSIApx, a fast and lightweight method for compressing the CSI of OFDM wireless links. We first prove that almost any sinusoid can be approximated as the linear combination of a small number of base sinusoids on constant frequencies. Exploiting the constant frequencies of the base sinusoids, CSIApx pre-computes key steps and finds a minimum square fit of the CSI vector with very few computations. We evaluate CSIApx with both experimental and synthesized CSI data, and the results show that CSIApx achieves very good compression ratios and approximation accuracy. We therefore believe CSIApx can be a very useful tool to be incorporated into the Wi-Fi protocol and will enable timely and accurate CSI feedback to improve the network performance.

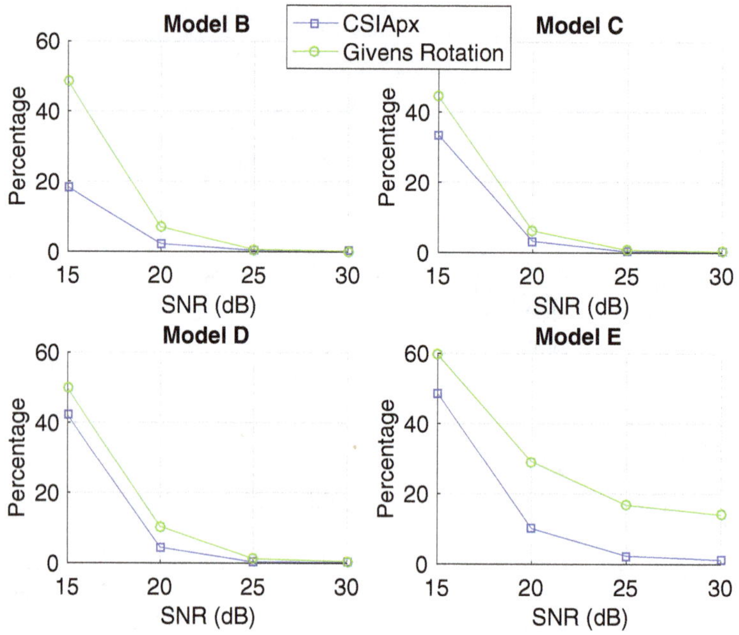

Fig. 16 MU-MIMO rate difference comparison with Givens rotation. The percentage of cases where the normalized rate differences are above 3% or lower than − 3% is shown when running CSIApx versus Givens rotation on the synthesized CSI data

Abbreviations

CSI: Channel state information; OFDM: Orthogonal frequency division multiplexing; MU-MIMO: Multi-user multiple-input-multiple-output; SVD: Singular value decomposition

Acknowledgments

The authors greatly appreciate the editor and reviewers for their work.

Funding

This research work was supported by the US National Science Foundation under Grant 1618358.

Authors' contributions

Both authors contributed to the work. Both authors read and approved the final manuscript.

Competing interests

The authors declare that they have no competing interests.

References

1. I. P. .-T. G. AC, Status of Project IEEE 802.11ac (2013). http://www.ieee802.org/11/Reports/tgac_update.htm. Accessed 15 Sept 2013
2. A. Mukherjee, Z. Zhang, in *2016 13th Annual IEEE International Conference on Sensing, Communication, and Networking (SECON)*. Channel state information compression for MIMO systems based on curve fitting (IEEE, London, 2016)
3. X. Wang, Channel feedback in OFDM systems (2014). Google Patents. US Patent No. 8,908,587. 9 Dec
4. X. Wang, S. B. Wicker, in *2013 IEEE Global Communications Conference (GLOBECOM)*. Channel estimation and feedback with continuous time domain parameters (IEEE, Atlanta, 2013), pp. 4306–4312
5. S. Ferguson, F. Labeau, A. M. Wyglinski, in *2010 IEEE 72nd Vehicular Technology Conference-Fall (VTC)*. Compression of channel state information for wireless OFDM transceivers (IEEE, Ottawa, 2010), pp. 1–5
6. V. P. G. Jiménez, T. Eriksson, A. G. Armada, M. J. F. García, T. Ottosson, A. Svensson, Methods for compression of feedback in adaptive multi-carrier 4G schemes. Wirel. Pers. Commun. **47**(1), 101–112 (2008)
7. U. K. Tadikonda, Adaptive bit allocation with reduced feedback for wireless multicarrier transceivers. PhD thesis (2007)
8. V. Pohl, P. H. Nguyen, V. Jungnickel, C. V. Helmolt, in *2003 IEEE Global Telecommunications Conference (GLOBECOM)*. How often channel estimation is needed in MIMO systems, vol. 2 (IEEE, San Francisco, 2003), pp. 814–818
9. K. Huang, R. W. Heath Jr., J. G. Andrews, Limited feedback beamforming over temporally-correlated channels. IEEE Trans. Sig. Process. **57**(5), 1959–1975 (2009)
10. X. Xie, X. Zhang, K. Sundaresan, in *19th Annual International conference on Mobile computing & networking (MOBICOM 2003)*. Adaptive feedback compression for MIMO networks (ACM, San Diego, 2013), pp. 477–488
11. R. Crepaldi, J. Lee, R. Etkin, S. Lee, R. Kravets, in *2012 Proceedings IEEE International Conference on Computer Communications INFOCOM*. CSI-SF: Estimating wireless channel state using CSI sampling & fusion (IEEE, Orlando, 2012), pp. 154–162
12. C. Carbonelli, S. Vedantam, U. Mitra, Sparse channel estimation with zero tap detection. IEEE Trans. Wirel. Commun. **6**(5), 1743–1763 (2007)
13. F. Wan, W. Zhu, M. Swamy, Semi-blind most significant tap detection for sparse channel estimation of OFDM systems. IEEE Trans. Circuits Syst. I. Reg. Papers. **57**(3), 703–713 (2010)
14. M. Duarte, Y. Eldar, Structured compressed sensing: from theory to applications. IEEE Trans. Signal Process. **59**(9), 4053–4085 (2011)
15. D. Vasisht, S. Kumar, H. Rahul, D. Katabi, in *Proceedings of the 2016 ACM SIGCOMM Conference*. Eliminating channel feedback in next-generation cellular networks (ACM, Florianopolis, 2016)
16. M. Guillaud, D. T. M. Slock, R. Knopp, in *2005 Proceedings of the Eighth International Symposium on Signal Processing and Its Applications, (ISSPA)*. A practical method for wireless channel reciprocity exploitation through relative calibration (IEEE, Sydney, 2005)
17. A. Mukherjee, Z. Zhang, in *2017 IEEE Global Communications Conference (GLOBECOM)*. Fast compression of OFDM channel state information with

constant frequency sinusoidal approximation (IEEE, Singapore, 2017)
18. Z. Zhang, A. Mukherjee, System and method for fast compression of OFDM channel state information (CSI) based on constant frequency sinusoidal approximation (2017). Google Patents. US Patent No. 9838104. 5 Dec
19. L. Schumacher, Implementation of the IEEE 802.11 TGn Channel Model (2006). http://www.info.fundp.ac.be/~lsc/Research/IEEE_80211_HTSG_CMSC
20. Y. Xie, Z. Li, M. Li, in *Proceedings of the 21st Annual International Conference on Mobile Computing and Networking (MOBICOM 2015)*. Precise power delay profiling with commodity WiFi (ACM, New York, 2015), pp. 53–64
21. N. Ravindran, N. Jindal, Multi-user diversity vs. accurate channel state information - MATLAB code (2009). http://www.ece.umn.edu/~nihar/mud_csi_code.html. Accessed 23 July 2012

11

Application of remote monitoring and management of high-speed rail transportation based on ZigBee sensor network

Jing Zhang

Abstract

Operational safety is the prerequisite for high-speed railway to carry out transportation work, and with the increasing mileage of high-speed railway in China, higher requirements are put forward for monitoring the operation environment of high-speed trains. From the point of view of the whole road, the wireless sensor network technology is fully applied to the monitoring of high-speed train operation environment. Based on the actual requirement of the high-speed train operating environment monitoring, the overall structure and logic frame of the high-speed train operating environment monitoring system based on wireless sensor network are put forward. By establishing the node structure model algorithm, and taking natural disaster and foreign object intrusion detection as an application example, the natural disaster and foreign object intrusion detection system based on wireless sensor network are designed.

Keywords: ZigBee sensor network, High-speed rail transport, Remote monitor and control

1 Introduction

Railway is the major artery of national economy, key infrastructure, and major livelihood projects, and is one of the backbones of the comprehensive transportation system and one of the main modes of transportation, which is of great importance and role in the economic and social development of our country (Xu D et al. 2016) [1]. Since the implementation of *Medium and long term railway network plan* in 2004, China's railway development has achieved remarkable results, which plays an important role in promoting economic and social development, ensuring and improving people's livelihood, supporting the implementation of the national major strategy, and strengthening our country's comprehensive strength and international influence (Zhu H., 2016) [2]. In recent 10 years, China's high-speed railway network has made remarkable achievements, and the scale of the road network has been expanding. By the end of 2015, the national railway business mileage reached 121,000 km, of which the speed railway was 19,000 km, accounting for more than 60% of the total mileage of the world high-speed railway (Gharghan S K et al. 2017) [3]. According to the newly revised *Medium and long term railway network plan*, by 2020, the scale of the railway network will reach 150,000 km, including 30,000 km of high-speed railway, and it will cover more than 80% of the large cities. By 2025, the scale of the railway network will reach 175,000 km, of which the high-speed railway is about 38,000 km, the network coverage is further expanded, and the road network structure is more optimized, the role of backbone is more significant, which can better play the role of railway in ensuring economic and social development (Mosleh M F et al. 2017) [4]. On the basis of the "four vertical and four horizontal" high-speed railway, the high-speed railway with passenger flow support, standard suitability, development requires should be increased, a high-speed railway network which takes the "eight vertical and eight horizontal" main channel as the skeleton, with connection of regional connection lines and intercity railway supplement is formed to realize the efficient and

Correspondence: navvnl210@163.com
School of Economics and Management, Beijing Jiaotong University, Beijing 100044, China

convenient connection between the provincial capital cities and the high-speed railways (Zhang Z et al. 2017) [5].

2 State of the art

Foreign scholars have been trying to apply wireless sensor networks in the railway field very early. While describing the application of wireless communication in railway, a railway monitoring system based on wireless sensor network is put forward by scholars (Alhmiedat T., 2017) [6]. In 2008, scholars put forward a railway monitoring system based on wireless sensor networks, which includes two aspects of train monitoring and line monitoring, and can be used to monitor the train running state in real time (Alhmiedat T et al. 2017) [7]. In orbit monitoring, Huo Hongwei and others put forward a railway track monitoring model based on multi gateway mobility, and when the specific node installed on the train is near the sensing node, the node covered currently is awakened, and nodes collect data to transmit the data collected to the gateway in a single hop or limited multi hop mode, when the mobile gateway leaves, the node enters a dormant state, which can store energy to achieve passing track and other carrying facilities monitoring information to the train, and it enables the train to understand the condition of the specific section in the course of running (Wu Q et al. 2018) [8]. To sum up, experts and scholars have carried out a preliminary study on wireless sensor networks for monitoring the operation environment of high-speed trains, including lines, trains, stations, natural environment along the way, transportation of dangerous goods, and so on, but most of studies remain in a specific area of research, various wireless sensor networks are self-contained and non-interference, resulting in a variety of complex problems in wireless sensor networks. The research on comprehensive monitoring of railway operation environment is in its infancy, and it is urgent to build a unified wireless sensor network to conduct overall monitoring of train operation environment.

3 Methodology

3.1 Node structure and typical energy consumption model algorithm

The node of wireless sensor network is a kind of micro embedded system, which consists of four parts: sensor module, processor module, wireless communication module, and energy supply module. Usually, the power consumption of the processor and sensor module is relatively low, and most of the energy consumption occurs in the wireless communication module. In view of the volume of sensing nodes, it is often impossible to use large capacity batteries to supply them, and because of the characteristics of wireless sensor networks, it is often not possible to replace batteries for sensing nodes. Therefore, how to improve the efficiency of communication and reduce energy consumption is one of the key problems to be solved in wireless sensor networks. A typical energy consumption model for wireless sensor networks is adopted.

$$E_{\mathrm{Tx}}\begin{cases} k \times E_{\mathrm{elec}} + k \times \varepsilon_{\mathrm{ft}} \times d^2, d < d_0 \\ k \times E_{\mathrm{elec}} + k \times \varepsilon_{\mathrm{mp}} d^4, d_0 \end{cases} \tag{1}$$

$$E_{\mathrm{Rx}} = k \times E_{\mathrm{elec}}. \tag{2}$$

The formula (1) indicates the loss of energy when sending k bit data to a receiver with a distance of d, including two parts: emission circuit loss and power amplification loss. E_{elec} is the loss energy of a transmitting circuit, and power amplification loss uses different models according to the different distances between the sender and receiver: free space model and multi path fading model. When the transmission distance is within a certain threshold d_0 (d_0 is the constant), when $d < d_0$, free space model is adopted, the power of the transmission distance is proportional to the square of the distance, ε_{ft} is the energy required for power amplification in the free-space channel model; when $d \geq d_0$, multi path fading model is adopted, the power consumption is proportional to the four times of the distance, $\varepsilon_{\mathrm{mp}}$ the energy required for power amplification in the multi path fading channel model. The formula (2) indicates the energy consumed by the receiver when receiving k bit data.

3.2 Energy consumption model algorithm for linear wireless sensor networks

According to the structure of wireless sensor network along the railway, the location of sink nodes and sensing nodes in wireless sensor network is determined, which is deployed in accordance with the actual monitoring requirements and relevant regulations along the railway line. The deployment of relay nodes in wireless sensor networks is directly related to the lifetime of wireless sensor networks. In general, a uniform deployment strategy is adopted by relay nodes, that is to say, all relay nodes located in all perceptual nodes are deployed with the same spacing. Under this deployment strategy, the distance between all relay nodes is equal, and the amount of energy consumed by single forwarding unit data is equal, but the nodes with close distance from the sink node need more data forwarding than those with far distance, therefore, it is easy to form an energy hot zone, resulting in the "energy hole" problem, which makes the wireless sensor network premature early to die.

In view of the above problems, a strategy of non-uniform optimization deployment is proposed, and the policy is described as follows: there is a uniform deployment between two adjacent sensing nodes or the sink node and the sensing node, that is the distance between two adjacent relay nodes is equal between two adjacent sensing nodes or the sink node and the sensing node. The relay nodes closer to the sink node assume larger data forwarding capacity, and a smaller deployment interval should be adopted; however, the relay nodes with far distance from the sink node assume smaller data forwarding capacity, and larger deployment spacing should be adopted, as far as possible, each relay node consumes the same energy in single data forwarding. Therefore, for the two adjacent sensing nodes which are closer to the sink node or between the sink node and the sensing node, the relay nodes should be more densely deployed. When the number of relay nodes that can be deployed is the same as that of the average deployment strategy, the unevenly optimized deployment strategy can make the energy consumption of all relay nodes in the network balanced, which can prolong the network life cycle and improve the network efficiency. Therefore, a mathematical model based on this strategy can be established to solve the problem of the total number of relay nodes; the number of relay nodes should be deployed between the two adjacent sensing nodes or the sink node and the sensing node, so that the network efficiency can be maximized.

The researched wireless sensor network is a typical linear network, and in order to avoid the problem of "energy hole," the energy balance of all nodes must be consumed. Then, the energy consumption problem of nodes will be analyzed in detail. According to the hypothesis, it can be known that the amount of data collected by each sensor node in 1 cycle is $l$bit, the amount of data each node accepts in 1 cycle is $r$bit, the amount of data that each node sends in 1 cycle is $j$bit, obviously, the amount of data received and sent by $l \leq j \leq l+r$ for relay nodes in 1 cycle is $j$bit. Because the energy consumed by the sensing data is negligible, it is known that the relay node consumes more energy than the sensing node. Therefore, the bottleneck of the life cycle of wireless sensor networks lies in relay nodes. According to the model of energy consumption, the energy consumed by the i relay node before the sensing node s_n is shown in formula (3).

$$E_i \begin{cases} j_i \times E_{\text{elec}} + j_i \times \varepsilon_{\text{ft}} \times d_i^2 + j_i \times E_{\text{elec}}, d_i < d_0 \\ j_i \times E_{\text{elec}} + j_i \times \varepsilon_{\text{mp}} \times d_i^4 + j_i \times E_{\text{elec}}, d_i \geq d_0 \end{cases} \quad (3)$$

j_i represents the amount of data received and sent by the relay node. In order to improve network efficiency and reduce "energy hole" problem, ideally, the energy consumed E_i by each node is equal, and all relay nodes consume all the energy at the same time, which is almost impossible in reality. In view of this problem, the model E_i is assumed to be near a definite value, and the constraint conditions are shown in formula (4).

$$\frac{\tilde{E}}{q} \leq E_i \leq q \cdot \tilde{E}, q \geq 1. \quad (4)$$

In the formula, q is a constant greater than 1 and infinitely close to 1.The lifetime of the network ζ is defined as the time used by the network when the energy of any node in the network is exhausted, as shown in formula (5).

$$\zeta = \frac{E_0}{\tilde{E}} \cdot T_p. \quad (5)$$

E_0 indicates the initial energy of nodes in a network, T_p indicates the time consumed by data acquisition and transmission. In general, the life cycle of the network increases with the number of nodes deployed, but the rate of increase will decrease. This is because as the size of the network increases, the contribution of a single node decreases. Therefore, the new parameters are selected as the objective function of the model, and network efficiency is defined as the ratio of the network life cycle to the number of network nodes, as shown in the formula (6).

$$\eta = \frac{\zeta}{N_s + N_r}. \quad (6)$$

N_s represents the number of perceived nodes and N_r represents the number of relay nodes. Network use efficiency indicates the change rate of network life cycle with network scale, which makes balance between the network lifetime and node deployment costs.

4 Result analysis and discussion

The railway operation environment monitoring system deploys a sensing node every 1000 m along the railway line, and the sink node is located at one end of the system. The length of the monitoring area is selected in four cases: 1000 m, 2000 m, 3000 m, and 4000 m. Because of the use of data fusion strategy, the amount of data transmitted by the sensing node is 8 bytes, 10 bytes, 12 bytes, and 15 bytes respectively. In the same number of relay nodes, two different deployment strategies are compared. When using the same number of relay nodes, the two different deployment strategies are adopted. With equally spaced deployment strategies, it can be seen that 14 relay nodes need to be deployed between two sensing nodes. The optimal deployment strategy is adopted to maximize the efficiency of the network, and MATLAB software is used to solve the problem. Figure 1 shows the comparison of the life cycle of the two strategies, and Fig. 2 shows the comparison of the use efficiency of the two strategies. It can be concluded that the

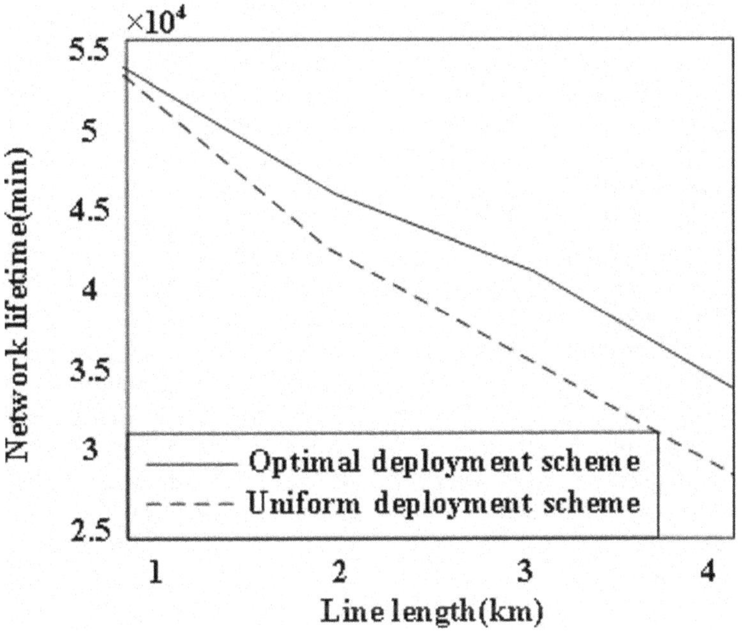

Fig. 1 Network use efficiency contrast diagram

network life cycle and use efficiency of the optimized deployment strategy are better than the equal distance deployment strategy. The life cycle of these two deployment strategies decreases with the increase of network length, but the reduction degree of network life cycle of the optimized deployment strategy is less. When the monitoring length is 4000 m, the network life cycle of the optimized deployment strategy is 20% higher than that of the equal distance deployment strategy. Therefore, the optimized deployment strategy proposed can effectively improve the life cycle and efficiency of the network.

When the monitoring area is 4000 m long, the network deploys 4 sensing nodes, and the energy consumption of relay nodes in the network using the node optimal deployment strategy is more balanced, when the network fails,

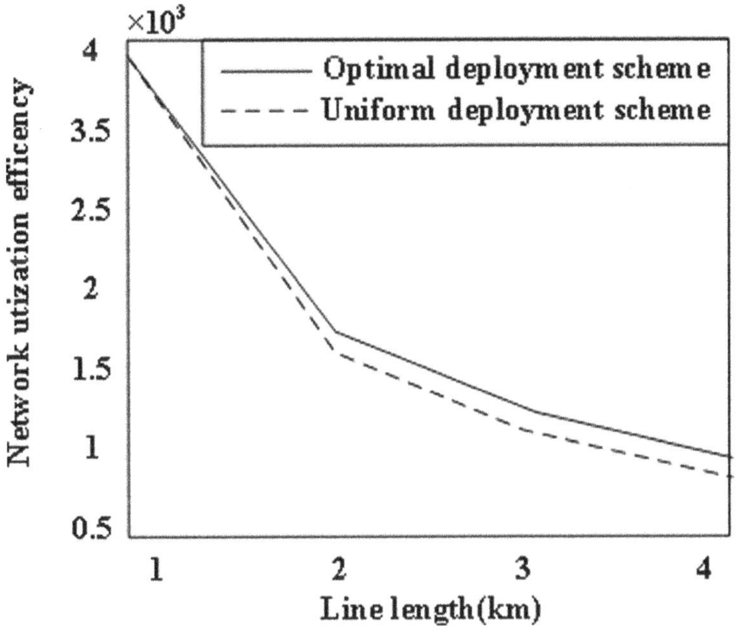

Fig. 2 Network use efficiency contrast diagram

the residual energy of nodes is 7.9% of the total energy of nodes. When the network with equal distance deployment strategy fails, the residual energy of nodes is 48.2% of the total energy of nodes. Therefore, the optimal deployment strategy designed can better save network energy and improve network efficiency. Without considering node failures, a network with only one row of relay nodes is simulated and analyzed. By simulating the linear wireless sensor networks with five relay nodes, the data packets with priority of 0, 1, 2, and 3 can be obtained in the network, and the number of data packets sent on average is shown in Fig. 3.

The routing protocol proposed is called mode I, which is compared with the following two data transmission modes. The mode II, which is a non-compulsory priority queuing system, is not adopted by the queueing form of full functional sensor node. The mode III, which is the priority of packet partition, is not considered, the intelligent forwarding mode is not adopted, and all data packets are transmitted by hopping step by step. Through simulation experiments, it can be seen that the total number of data packets transmitted by linear wireless sensor networks under these three data transmission modes is shown in Fig. 4. Among them, the total number of transmission data packets of modes I and II is the same, so the network life of these two ways is the same, and the total number of data packets transmitted by III is less, so the network life is slightly smaller in this way.

By the simulation of a network with five relay nodes, the average transmission delay of data packets with four priorities as 0, 1, 2, and 3 can be obtained under the two transmission modes of mode I and mode II. The delay comparison of the two transmission modes through simulation is shown in Fig. 5.

From the results of the graph, it can be seen that the method of non-mandatory priority queueing system is adopted, although the transmission delay of data packets with lower priority is increased, the transmission delay of the highest priority packet is effectively reduced, which improves the timeliness of packet transmission with higher priority. In this way, the queuing delay of the data packet is allocated to data packets with different priority levels, which makes $W_K < W_{K-1} < \ldots < W_0$, while in general $\lambda_0 >> \lambda_K$, that is, the number of data packets with high priority is small. Therefore, on average, the transmission delay of data packets with high priority is reduced a lot, while on average the transmission delay of each data packet with low priority level is not much increased, which is a good way to reduce the delay. Therefore, the routing protocol can effectively reduce the transmission delay of high priority packets in the case of ensuring the network lifetime. On the basis of the existing hardware conditions, five communication technologies of the above eight modules are tested on the number of packets sent successfully through the instant uplink and downlink of the train at high speed.

When the train is running within the speed range of 300–350 km/h, the single output of the four modules' downlink and uplink packets are tested respectively. When the ground sends data packets to the train, the number of data packets successfully received by the vehicle test platform is instantaneous after the train passes the ground test platform. According to the test results, the number of data packets sent successfully decreases as the speed of the train increases, and the number of

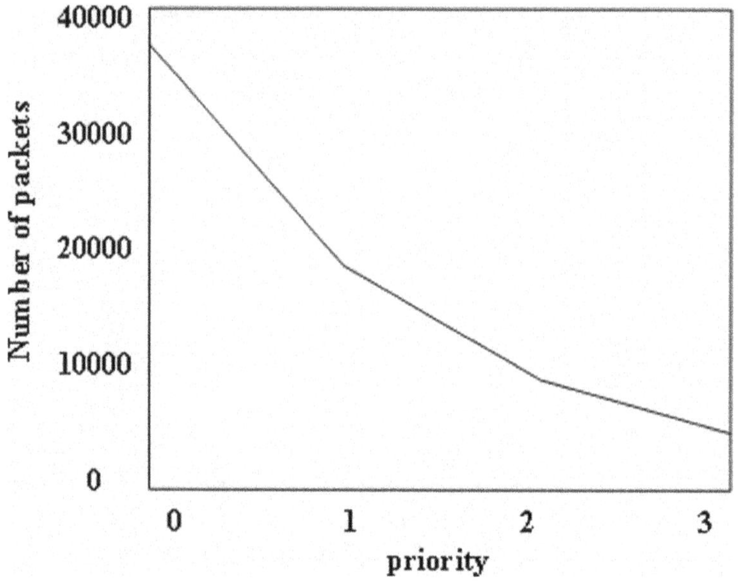

Fig. 3 Number of packets with different priorities on average

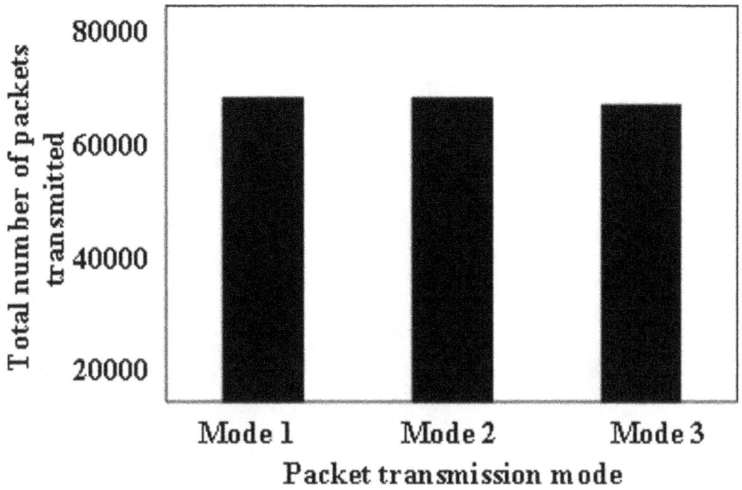

Fig. 4 Data packet transmission quantity contrast diagram

data packets sent successfully by the communication module based on ZigBee technology is the largest. When the train sends data packets to the ground, the number of data packets successfully received by the ground test platform is instantaneous after the train passes the ground test platform. According to the test results, the number of data packets sent by the communication module based on ZigBee technology is also the largest. Through several tests, the average packet loss rate of uplink and downlink of each communication module is calculated. According to the test results, when the communication module based on RFID technology is set to 250 kbps at the air speed and the baud rate is 9600, the rate of packet loss in both uplink and downlink is the lowest. The packet loss rate of communication module based on ZigBee technology is relatively high. By adding the delay of the uplink and the delay of the downlink, the problem of two laptops' time asynchrony can be eliminated and the more accurate round-trip delay of the packet transmission can be obtained.

Figure 6 shows the relationship between the data packet transmission delay of four modules and the speed of the vehicle, and Fig. 6–9 shows the average delay of packet transmission of four modules. According to the test results, there is no obvious linear relationship between data packet transmission delay and vehicle speed. Among them, the third set of module based on RFID technology, because of its baud rate is 38,400, which is larger than other modules' baud rate of 9600, so the data packet transmission delay of it is the smallest, which concentrating between 149 ms and 195 ms, and the average transmission delay is 167.75 ms. Transmission delay

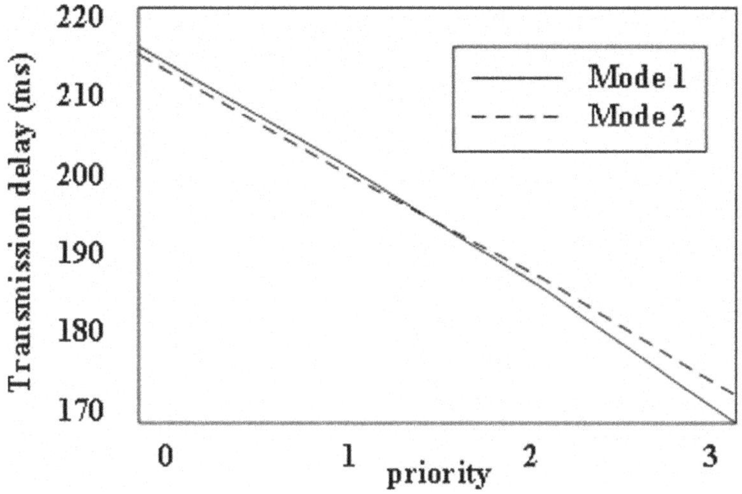

Fig. 5 Data packet transmission delay contrast diagram

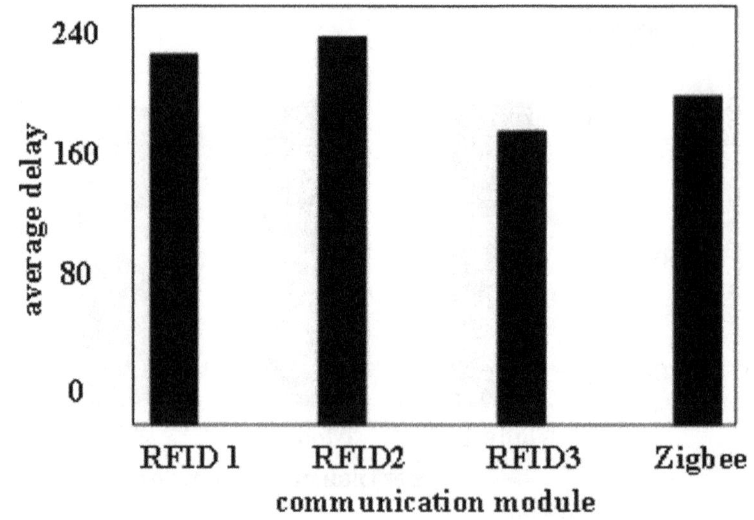

Fig. 6 Packet transmission average delay

time of module based on ZigBee technology is the second, which concentrating between 170 ms and 193 ms, and the average transmission delay is 186.89 ms. By the analysis of the above test results, the communication models based on ZigBee technology, the single transmission volume of uplink and downlink data packets is the largest, and the transmission delay is moderate to be recommended for building wireless sensor networks.

5 Conclusion

Based on the actual situation of high-speed train operation environment monitoring, the current situation and business requirements of the existing high-speed train running environment monitoring are fully investigated and analyzed, the advantages of wireless sensor networks is exerted, and the new idea of applying wireless sensor network technology in the field of high-speed train operation environment monitoring is put forward; the overall framework and logical framework of high-speed train operation environment monitoring system based on WSN are proposed; a topology structure for ground wireless sensor networks is designed; the node deployment strategy and routing protocol for linear wireless sensor networks are studied; the performance of ground to ground communication for various wireless communication technologies is tested in real high-speed rail environment, a transmission scheme of vehicle ground wireless sensor network based on relay transmission is designed, and it has been verified in field test; a natural disaster monitoring system based on wireless sensor network is designed. The overall framework of high-speed train operation environment monitoring system based on SOA and wireless sensor network is proposed. The framework runs through railway computer networks and wireless sensor networks, which can integrate all the sensor resources distributed in the bureaux, stations, and routes of all roads to form a unified sensor network. The comprehensive application of wireless sensor network in railway-related professional fields is realized to form an innovative application mode with internal and external penetration, and the integration of service and production.

Funding
No Funding.

Authors' contributions
JZ has done a lot of research and analysis on ZigBee sensor network and contributed to the summary of the paper. The author read and approved the final manuscript.

Competing interests
The author declares that she has no competing interests.

References

1. D. Xu, P. Wang, B. Shen, Food pollution remote intelligent monitoring system based on ZigBee technology. Adv. J. Food Sci. Technol. **11**(4), 308–314 (2016)
2. H. Zhu, Analysis of a wireless local area network based on the ZigBee technology applied to agro-food safety production and monitoring. Adv. J. Food Sci. Technol. **10**(3), 204–208 (2016)
3. S.K. Gharghan, R. Nordin, M. Ismail, Development and validation of a track bicycle instrument for torque measurement using the ZigBee wireless sensor network. Int. J. Smart Sensing Intell. Syst. **10**(1), 124–145 (2017)
4. M.F. Mosleh, D.S. Talib, Implementation of active wireless sensor network monitoring using ZigBeeprotocol. J. Eng. Sc. Technol. **12**(11), 3082–3091 (2017)
5. Z. Zhang, P. Wu, W. Han, et al., Remote monitoring system for agricultural information based on wireless sensor network. J. Chin. Inst. Eng. **40**(1), 75–81 (2017)
6. T. Alhmiedat, Low-power environmental monitoring system for ZigBee wireless sensor network. Ksii Trans. Internet Inform. Syst. **11**(10), 4781–4803 (2017)
7. T. Alhmiedat, G.A. Samara, Low cost ZigBee sensor network architecture for indoor air quality monitoring. Int. J. Comput. Sci. Inform. Security **15**(1), 140–144 (2017)
8. Q. Wu, J. Cao, C. Zhou, et al., Intelligent smoke alarm system with wireless sensor network using ZigBee. Wireless Commun. Mobile Comput. **2018**(3), 1–11 (2018)

12

Modeling the UE-perceived cellular network performance following a controller-based approach

Jessica Mendoza* , David Palacios, Isabel de-la-Bandera, Eduardo Baena, Emil J Khatib and Raquel Barco

Abstract

During the last few years, mobile communication networks have experienced a huge evolution. This evolution culminates with the arrival of the fifth generation (5G) of mobile communication networks. As a result, the complexity of network management tasks has been increasing and the need to use automatic management algorithms has been demonstrated. However, many mobile network operators (MNOs) are reluctant to evaluate these algorithms in their networks. To address this issue, in this paper, a modeling approach is proposed. In this sense, the behavior of a commercial network, as it is perceived by user equipments (UEs), has been replicated in a research testbed using a three-step modeling process. The first step consists on performing a measurement campaign in several external networks. The second step is composed of the measurement campaign result analysis and the classification of the results in different types of scenarios. Finally, the third step is related to the application of a modeling algorithm in a research testbed. In order to perform the last step, the use of a method based on a controller is proposed. The modeling process presented in this paper allows to replicate the network behavior from users located in different areas and with different conditions point of view. Moreover, the use of a testbed environment can help to avoid downtime in commercial networks caused by possible algorithm bugs.

Keywords: Modeling, UE-perceived, Testbed, Measurement campaign, Mobile communication networks

1 Introduction

Mobile communication networks have experienced a large development over the last few years. According to studies conducted by Cisco [1], *"Global mobile data traffic was 7% of total IP traffic in 2016 and will be 17% of total IP traffic by 2021"*. This growth in traffic will be especially pronounced with the arrival of the fifth generation (5G) of mobile communication networks. In these networks, an increment in the number of users and offered services is also expected, resulting in an increase in the number of configuration parameters of the networks and possible scenarios. In this context, the development of automatic algorithms that allow the management of networks in a faster and more efficient way become a priority for mobile network operators (MNOs). However, MNOs are increasingly reluctant to implement in their own network algorithms that have not been assessed in prototypes of real networks. The reason is to avoid a possible service degradation due to a suboptimal parameter configuration change made by a certain algorithm. This problem is getting worse in 5G networks, since in these networks, a reduction of the downtime is sought. Thus, many automatic management network algorithms developed by researchers and tested in simulated environments are never implemented in real networks. In the cases in which the MNOs decide to assess an algorithm in their own networks, the tests will be carried out in limited areas of the network before extending them to the complete network. This methodology presents some important limitations, since different network areas may behave in a different way and, therefore, the conclusions obtained from a specific area of the network might not be generalizable to the entire network.

To deal with this problem, a modeling approach could be followed. Modeling is the act of representing something. This representation is often mathematical. Thus, in

*Correspondence: jmr@ic.uma.es
[1] Department of Communication Engineering, University of Malaga, Andalucía Tech., 29071 Malaga, Spain

the literature, there are many works focused on the development of analytical models that describe the behavior of different types of phenomena. In relation to mobile communication networks, in [2] and [3], the authors present analytical models of the behavior of different metrics associated with the quality of experience (QoE) of video services. Download throughput models are proposed in [4] and [5]. In the first case, the authors use neural networks, whereas in the second case, the authors use linear regression. However, modeling can also be the act of imitating the behavior of a specific situation. In mobile communication network scope, to the authors' knowledge, in the literature, there are no other works related to the modeling in this line.

On the other hand, with the aim of satisfying the needs of a growing number of users, the goal of network management tasks has been changing over time, from management focused on the performance of a certain section of the network (typically the radio access network), such as in [6], to a management that is increasingly closer to the user, and to the user's perception of the performance or behavior of the network [7], obtaining in this way an end-to-end (E2E) vision of the network behavior.

In line with the latter, this paper aims at showing how the behavior of a commercial cellular network could be replicated from a user equipment (UE) point of view by using a research testbed (Fig. 1). In this sense, by the adjustment of the configuration parameters of the testbed network, it is intended to obtain a network equivalent to the commercial one in terms of user perception, that is, to ensure that users connected to the testbed network report the same values of users' performance indicators (measured user performance indicators in Fig. 1) than the users connected to the commercial network (target user performance indicators in Fig. 1), thus providing an environment in which testing automatic management network algorithms before their implementation in MNO commercial networks. One of the main advantages of the use of a network model is that different areas, with different behaviors, could be independently modeled and optimized. In addition, the use of a testbed environment allows detecting possible failures or bugs in the algorithms before being implemented in the real network, avoiding possible configuration failures of the network parameters and the consequent service degradations.

To achieve the global objective, the modeling of a cellular network behavior from a UE perspective, a three-step process is proposed. The first step consists of performing a measurement campaign. For this purpose, similarly to the measurement campaign conducted in [8], experiments have been carried out in commercial networks of some of the main MNOs in Europe. The second step consists in the analysis and classification of the data collected in the measurement campaign, obtaining different behavior patterns of the measured networks. The output of the second step is taken as the input to perform the last step. Such step refers to the modeling of a commercial network performance from the UE point of view. In order to perform the modeling, a method based on a controller has been used. In particular, the use of the Taguchi method (TM) is proposed, in a similar way to the one proposed in [9], where antenna azimuths and tilts are optimized in order to improve the user throughput. In this paper, the TM controller aims at adjusting the value of specific network configuration parameters depending on the behavior to be replicated that will be represented by user performance indicators.

The rest of the paper is organized as follows. Firstly, the methodology developed in the modeling process is presented, indicating how the design of the measurement campaign has been made and providing the main characteristics of the proposed method to carry out the modeling. Secondly, an overview of the testbed network is provided, and the result of the measurement campaign, as well as the result of the modeling algorithm application are shown. Finally, some conclusions are drawn from the previous study and tests.

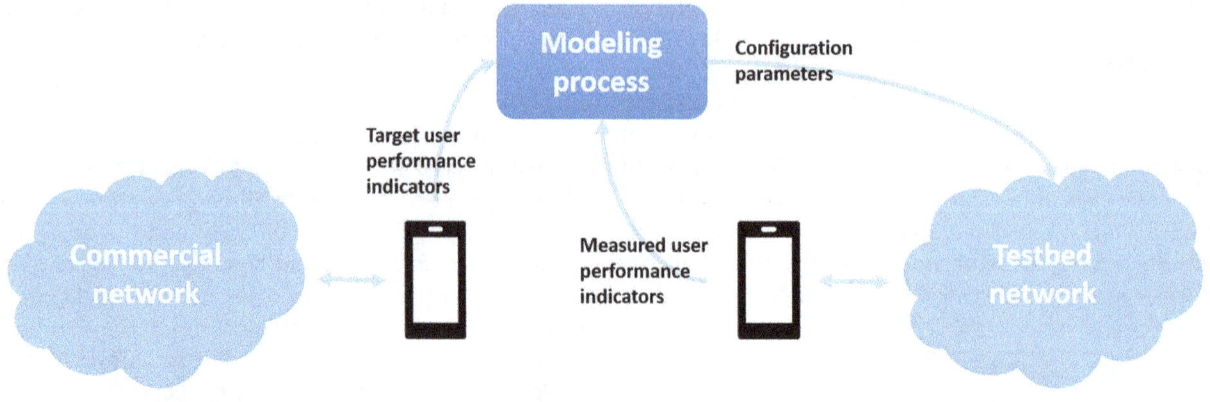

Fig. 1 Modeling scheme

2 Modeling methodology

This section presents the modeling process presented in this study. To model the UE perception of a commercial network behavior, a three-step approach is proposed as follows (Fig. 2):

- Measurement campaign. This first phase consists in taking measurements in commercial networks behavior as perceived by the UEs. In this sense, the values of some user performance indicators such as user download throughput has been collected.
- Analysis results and classification. By inspecting the resulting measurements, these are grouped, resulting in several scenarios that represent the behavior of the studied networks.
- Modeling. The objective of this task is to obtain a set of configuration parameters, such as the eNodeB transmission power or the bandwidth, to be applied to the testbed network in order to obtain similar user performance indicators that those obtained from measurement campaign for different considered scenarios.

In the next subsections, an overview of the measurement campaign motivation and methodology as well as the main characteristics of the algorithm used to perform the modeling is provided.

2.1 Measurement campaign

For a model to be built, a number of observations of the entity to be modeled are needed. These observations will allow extracting the underlying relations among the involved variables as well as providing a general view of the entity behavior and neglecting punctual deviations. In the context of cellular networks, a model may be extracted from its performance indicators, which take different shapes and quantify different aspects of the same system. In particular, performance information is usually gathered from network elements, such as the base stations, as counters, which count the number of times an event occurred, or key performance indicators (KPIs), which are a type of performance measurement computed from the first. Performance information may also be gathered from UEs, either by directly collecting data from radio channel measurements (such as reference signal received power, RSRP; or reference signal received quality, RSRQ) or after a complex event processing (CEP) phase is applied on signaling messages between the UE and the network.

Counters and KPIs are often collected as aggregate magnitudes (not carrying information about specific UEs) within a certain time period; typically, with a time resolution in the order of several minutes. They are periodically stored in databases belonging to the operations support systems (OSS), from which they can be retrieved for management tasks. These indicators quantify the performance of a network segment, either the radio access network (RAN) or the core network (CN), and as a result, do not allow MNOs to obtain an E2E vision of the network.

On the other hand, performance information gathered from the UEs is registered with a much higher resolution; in the order of seconds. This information may be retrieved either from the UEs themselves, by means of driving tests with UEs modified ad hoc, or from network elements. This source of performance information provides an E2E vision, allowing MNOs to follow user-centric approaches for management purposes, which is a trend nowadays [10, 11]. This additional information, which is crucial for novel network management, comes at the expense of increased processing and storing needs for MNOs. This is the type of information used to perform the modeling tasks in this paper.

2.2 Taguchi method

TM is one of the most widely used tools in the engineering experiment design scope. TM computes the values that the inputs (called factors or parameters) of a system should take in order to obtain a specific value on a target response metric. In the mobile network modeling case, these factors will be network configuration parameters, and the response metric, the UE's metric selected to perform the modeling. To this end, the method executes successive iterations until a specified stop criterion is met. In each iteration, a set of experiments with different combinations of input factors or parameter values are performed. Each experiment consists on the use of a service by a UE (e.g., file download, video streaming). Experiment results are metrics that can represent the performance of the networks in different levels, such as radio, KPI, and key quality indicators (KQI). These metric values depend on the type of service that is being used by the user as well as the service traffic profile. Therefore, modeling results in terms of configuration parameters will depend on the service used to perform the experiments.

For each factor, TM defines a set of levels (possible values of the factors to be tested in each iteration) and an initial range of values (maximum and minimum values that could be adopted during the process).

In order to decide which combination of input factors values is selected in each experiment, TM proposes the use of an orthogonal array (OA) [12]. Thus, given a response metric and with the aim of reducing the number of experiments needed to carry out the process, an OA represents the configuration of input factors that should be used in each experiment. TM can be used with three objectives: (a) to minimize a response metric value: the smaller-the-better (STB); (b) to maximize a response metric value: the larger-the-better (LTB); and (c) to achieve a target value of a response metric: the nominal-the-best

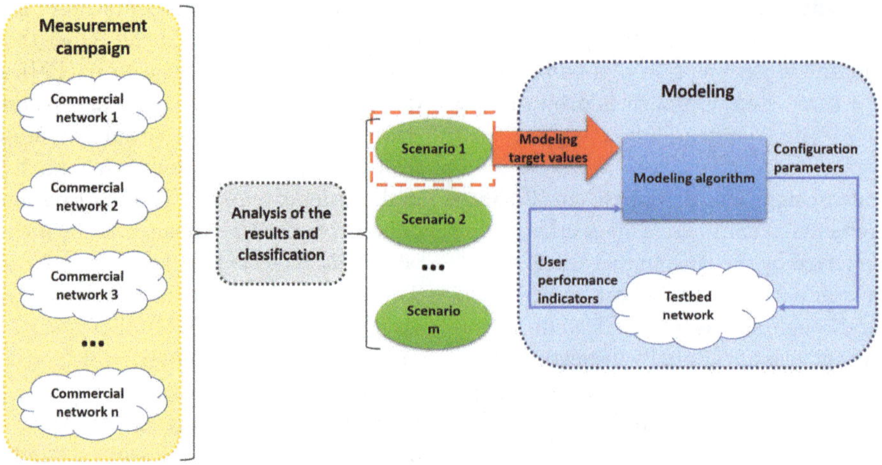

Fig. 2 Modeling process steps

(NTB). Since the modeling purpose is to replicate the behavior of a system, the TM approach that better fits the modeling problem is the NTB.

To obtain the optimal values of the factors, TM uses a metric called signal to noise (SN). The SN metric is computed after each iteration for each factor level. The optimal level of a factor is the one that produces the highest value of the SN metric. Depending on the objective of the method: STB, LTB, and NTB, the SN metric will be calculated differently. In the NTB case, the SN metric is computed as follows:

$$SN = -10 \cdot \log_{10}\left(\frac{\mu^2}{\sigma^2}\right) \tag{1}$$

where μ is the response metric mean of the experiments performed with a certain factor level and σ is their standard deviation.

Figure 3 shows the steps to follow during the process. The first step is the selection of an appropriate OA and an objective SN metric that fits the problem. Then, a mapping between the levels of each factor and the real values of these are made. To this end, it is necessary to define the maximum and minimum values to be tested for each factor. Afterwards, the experiments are carried out and, through the calculation of the SN metric, the combination of factors that provide the maximum response metric result is sought. Once the optimum values have been obtained in the iteration, the stop criterion, defined as the difference between two adjacent values of a factor, is checked. If the stop criterion is met, the process is finished; in the opposite case, the maximum and minimum values to be tested of each factor are reduced, taking as central value the optimal value in the previous iteration. Then, the mapping between the levels of each factor and the measurements of these, as well as the experiments, is made again.

One limitation of the TM is that only one metric can be used as a response metric. However, in some cases, such as the one presented in this work, the objective of the process (maximize, minimize, or achieve a target value) is defined as a combination of different metrics. Such a situation is called the multi-response problem in TM. An approach to address this limitation is proposed in [13] and is the one

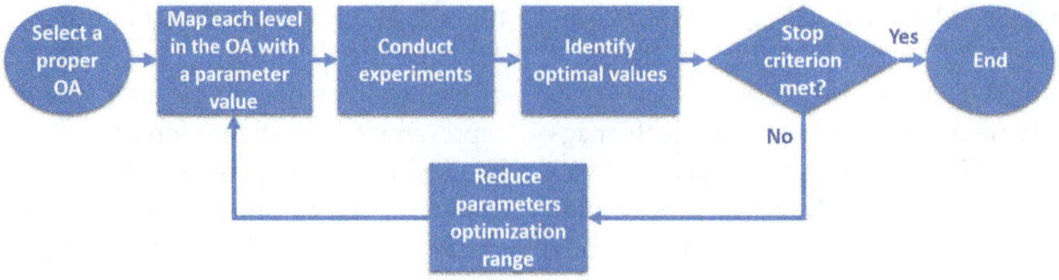

Fig. 3 Operation diagram of Taguchi method

followed in this paper. Specifically, the proposed solution consists of four steps: first, the SN metric is calculated for each response metric and for each combination of factors. Then, the average SN for each response metric and for each factor level and a certain weight is given at each level in relation to the maximum SN calculated for that factor and response. Next, the average weight for each factor level is calculated, considering the weight obtained in all the responses. Finally, for each factor, the factor level with the highest average weight is taken as the optimum factor level.

3 Proof of concept

In this section, first an overview of the main characteristics of the testbed used to perform the modeling tasks is provided. Then, the results obtained from the three steps that make up the modeling process, explained in the previous section, are presented.

3.1 Testbed description

The University of Malaga Heterogeneous Network (UMAHetNet) [14] is a full indoor long-term evolution (LTE) network deployed at the Telecommunication Engineering School. The RAN consists of 12 base stations in the shape of picocells. The evolved packet core (EPC) is integrated in a single compact equipment, namely, the evolved Core Network Solution (eCNS), which is composed of all core network elements (home subscriber server, HSS; mobility management entity, MME; serving gateway, S-GW; packet data network gateway, P-GW; and policy and charging rules function, PCRF), as shown in Fig. 4. On top of this, UMAHetNet is monitored and coordinated by a network management system (NMS), the Huawei solution iManager U2000.

These network elements are interconnected by means of a 24-port Gbit switch providing all the equipment access to the Internet.

Both the picocells and the eCNS are fully configurable. The picocells can be managed either directly through a local management terminal (LMT) client or by means of the U2000 and eCNS clients. These two latter allow the user to fully customize all the network parameters, as well as to monitor the network status using built-in and user-defined KPIs. Furthermore, this testbed enables the user to integrate his own self-organizing network (SON) applications and algorithms, from self-configuration techniques to any self-optimizing and self-healing application.

In order to enable the integration of new management mechanisms into the NMS, as well as to allow other researchers deploy their own management functionalities in the testbed, a web service application programming interface (API) has been developed. Among the different types of web service APIs that exist (simple object access protocol, SOAP, representational state transfer, REST, common object request broker architecture, CORBA, etc.), it has been decided to implement an API of type REST. Whereas SOAP, CORBA, and other remote procedure calls (RPC) aim at defining a communication protocol between the client and the web server [15], REST is a web architecture style that uses the hypertext

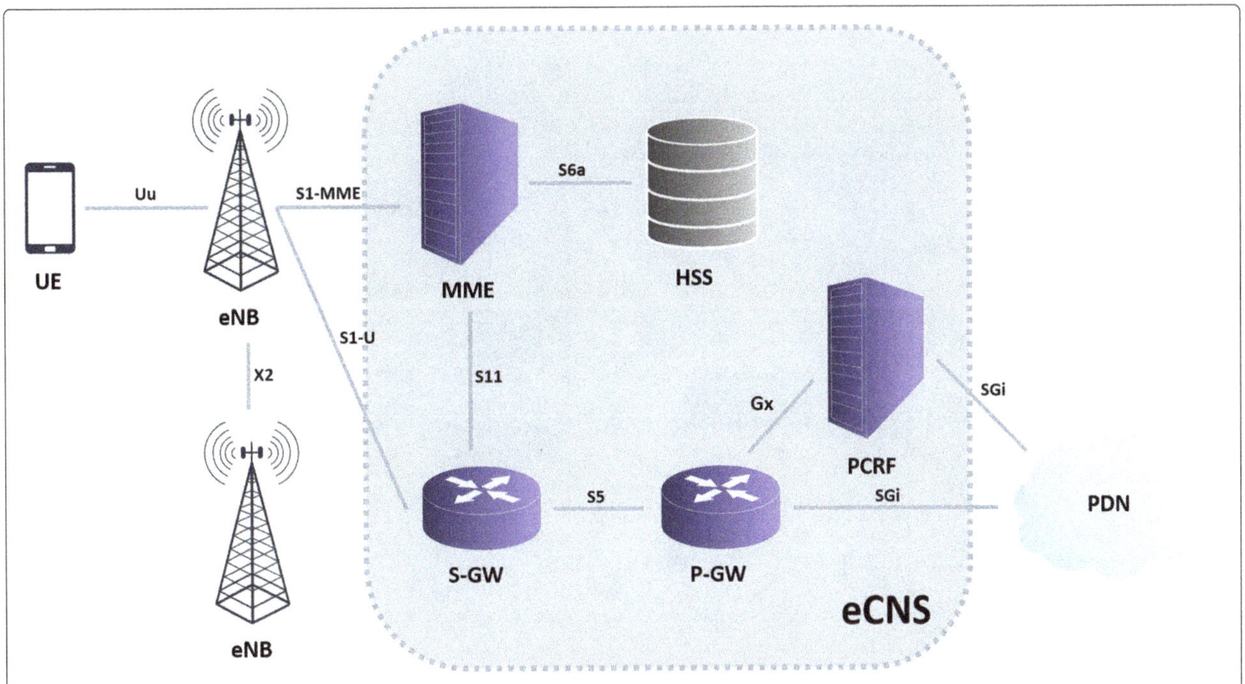

Fig. 4 UMAHetNet general scheme

transfer protocol (HTTP) in the communication between these two parts. This makes the use of RPC systems more complex and inflexible. Moreover, REST allows the transmission of a greater variety of data formats (e.g. plain text, JavaScript object notation, JSON or extensible markup language, XML).

Thus, on the one hand, users communicate with the REST API using a RESTful interface. This interface is accessible using standard hypertext transfer protocol (HTTP) libraries, being easily integrable with algorithms implemented on any programming language for SON functions, monitoring, etc. On the other hand, the REST API will traduce users' HTTP queries in (a) man-machine language (MML) commands that allow users to perform both RAN and CN configuration parameters changes and (b) structured query language (SQL) queries that allow users to consult network configuration parameters as well as performance indicators. In Fig. 5, a diagram of the REST API is shown.

In UMAHetNet, UEs are Linux-based devices, which are connected to the UMA picocells and which gather user's measured performance indicators. Specifically, UEs are MONROE nodes, which are further described in [16]. The MONROE project aims at establishing a collaboration framework for European researchers in the field of mobile broadband (MBB) communications. To that end, the project allows each partner to execute MBB-related experiments in another partner's facilities, so that information from several cellular commercial networks can be gathered and compared [17]. The experiments as well as the data collection are performed by MONROE nodes, which are permanently connected to the partner's facilities. To manage the execution of experiments, a platform has been developed in the scope of this project [18].

Figure 6 shows the hardware structure used to perform the modeling in the UMAHetNet and the connections between its elements. This structure consists in a management device, a MONROE node and the UMAHetNet itself. The intelligent of the system is centralized in the management device (a computer, in this case). On the one hand, this element communicates with the MONROE node through a secure shell (SSH) connection. The management device sends commands to the MONROE node to perform a certain experiment and to retrieve the resulting performance measurements collected by the node during the experiments. The results of the experiments are processed by the management device, which will decide the adjustment to be made in the network configuration parameters. On the other hand, the management device communicates with the UMAHetNet through the REST API described above. The management device send to the REST API the new values of the configuration parameters, and the REST API transforms the request of the management device in MML commands and sends them to the UMAHetNet to make the configuration changes.

3.2 Results

In this subsection, the results of the different tasks of the modeling process are shown, which are the measurement campaign, the subsequent data analysis and classification, and the use of the modeling algorithm. To carry out the experiments related to the measurement campaign

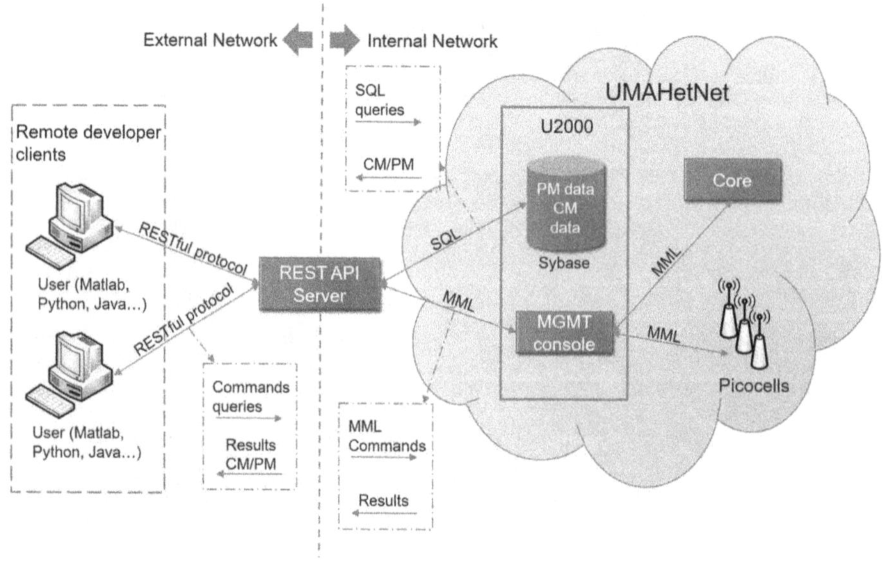

Fig. 5 REST API diagram

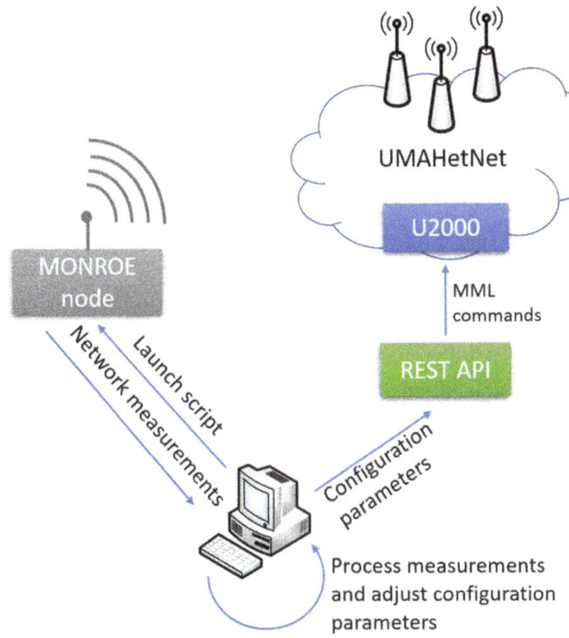

Fig. 6 Integration of hardware extension

and the modeling steps, a file transfer service has been selected. File transfer service is the base of most of the services demanded by UEs, especially in the MBB service category, which carries most of the cellular traffic nowadays [19] and whose most relevant performance indicator is throughput.

3.2.1 Measurement campaign

In order to validate the model of the network performance, a measurement campaign has been devised. The measurement campaign aims at obtaining a set of measurements of the UE perception of the behavior of commercial networks. These measurements will be processed and taken as the target value of the modeling algorithm. To have a broader view of the commercial networks' behavior, different networks of some of the main European MNOs (Yoigo, Vodafone, Orange, Telia, Telenor, and TIM) have been studied. As said before, the service used for the campaign is file transfer service. By means of a battery of tests in which files of different sizes have been proved (100MB, 50MB, 10MB, 1MB, and 0.5MB), it has been determined that the optimal file size for the measurements is 1MB, achieving a tradeoff between the time to perform a complete download and the statistical significance of the results.

The measurement campaign has been performed using the MONROE platform (see Section 3.1) to manage the experiments and download the results and the MONROE nodes connected to MNOs networks to execute the experiments and gather UE metrics. The UE metrics used in this paper are collected by the nodes in two different ways. On the one hand, the nodes have installed the cURL tool [20] that provides metrics such as the setup time, the total download time, and the average achieved throughput. On the other hand, the metadata of the nodes, related to the radio signal, are collected during the execution of the experiments by means of the LTE modem available at MONROE nodes: a ZeroMQ socket. These data are the RSRP, the RSRQ, and the received signal strength indicator (RSSI).

3.2.2 Data analysis and classification of scenarios

The metrics chosen to model an external commercial network are the RSRP (in dBm) and the file transfer protocol (FTP) user throughput, averaged along the duration of each experiment. Figure 7 shows these metrics with a scatter plot, where two well differentiated behaviors can be seen. The left group of samples stands for low values of RSRP, and consequently, low throughput values. These samples may have been gathered from nodes experiencing low-signal quality, possibly due to high propagation losses or interferences. The right group, however, presents higher values of RSRP and throughput. Thus, two behaviors may be distinguished, each one representing one of the groups of samples:

1. Low-performance behavior. According to the samples gathered in the measurement campaign, this scenario may be characterized by an average RSRP of – 102 dBm and an average throughput of 2 Mbps.

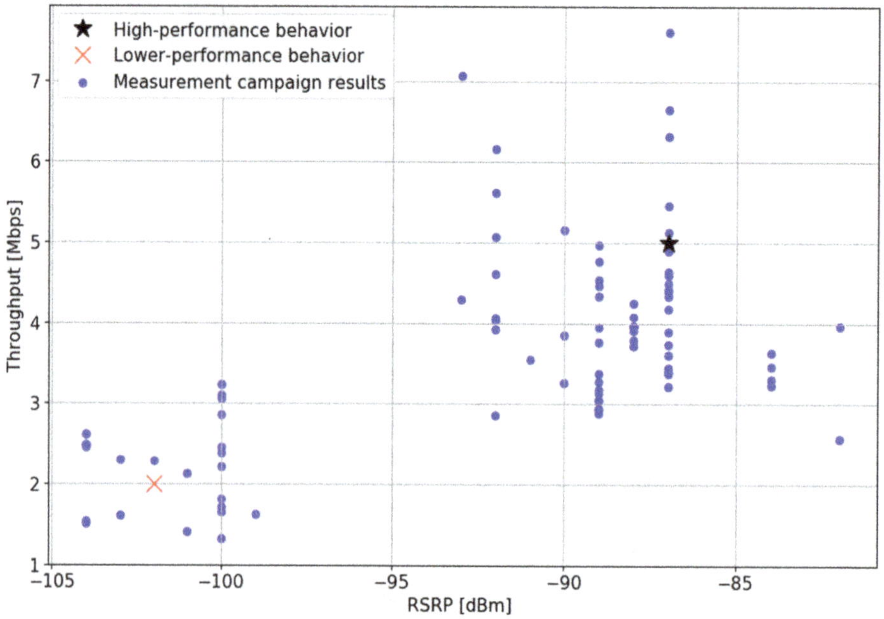

Fig. 7 Classification of measurement campaign results. Centroids representing the average behavior of each cluster are represented with a red cross (low-performance behavior) and a black star (high-performance behavior)

2. High-performance behavior. This behavior may be characterized by an average RSRP of – 87 dBm and an average throughput of 5 Mbps.

3.2.3 Modeling algorithm application

The modeling algorithm application step have been carried out in the UMAHetNet network. The file size used in the file transfer service is 1 MB as in the measurement campaign. The scenario in which the modeling have been performed consists of two terminals: a MONROE node and a smartphone, each one connected to a different picocell. The picocell in which the MONROE node is connected is considered as the serving cell and the other one as the interfering cell. As said before, in this case, the metrics selected to model an external commercial network behavior are the RSRP and the UE throughput. These measurements are provided by the MONROE node as the results of the experiments. The configuration parameters that will be modified to achieve the target behavior (the input factors of TM algorithm) are the transmission power of the serving and the interfering cell, and the serving cell available bandwidth. For each configuration parameter, three levels are defined. The OA used to perform the modeling, which depends on the number of factors as well as the levels of each factors, can be found in [21]. Finally, the TM stop criterion is set to 0.2 since with this value a tradeoff between modeling accuracy and processing time is achieved. The main configuration parameters of the UMAHetNet are summarized in Table 1.

Due to the fact that most of the samples obtained in the measurement campaign belong to the second of the scenarios identified in Section 3.2.2, this scenario has been selected for modeling. Thus, the target values of throughput and RSRP are 5 Mbps and – 87 dBm, respectively.

Figures 8, 9, 10, 11, and 12 show the results of the network modeling. In Figs. 8 to 10, the convergence of the configuration parameters (serving cell transmission power, interfering cell transmission power, and serving cell available bandwidth, respectively) is depicted. In these figures, it can be seen how, in each iteration, the TM controller is reducing the search range, focusing on those

Table 1 Configuration of the UMAHetNet

Scenario	2 UEs. Each one connected to a different picocell
System bandwidth	5, 10, 15 and 20 MHz
Serving cell transmission power	– 4 to – 12 dBm
Interfering cell transmission power	– 6 to – 20 dBm
Mobility	Static user
Transmission direction	Download
Service	File transfer service (file size: 1MB)
e-NodeB (picocell)	Omnidirectional antennas. MIMO
Scheduler	Enhanced proportional fair
Time Transmission Interval (TTI)	1 ms

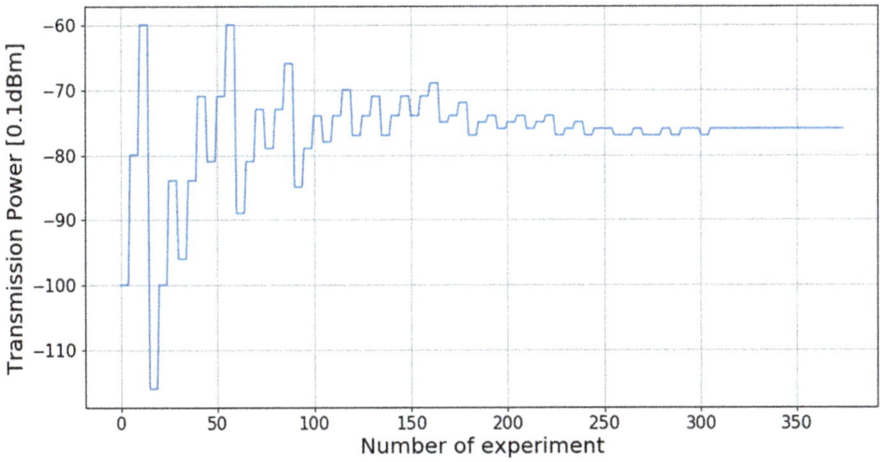

Fig. 8 Serving cell transmission power adjustment

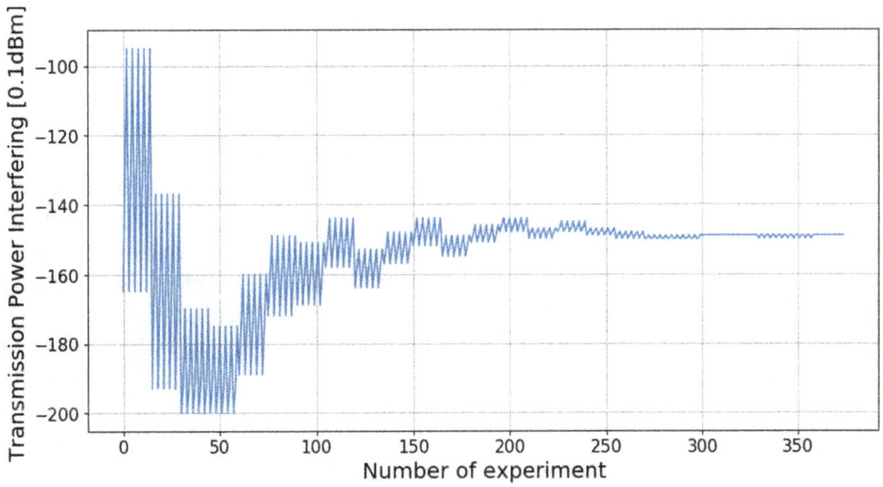

Fig. 9 Interfering cell transmission power adjustment

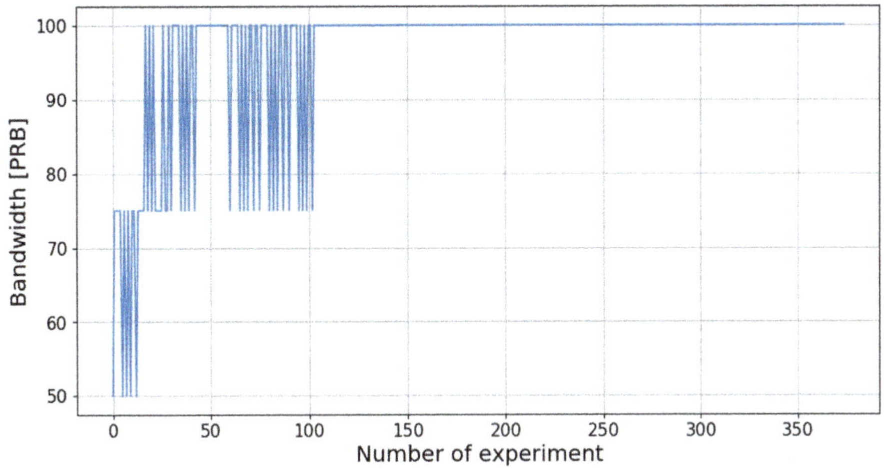

Fig. 10 Serving cell bandwidth adjustment

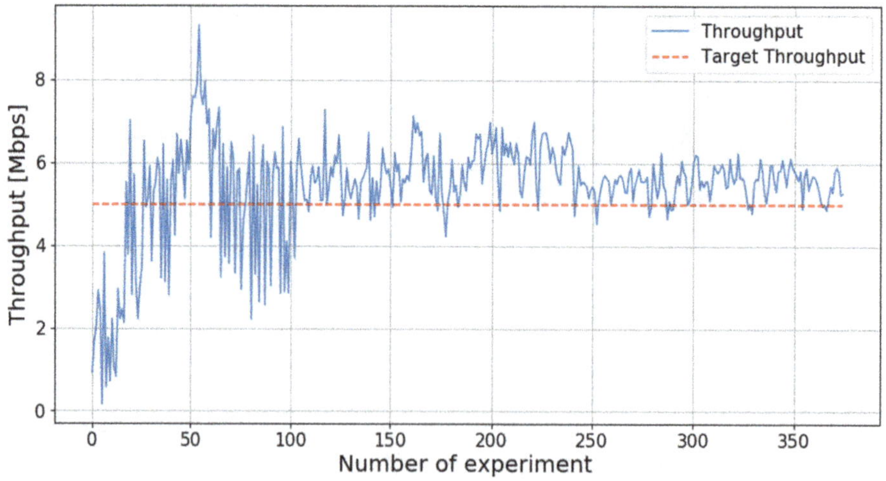

Fig. 11 FTP throughput

values of the configuration parameters for which the throughput and RSRP obtained in the simulations are closer to the target values. Finally, the values of configuration parameters, for which the target values of throughput and RSRP are achieved are in the serving cell transmission power case, – 7.6 dBm; in the interfering cell transmission power case – 14.9 dBm, and in the serving cell, the available bandwidth is 100 physical resource blocks (PRBs) (20 MHz).

Figures 11 and 12 show the resulting values of throughput and RSRP obtained in each experiment (represented in blue) as well as the target values of these metrics (represented in red). It can be observed how the values of throughput and RSRP resulting from the successive experiments get closer to the throughput and RSRP target values along the iterations. So, as a result of the modeling process, specific values for the different considered configuration parameters, with which the target values of the measured user performance indicators are achieved, are obtained.

On the one hand, in Table 2, the results of the modeling process proposed in this paper are shown (testbed configuration parameters). As it is explained before, these values are the ones that should be use in the testbed network in order to achieve the target values of user performance indicators. On the other hand, the final values obtained during the modeling process for the user performance indicators: RSRP and throughput and the target values of these indicators are also listed in Table 2. Results obtained for the modeling process has been computed as the mean values of the indicators once the TM has converged. Regarding these results, it can be appreciated in a

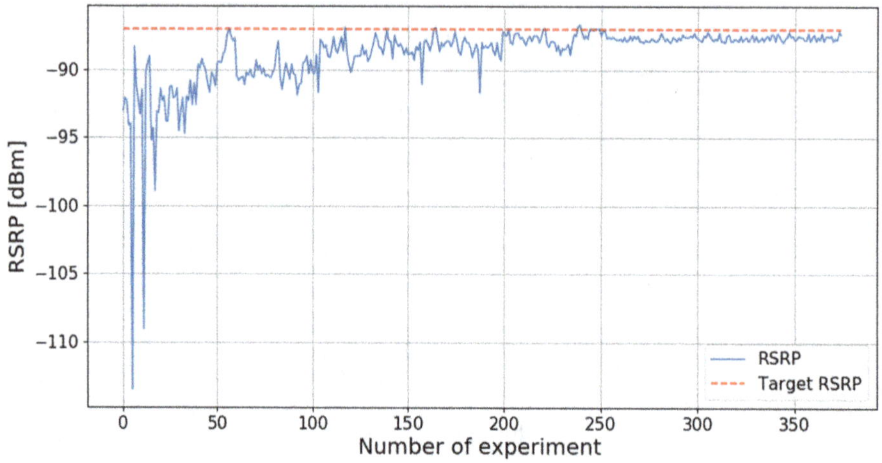

Fig. 12 RSRP

Table 2 Summary of the target values and the results obtained

Testbed configuration parameters	Serving cell transmission power	– 7.6 dBm
	Interfering cell transmission power	– 14.9 dBm
	Serving cell bandwidth	20 MHz
Measured user performance indicators	RSRP	– 87.62 dBm
	Throughput	5.38 Mbps
Target user performance indicators	RSRP	– 87 dBm
	Throughput	5 Mbps

quantitative manner, that the final user performance indicators values are very close to the target ones. The final values of RSRP and throughput achieved in the modeling process are – 87.62 dBm and 5.38 Mbps, respectively, while the target values are – 87 dBm and 5 Mbps.

4 Conclusion

New management network algorithms developed by researchers are rarely tested on commercial networks due to the reluctance of most MNOs. To deal with this problem, in this paper, the modeling of the UE perception of a commercial network performance using a testbed network is proposed. Firstly, a measurement campaign has been developed. To this end, several user performance indicators have been measured in different types of scenarios in commercial networks of several European operators. Secondly, by analyzing the results of the measurement campaign, it has been possible to differentiate two groups that represent the behavior of the studied networks. Finally, taking one of the previous groups as input, the modeling over a testbed network has been carried out using a TM controller. This method can be used by operators to have a preliminary assessment of the effects of a certain optimization action over their network, without actually interrupting or affecting services. It can also be used by researchers to draw better conclusions on how their algorithms will behave in a commercial network; as well as debug them for different scenarios.

As future work, the development of a more complex commercial network model is proposed. In particular, a wider range of user performance indicators will be considered, so that the different services could be better represented. New scenarios will also be considered to allow the modeling of wider range of behaviors that can be found in a commercial network with UMAHetNet. Finally, the use of new configuration parameters will be considered, to be able to model the more complex scenarios discussed above.

Abbreviations

5G: 5th generation; API: Application programing interface; CEP: Complex event processing; CN: Core network; CORBA: Common object request broker architecture; E2E: End-to-end; eCNS: Evolved Core Network Solution; EPC: Evolved packet core; FTP: File transfer protocol; HSS: Home subscriber server; HTTP: Hypertext transfer protocol; JSON: JavaScript object notation; KPI: Key performance indicator; KQI: Key quality indicator; LMT: Local management terminal; LTB: The larger-the-better; LTE: Long-Term Evolution; MBB: Mobile broadband; MME: Mobility management entity; MML: Man-machine language; MNO: Mobile network operator; NMS: Network management system; NTB: The nominal-the-better; OA: Orthogonal array; OSS: Operations support system; P-GW: Packet data network gateway; PCRF: Policy and charging rules function; PRB: Physical resource block; QoE: Quality of experience; RAN: Radio access network; REST: Representational state transfer; RPC: Remote procedure calls; RSRP: Reference signal received power; RSRQ: Reference signal received quality; RSSI: Received signal strength indicator; S-GW: Serving gateway; SN: Signal to noise; SOAP: Simple object access protocol; SON: Self-organizing network; SQL: Structured query language; SSH: Secure shell; STB: The smaller-the-better; TM: Taguchi method; TTI: Time transmission interval; UE: User equipment; UMAHetNet: University of Malaga Heterogeneous Network; ŞXML: Extensible markup language

Acknowledgements

Not applicable.

Authors' contributions

EB designed and performed the measurement campaign. JMR, DP, and IB analyzed the measurement campaign results and performed the modeling tasks. JMR and EJK developed the API REST used in this work. JMR and EB designed and implemented the hardware structure used to perform the modeling tasks in the UMAHetNet. JMR was the major contributor in writing the manuscript. RB was the responsible of supervise all tasks and for the funding acquisition. All authors read and approved the final manuscript.

Funding

This work has been partially funded by the European Union's Horizon 2020 research and innovation program under both the grant agreement No. 644399 (MONROE) through a second open call project (eSON) and the project ONE5G (ICT-760809), the Junta de Andalucía (Research Project of Excellence P12-TIC-2905), and the Spanish Ministry of Economy and Competitiveness (project TEC2015-69982-R). The authors would like to acknowledge the contributions of their colleagues in the project, although the views expressed in this contribution are those of the authors and do not necessarily represent the project.

Competing interests

The authors declare that they have no competing interests.

References

1. Cisco, Cisco Visual Networking Index: Global Mobile Data Traffic Forecast Update, 2016-2021 (2017). https://www.cisco.com/c/en/us/solutions/collateral/service-provider/visualnetworking-index-vni/complete-white-paper-c11-481360.html. Accessed Dec 20 2018
2. Z. Xiao, Y. Xu, H. Feng, T. Yang, B. Hu, Y. Zhou, in *2015 IEEE Global Communications Conference (GLOBECOM)*. Modeling streaming QoE in wireless networks with large-scale measurement of user behavior, (2015), pp. 1–6. https://doi.org/10.1109/GLOCOM.2015.7417690
3. X. Zhang, Y. Xu, H. Hu, Y. Liu, Z. Guo, Y. Wang, Modeling and analysis of Skype video calls: rate control and video quality. IEEE Trans Multimed. **15**(6), 1446–57 (2013). https://doi.org/10.1109/TMM.2013.2247988
4. T. Rehman, M. A. I. Baig, A. Ahmad, in *2017 IEEE 8th Annual Ubiquitous Computing, Electronics and Mobile Communication Conference (UEMCON)*. LTE downlink throughput modeling using neural networks, (2017), pp. 265–70. https://doi.org/10.1109/UEMCON.2017.8249044
5. J. Cainey, B. Gill, S. Johnston, J. Robinson, S. Westwood, in *39th Annual IEEE Conference on Local Computer Networks Workshops*. Modelling download throughput of LTE networks, (2014), pp. 623–8. https://doi.org/10.1109/LCNW.2014.6927712
6. P. Muñoz, R. Barco, I. de la Bandera, Load balancing and handover joint optimization in LTE networks using fuzzy logic and reinforcement learning. Comput. Netw. **76**, 112–25 (2015)
7. 5G Infrastructure PPP Association and others, 5G Vision - The 5G

Infrastructure Public Private Partnership: the next generation of communication networks and services (2015)

8. F. Mah-Rukh, K. Kousias, A. Lutu, M. Rajiullah, Ö. Alay, A. Brunström, A. Argyriou, in *11th Workshop Wireless Network Testbeds, Expeimental Evaluation & Characterization*. FLEX-MONROE: aunified platform for experiments under controlled and operational LTE settings, (2017), pp. 1–8. https://doi.org/10.1145/3131473.3131477
9. A. Awada, B. Wegmann, I. Viering, A. Klein, in *2011 IEEE 73rd Vehicular Technology Conference (VTC Spring)*. A joint optimization of antenna parameters in a cellular network using Taguchi's method, (2011), pp. 1–5. https://doi.org/10.1109/VETECS.2011.5956217
10. A. Gómez-Andrades, R. Barco, I. Serrano, P. Delgado, P. Caro-Oliver, P. Muñoz, Automatic root cause analysis based on traces for LTE self-organizing networks. IEEE Wirel. Commun. **23**(3), 20–28 (2016). https://doi.org/10.1109/MWC.2016.7498071
11. V. Buenestado, M. Toril, S. Luna-Ramírez, J. M. Ruiz-Avilés, A. Mendo, Self-tuning of remote electrical tilts based on call traces for coverage and capacity optimization in LTE. IEEE Trans. Vehicular Technol. **66**(5), 4315–4326 (2017). https://doi.org/10.1109/TVT.2016.2605380
12. G. Taguchi, *Taguchi Methods. Research and Development*, 1st Edition. (American Supplier Institute, S. Konishi, eds.) (Taguchi Methods Series, 1993)
13. A. Al-Refaie, T.-H. Wu, M.-H. Li, An effective approach for solving the multi-response problem in Taguchi method. Jordan J. Mech. Industr. Engineer. **4**(2), 314–23 (2010)
14. S. Fortes, S.-R. J. Antonio, D. Palacios, E. Baena, R. Mora-García, M. Medina, P. Mora, R. Barco, The Campus as a Smart City: University of Málaga environmental, learning, and research approaches. Sensors. **19**(6) (2019). https://doi.org/10.3390/s19061349
15. J. Tihomirovs, J. Grabis, Comparison of SOAP and REST based web services using software evaluation metrics. Information Technology and Management Science. **19**(1), 92–97 (2016)
16. M. Peón-Quirós, T. Hirsch, S. Alfredsson, J. Karlsson, A. S. Khatouni, Ö. Alay, Deliverable D1.3 Final Implementation. https://www.monroe-project.eu/resources/projectdeliverables/. Accessed Nov 3 2018
17. MONROE Measuring Mobile Broadband Networks in Europe. https://www.monroe-project.eu/. Accessed Feb 8 2019
18. Ö. Alay, A. Lutu, M. Peón-Quirós, V. Mancuso, T. Hirsch, K. Evensen, A. Hansen, S. Alfredsson, J. Karlsson, A. Brunström, A. Safari Khatouni, M. Mellia, M. A. Marsan, in *Proceedings of the 23rd Annual International Conference on Mobile Computing and Networking*. Experience: an open platform for experimentation with commercial mobile broadband networks (ACM, New York, 2017), pp. 70–80. http://doi.acm.org/10.1145/3117811.3117812. https://doi.org/10.1145/3117811.3117812
19. Ericsosn, Ericsson Mobility Report (2018). https://www.ericsson.com/491e34/assets/local/mobilityreport/documents/2018/ericsson-mobility-report-november-2018.pdf. Accessed Dec 20 2018
20. D. Stenberg, *Everything curl*, (2018). https://curl.haxx.se/book.html. Accessed Apr 22 2018
21. N. J. A. Sloane, A library of orthogonal arrays. http://neilsloane.com/oadir/. Accessed Jul 18 2018

13

Station green lighting system based on sensor network

Min Wang[1], Zhongbo Wu[1*] and Hang Qin[2]

Abstract

Aiming at the problems of high energy consumption, chaotic lines, and inconsistent lighting sources in many stations, this paper presents a simple, reliable, and convenient station green lighting system based on sensor network. According to the arrival and departure time of the train, the lighting equipment of the corresponding track is automatically closed, which saves the power resources and human resources, and improves the service life of the lighting equipment. In addition, an intelligent illumination control method based on BP neural network is proposed. The test is carried out in the real environment of the station. The designed system can intelligently adjust the working conditions of the lighting system, save energy to the greatest extent and meet the practical requirements.

Keywords: Station green lighting system, Sensor network, BP neural network

1 Introduction

Railway passenger station is a public place for passengers to get on and off the train. The lighting of its platform, waiting room and other places can provide good visual conditions and comfortable visual environment for passengers and station staff to ensure the service quality of the station and passenger safety. However, at present, there are many shortcomings in station lighting, such as large power consumption, short light source life, backward control mode, and so on [1].

At present, although China's electric power industry is developing rapidly, the situation of insufficient power supply and low power efficiency is still serious, which will continue to exist for a considerable period of time in the future. At the same time, the waste of power resources in the operation and management of stations is quite serious, and the power consumption funds are still high. In practice, the operation and maintenance cost of the station lighting system will even exceed the construction cost, among which the water and electricity cost, management cost, and maintenance cost of mechanical and electrical equipment account for a considerable proportion of the total cost [2].

* Correspondence: wuzhongbo@hbuas.edu.cn
[1]Computer Engineering School, Hubei University of Arts and Science, Xiangyang, China
Full list of author information is available at the end of the article

In order to realize the intelligence of a lighting system, some researchers designed an intelligent lighting system based on ZigBee. There are also some intelligent lighting systems based on the Internet of Things architecture. Airports, subway, and other occasions have begun to use intelligent systems to improve lighting efficiency. Chun et al. presents a real-time smart lighting control system using human location estimation [3]. Park et al. put forward a novel model-based PID gain design method to improve the performances of dynamic bending light (DBL) module that change horizontal angle [4]. Some researchers have studied outdoor lighting systems [5, 6], while others have studied indoors such as classrooms and homes [7–10]. Wireless control technology was added to the lighting system, but they still need people to participate in and control and have not achieved an energy-saving effect. Some researchers began to use large data and clustering algorithm to analyze the illumination area in order to control the illumination area more intelligently [11, 12].

We realize a green lighting system of railway station based on a sensor network. Controlling the lighting equipment automatically according to the arriving and leaving time of the train, the system can economize electric and human resource and improve the using life of the lighting equipment.

2 System architecture

A station green lighting system is mainly composed of a lighting unit, scene controller, and monitoring host, as shown in Fig. 1. The system staff can check, manage, and control the working status of each lighting unit of the whole lighting system by monitoring the host computer. A monitoring host is set up in the system. The host computer is connected to the Internet and installed with the monitoring software of the lighting system. The scene controller and its controlled lighting unit are the basic components of the system. The monitoring host exchange information with the system through the Internet and a GPRS wireless network. The number of scene controllers is determined according to the scale of landscape lighting and application environment. Each scene controller controls 1–127 lighting units. Local communication of the system adopts the ZigBee wireless sensor network, the lighting unit completes the WSN sensor network device function, while the scene controller realizes WSN gateway function and acts as respective sensor network. The coordinator is responsible for the networking and data transmission management of sensor devices. In addition to completing the function of sensor equipment, the lighting unit in the system needs to complete the work including collecting the detection data of the lighting unit, sending data according to the system requirements, battery-charging management, lighting control, and so on.

The automatic control and manual control functions of the lighting system are implemented in the application program of the monitoring server. The main purpose is to control the scene controller according to different conditions, so as to achieve the purpose of controlling the lighting equipment. Its functions include the following:

1. According to illumination, automatic control of lighting equipment is realized. Illuminometers are installed in waiting rooms, train tracks, and squares. Illuminometers are connected with servers. We use T to represent the sample value of the illuminometer, and server-side applications periodically read T. For waiting rooms and squares, when T is less than the pre-set threshold V of the program, the corresponding lighting equipment will be closed, when T is greater than V, the corresponding lighting equipment will be disconnected; for lanes, when T is less than V, the automatic control function of lanes will be activated, and when T is greater than V, the automatic control function of lanes will be closed.
2. Controlling scene controllers according to arrival and departure time of trains in the train information table.

The server application program periodically reads the train information in the database. If a train is about to arrive, it opens the lighting equipment of the corresponding controller. If the train on the track leaves, it disconnects the lighting equipment of the corresponding controller.

3 System implementation

3.1 Lighting unit

The lighting unit mainly consists of a solar panel, power management module, storage battery, LED lamp control

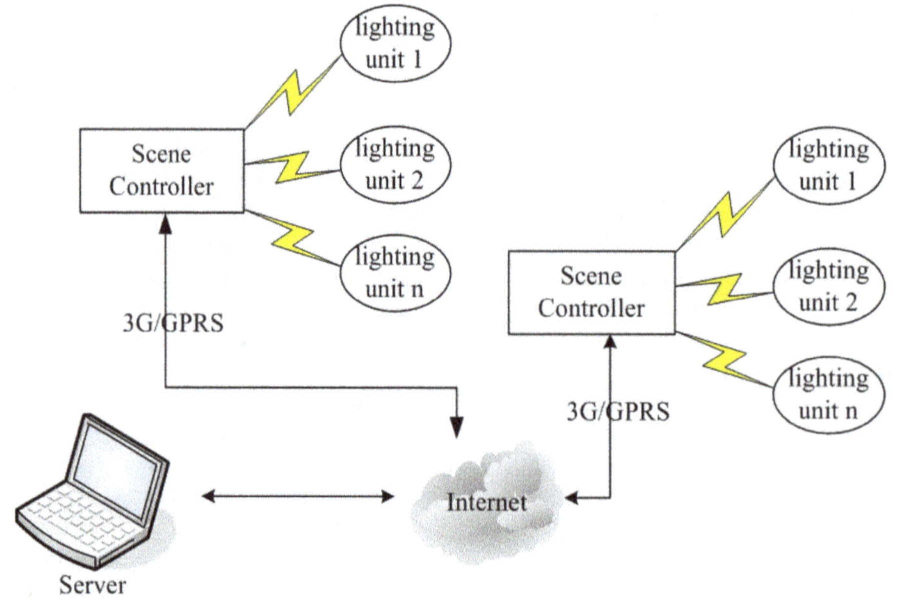

Fig. 1 System architecture

module, and wireless transceiver module. Solar panels convert light energy into electricity and charge batteries through the power management module. When the lighting system is turned on, the power management module converts the stored energy of the battery into 12 V DC required for LED lighting. The power module detects the voltage of the battery in real time. When the voltage of the battery is below the threshold, the module automatically transfers the power of the LED to the market and completes the conversion of 220 V AC to 12 V DC.

The control module of the LED lamp completes the switch, color adjustment, and dimming of the LED lamp according to the need of the scene setting. At present, LED lamps are usually packaged with 1-W or 3-W beads, which emit different colors of light through different phosphor LED beads. There are three kinds of packaging modes of LED beads: series, parallel, and hybrid. According to the color and brightness requirements of lighting, the packaging mode of LED beads can be selected. Controlling the brightness of LED beads can be achieved by changing the current of the LED beads and adjusting the time of lighting the LED beads. Relatively changing the current adjustment method, using the characteristics of a high flash of the LED to change the lighting time of the LED, is simpler and easier to achieve. It is the main way to adjust the brightness of the LED beads.

3.2 Monitoring server

The monitoring server in the system is the information center of the whole lighting system. When the system runs, the upper computer software receives the status information of the lighting unit transmitted from the scene controller via the Internet.

The monitoring server sends queries and setting instructions to the scene controller according to the scene setting requirements, and then forwards them to the corresponding lighting unit by the scene controller.

3.3 Scene controller

The scene controller's built-in GPRS module communicates with PC after accessing Internet through GPRS network. At the same time, in the ZigBee wireless sensor network, its role is as a coordinator, responsible for the networking of wireless sensors and the management of sensor equipment. In the system design, the maximum number of communication nodes in each sensor network is set to 128, that is, one coordinator and 127 devices. The number of lighting units in a lighting system may exceed 127. There are more than two coordinators and their responsible networks in a system at the same time. A unique 16-bit network PANID is set for each coordinator in the system. The embedded ZigBee terminal module in the lighting unit managed by the system needs the same PANID as the network coordinator. In this way, the coordinator located in the scene controller can accept the request to join the network of the same PANID terminal within its network coverage and then add the information of the new lighting unit node.

In the operation of the system, the scene controller does not process and save the information sent by the monitoring host and the lighting unit. It directly sends the state detection information sent by the lighting unit to the monitoring host through the local area network for processing and, at the same time, sends the instructions issued by the monitoring host to the lighting units. The monitoring host is responsible for the information processing and judgment of multiple scenes and lighting units in the whole system. Scene control in the system acts as a sensor gateway, responsible for communication with various devices and Internet network.

The hardware composition of the sensor gateway includes an MCU unit, GPRS module unit, ZigBee module unit, power management unit, and clock unit. The input voltage of the power management unit converts the battery voltage to 4.1 V required by the GPRS module, 5 V required by MCU, and 3.3 V required by the MCU module. UART0 and UART1 of MCU module are connected with GPRS module and ZigBee module respectively to realize network control and communication. In circuit design, attention should be paid to the fact that the high current at the start of the GPRS module will result in a voltage drop of 0.6–0.7 V. It is necessary to design 1–2100 uF tantalum capacitors between the 4.1-V output terminal and the ground to avoid the restart caused by the protection of GPRS module due to the voltage drop to 3.0 V. The scene controller uses NXPLPC1766 MCU, and its two UART ports are connected with GPRS module and ZigBee transceiver module, respectively. UDP and IP protocol stack are implemented on software based on embedded operating system uC/OS II. The monitoring host in the system can exchange information with a gateway through UDP protocol.

3.4 Networking process of sensor networks

The ZigBee wireless sensor network is used for local area communication of lighting control system. It is a low-speed, low-power, and short-distance wireless communication technology. ZigBee supports a variety of networking modes. Considering efficiency and reliability, the system uses star topology networking. Lighting system deploys one or more scene controllers and each scene controller communicates directly with the lighting unit.

The system predefines a PANID for each scenario controller as the network identifier. The scene controller broadcasts broadcast frames 60 s after starting and opens lighting unit's request response to join the network. Once the illumination unit is activated or reset, channel scanning is performed regularly. Once it finds that there are scene controllers available in the network, it sends a request. After the scene controller detects the request, it decides to accept or reject the device to join the network and update its own network table.

In the system, data transmission between scene controller and lighting unit of sensor network adopts direct transmission mode (no intermediate device forwarding), that is, scene control transmits data directly to the lighting unit, and sends confirmation information to the scene controller when the lighting unit receives data. The data transmission mode requires that the terminal node equipment be in the data-receiving state at any time, that is to say, it should be in the wake-up state at any time.

Scene controller uses unicast mode to send information to poll sensor nodes. The scene controller starts up and sends a data transmission request frame to the lighting unit periodically according to the order of lighting units in the network table. The lighting unit receives a request frame, returns a reply frame, and contains its status information in the reply frame.

Scene controller starts to get the IP address and establish a network table. It periodically reports the status information of the illumination unit in the sensor network to the host computer. The default is 5 min, which can be set up. The host computer sets the polling interval of the scene controller through the network and calibrates the local clock of the scene controller. The system adopts a two-level synchronization mechanism to solve the synchronization problem. The upper computer software and the scene controller communication protocol use the checking time frame. The upper computer sends the checking time frame regularly. The scene controller obtains the upper computer time through the frame and checks the local time. In sensor networks, scene controllers send broadcasting pulse frames every 60 s to synchronize the nodes of the network they manage. The pulse frames contain the updated data of the counter in seconds. After receiving the broadcast pulse frames, the illumination unit updates the counting value of the local timer. The internal timer of the illumination unit updates the counting value of the timer at 1 s. The sensor gateway broadcasts the current time information every 10 s. The sensor gateway contains a clock chip. The internal time counting unit of the sensor network is seconds. The sensor gateway converts the clock chip's H:MM:SS into a seconds counting. Each sensor device receives the time data and updates the internal time counter. The timer of each sensor device is interrupted once in 1 s, and the time counter in interruption service is added by 1.

3.5 Database design

The database contains train information table, equipment status table, operation log table, and so on. The most important table is the train information table. The information in this table is the main basis for automatic control of server applications program (Table 1).

The administrator can modify the information of this table through the application program, such as adding, deleting, and modifying the information of the number of trains. Normally, late = 0, arriveActual = arrival, leaveActual = leave; when the train is late, the administrator modifies the corresponding number of late through the application program. The program automatically modifies arriveActual and leaveActual, arriveActual = arrive + late, and leaveActual = leave + late. When the train leaves, the program automatically modifies late = 0, arriveActual = arrive, and leaveActual = leave

4 System improvement

General lighting for waiting rooms, indoor, and station platforms is generally designed for lighting purposes. Through lighting calculation and scheme design, the illumination of buildings can reach the national lighting standards. But in practice, we all hope that the lighting of various places can not only satisfy people's visual condition of work, but also their study and life. It also hopes to respond to the initiative of national green lighting and save energy consumption (Figs. 2 and 3). Therefore, in the design of the green lighting automatic control system, the first consideration is to combine the illumination and energy saving very well, which can meet its illumination in real time according to its needs, so that the comprehensive index of illumination and energy saving can reach the optimum.

There are some limitations according to the feedback control of illumination. On the one hand, it is difficult to establish an exact mathematical model, and the model is often nonlinear. On the other hand, the open-closed-loop parameter control is more complex which has a large amount of calculation and does not have self-learning

Table 1 Database design

Attribute	Description
trainid	Train number
begin	Departure Station
end	Terminus
track	Track
arrive	Prescribed arrival time
leave	Prescribed departure time
arriveActual	Actual arrival time
leaveActual	Actual departure time
late	Delay time

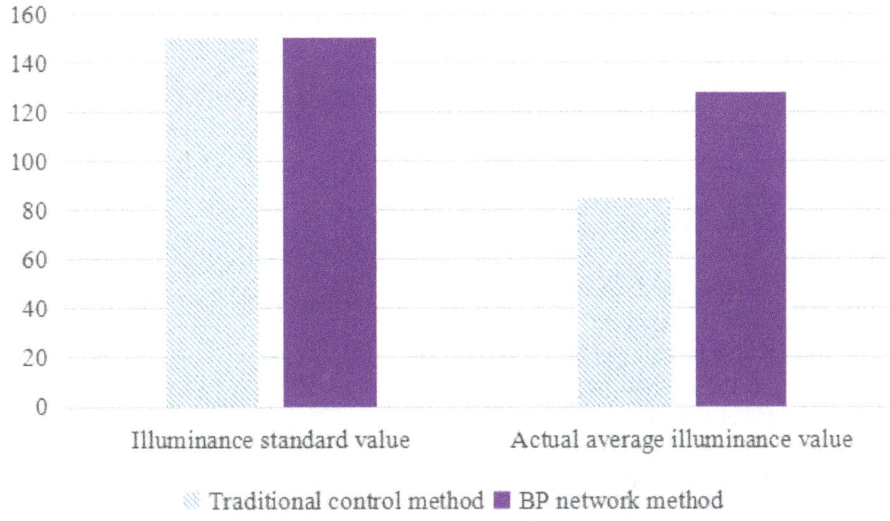

Fig. 2 Comparison of illuminance

function. Real time and accuracy cannot be well guaranteed.

In recent years, artificial neural network (ANN) technology has been developed to the practical stage and has achieved fruitful results in many fields [13, 14]. A neural network is a large-scale information parallel processing system. It imitates the function of human brain neurons. It has the characteristics of distributed information storage, parallel processing, self-adapting, and self-learning. It can map any function relationship, which is an important function of a neural network.

The illumination control in the green lighting control system can be regarded as a "black box" by combining illumination and power organically according to the variables transmitted by sensors (illumination, voltage, and current). Only the relationship between input and output is studied, and the network is trained by the accumulated experience. In the process of illumination control, as long as the required parameters are provided to the network as input, the intelligent control of illumination can be realized.

BP (back-propagation) network is the most widely used artificial neural network model at present, because its learning algorithm is based on the back-propagation learning algorithm named.

Under certain conditions, there exists a three-layer back-propagation network with input, output, and a nonlinear hidden layer for any given $\varepsilon > 0$, which can be used to approximate any nonlinear function with the mean square deviation not exceeding ε.

$P_1, P_2, P_3, \cdots, P_R$ is input which can be written in matrix form $P = [P_1, P_2, P_3, \cdots, P_R]$. Their corresponding

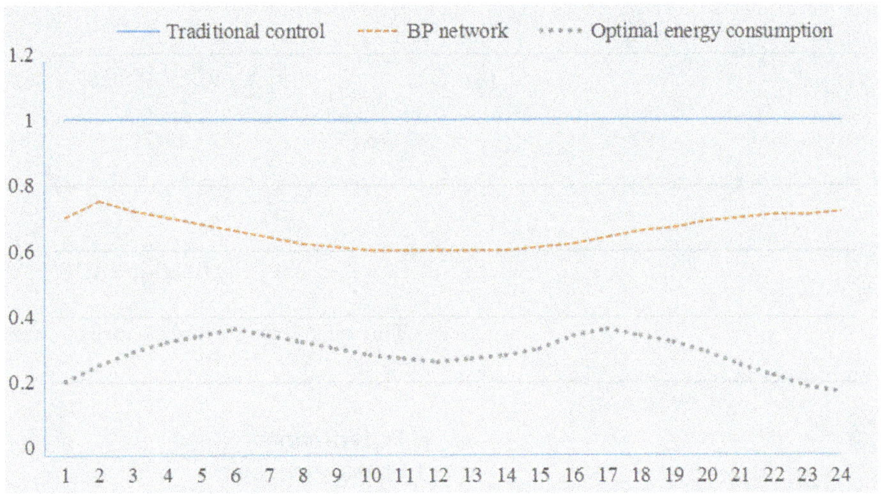

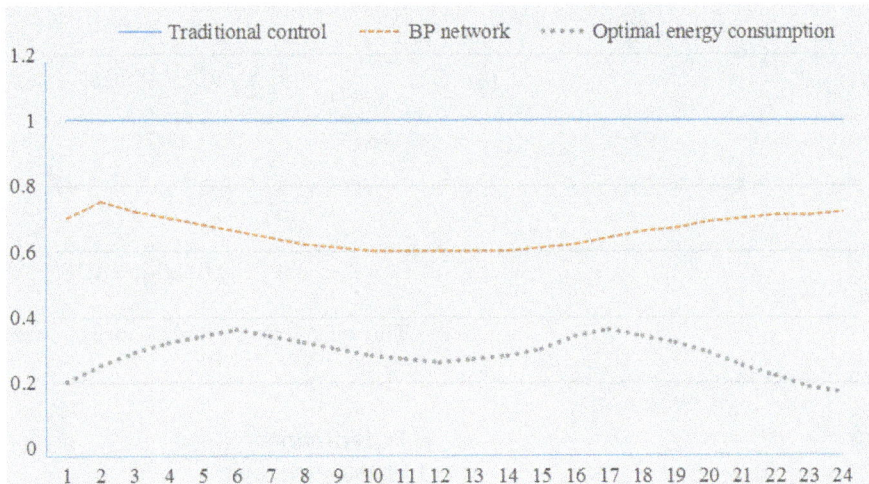

Fig. 3 Comparison of energy consumption

weights are W_1, W_2, W_3, $\cdots$, W_R which can be written in matrix form $W = [W_1, W_2, W_3, \cdots, W_R]$. Net input of neurons is $n = W_p + b$. Output of neurons is $q = f \cdot W_p + b$.

The weights W and threshold b are all adjustable parameters of neurons, which are adjusted by designers according to learning rules. The input/output relationship is consistent with the design objective. f is a transfer function, which is selected by the designer according to the design objective.

The BP network adopts a supervised learning algorithm. Firstly, we provide a set of input/output data pairs $\{x_1, t_1\}$, $\{x_2, t_2\}$, $\{x_h, t_h\}$ as training sample sets. x_h is the input of the network, and t_h is the desired target output response. The training process of the network is to supply each input to the network. The actual output of the network is compared with the target output. The error between the two is regarded as the performance of the network.

We take mean square error as the performance target of BP algorithm.

$$F(x) = E\left[e^T e\right] = E\left[(t - \lambda)^T (t - \lambda)\right] \tag{1}$$

The square error of the K iteration can be used instead of the upper one.

$$F = (t(k) - \lambda(k))^T (t(k) - \lambda(k)) = e^T(k) e(k) \tag{2}$$

Different numerical methods, such as the fastest descent method, Newton method, and conjugate gradient method, can be used to modify the weights and thresholds to reduce errors.

We adopt the fastest descent method.

$$W_{i,j}^m(k+1) = W_{i,j}^m(k) - \lambda \frac{\partial \hat{F}}{\partial W_{i,j}^m} \tag{3}$$

$$b_i^m(k+1) = w_i^m(k) - \lambda \frac{\partial \hat{F}}{\partial w_i^m} \tag{4}$$

λ is the learning rate $\frac{\partial \hat{F}}{\partial W_{i,j}^m} = \frac{\partial \hat{F}}{\partial n_i^w} \times \frac{\partial n_i^m}{\partial W_{i,j}^m}$.

$$\frac{\partial \hat{F}}{\partial b_i^m} = \frac{\partial \hat{F}}{\partial n_i^w} \times \frac{\partial n_i^m}{\partial b_i^m} \tag{5}$$

n_i^m is the net input of layer m $n_i^m = \sum_{j-1}^{T^{m-1}} W_{i,j}^m \lambda_i^{m-1} + b_i^m$

So $\frac{\partial n_i^m}{\partial W_{i,j}^m} = \lambda_j^{m-1}, \frac{\partial n_i^m}{\partial b_i^m} = 1$.

Definition sensitivity $C_i^m = \frac{\partial \hat{F}}{\partial n_i^m}$ (That is the change rate of the ith element to the net input of m level by $\hat{F}$.)

$$\frac{\partial \hat{F}}{\partial W_{i,j}^m} = T_i^m \alpha_j^{m-1}, \frac{\partial \hat{F}}{\partial b_i^m} = C_i^m \tag{6}$$

Now the fastest descent method can be written as

$$W_{i,j}^m(k+1) = W_{i,j}^m(k) - \lambda C_i^m \lambda_j^{m-1}, b_i^m(k+1) = b_i^m(k) - \lambda C_i^m \tag{7}$$

Its matrix form can be written as

$$W^m(k+1) = W^m(k) - \lambda C^m \left(\lambda^{m-1}\right)^T, b^m(k+1) = b^m(k) - \lambda C^m \tag{8}$$

$$C^m = \frac{\partial \hat{F}}{\partial n^m} = \begin{bmatrix} \frac{\partial \hat{F}}{\partial n_1^m} \\ \frac{\partial \hat{F}}{\partial n_2^m} \\ \cdots \\ \frac{\partial \hat{F}}{\partial n_n^m} \end{bmatrix} \tag{9}$$

Sensitivity should be calculated from the last formula to the first layer by the recursive formula.

$$C^m \to C^{m-1} \cdots C^2 \to C^1, C^m = \left(\frac{\partial n^{m+1}}{\partial n^m}\right)^T \frac{\partial \hat{F}}{\partial n^{m+1}} \tag{10}$$

We use the Jacobi matrix to derive the recurrence relation of sensitivity.

$$\frac{\partial n^{m+1}}{\partial n^m} = \begin{bmatrix} \frac{\partial n_1^{m+1}}{\partial n_1^m} & \frac{\partial n_1^{m+1}}{\partial n_2^m} & & \frac{\partial n_1^{m+1}}{\partial n_s^m} \\ \frac{\partial n_2^{m+1}}{\partial n_1^m} & \frac{\partial n_2^{m+1}}{\partial n_2^m} & \cdots & \frac{\partial n_2^{m+1}}{\partial n_s^m} \\ \vdots & \vdots & & \vdots \\ \frac{\partial n_s^{m+1}}{\partial n_1^m} & \frac{\partial n_s^{m+1}}{\partial n_2^m} & & \frac{\partial n_s^{m+1}}{\partial n_s^m} \end{bmatrix} \tag{11}$$

The elements (i,j) in the matrix can be calculated by the following formula.

$$\begin{aligned} \frac{\partial n_i^{m+1}}{\partial n_j^m} &= \frac{\partial \left[\sum_{i-1}^{C^m} W_{i,j}^{m+1} \lambda_j^m + b_i^{m+1}\right]}{\partial n_j^m} \\ &= W_{i,j}^{m+1} \frac{\partial \lambda_j^m}{\partial n_j^m} = W_{i,j}^{m+1} \frac{\partial f^m\left(n_j^m\right)}{\partial n_j^m} \\ &= W_{i,j}^{m+1} f^m\left(n_j^m\right) = W^{m+1} F^m(n^m) \end{aligned} \tag{12}$$

The sensitivity formula can be rewritten as $C_j^m = -2(t_j - \lambda_j) f^m(n_j^m)$.

5 Experiments

5.1 Simulation setup

The neural network selects a three-layer BP network with a hidden layer. The input and output of the

Table 2 Eight neurons

Symbol	Definition
e	The actual illuminance value measured by sensors
u	Voltage measured by sensor
i	Current measured by sensor
E	Illuminance standard value
O	Choosing luminous flux of light source
L	The length of the platform
W	The width of the platform
H	The height of the platform

network depend on the definition of the external problem. For the lighting system which needs to adjust the illumination on-line in real time, in order to satisfy the illumination conditions and achieve the optimal energy saving, even if the comprehensive index of illumination and power reaches the optimum, the input layer selection of the network depends on the definition of the external problem. The eight neurons are as follows (Table 2).

With each group of lights as the output (there are eight groups of lights here), according to the value of illumination, automatically realize the switch of each group of lights, so that the illumination remains at a predetermined value, so that only eight neurons in the output layer, the number of neurons in the hidden layer using the gradual growth method to determine, that is, starting with a simple network, if it does not meet the requirements, gradually increase the number of neurons in the hidden layer until appropriate.

5.2 Results and discussion

With the intelligent lighting control system, the lighting system can work in the full automatic state system and automatically adjust the illumination to the most suitable level according to the preset value. This can make good use of natural light, and when the weather or season changes, the system can ensure that the light in the waiting room is maintained at a predetermined level. From the simulation results, it can be seen that the BP network algorithm in the neural network can not only make the illumination satisfactory and make the fatigued passengers be in a comfortable environment, but also save considerable energy. It can save 25–45% of the energy, which is in line with the concept of green lighting.

Figure 4 shows our green lighting system, which turns on the traffic lights of the corresponding lanes 5 min before the train arrives and turns off the traffic lights of the corresponding platforms 10 min after the train leaves. When you click the manual control button, an open and closed control panel will be displayed on the interface. You can click on or off the button to control the corresponding track lights. The system is widely used in Xiangyang, Luoyang, Sanmenxia, and other stations, and each station can save 30–50% of the electricity cost each year.

6 Conclusion

This paper puts forward the concept of "green lighting" for station lighting, which not only has great significance for saving electric energy and providing a healthy and comfortable environment for passengers, but also promotes the implementation of a green lighting project in China. In the automatic control system of green lighting, the intelligent control of illumination based on neural network theory is proposed, which has certain theoretical research significance and practical application value.

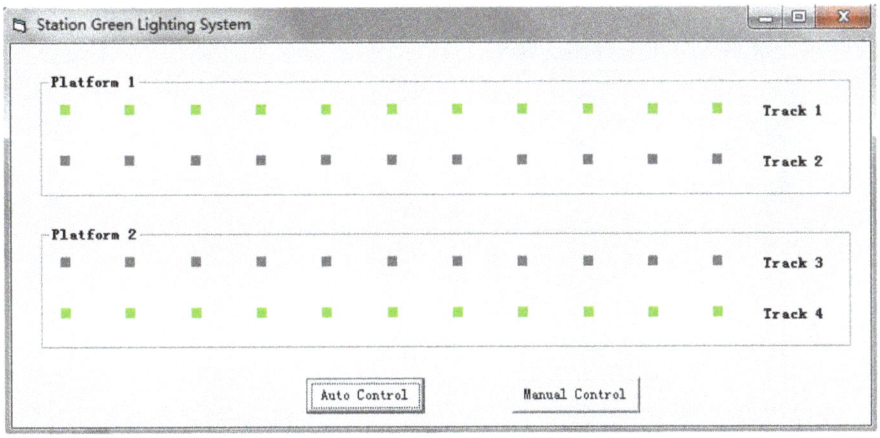

Fig. 4 Interface of actual system

Funding

The authors acknowledge the Hubei Ministry of Education Foundation (Grant: B2016176) and Open Fund of Hubei Superior and Distinctive Discipline Group of "Mechatronics and Automobiles".

Authors' contributions

MW carried out the design of system architecture and drafted the manuscript. ZW carried out the development of prototype system. HQ participated in the design of the study and performed the statistical analysis. ZW conceived of the study, participated in its design and coordination, and helped to draft the manuscript. All authors read and approved the final manuscript.

Competing interests

The authors declare that they have no competing interests.

Author details

[1]Computer Engineering School, Hubei University of Arts and Science, Xiangyang, China. [2]Computer School, Yangtze University, Jingzhou, China.

References

1. R. Fiederling, "Smartrix" a new chapter in the smart matrix LED lighting. ATZelektronik Worldw. **12**(4), 50–55 (2017)
2. X. Ye, X. Xia, Optimal lighting project maintenance planning by a control system approach. IFAC Proc. Vol. **47**(3), 3152–3157 (2014)
3. S.Y. Chun, C. Lee, J. Jang, et al., Real-time smart lighting control using human motion tracking from depth camera. J. Real-Time Image Proc. **10**(4), 805–820 (2015)
4. S.H. Park, B.U. Im, D.K. Park, et al., Model based optimum Pid gain design of adaptive front lighting system. Int. J. Automot. Technol. **19**(5), 923–933 (2018)
5. I. Wojnicki, S. Ernst, L. Kotulski, et al., Economic impact of intelligent dynamic control in urban outdoor lighting. Energies **9**(5), 1–14 (2016)
6. L. Tahkamo, R.S. Rasanen, L. Halonen, et al., Life cycle cost comparison of high-pressure sodium and light-emitting diode luminaires in street lighting. Int. J. Life Cycle Assess. **21**(2), 137–145 (2016)
7. H. Lee, S. Kwon, J. Lim, et al., A development of a lighting control system based on context-awareness for the improvement of learning efficiency in classroom. Wirel. Pers. Commun. **86**(1), 165–181 (2016)
8. M.K. Jha, N. Kumar, Energy efficient lighting system for indoor parking with ubiquitous communication. Wirel. Pers. Commun. **100**(2), 379–389 (2018)
9. A. Yano, K. Fujiwara, Plant lighting system with five wavelength-band light-emitting diodes providing photon flux density and mixing ratio control. Plant Methods **8**(1), 46–46 (2012)
10. S. Moon, S. Kwon, J. Lim, et al., Implementation of smartphone-based color temperature and wavelength control LED lighting system. Clust. Comput. **19**(2), 949–966 (2016)
11. F. He, L. Zhang, Mold breakout prediction in slab continuous casting based on combined method of GA-BP neural network and logic rules. Int. J. Adv. Manuf. Technol. **95**(95), 4081–4089 (2018)
12. C. Zhang, C. Liu, X. Zhang, et al., An up-to-date comparison of state-of-the-art classification algorithms. Expert Syst. Appl. **82**(82), 128–150 (2017)
13. A.H. Moghaddam, M.H. Moghaddam, M. Esfandyari, et al., Stock market index prediction using artificial neural network. J. Econ. Finance Adm. Sci. **21**(41), 89–93 (2016)
14. H. Zhenlong, Z. Qiang, W. Jun, et al., The prediction model of cotton yarn intensity based on the CNN-BP neural network. Wirel. Pers. Commun. **102**(2), 1905–1916 (2018)

14

Performance analysis and optimization for non-uniformly deployed mmWave cellular network

Xuefei Zhang[1*], Yinjun Liu[1], Yue Wang[1] and Juan Bai[2]

Abstract

In this paper, we propose a multi-tier mmWave cellular framework where sub-6 GHz macro BSs (MBSs) are assumed as a Poisson point process (PPP) and small-cell BSs (SBSs), operating on either mmWave or sub-6 GHz, follows non-uniform Poisson cluster point (PCP) model. This paper proposes both centralized and distributed user association algorithms. For the centralized two-step algorithm, we aim to maximize the sum rate while satisfying quality of service (QoS) and power consumption constraints based on eigenvalue analysis. Then, we derive the association probability, the coverage probability, and the average achievable rate, cosidering directivity and blockage effect, by stochastic geometry. On this basis, a distributed user association algorithm is proposed. The simulation results demonstrate the accuracy of our theoretical analysis and also reveal the effect of some parameters on the network performance. In addition, the proposed centralized algorithm can achieve near-optimal sum rate with a low complexity.

Keywords: MmWave cellular network, Non-uniform distribution, Stochastic geometry

1 Introduction

Nowadays, the fifth generation mobile communication network (5G) are undoubtedly one of the most attractive topics in both academic and industrial fields. Many res earchers engage themselves to explore the new advanced methodologies and technologies in 5G, to support the booming data traffic with reduced energy consumption and improved quality of service (QoS) provision. Some enabling technologies, such as heterogeneous networks (HetNets), massive multiple-input multiple-output (MIMO), and millimeter wave (mmWave) techniques, have been identified to bring 5G to fruition [1].

MmWave communications, which operate from 30 GHz to 300 GHz, have attracted much attention due to the shortage of microwave frequency [2]. It can provide the larger bandwidth, which can support the higher data rate transmission. Besides, mmWave communications have no effect on the traditional cellular communication in sub-6 GHz, since they operate at the higher and different frequencies. Therefore, the deployment of mmWave BSs not only offloads the data traffic of existing cellular frequency, but also reduces the interference. However, a major channel characteristic of mmWave communications is its sensitivity to the blockages, which is a serious problem due to the poor performance of mmWave links at penetrating into or diffracting around solid objects [3]. Therefore, mmWave communication is commonly used in short-range directional transmission with directional antennas instead of isotropic antennas. On the other hand, the rigorous directivity and high penetration losses lead to the fact that mmWave systems are noise-limited. Therefore, the user association algorithms designed for the interference-limited networks cannot be directly applied in the mmWave systems [4].

Due to the two mentioned fundamental differences from conventional sub-6 GHz cellular systems, the channel models of mmWave communications should be modeled carefully. Some research proposed the mmWave-pattern models including the effects of blocking, directional beamforming, small-scale fading, and path loss for line-of-sight (LOS) and non-line-of-sight (NLOS) links [3–5]. Specifically, the blockages, resulting in the serious power loss at the high mmWave frequencies, are assumed stationary and isotropic [5, 6]. For mathematical

*Correspondence: zhangxuefei@bupt.edu.cn
[1] National Engineering Laboratory for Mobile Network Technologies, Beijing University of Posts and Telecommunications, 100876 Beijing, China
Full list of author information is available at the end of the article

tractability, a LOS ball model was proposed for simplifying blockage modeling where BSs inside the LOS ball were considered to be in LOS whereas any BS outside of the LOS ball was treated as NLOS [6]. In [7], this blocking model was modified by adding a LOS probability within the LOS ball, and this approach was shown to reflect several realistic blockage scenarios. Meanwhile, the pathloss exponent is considered higher in NLOS links (e.g., commonly about 4) than that in LOS links (e.g., around 2) [8]. Furthermore, some real data is obtained to verify the parameters of some blocking model (e.g., real building data in the UT Austin, downtown LA regions, NYC, and Chicago [4, 6]). It is revealed that the environment has the effect on the block behaviors, and the blocking models should be carefully selected according to the propagation environment (e.g., 3GPP-like models could be sufficient to fit the urban regions; generalized LOS ball model gives a good fit for dense random deployment of BSs, etc. [4]).

Meanwhile, the directional beamforming, for compensating the increased path loss at mmWave frequencies and for overcoming the more serious noise due to the large transmission bandwidth, is approximated by considering a sectored antenna model [9, 10]. In [9], mmWave BSs are equipped with directional antennas, and UEs perform the perfect beam alignment, whereas in [10], both the users and mmWave BSs are assumed to estimate the angles of arrival and to adjust their antenna orientations accordingly. Moreover, large antenna arrays, enabling massive MIMO and hybrid beamforming techniques, can contribute to the directive transmission. The rate of multi-user MIMO with hybrid beamforming [11] or with user pairing [12] or joint spatial division and multiplexing [13] in mmWave network are analyzed.

2 Methods/experimental

Motivated by the mathematical tractability, stochastic geometry has been accepted as a popular and useful tool to analyze performance of HetNets, and some significant results and conclusions are obtained [14–19]. Stochastic geometry is introduced into the mmWave network analysis where locations of both sub-6 GHz and mmWave BSs are considered as Poisson point process (PPP), and block effects are incorporated into the system model [20]. The expressions, or upper and lower bound of some performance metrics (e.g., coverage probability, outage probability, average rate, etc.) are derived mostly under PPP-based mmWave network scenario [3, 4, 6, 9, 10]. Actually, as one of capacity-increasing techniques, mmWave BSs commonly work along with the cellular network and are distributed only on some hot spot areas. Therefore, non-uniform location distribution of mmWave BS is more appropriate to model the realistic scenario. To the best of our knowledge, few works adopt the non-uniform mmWave BS location distribution model, which will be the main difference of our paper from previous papers. In this paper, we make centralized and a distributed user association analysis for mmWave cellular system. The contributions of this paper can be summarized as

- We propose a centralized two-stage user association and power control algorithm to maximize the network sum rate while satisfying SINR and power consumption constraints. The optimization problem is divided into user association and power control subproblems. Specifically, for user association, the eigenvalue analysis is performed to drop communication links successively that cause the maximum sum of the interference power in the network until the feasibility condition is satisfied. In addition, binary power (BP) allocation is applied to solve the optimal power allocation.
- We derive analytical expressions on the association probability, the coverage probabilities, and the average achievable rate, considering directivity and blockage effect, using stochastic geometry. On this basis, a distributed distance-based user association method is proposed to decide the pattern of BSs (conducting in either sub-6 GHz or mmWave).

The remainder of this paper is organized as follows. In Section 3, the system model for mmWave cellular networks is described. The proposed centralized algorithms are presented in Section 4. In Section 5, the analytical expressions for the association probability, coverage probability, and the average achievable rate are derived. Further, a distributed distance-based user association method is proposed to decide the patterns of BSs. Simulation results are provided in Section 6 to compare the performance of the proposed algorithms, which are followed by our conclusions in Section 7.

3 System model

We consider a downlink mmWave cellular network. Specifically, macro base stations (MBSs) are uniformly distributed in $\mathbb{R}^2$ according to Poisson point process (PPP) with density λ_{mu}; small-cell base stations (SBSs) are generated by Poisson cluster process (PCP) Φ_s where the parent points following PPP with density λ_p forms the centers of these clusters and the daughter points (i.e., SBSs) are uniformly distributed around the cluster center within the radius R [19]. The number of daughter points within each cluster is c. In a hybrid deployment, MBSs, under sub-6 GHz, are essential to provide wide coverage to guarantee a consistent service for UEs whereas mmWave may be used in SBSs mainly to deliver high rate to individual UEs. Specifically, all MBSs and some SBSs share sub-6 GHz, whereas the rest of SBSs operate at the mmWave bands. It is worth noting that SBSs in one cluster will choose the same pattern, i.e., either sub-6 GHz or mmWave. Without loss of generality, we assume that a SBS cluster is under sub-6 GHz with probability p_{su}, i.e, the density of sub-6 GHz SBSs $\lambda_{su} = cp_{su}\lambda_p$; otherwise, it conducts under mmWave, i.e, the density of mmWave SBSs $\lambda_{sm} = cp_{sm}\lambda_p$ where $p_{sm} = 1-p_{su}$. Thus, we regard BS location distribution as three independent tiers, i.e., sub-6 GHz MBS tier, sub-6 GHz SBS tier, and mmWave SBS tier, as is shown in Fig. 1.

The two main characteristics of mmWave transmission are blockage effect and directivity at transmitter and/or receiver.

(1) The actual mmWave transmission directivity is approximated by a sectored model [3, 4, 6, 9, 10], where a UE receives a signal with directivity gain $G_{\max}$ if the UE's angle θ with respect to the best beam alignment is within the main beamwidth 2ω of the serving mmWave cell and with directivity gain $G_{\min}$ otherwise. This is formulated by

$$G(\theta) = \begin{cases} G_{\max} & |\theta| \le \omega \\ G_{\min} & \text{otherwise} \end{cases} \quad (1)$$

whereas $G(\theta) = 1$ under sub-6 GHz cell. Due to the fact that θ is independent of all other variables, we simplify the $G(\theta)$ as G_j for the j-tier directivity gain.

(2) We approximate the blockage effect by a modified LOS ball model under mmWave link [4], i.e., a UE within a distance D to an mmWave SBS is LOS link, otherwise is considered NLOS link.

Thus, the received SINR at the typical user (at origin o) is

$$\mathrm{SINR}_k(x) = \frac{P_k G_k h_x \|x\|^{-\alpha_k}}{I_{\Phi\backslash x}(x) + N_0}, \quad (2)$$

where x is the location of the associated BS, Φ is the set of all BSs' locations, h_x is the small-scale fading coefficient between UE at origin and BS at x and is assumed to be independent and identically distributed (i.i.d.) Rayleigh fading [9]. The path loss exponent of an mmWave link depends on the link distance whereas the path loss exponent of a sub-6 GHz link is independent of the link distance. Specifically, the path loss exponent a_k equals a_s if the link is under sub-6 GHz, equals a_l if the link is LOS under mmWave and equals a_n otherwise. The additive noise is complex Gaussian distributed [19]. The transmit power of MBSs, SBSs operating at sub-6 GHz, and SBSs operating at mmWave frequency are P_{mu}, P_{su}, P_{sm}, respectively. The aggregated interference for UE at x can be expressed as [4, 21]

$$I_{k,\Phi\backslash x}(x) = \begin{cases} \sum\limits_{j\in\{\mathrm{mu,su}\}} \sum\limits_{z\in\Phi_j\backslash x} P_j G_j h_z \|z\|^{-\alpha_j} & k \in \{\mathrm{mu,su}\} \\ 0 & k \in \{\mathrm{sm}\} \end{cases} \quad (3)$$

where Φ_j represents the set of j-tier BS locations and

- k = mu means that k tier is sub-6 GHz macro BS tier.
- k = s means that k tier is sub-6 GHz SBS tier.
- k = sm means that k tier is mmWave SBS tier.

The conditions that UE associates with MBS tier, sub-6 GHz SBS tier and mmWave SBS tier are illustrated in

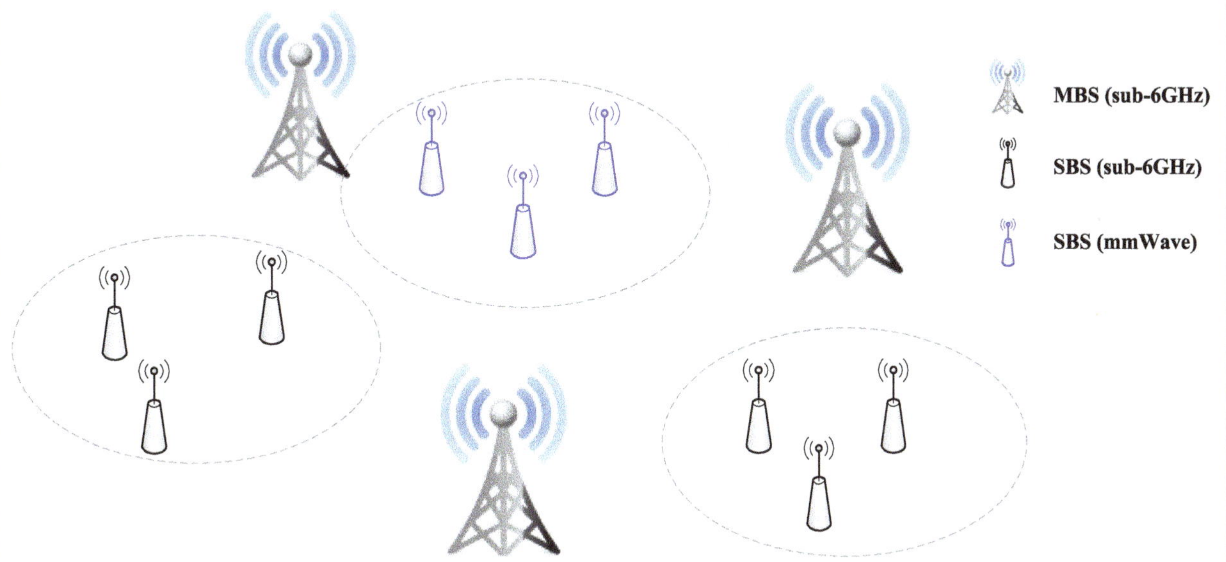

Fig. 1 System model for mmWave cellular networks

Figs. 2, 3, and 4, respectively. Unlike sub-6 GHz networks, mmWave cellular systems will be noise-limited due to the directivity and blocking effects discussed earlier, in conjunction with the large bandwidth which brings in much more noise power [4, 9]. In such cases, the SNR distribution can be used as an approximation of the SINR. This is in contrast to sub-6 GHz cellular networks, which are often interference-limited, meaning SIR ≈ SINR instead.

In this paper, we consider an mmWave cellular network in which users intend to communicate with the BS in either mmWave or sub-6 GHz band. In this mmWave cellular system, we propose a centralized and a distributed user association algorithm, respectively. In Section 4, we propose a centralized user association and power control algorithm. The main idea of the centralized algorithm is to design the user association and transmit power of users to maximize the network sum rate while satisfying the individual target SINR constraints for all links. Note that the centralized algorithm depends on the global channel state information (CSI) possibly at a centralized controller, which is significant to the power design. Note that the centralized algorithm requires global channel state information (CSI) possibly at a centralized controller, which may incur high CSI feedback overhead. To resolve this issue, we propose a distributed user association method in Section 5, which requires distance about the link between the transmitter and its corresponding receiver only. The main goal of the distributed algorithm is to maximize the coverage probability, thereby the basis of which is to calculate the coverage probability. Therefore, we derive the coverage probability as well as average achievable rate in Section 5. Moreover, the probability distribution function (pdf) of associated distance and the association probability are also derived in Section 5, which are the necessary components of the expression of coverage probability. Thus, we analyze the mmWave cellular network performance from the perspectives of centralized and distributed control.

4 A centralized two-stage user association and power control algorithm

A main feature of mmWave cellular networks is that the mmWave links are managed by MBSs in a centralized manner. Commonly, the centralized algorithm can provide the upper performance bound for its distributed counterpart. In this section, we suppose that MBS is able to acquire global channel state information (CSI). Based on the system model in Section 2, we generate a mmWave cellular network of N sub-6 GHz MBSs and M mmWave SBSs. On this basis, we would like to maximize the network sum rate under SINR and power consumption constraints. The optimization problem can be formulated as

$$\max_{\mathbf{P}} \sum_{i=1}^{M+N} \log\left(1 + \mathrm{SINR}_i(x)\right), \tag{P1}$$

$$\text{s.t.}\ (\mathbf{I} - \mathbf{F})\,\mathbf{P} \geq \mathbf{b}, \tag{4}$$

$$0 \leq P_i \leq P_{\mathrm{mu}}^{\max}, i = 1, 2, \cdots, N, \tag{5}$$

$$0 \leq P_i \leq P_{\mathrm{sm}}^{\max}, i = N+1, N+2, \cdots, N+M. \tag{6}$$

where (4) corresponds to the SINR requirement to be satisfied at all $M+N$ receivers. All $M+N$ inequalities are put together in a matrix-form inequality, as is shown in (4). The element in the normalized gain matrix $\mathbf{F}$ is

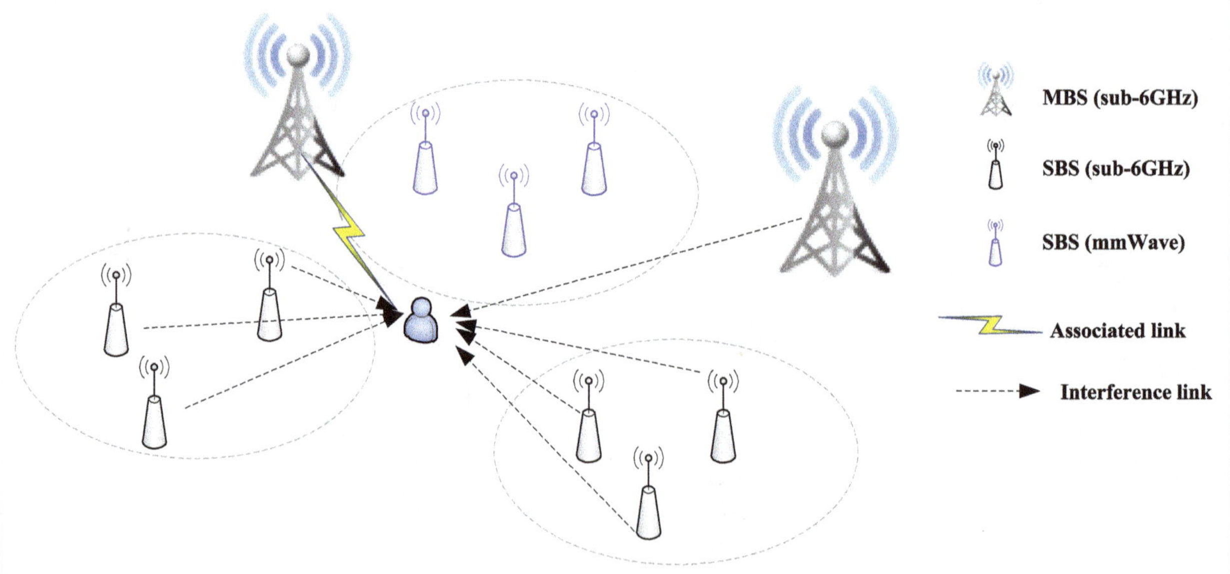

Fig. 2 The condition that UE associates with MBS tier

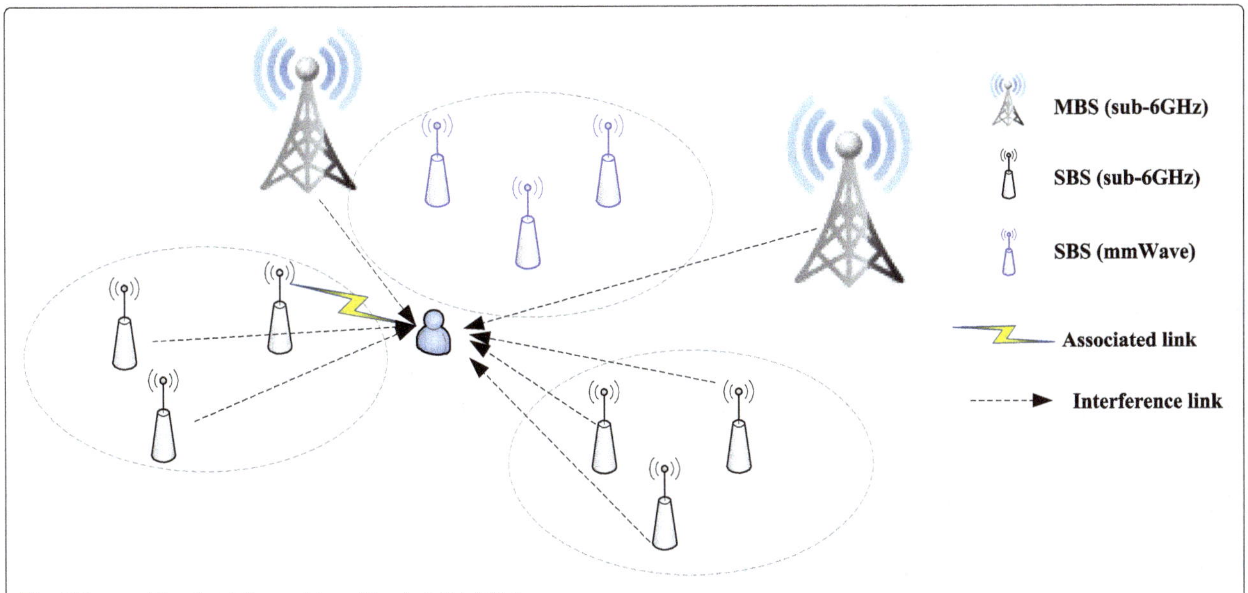

Fig. 3 The condition that UE associates with sub-6 GHz SBS tier

$F_{ki} = \begin{cases} \frac{\gamma_T \beta_{ki}}{\beta_{kk}}, & k \neq i, k \leq N \\ 0, & \text{otherwise} \end{cases}$, where $\beta_{ki} = G_{ki} h_{ki} \|x_{ki}\|^{-\alpha_k}$ consists of the antenna gain, channel gain, and path loss from transmitter k to receiver i. The elements in $\mathbf{b}$ are auxiliary variables which can be considered as the normalized SINR threshold for every receiver and $\mathbf{b} = \left[\frac{\gamma_T N_0}{\beta_{11}}, \frac{\gamma_T N_0}{\beta_{22}}, ..., \frac{\gamma_T N_0}{\beta_{(N+M)(N+M)}}\right]$. The power control vector $\mathbf{P} = [P_1, P_2, ..., P_N, P_{N+1}, ..., P_{N+M}]^T$. $P_{\text{mu}}^{\max}$ and $P_{\text{sm}}^{\max}$ are the maximum power of MBS and mmWave SBS, respectively. $\mathbf{I}$ is the identity matrix. In this section, we focus on the condition $p_{\text{sm}} = 1$, which means that all SBSs operate at the mmWave frequency. Without loss of generality, the results can be extended to the condition $0 \leq p_{\text{sm}} < 1$, i.e., the sub-6 GHz SBSs are included in the optimization.

We can see that the problem in (P1) is non-convex, which is difficult to obtain the optimal solution. Therefore, we decouple this problem into two subproblems, i.e., user association subproblem and power control subproblem. On this basis, we propose a centralized two-stage user association and power control algorithm, denoted by TS algorithm to get the closely local optimum. In the following, we provide the solution to the two subproblems separately.

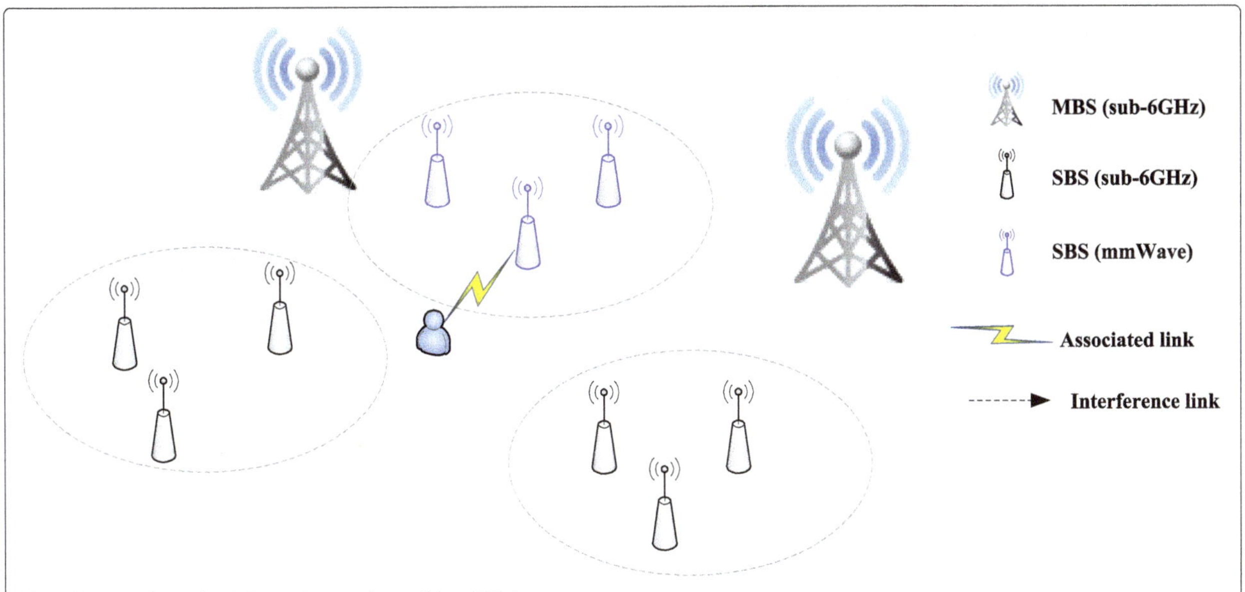

Fig. 4 The condition that UE associates with mmWave SBS tier

4.1 The first stage: user association

The first stage is to find the feasible user association set to satisfy (4) that can be considered as the prerequisite of the next power control. The matrix **F** is comprised of non-negative elements and is irreducible because all the links interfere with each other. By the Perron-Frobenious theorem [22, 23], the SINR constraint set in (4) is non-empty if and only if the maximum modulus eigenvalue of **F** is smaller than 1, i.e., $\rho(\mathbf{F}) < 1$. Otherwise, the power control solution is infeasible. The explanation for this conclusion is that the network prefer to support more transmissions, but sometimes it has to drop some links to satisfy SINR constraints. The key idea to solve this problem is to drop communication links successively that cause the maximum sum of the interference in the network until the feasibility conditions are satisfied. It is worthy noting that since the mmWave BSs will not cause the interference due to the high attenuation loss, none of mmWave BSs will be dropped in the first stage.

4.2 The second stage: power allocation

After the user association, the original optimization problem (P1) can be reformulated as P2, i.e., a tractable power-constrained sum rate maximization problem.

$$\max_{\mathbf{P}} \sum_{i=1}^{M+N'} \log\left(1+\mathrm{SINR}_i(x)\right), \tag{P2}$$

$$\text{s.t.} 0 \le P_i \le P_{\mathrm{mu}}^{\max}, i = 1, 2, \cdots N', \tag{7}$$

$$0 \le P_i \le P_{\mathrm{sm}}^{\max}, i = N'+1, N'+2, \cdots N'+M. \tag{8}$$

where N' is the number of the updated sub-6 GHz BSs set after the user association stage.

Geometric programming (GP) is a widely adopted method to obtain the optimal solution to the power-constrained sum rate maximization problem [24] and binary power (BP) can provide the near optimal performance but with low complexity [25]. Therefore, we tackle the power control subproblem by BP. In addition, the optimal mmWave BS transmit power is $P_i = P_{\mathrm{sm}}^{\max}, i = N'+1, N'+2, ..., N'+M$ due to the fact that mmWave link is interference-free.

The details of TS algorithm are describe below.

Centralized two-stage user association and power control algorithm

Initialization: Set gain matrix $\mathbf{F}^t$ for $t = 0$ according to the global CSI. All cellular and mmWave links are active.

Step 1: Calculate $\rho\left(\mathbf{F}^t\right)$. If $\rho\left(\mathbf{F}^t\right) < 1$, solve the sum rate maximization by BP to obtain the power allocation method. Then, the algorithm stops. Otherwise, go to step 2.

Step 2: Calculate $q = \arg \max_{q \in \{1,2,...,N+M\}} \left\|\mathbf{f}_q\right\|_2$, where $\mathbf{f}_q$ is the column of the matrix $\mathbf{F}^t$, i.e., $\mathbf{F}^t = \left[\mathbf{f}_1^t, \mathbf{f}_2^t, ..., \mathbf{f}_{N+M}^t\right]$.

Step 3: Remove the qth column and row in $\mathbf{F}^t$. $t \leftarrow t+1$. The updated matrix $\mathbf{F}^t$ is the reduced matrix. Return to step 1.

5 Distributed user association analysis of mmWave cellular network

In the previous section, we propose a centralized two-stage user association and power control algorithm. Although centralized algorithm can achieve a good performance, the main challenge is the frequent coordination between transmitters and the high signaling overheads for sharing CSI. In order to solve this problem, we conduct a distributed user association analysis for mmWave cellular network. Specifically, we derive some key performance metrics considering the directivity and blockage effect. On this basis, we propose a distributed distance-based user association method to decide the pattern of BSs (either sub-6 GHz or mmWave pattern).

In order to enable the derivations and the theorems in this section easier to understand. Table 1 lists the symbols and parameters for distributed user association.

For the parameters k, i, j, and l, they represent k-tier, i-tier, j-tier, and l-tier. The following theorem provides the per tier association probability, which is essential for deriving the main results in the sequel.

5.1 Performance analysis

In this subsection, we derive the association probability, coverage probability, and the average achievable rate under mmWave cellular network.

Table 1 Symbols and parameters for distributed user association

Symbol and parameter	Description
A_k	The probability that a typical user receives the maximum received power from a k-tier BS.
r_k	The distance from UE to the nearest k-tier BS.
P_k	The transmission power of k-tier BS.
P_k	The directivity gain of a k-tier link.
α_k	The path loss exponent in the k-th tier.
$P_{r,k}(r_k)$	The long-term received power from a k-tier BS at UE.
$F_{R_l}(r_l)$	The cumulative distribution function (cdf) of the distance to the nearest l-tier BS.
$f_{X_k}(x_k)$	The probability distribution function (pdf) of k-tier associated distance x_k.
$\mathrm{SINR}_k(x)$	The received SINR at the typical user (at origin o) conditioned on the associated k-tier BS at x.
$\mathfrak{R}_k$	The average achievable rate on the condition that user is associated with k-tier.
$\mathfrak{R}$	The average achievable rate.
φ_{sm}	The sm-tier coverage probability conditioned on the associated distance.
φ_{mu}	The mu-tier coverage probability conditioned on the associated distance.

The following theorem provides the per tier association probability, which is essential for deriving the main results in the sequel.

Theorem 1 *The association probability in the kth tier is*

$$A_k = \begin{cases} \mathbb{E}_{r_k}\Big[\exp\Big(-\pi\lambda_j\Big(\frac{P_j}{P_k}\Big)^{\frac{2}{\alpha_s}} r_k^2\Big)\Big(\frac{\omega}{\pi}\Big(1-\mathcal{E}_{Y_l}\Big(\Big(\frac{P_l G_{\max}}{P_k}\Big) r_k^{\alpha_s}\Big)\Big) \\ \quad +\frac{\pi-\omega}{\pi}\Big(1-\mathcal{E}_{Y_l}\Big(\Big(\frac{P_l G_{\min}}{P_k}\Big) r_k^{\alpha_s}\Big)\Big)\Big)\Big], \\ \qquad j,k \in \{\mathrm{mu},\mathrm{su}\}, j \neq k, l = \mathrm{sm} \\ \mathbb{E}_{r_k}\Big[\frac{\omega}{\pi}\exp\Big(-\pi\sum\limits_{j\in\{\mathrm{mu},\mathrm{su}\}}\lambda_j\Big(\frac{P_j}{P_k G_{\max}}\Big)^{\frac{2}{\alpha_s}} r_k^2\Big) \\ \quad +\frac{\pi-\omega}{\pi}\exp\Big(-\pi\sum\limits_{j\in\{\mathrm{mu},\mathrm{su}\}}\lambda_j\Big(\frac{P_j}{P_k G_{\min}}\Big)^{\frac{2}{\alpha_s}} r_k^2\Big)\Big], \\ \qquad k = \mathrm{sm} \end{cases} \tag{9}$$

where $\mathcal{E}_{Y_l}(y)$ is expressed as

$$\mathcal{E}_{Y_l}(y) = \begin{cases} F_{R_l}\left(\sqrt[\alpha_n]{y}\right) & D < \sqrt[\alpha_l]{y} < \sqrt[\alpha_n]{y} \text{ or } D < \sqrt[\alpha_n]{y} < \sqrt[\alpha_l]{y} \\ F_{R_l}\left(\sqrt[\alpha_l]{y}\right) & \sqrt[\alpha_l]{y} < \sqrt[\alpha_n]{y} < D \text{ or } \sqrt[\alpha_n]{y} < \sqrt[\alpha_l]{y} < D \\ F_{R_l}\left(\sqrt[\alpha_l]{y}\right) + F_{R_l}\left(\sqrt[\alpha_n]{y}\right) - F_{R_l}(D) & \sqrt[\alpha_l]{y} < D < \sqrt[\alpha_n]{y} \\ F_{R_l}(D) & \sqrt[\alpha_n]{y} < D < \sqrt[\alpha_l]{y} \end{cases} \tag{10}$$

where $F_{R_l}(r_l) = 1 - \exp\left(-\pi\lambda_l r_l^2\right)$ is the cumulative distribution function (cdf) of the distance of the nearest l-tier BS.

Proof See Appendix A. □

From Theorem 1, we further derive the the distance between a user and its serving BS, of which the premise is that the user is associated with the kth tier.

Corollary 1 *The probability distribution function (pdf) of k-tier associated distance*

$$f_{X_k}(x_k) = -\frac{\partial \Pr\left[r_k > x_k, P_{r,k} > \max\limits_{j,j\neq k} P_{r,j}\right]}{A_k \partial x_k} \tag{11}$$

where $\Pr\left[r_k > x_k, P_{r,k} > \max\limits_{j,j\neq k} P_{r,j}\right]$ is obtained through the method to calculate A_k, in which the only difference is the integral interval over r_k is x_k to infinity instead of zero to infinity.

Depending on the association probability and the pdf of the associated distance, we further obtain the coverage probability and average achievable rate, which are the two important indicators to evaluate the network performance.

In this paper, a UE is said to be in coverage if it is able to connect to at least one BS with SINR above its threshold.

Theorem 2 *The coverage probability is*

$$\Pr\left(\bigcup_{k\in\{mu,su,sm\}} SINR_k(x_k) > \gamma_k\right) = \mathbb{E}_{x_k}\left[\frac{\omega}{\pi}\exp\left(-\frac{\gamma_k N_0}{P_k G_{\max} x_k^{-\alpha_k}}\right) + \frac{\pi-\omega}{\pi}\exp\left(-\frac{\gamma_k N_0}{P_k G_{\min} x_k^{-\alpha_k}}\right)\right] + \sum_{k\in\{mu,su\}} \mathbb{E}_{x_k}\left[\exp\left(-\frac{\gamma_k \sum\limits_{z\in\Phi_j} P_j h_z z^{-\alpha_s}}{P_k x_k^{-\alpha_s}}\right)\right] \mathbb{E}^!_{x_k}\left[\exp\left(-\frac{\gamma_k \sum\limits_{z\in\Phi_j} h_z z^{-\alpha_s}}{x_k^{-\alpha_s}}\right)\right] \tag{12}$$

where the first term can be calculated by averaging over x_k, and the pdf of x_k is given by Corollary 1. $\mathbb{E}_{x_k}\left[\exp\left(-\frac{\gamma_k \sum_{z\in\Phi_j} P_j h_z z^{-\alpha_s}}{P_k x_k^{-\alpha_s}}\right)\right]$ and $\mathbb{E}^!_{x_k}\left[\exp\left(-\frac{\gamma_k \sum_{z\in\Phi_j} h_z z^{-\alpha_s}}{x_k^{-\alpha_s}}\right)\right]$ have been derived in [19].

Proof See Appendix B. □

Last, we derive the average achievable rate of mmWave cellular network. The average achievable rate $\Re$ can be derived by the method to analyze the coverage probability in Appendix B.

Theorem 3 *The average achievable rate is*

$$\Re = \sum_{k\in\{mu,su,sm\}} A_k \Re_k = \sum_{k\in\{mu,su,sm\}} A_k \mathbb{E}_{x_k}\left[\int_0^\infty \Pr\left(SINR_k(x_k) > e^t - 1\right) dt\right] \tag{13}$$

where $\Re_k$ is the average achievable rate on the condition that user is associated with k-tier and $\Pr\left(SINR_k(x_k) > e^t - 1\right)$ can be easily deduced according to (25) and (26) in Appendix B.

5.2 Distributed distance-based user association method

On the basis of the expression of coverage probability, together with the known associated distance, we propose a simple method to judge which pattern of BS (sub-6 GHz or mmWave) are preferred by comparing the conditional coverage probabilities of sub-6 GHz BS and mmWave BSs. In order to make the problem mathematically trackable, we set $p_{\mathrm{sm}} = 1$ in the following analysis.

The sm-tier coverage probability conditioned on the associated distance is

$$\varphi_{\mathrm{sm}} = \frac{\omega}{\pi}\exp\left(-\frac{\gamma_{\mathrm{sm}} N_0}{P_{\mathrm{sm}} G_{\max} x_{\mathrm{sm}}^{-\alpha_{\mathrm{sm}}}}\right) + \frac{\pi-\omega}{\pi}\exp\left(-\frac{\gamma_{\mathrm{sm}} N_0}{P_{\mathrm{sm}} G_{\min} x_{\mathrm{sm}}^{-\alpha_{\mathrm{sm}}}}\right) \tag{14}$$

The mu-tier coverage probability conditioned on the associated distance is [16]

$$\varphi_{\text{mu}} = \exp\left\{ - 2\pi\lambda_{\text{mu}} x_{\text{mu}}^2 \gamma_{\text{mu}}^{\frac{2}{\alpha_s}} \left[B\left(1; \frac{2}{\alpha_s}, 1-\frac{2}{\alpha_s}\right) - B\left(\frac{1}{1+\gamma_{\text{mu}}}; \frac{2}{\alpha_s}, 1-\frac{2}{\alpha_s}\right)\right]\right\} \quad (15)$$

where $B(x; y, z)$ is the incomplete beta function.

The user association decision is determined by the coverage probability conditioned on the associated distance, that is, the user will associate to the nearest mmWave BS if $\varphi_{\text{sm}} \geq \varphi_{\text{mu}}$, otherwise associates to the nearest MBS.

Remark 1 *According to Jensen inequality,* [26] $a_1 \exp\left(-b_1 x^{\alpha_{\text{sm}}}\right) + a_2 \exp\left(-b_2 x^{\alpha_{\text{sm}}}\right) \geq \exp\left(-\left(a_1 b_1 + a_2 b_2\right) x^{\alpha_{\text{sm}}}\right)$*, the sufficient condition for the inequality* $\varphi_{\text{sm}} \geq \varphi_{\text{mu}}$ *is* $\exp\left(-\left(a_1 b_1 + a_2 b_2\right) x^{\alpha_{\text{sm}}}\right) \geq \exp\left(-cx^2\right)$. *Therefore, the user will associate with nearest mmWave BS if* $D \leq x \leq \left(\frac{c}{a_1 b_1 + a_2 b_2}\right)^{\frac{1}{\alpha_n - 2}}$ *or* $x < \min\left\{D, \left(\frac{c}{a_1 b_1 + a_2 b_2}\right)^{\frac{1}{\alpha_l - 2}}\right\}$.

6 Results and discussion

In this section, we present numerical and simulation results on association probability, coverage probability, and average rate for mmWave cellular network and then evaluates the performance of the proposed centralized TS algorithm. The simulation parameters are listed in Table 2.

We start by looking into the association probabilities and the different factors that affect the probability. Figure 5 illustrates the association probability of sub-6 GHz MBS, sub-6 GHz SBS, and mmWave SBS against the mmWave probability p_{sm}. Figure 5 shows that the simulation results match the analytic results well, which validates our analysis. It is also seen in Fig. 1 that more UEs associate with mmWave BS with the growth of p_{sm}, which is in agreement with our intuition. Moreover, it can be seen that MBSs still undertake more UEs due to higher transmission power.

Figure 6 shows that coverage probability decreases with the larger c value whereas the gap between different c diminishes with the increasing mmWave probability p_{sm}. That is because larger c value under low p_{sm} results in more sub-6 GHz SBSs that cause more interference. However, the effect of c on coverage probability becomes weaker with the increasing p_{sm} due to the fact that more SBSs conduct in interference-free mmWave pattern with the increase of p_{sm}. The extreme case is that the gap disappears for $p_{\text{sm}} = 1$, as is shown in Fig. 2. Thus, the results in Fig. 2 demonstrates that mmWave links can contribute to coverage probability.

Similarly, the higher achievable average rate is along with the larger p_{sm} in Fig. 7. It is worth noting that a cross point appears with the increasing p_{sm} under different c. That means less sub-6 GHz SBSs (i.e., lower interference) or more mmWave SBSs (i.e., better channel state) can achieve higher achievable average rate.

From Fig. 8, we can see that the cluster radius R has no strong effect on UE association when SBS density is small, which further verifies that a PCP can be approximated as a PPP under small value c. Moreover, it is noticeable that mmWave support highest data rate even under the least number of associated UEs by comparing Figs. 8 and 9. Therefore, mmWave communications outperforms sub-6 GHz communications on rate delivery for high directivity and interference-free transmission environment. It can be seen in Fig. 10 that the coverage probability grows in accordance with the increasing cluster radius whereas it tends to keep stable for small values of c. Higher c value results in stronger non-uniform distribution, which can implicitly confirm that non-uniformity can reduce coverage area.

This finding obtained in Fig. 11 is in agreement with the published paper [6] that the coverage probability

Table 2 Simulation parameters

Parameters	Value
Side length of simulation area	$L = 1500\,m$
MBS density	$\lambda_m = 5/\pi L^2$
SBS parent density	$\lambda_p = 2\lambda_m$
The number of SBS in a cluster	$c = [4:10]$
The probability of mmWave BS among SBSs	$p_{\text{sm}} = [0:0.1:1]$
Cluster range of SBS	$R = [50:100:750]$
Transmit power of MBS	$P_m = 46$ dBm $= 39$ W [10]
Transmit power of SBS	$P_{\text{su}} = P_{\text{sm}} = 30$dBm $= 1$W [10]
SINR threshold	$\gamma_T = [-10:2:10]$dB
Antenna gain	$\omega = 5^o, G_{\max} = 18$dBi $= 63, G_{\min} = -2$ dBi $= 0.63$ [10]
Path loss	$\alpha_s = 3, \alpha_l = 2, \alpha_n = 4$ [27]
Small-scale fading	$h \sim \exp(1)$
mmWave bandwidth	$W_s = 1$ GHz [9]
Noise power of mmWave	-174 dBm/Hz $+ 10\log(W_s) + 10$ dB [9]

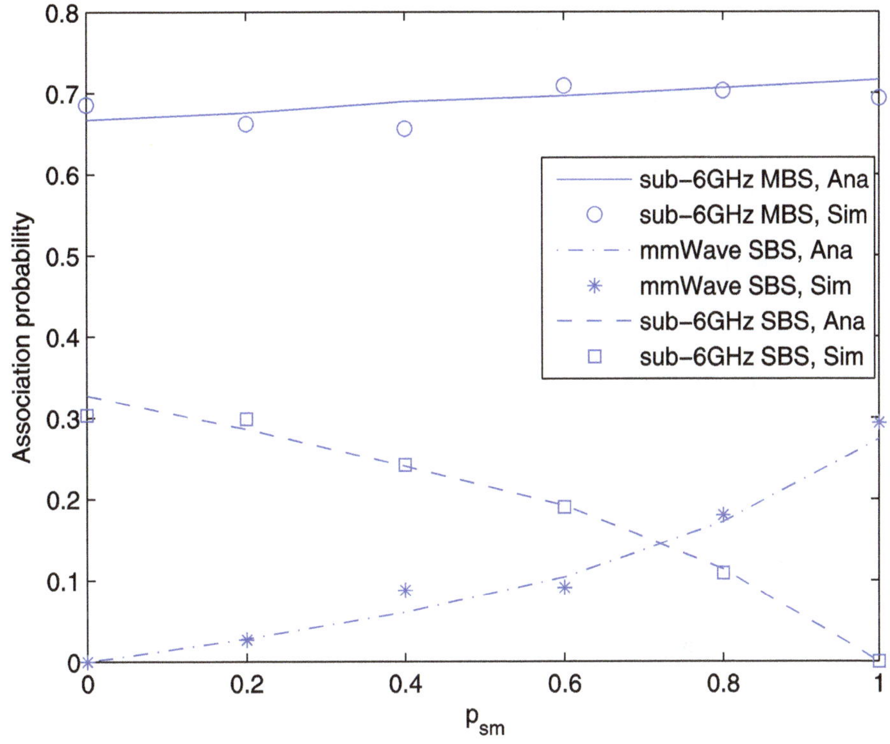

Fig. 5 Association probability vs. the mmWave probability p_{sm}

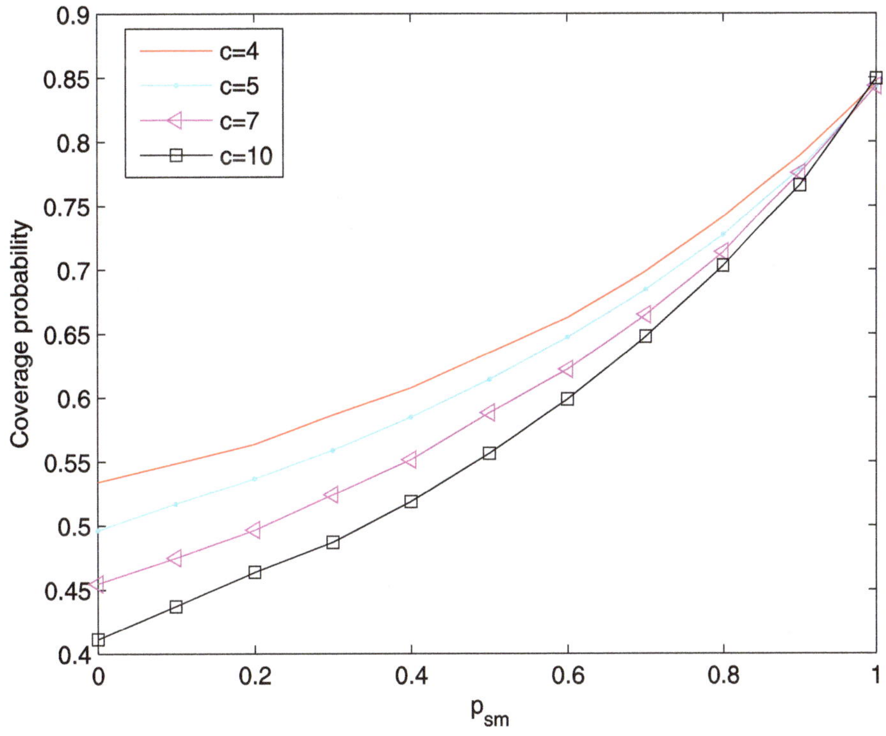

Fig. 6 Coverage probability vs. the mmWave probability p_{sm}

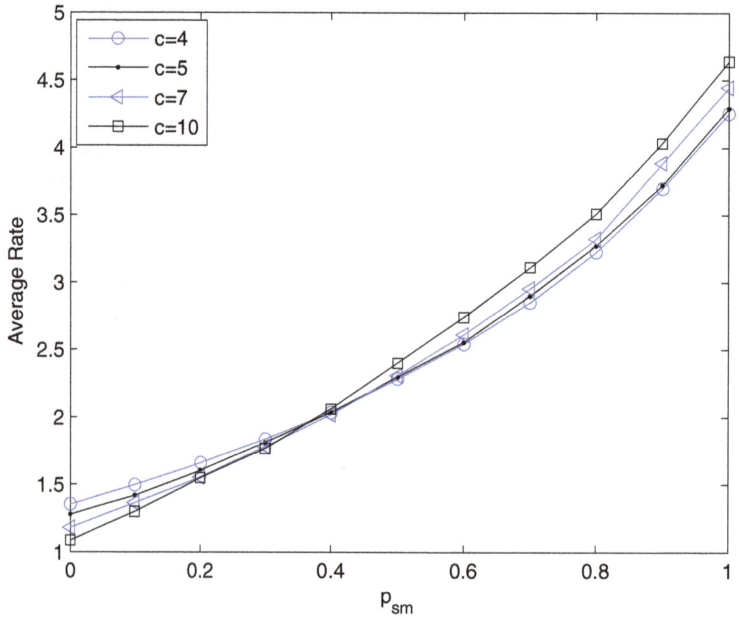

Fig. 7 Average achievable rate vs. the mmWave probability p_{sm}

decreases with the growing SINR threshold. Also, it illustrates that more mmWave links are able to enhance the coverage probability. This is mainly due to interference-free transmission environment under mmWave communications.

As expected, the sum rate decreases with the increasing SINR threshold in Fig. 8. Meanwhile, we compare the performance of the proposed centralized TS algorithm (using BP) and geometric programming (GP) algorithm where the performance of GP is commonly adopted as a

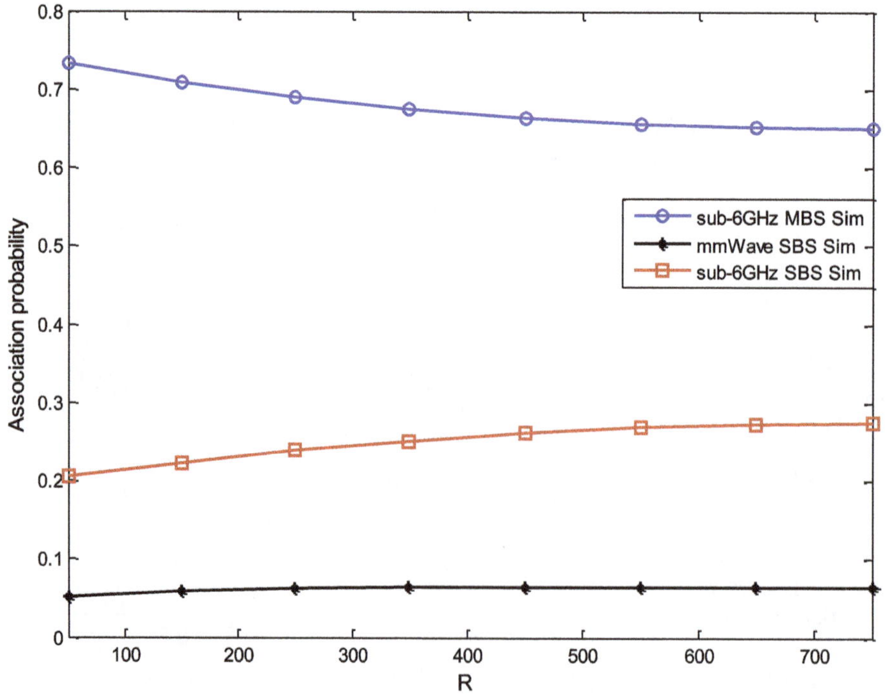

Fig. 8 Association probability vs. the cluster radius R

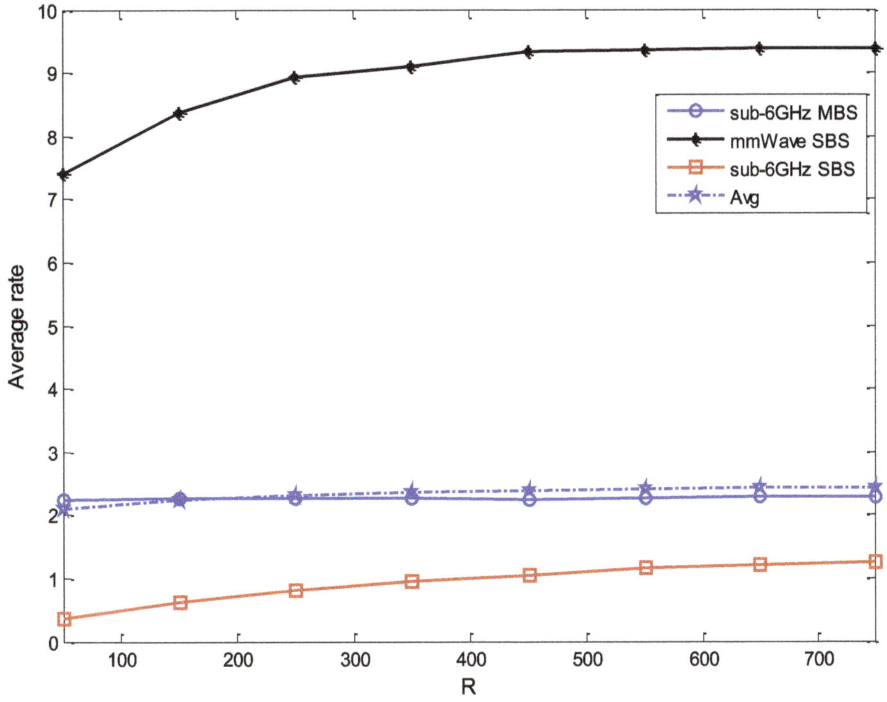

Fig. 9 Average achievable rate vs. the cluster radius R

benchmark. The observation in Fig. 12 is that TS algorithm is always approximate to the GP for $c = 1$. Although the gap between GP and BP becomes obvious with the larger c, BP is near to the optimal results under low SINR. The results demonstrate near-optimality of BP under low SINR constraint. Moreover, the complexity of the BP-based transmitter design is reduced since only a two-level power control is required, which is a potential in the design and analysis of wireless networks.

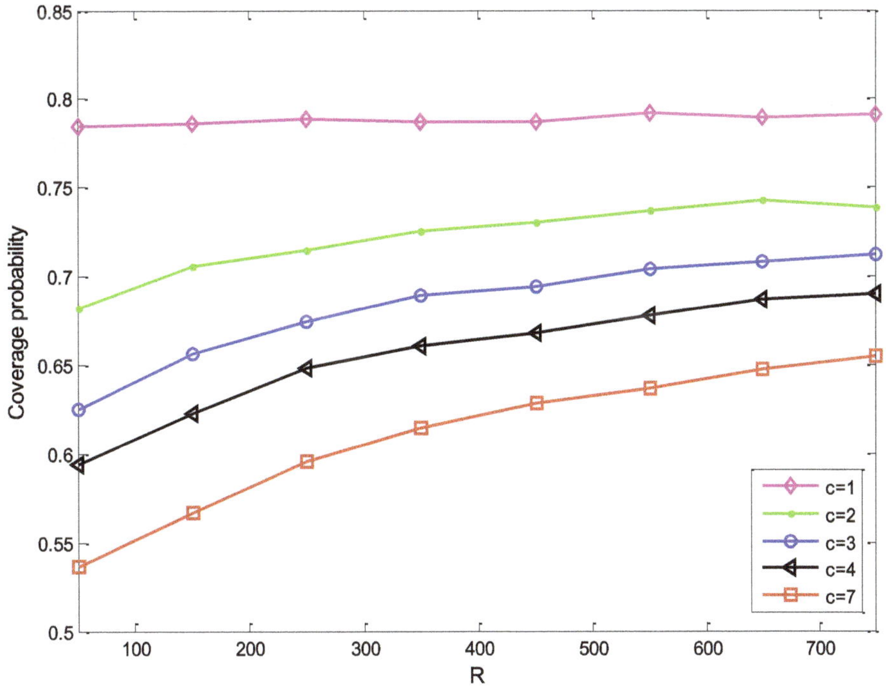

Fig. 10 Coverage probability vs. the cluster radius R

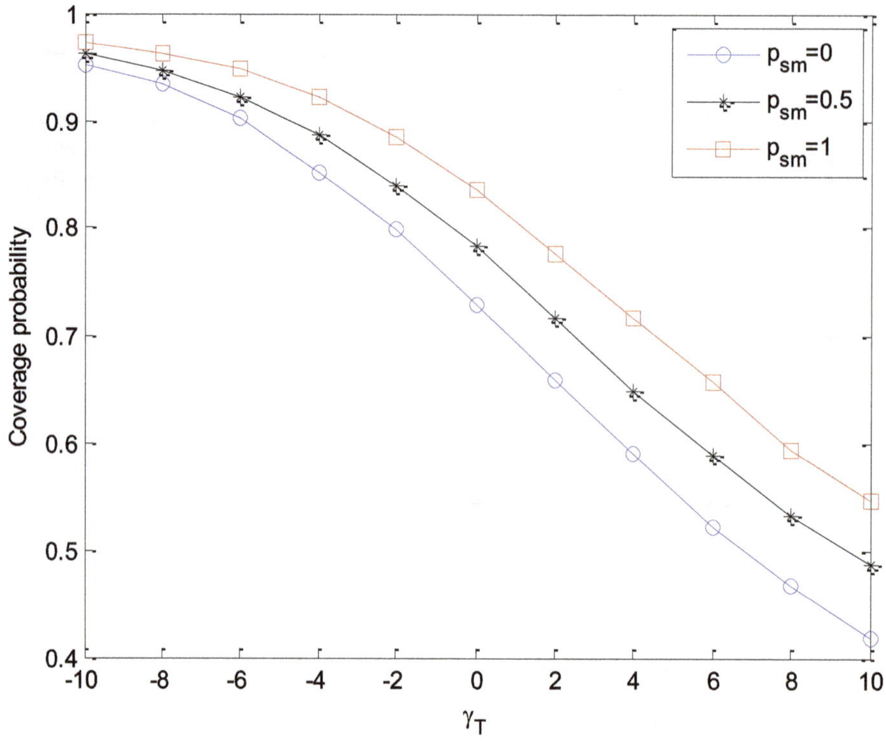

Fig. 11 Coverage probability vs. SINR threshold

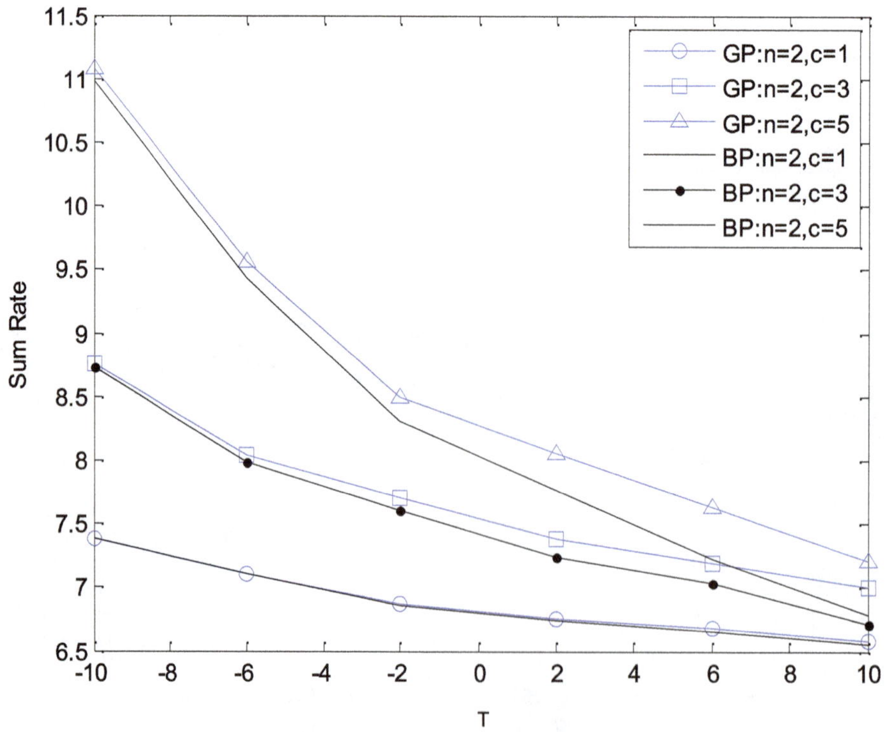

Fig. 12 Sum rate of the centralized algorithms vs. SINR threshold

7 Conclusion

This paper proposes a realistic multi-tier mmWave cellular framework where sub-6 GHz MBS deployment is assumed as a PPP, and SBS operating on either mmWave or sub-6 GHz follows PCP model. In this system, we propose both centralized and distributed user association algorithms. For the centralized two-step algorithm, we aim to maximize the sum rate while satisfying QoS and power consumption constraints based on eigenvalue analysis. Then, we derive the association probability, the coverage probability, and the average achievable rate, incorporating directivity and blockage effect, by stochastic geometry. On this basis, the distributed user association algorithm is proposed. The simulation results demonstrate the accuracy of our theoretical analysis and also reveal that the effect of some parameters on the network performance. In addition, the proposed centralized algorithm can achieve near-optimal sum rate with a low complexity under non-dense scenario.

Appendix

A. Proof of Theorem 1

The associated probability in the k-tier is given by

$$A_k = \Pr\left[P_{r,k} > \max_{j,j\neq k} P_{r,j}\right] = \mathbb{E}_{r_k,\mathbf{G}}\left[\Pr\left(P_k G_k r_k^{-\alpha_k} > \max_{j,j\neq k} P_j G_j r_j^{-\alpha_j}\right)\right]$$
$$\stackrel{(a)}{=} \mathbb{E}_{r_k,\mathbf{G}}\left[\prod_{j\in\{\text{mu,su,sm}\},j\neq k} \Pr\left(P_k G_k r_k^{-\alpha_k} > P_j G_j r_j^{-\alpha_j}\right)\right] \tag{16}$$

where a is given from the independence of PPP. The pdf of r_k is given by [16]. $\mathbf{G}$ is a vector involving the directivity gains of all links.

First we calculate $\Pr\left(P_k G_k r_k^{-\alpha_k} > P_j G_j r_j^{-\alpha_j}\right)$

1. If $k,j \in \{\text{mu,su}\}$ and $j \neq k$

$$\Pr\left(P_k G_k r_k^{-\alpha_k} > P_j G_j r_j^{-\alpha_j}\right) = \Pr\left(r_j > \left(\frac{P_j}{P_k}\right)^{\frac{1}{\alpha_s}} r_k\right)$$
$$= \exp\left(-\pi\lambda_j\left(\frac{P_j}{P_k}\right)^{\frac{2}{\alpha_s}} r_k^2\right) \tag{17}$$

2. If $k \in \{\text{mu,su}\}$ and $j = \text{sm}$

$$\Pr\left(P_k G_k r_k^{-\alpha_k} > P_j G_j r_j^{-\alpha_j}\right) = \Pr\left(r_j^{\alpha_j} > \left(\frac{P_j G_j}{P_k G_k}\right) r_k^{\alpha_k}\right)$$
$$= 1 - \int_0^{\left(\frac{P_j G_j}{P_k}\right) r_k^{\alpha_s}} \ell_{Y_j}(y)\,dy = 1 - \mathcal{E}_{Y_j}\left(\left(\frac{P_j G_j}{P_k}\right) r_k^{\alpha_s}\right) \tag{18}$$

where the variable $Y_j = r_j^{\alpha_j}$. $\ell_{Y_j}(y)$ and $\mathcal{E}_{Y_j}(y)$ are the pdf and cdf of variable Y_j, respectively. The cdf of variable Y_j is given by

$$\mathcal{E}_{Y_j}(y) = \Pr\left(Y_j \leq y\right) = \Pr\left(r_j \leq \sqrt[\alpha_j]{y}\right) \tag{19}$$

where the last term in (19) can be calculated based on the pdf of distance between user and the nearest j-tier BS. Therefore, $\mathcal{E}_{Y_j}(y)$ is expressed as

$$\mathcal{E}_{Y_j}(y) =$$
$$\begin{cases} F_{R_j}\left(\sqrt[\alpha_n]{y}\right) & D < \sqrt[\alpha_l]{y} < \sqrt[\alpha_n]{y} \quad \text{or } D < \sqrt[\alpha_n]{y} < \sqrt[\alpha_l]{y} \\ F_{R_j}\left(\sqrt[\alpha_l]{y}\right) & \sqrt[\alpha_l]{y} < \sqrt[\alpha_n]{y} < D \quad \text{or } \sqrt[\alpha_n]{y} < \sqrt[\alpha_l]{y} < D \\ F_{R_j}\left(\sqrt[\alpha_l]{y}\right) + F_{R_j}\left(\sqrt[\alpha_n]{y}\right) - F_{R_j}(D) & \sqrt[\alpha_l]{y} < D < \sqrt[\alpha_n]{y} \\ F_{R_j}(D) & \sqrt[\alpha_n]{y} < D < \sqrt[\alpha_l]{y} \end{cases} \tag{20}$$

where $F_{R_j}(r_j) = 1 - \exp\left(-\pi\lambda_j r_j^2\right)$ is the cdf of the distance of the nearest j-tier BS.

3. If $k = \text{sm}$ and $j \in \{\text{mu,su}\}$

$$\Pr\left(P_k G_k r_k^{-\alpha_k} > P_j G_j r_j^{-\alpha_j}\right) =$$
$$\begin{cases} \Pr\left(r_j > \left(\frac{P_j G_j}{P_k G_k}\right)^{\frac{1}{\alpha_s}} r_k^{\frac{\alpha_l}{\alpha_s}}\right) = \exp\left(-\pi\lambda_j\left(\frac{P_j}{P_k G_k}\right)^{\frac{2}{\alpha_s}} r_k^{\frac{2\alpha_l}{\alpha_s}}\right) & r_k \leq D \\ \Pr\left(r_j > \left(\frac{P_j G_j}{P_k G_k}\right)^{\frac{1}{\alpha_s}} r_k^{\frac{\alpha_n}{\alpha_s}}\right) = \exp\left(-\pi\lambda_j\left(\frac{P_j}{P_k G_k}\right)^{\frac{2}{\alpha_s}} r_k^{\frac{2\alpha_n}{\alpha_s}}\right) & r_k > D \end{cases} \tag{21}$$

Thus, the k-tier associated probability is

$$A_k = \begin{cases} \mathbb{E}_{r_k,\mathbf{G}}\left[\exp\left(-\pi\lambda_j\left(\frac{P_j}{P_k}\right)^{\frac{2}{\alpha_s}} r_k^2\right)\left(1 - \mathcal{E}_{Y_l}\left(\left(\frac{P_l G_l}{P_k}\right) r_k^{\alpha_s}\right)\right)\right], & j,k \in \{\text{mu,su}\}, j \neq k, l = \text{sm} \\ \mathbb{E}_{r_k,\mathbf{G}}\left[\exp\left(-\pi \sum_{j\in\{\text{mu,su}\}} \lambda_j\left(\frac{P_j}{P_k G_k}\right)^{\frac{2}{\alpha_s}} r_k^2\right)\right] & k = \text{sm} \end{cases}$$
$$\stackrel{(a)}{=} \begin{cases} \mathbb{E}_{r_k}\left[\exp\left(-\pi\lambda_j\left(\frac{P_j}{P_k}\right)^{\frac{2}{\alpha_s}} r_k^2\right)\left(\frac{\omega}{\pi}\left(1 - \mathcal{E}_{Y_l}\left(\left(\frac{P_l G_{\max}}{P_k}\right) r_k^{\alpha_s}\right)\right) + \frac{\pi-\omega}{\pi}\left(1 - \mathcal{E}_{Y_l}\left(\left(\frac{P_l G_{\min}}{P_k}\right) r_k^{\alpha_s}\right)\right)\right)\right], \\ \qquad j,k \in \{\text{mu,su}\}, j \neq k, l = \text{sm} \\ \mathbb{E}_{r_k}\left[\frac{\omega}{\pi}\exp\left(-\pi \sum_{j\in\{\text{mu,su}\}} \lambda_j\left(\frac{P_j}{P_k G_{\max}}\right)^{\frac{2}{\alpha_s}} r_k^2\right) + \frac{\pi-\omega}{\pi}\exp\left(-\pi \sum_{j\in\{\text{mu,su}\}} \lambda_j\left(\frac{P_j}{P_k G_{\min}}\right)^{\frac{2}{\alpha_s}} r_k^2\right)\right], \\ \qquad k = \text{sm} \end{cases} \tag{22}$$

where a averages the directivity gain which is a trivial work due to the fact that the directivity gain of each link is independent of other variables. Additionally, averaging over r_k is also a simple work once we know the pdf of r_k [10].

B. Proof of Theorem 2

The coverage probability can be expressed as

$$\Pr\left(\bigcup_{k\in\{\text{mu,su,sm}\}} \text{SINR}_k(x_k) > \gamma_k\right) = \sum_{k\in\{\text{mu,su,sm}\}} \mathbb{E}_{x_k,\mathbf{G}}\left[\Pr\left(\text{SINR}_k(x_k) > \gamma_k\right)\right] \tag{23}$$

where $\Pr\left(\text{SINR}_k(x_k) > \gamma_k\right)$ is given by

$$\Pr\left(\text{SINR}_k(x_k) > \gamma_k\right) = \Pr\left(\frac{P_k G_k h_{x_k} x_k^{-\alpha_k}}{\sum_{j\in\{\text{mu,su}\}} I_j + N_0} > \gamma_k\right) \approx \begin{cases} \Pr\left(\frac{P_k G_k h_{x_k} x_k^{-\alpha_k}}{N_0} > \gamma_k\right) & k = \text{sm} \\ \Pr\left(\frac{P_k G_k h_{x_k} x_k^{-\alpha_k}}{\sum_{j\in\{\text{mu,su}\}} I_j} > \gamma_k\right) & k \in \{\text{mu,su}\} \end{cases} \tag{24}$$

where the interference can be ignored when UE is in the coverage of mmWave SBS, i.e., $k = sm$ and the interference cannot be neglected when UE is in the coverage of sub-6 GHz SBS or MBS, i.e., $k \in \{\text{mu,su}\}$. The following is the discussion about the associated distance-based conditional k-tier coverage probability,

(1) If $k = \text{sm}$,

$$\Pr\left(\frac{P_k G_k h_{x_k} x_k^{-\alpha_k}}{N_0} > \gamma_k\right) = \Pr\left(h_{x_k} > \frac{\gamma_k N_0}{P_k G_k x_k^{-\alpha_k}}\right) = \exp\left(-\frac{\gamma_k N_0}{P_k G_k x_k^{-\alpha_k}}\right) \tag{25}$$

(2) If $k = \text{mu,su}$,

$$\Pr\left(\frac{P_k G_k h_{x_k} \|x_k\|^{-\alpha_k}}{\sum_{j\in\{\text{mu,su}\}} I_j} > \gamma_k\right) = \mathbb{E}_{x_k}\left[\exp\left(-\frac{\gamma_k \sum_{z\in\Phi_j} P_j h_z z^{-\alpha_s}}{P_k x_k^{-\alpha_s}}\right)\right] \mathbb{E}^!_{x_k}\left[\exp\left(-\frac{\gamma_k \sum_{z\in\Phi_j} h_z z^{-\alpha_s}}{x_k^{-\alpha_s}}\right)\right] \tag{26}$$

where $\mathbb{E}_{x_k}\left[\exp\left(-\frac{\gamma_k \sum_{z\in\Phi_j} P_j h_z z^{-\alpha_s}}{P_k x_k^{-\alpha_s}}\right)\right]$ and $\mathbb{E}^!_{x_k}\left[\exp\left(-\frac{\gamma_k \sum_{z\in\Phi_j} h_z z^{-\alpha_s}}{x_k^{-\alpha_s}}\right)\right]$ of PPP and PCP have been derived in [19], respectively.

Thus, the coverage probability is given by

$$\begin{aligned}\Pr\left(\bigcup_{k\in\{\text{mu,su,sm}\}} \text{SINR}_k(x_k) > \gamma_k\right) &= \mathbb{E}_{x_k}\left[\frac{\omega}{\pi}\exp\left(-\frac{\gamma_k N_0}{P_k G_{\max} x_k^{-\alpha_k}}\right) + \frac{\pi-\omega}{\pi}\exp\left(-\frac{\gamma_k N_0}{P_k G_{\min} x_k^{-\alpha_k}}\right)\right] \\ &+ \sum_{k\in\{\text{mu,su}\}} \mathbb{E}_{x_k}\left[\exp\left(-\frac{\gamma_k \sum_{z\in\Phi_j} P_j h_z z^{-\alpha_s}}{P_k x_k^{-\alpha_s}}\right)\right] \mathbb{E}^!_{x_k}\left[\exp\left(-\frac{\gamma_k \sum_{z\in\Phi_j} h_z z^{-\alpha_s}}{x_k^{-\alpha_s}}\right)\right]\end{aligned} \tag{27}$$

where the first term can be calculated by averaging over x_k. The pdf of x_k is given by Corollary 1. $\mathbb{E}_{x_k}\left[\exp\left(-\frac{\gamma_k \sum_{z\in\Phi_j} P_j h_z z^{-\alpha_s}}{P_k x_k^{-\alpha_s}}\right)\right]$ and $\mathbb{E}^!_{x_k}\left[\exp\left(-\frac{\gamma_k \sum_{z\in\Phi_j} h_z z^{-\alpha_s}}{x_k^{-\alpha_s}}\right)\right]$ have been derived in [19].

Abbreviations

5G: the fifth generation mobile communication network; BP: Binary power; CSI: Channel state information; GP: Geometric programming; HetNets: Heterogeneous networks; LOS: Line-of-sight; MBS: Macro base station; MIMO: Multiple-input multiple-output; mmWave: Millimeter wave; NLOS: Non-line-of-sight; PCP: Poisson cluster point; PGF: Probability generating function; PPP: Poisson point process; QoS: Quality of service; SBS: Small-cell base station; SINR: Signal to interference plus noise ratio; UE: User

Funding

This work is supported by the National Natural Science Foundation of China (No. 61701037 and 61601503), Beijing Natural Science Foundation (No. L172033) and the 111 Project of China (B16006).

Authors' contributions

The idea of this work was proposed by XZ and YL. YW performed the simulation. XZ wrote the manuscript. JB has given critical revision and helped revise the manuscript. All authors read and approved the final manuscript.

Competing interests

The authors declare that they have no competing interests.

Author details

[1]National Engineering Laboratory for Mobile Network Technologies, Beijing University of Posts and Telecommunications, 100876 Beijing, China. [2]Air Force Engineering University, 710051 Xi'an, China.

References

1. J.G. Andrews, et al., What will 5G be? IEEE J. Sel. Areas Commun. **32**(6), 1065–1082 (2014)
2. F. Boccardi, et al., Five disruptive technology directions for 5G. IEEE Commun. Mag. **52**(2), 74–80 (2014)
3. M. Di Renzo, Stochastic geometry modeling and analysis of multi-tier millimeter wave cellular networks. IEEE Trans. Wirel. Commun. **14**(9), 5038–5057 (2015)

4. J.G. Andrews, et al., Modeling and analyzing millimeter wave cellular systems. IEEE Trans. Commun. **65**(1), 403–430 (2017)
5. L. Wei, et al., Key elements to enable millimeter wave communications for 5G wireless systems. IEEE Wirel. Commun. **21**(6), 136–143 (2014)
6. T. Bai, et al., Coverage and rate analysis for millimeter-wave cellular networks. IEEE Trans. Wirel. Commun. **14**(2), 1100–1114 (2015)
7. S. Singh, et al., Tractable model for rate in self-backhauled millimeter wave cellular networks. IEEE J. Sel. Areas Commun. **33**(10), 2196–2211 (2015)
8. A. Ghosh, et al., Millimeter-wave enhanced local area systems: a high-data-rate approach for future wireless networks. IEEE J. Sel. Area Commun. **32**(6), 1152–1163 (2014)
9. H. Elshaer, et al., Downlink and uplink cell association with traditional macrocells and millimeter wave small cells. IEEE Trans. Wirel. Commun. **15**(9), 6244–6258 (2016)
10. M. Di Renzo, in *IEEE International Conference on Communications (ICC)*. Stochastic geometry modeling and performance evaluation of mmWave cellular communications, (London, 2015), pp. 5992–5997
11. M. N. Kulkarni, A. Ghosh, J. G. Andrews, A comparison of MIMO techniques in downlink millimeter wave cellular networks with hybrid beamforming. IEEE Trans. Commun. **64**(5), 1952–1967 (2016)
12. F. W. Vook, et al., in *IEEE 48th Asilomar Conference on Signals, Systems and Computers*. Massive MIMO for mmWave systems, (Pacific Grove, 2014), pp. 820–824
13. A. Adhikary, et al., Joint spatial division and multiplexing for mm-wave channels. IEEE J. Sel. Areas Commu. **32**(6), 1239–1255 (2014)
14. M. Haenggi, et al., Stochastic geometry and random graphs for the analysis and design of wireless networks. IEEE J. Sel. Areas Commu. **27**(7), 1029–1046 (2009)
15. R. W. Heath, M. Kountouris, T. Bai, Modeling heterogeneous network interference using Poisson point processes. IEEE Trans. Signal Pro. **61**(16), 4114–4126 (2013)
16. H. S. Dhillon, et al., Modeling and analysis of K-tier downlink heterogeneous cellular networks. IEEE J. Sel. Area Commu. **30**(3), 550–560 (2012)
17. S. Mukherjee, Distribution of downlink SINR in heterogeneous cellular networks. IEEE J. Sel. Area Commu. **30**(3), 575–585 (2012)
18. N. Deng, W. Zhou, M. Haenggi, Heterogeneous cellular network models with dependence. IEEE J. Sel. Area Commu. **33**(10), 2167–2181 (2015)
19. Y. J. Chun, M. O. Hasna, A. Ghrayeb, Modeling heterogeneous cellular networks interference using poisson cluster processes. IEEE J. Sel. Area Commu. **33**(10), 2182–2195 (2015)
20. T. Bai, et al., in *Global Conference on Signal and Information Processing (GlobalSIP)*. Coverage analysis for millimeter wave cellular networks with blockage effects, (Austin, 2013), pp. 727–730
21. R. K. Ganti, M. Haenggi, Interference and outage in clustered wireless ad hoc networks. IEEE Trans. Inform. Theory. **55**(9), 4067–4086 (2009)
22. G. J. Foschini, Z. Miljanic, A simple distributed autonomous power control algorithm and its convergence. IEEE Trans. Vel. Technol. **42**(4), 641–646 (1993)
23. N. Lee, et al., Power control for D2D underlaid cellular networks: modeling, algorithms, and analysis. IEEE J. Sel. Area Commun. **33**(1), 1–13 (2015)
24. C. Guo, et al., -Fair power allocation in spectrum-sharing networks. IEEE Trans. Veh. Technol. **65**(5), 3771–3777 (2016)
25. A. Gjendemsj, et al., Binary power control for sum rate maximization over multiple interfering links. IEEE Trans. Wirel. Commun. **7**(8), 3164–3173 (2008)
26. H. Alzer, On some inequalities for the incomplete gamma function. Math. Comput. Am. Math. Soc. **66**(218), 771–778 (1997)
27. A. Thornburg, T. Bai, R. Heath, Performance analysis of mmWave ad hoc networks. IEEE Trans. Signal. Process. **64**(15), 4065–4079 (2016)

15

Q-learning-enabled channel access in next-generation dense wireless networks for IoT-based eHealth systems

Rashid Ali[1], Yazdan Ahmad Qadri[1], Yousaf Bin Zikria[1], Tariq Umer[2], Byung-Seo Kim[3] and Sung Won Kim[1*]

Abstract

One of the key applications for the Internet of Things (IoT) is the eHealth service that targets sustaining patient health information in digital environments, such as the Internet cloud with the help of advanced communication technologies. In eHealth systems, wireless networks, such as wireless local area networks (WLAN), wireless body sensor networks (WBSN), and wireless medical sensor networks (WMSNs), are prominent technologies for early diagnosis and effective cures. The next generation of these wireless networks for IoT-based eHealth services is expected to confront densely deployed sensor environments and radically new applications. To satisfy the diverse requirements of such dense IoT-based eHealth systems, WLANs will have to face the challenge of assisting medium access control (MAC) layer channel access in intelligent adaptive learning and decision-making. Machine learning (ML) offers services as a promising machine intelligence tool for wireless-enabled IoT devices. It is anticipated that upcoming IoT-based eHealth systems will independently access the most desired channel resources with the assistance of sophisticated wireless channel condition inference. Therefore, in this study, we briefly review the fundamental models of ML and discuss their employment in the persuasive applications of IoT-based systems. Furthermore, we propose Q-learning (QL) that is one of the reinforcement learning (RL) paradigms as the future ML paradigm for MAC layer channel access in next-generation dense WLANs for IoT-based eHealth systems. Our goal is to contribute to refining the motivation, problem formulation, and methodology of powerful ML algorithms for MAC layer channel access in the framework of future dense WLANs. This paper also presents a case study of next-generation WLAN IEEE 802.11ax that utilizes the QL algorithm for intelligent MAC layer channel access. The proposed QL-based algorithm optimizes the performance of WLAN, especially for densely deployed devices environment.

Keywords: Internet of Things, eHealth systems, Machine learning, Next-generation dense WLANs, MAC layer channel access

1 Introduction

Internet of Things (IoT) technology connects physical objects with the help of sensors and actuators by utilizing the existing infrastructure of communication networks, specifically with the help of unlicensed wireless networks [1]. Therefore, IoT technology uses the existing network infrastructure and communication technologies to ensure its strength. Sensors and actuators play a vital role in connecting the physical world to the digital world [2, 3]. The applications of IoT technology such as smart-cities, smart-industries, smart-metering, smart-grid, and smart-healthcare systems (IoT-based eHealth) are continuously increasing. It is expected that by the end of 2020, wireless-enabled devices will increase to 36.5 billion, and 70% of those would comprise sensor devices [1].

One of the key applications for the IoT is the eHealth service that targets sustaining patient health information in digital environments such as the Internet cloud with the help of advanced communication technologies. The World Health Organization (WHO) conducted a survey in 2013 and highlighted that upcoming decades would face the challenge of shortage of global health workforce, which would reach 12.9 million [4]. The main reasons of

*Correspondence: swon@yu.ac.kr
[1]Department of Information and Communication Engineering, Yeungnam University, Gyeongsan, 38541, Republic of Korea
Full list of author information is available at the end of the article

the decline are decreased interest in young people to pursue this profession, aging of current workforce, and the growing risk of non-infectious diseases such as cancer and heart stroke [4]. However, nowadays, health-related information can be easily monitored and tracked with the help of smart sensors and devices. This IoT-based eHealth enables people to allow emergency services/hospitals, doctors, and relatives to access their health-related data through different applications for immediate and efficient treatment. Handheld devices, such as smartphones and fitness bands, can act as on-body coordinators for personalized health monitoring because they are equipped with a variety of sensors, such as heart rate measurement sensor, blood glucose and pressure sensors, temperature sensors, humidity measurement sensors, accelerometers, magnetometers, and gyroscope (Fig. 1) [5]. There exist several built-in applications in such handheld smartphones, such as S-Health, to keep track of daily body fitness. However, there are always concerns regarding data privacy and security, reliability, and trustworthiness in the extensive usage of wearable smart devices [5].

One of the key issues in IoT-based eHealth systems is the requirement of appropriate communication technologies for efficient information sharing [6, 7]. Particularly, reliable connectivity is essential for real-time health-related information sharing. Wireless communication technologies are flexible and cost-effective for IoT-based information sharing. As shown in Fig. 1, a combination of both short-range wireless communication technologies, such as Bluetooth Low Energy (BLE), ZigBee, and IEEE 802.11 wireless local area network (WLAN), and long-range wireless communication technologies, such as Sub-1 Giga, LoRaWAN, and 4G/5G/LTE cellular systems, are typically considered [8, 9]. Both academic and industrial communities have recognized the significant attention given to future WLANs (IEEE 802.11) for IoT-based eHealth systems. One of their motivating services is the promisingly high throughput to support extensively advanced technologies even in densely deployed devices environment [10, 11]. However, unlicensed WLAN would face huge challenges in the future to access the shared channel resources, especially for highly dense IoT device deployments. The use of small cells and information-centric sensor networks in forthcoming IoT-based system may help to reduce the performance degradation issues [3, 12]. The most popular wireless channel resource utilization technique utilized by the WLAN medium access control (MAC) protocol is known as carrier sense multiple access with collision avoidance (CSMA/CA). To achieve maximum channel resource utilization through fair channel access in the WLANs with the ever-increasing density of contending IoT devices, the CSMA/CA scheme is very important as a part of IoT-based systems. CSMA/CA uses a binary exponential backoff (BEB) as its typical and

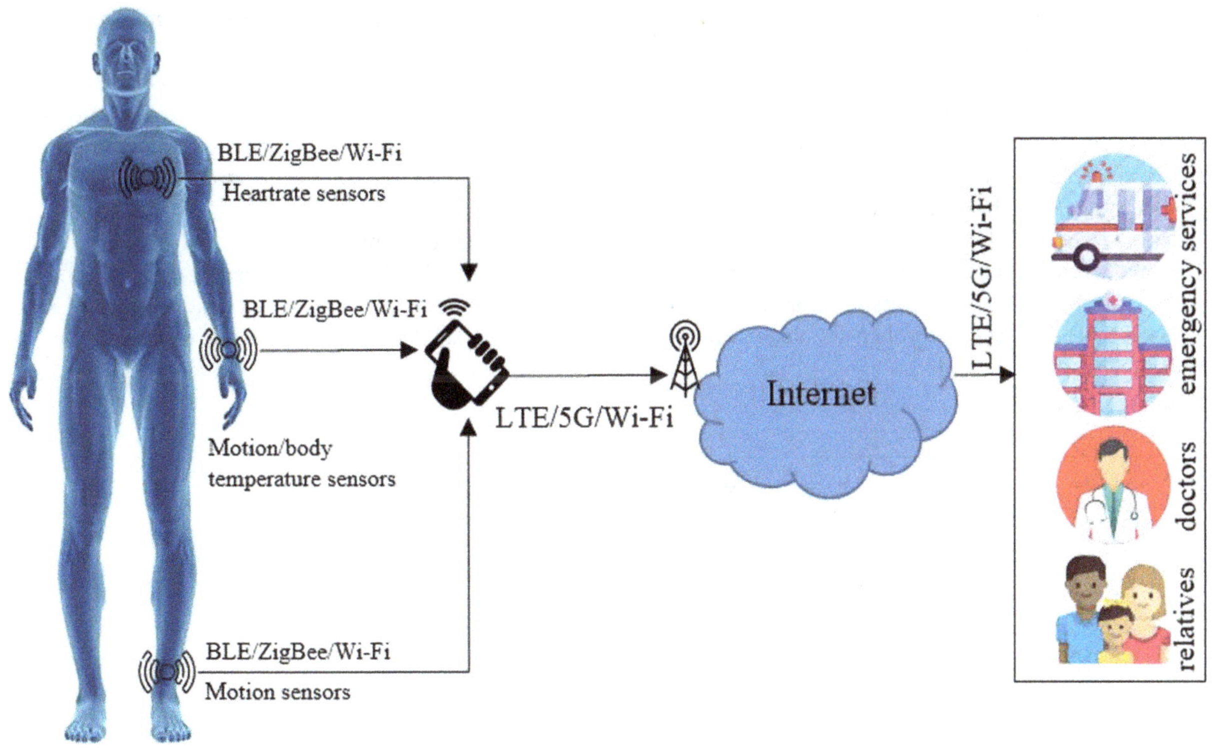

Fig. 1 Role of communication technologies in IoT-based eHealth system

traditional channel contention mechanism [11]. In BEB, a backoff value for contention is generated randomly from a specified contention window (CW). The CW size is exponentially increased for each unsuccessful transmission and reset to its initial size once transmitted successfully. For a network with a heavy load, resetting CW to its minimum size after successful transmission will result in more collisions and poor network performance. Similarly, for fewer contending devices, the blind exponential increase of CW for collision avoidance causes an unnecessary long delay. Besides, this blind increase/decrease of the backoff CW is more inefficient in highly dense networks proposed for IoT-based systems. Thus, the current CSMA/CA mechanism does not allow wireless networks to achieve high efficiency in highly dense environments.

Future dense WLANs are anticipated to infer the diverse and interesting features of both the devices' environments and their behavior to spontaneously optimize the reliability and efficiency of communication. Machine learning (ML), which is one of the prevailing machine intelligence tools, establishes an auspicious paradigm for optimization of the performance of WLANs [13]. As illustrated in Fig. 2, we can imagine an intelligent IoT device that is capable of accessing channel resources with the aid of ML. Therefore, an intelligent device would observe and learn the performance of a specific action with the objective of preserving a specific performance metric. Further, based on this learning, the intelligent device aims to reliably improve its performance while executing future actions by exploiting previous experience. ML algorithms are typically categorized into supervised [14] or unsupervised [15] learning algorithms. The supervised and unsupervised algorithms specify whether there are categorized samples in the available data (usually known as training data). Recently, another class of ML, known as reinforcement learning (RL), has emerged. It is encouraged by behavioral psychology [16, 17]. RL is concerned with a certain form of reward for a learner (such as an intelligent IoT device) that is associated with its environment (such as IoT-based eHealth system) through its observations and actions.

In this study, we briefly assess the fundamental perceptions of ML and propose services in persuasive applications for IoT-based systems based on the supervised, unsupervised, and RL categories. ML can be used extensively for revealing numerous practical problems in the future dense WLANs of the IoT-based application like eHealth systems. Examples include massive multiple-input multiple-output (MIMO), device-to-device (D2D) communications, femto/small cell-based heterogeneous networks, and high contention in dense WLAN environments. Following are the contributions of this paper:

- We briefly present the fundamental insights of ML in persuasive applications for IoT-based systems.
- Furthermore, we propose Q-learning (QL) that is one of the prevailing algorithms of RL as the future ML paradigm for channel access in contention-based dense WLANs for IoT-based systems.

The goal of this paper is to aid readers in refining the enthusiasm for problem devising and the approach to powerful ML algorithms for channel access in the framework of future dense WLANs to tap into previously unexplored applications of IoT-based systems. Table 1 shows list of acronyms used in this paper.

2 Machine learning in WLANs for IoT-based systems

As aforementioned, ML is usually categorized as supervised, unsupervised, and the most recently evolved RL algorithms. In this section, we elaborate the role of these categories in wireless communication networks for IoT-based systems. Figure 3 summarizes the family architecture of ML techniques, models, and their potential applications in dense IoT-based systems.

2.1 Supervised learning

In supervised ML, the learning agent learns from a labeled training dataset supervised by an erudite exterior supervisor. Each labeled training dataset is a depiction of a state comprising a specification, label, particular action,

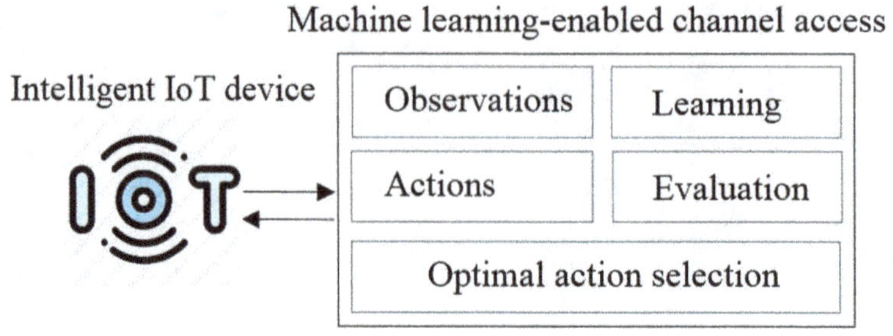

Fig. 2 Intelligent channel access mechanism for IoT-based systems

Table 1 List of acronyms used in this paper

Acronyms	Full description
AP	Access point
BL	Bayesian learning
BLE	Bluetooth Low Energy
CRNs	Cognitive radio networks
CSMA/CA	Carrier sense multiple access/collision avoidance
D2D	Device-to-device
HMM	Hidden Markov model
ICA	Independent component analysis
IoT	Internet of Things
LoRaWAN	Long-range wireless area network
LTE	Long-Term Evolution
MAC	Medium access control
MDP	Markov decision process
MIMO	Multiple-input multiple-output
MIP	Mixed integer programming
ML	Machine learning
PCA	Principle component analysis
POMDP	Partially observed MDP
QL	Q-learning
RA	Regression analysis
RL	Reinforcement learning
SVM	Support vector machine
WBSN	Wireless body sensor network
WHO	World Health Organization
WLAN	Wireless local area network
WMSN	Wireless medical sensor network

and class to which that particular action belongs. The objective of supervised ML is to make the system infer its retorts so that it acts intelligently in states not present in the labeled training dataset [18]. Although supervised ML is a significant type of ML, it is not suitable for a learner to learn the environment without the help of a supervisor and the available training dataset in it. Therefore, for the systems that need to deal interactively, it is often impractical to obtain a sample training dataset of anticipated behavior that is equally precise and descriptive regarding all the states in which the device has to perform actions in the future. In an unexplored environment, wherein ML is expected to be most valuable, a device must be able to learn from its own experience of interaction with the environment [18, 19].

Examples of supervised ML algorithms are regression models [20], k-nearest neighbor (KNN) [21], support vector machine (SVM) [22], and Bayesian learning (BL) [18]. Regression analysis (RA) depends on a statistical method for assessing the relations among input parameters. The objective of RA is to envisage the assessment of one or more continuously valued estimation objectives, given the assessment of a vector of input parameters. The estimation objective is a function of the independent parameters. The KNN and SVM techniques are mostly employed to categorize different objects in the system. In the KNN technique, an agent/device is categorized according to the votes of the neighbor agents. The agent is associated with the category that is most common among its k-nearest neighbors. On the contrary, the SVM algorithm uses non-linear mapping for object classification. First, it converts the original training dataset into a higher measurement, where it befits distinguishability. Later, it explores for the optimized linearly separating hyperplane that is accomplished by distinguishing one category of agents from another [18]. On the contrary, the idea of BL is to estimate a posterior distribution of the target variables, given some inputs and the available training datasets. The hidden Markov model (HMM) is a simple example of reproductive paradigms that can be learned with the help of BL [19]. HMM is a tool for expressing probability distributions of the trail of observations in the system. More specifically, it is a generalization method, where the unseen (hidden) variables of the system are associated with each other through a Markov decision process (MDP) [23]. These hidden variables control the particular constituent to be selected for each observation, while being relatively independent of each other.

These examples of supervised ML paradigms can be used for estimating wireless radio parameters that are related to the quality of service and quality of experience requirements of a particular user/device. Similar to a massive MIMO system of hundreds of radio antennas, the available channel estimation may lead to optimal dimensional search problems, which can be easily learned using any of the abovementioned supervised learning models. The SVM functions are cooperative for data classification problems. A hierarchical SVM (H-SVM), in which each hierarchical level is comprised of a fixed number of SVM classifiers, was proposed in [23]. H-SVM is used to intelligently estimate the Gaussian channel's noise level in a MIMO system by exploiting the training data. KNN and SVM can be pragmatic in finding the optimum handover solutions in wireless networks. Similarly, the BL model can be invoked for wireless channel characteristics learning and estimation in future generation ultra-dense wireless networks. For example, Wen et al. [24] estimated both the radio channel parameters in a specific radio cell and those of the intrusive links of the neighboring radio cells using BL techniques to deal with the pilot contamination problem faced by massive MIMO systems. Another application of BL was proposed in [25], where a Bayesian inference model was proposed for considering and statistically describing a variety of methods that are

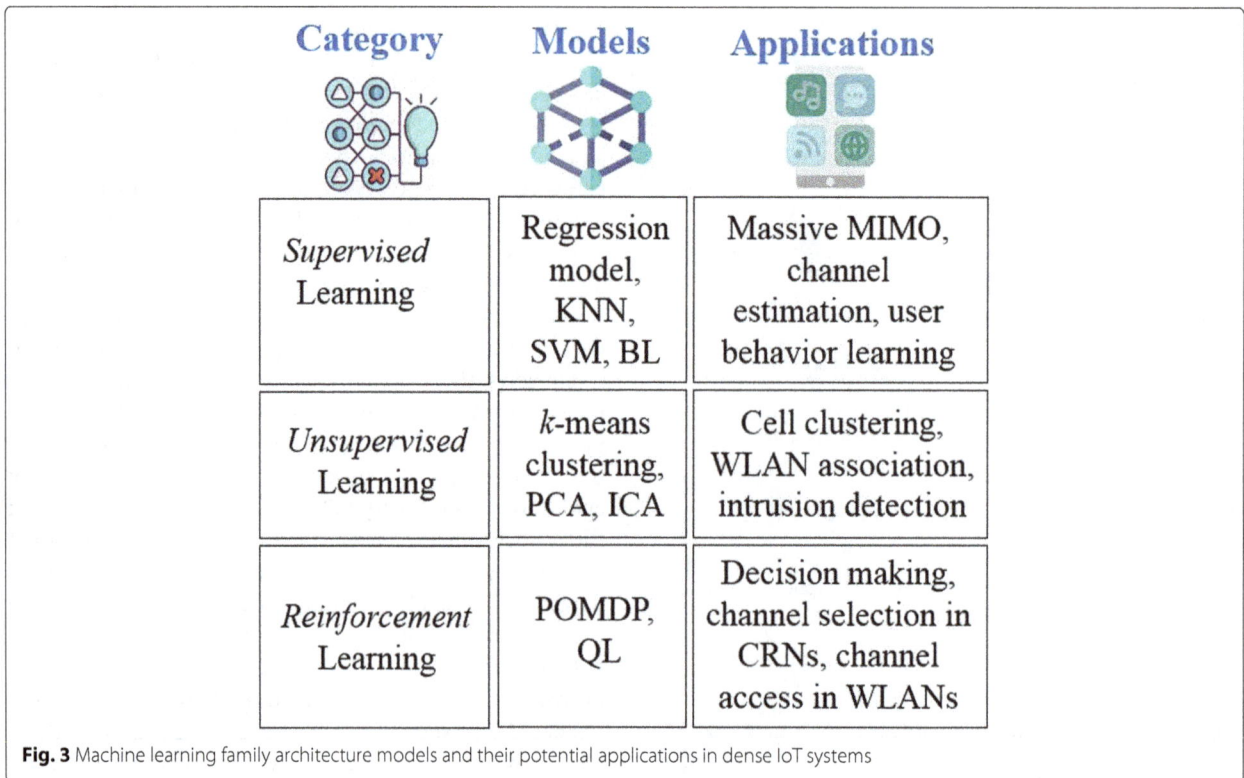

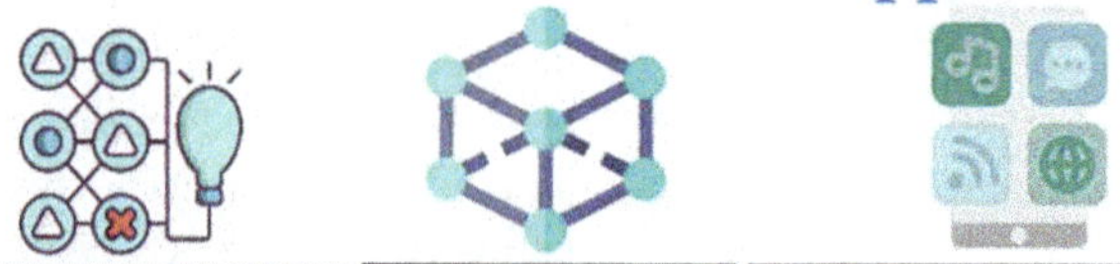

Category	Models	Applications
Supervised Learning	Regression model, KNN, SVM, BL	Massive MIMO, channel estimation, user behavior learning
Unsupervised Learning	*k*-means clustering, PCA, ICA	Cell clustering, WLAN association, intrusion detection
Reinforcement Learning	POMDP, QL	Decision making, channel selection in CRNs, channel access in WLANs

Fig. 3 Machine learning family architecture models and their potential applications in dense IoT systems

proficient at learning the predominant factors for cognitive radio networks (CRNs). Their proposed mechanism covers both the MAC and the network layers of a wireless network.

2.2 Unsupervised learning

Unsupervised ML is usually regarding the verdict structure veiled in a collection of unlabeled training datasets. The terms supervised ML and unsupervised ML would appear to profoundly categorize most ML-based paradigms; however, they are not accurate. The aim of supervised ML is to learn the mapping from an input dataset to an output result where accurate values are provided by a supervisor. On the contrary, in unsupervised learning, there is no external supervisor but only the available input dataset. The objective is to find symmetries in the dataset. There is an edifice of the available dataset space, e.g., that certain patterns occur often, such patterns can help understand the action to be performed in the future for any unknown input. In the statistical context, this is also known as density estimation [18].

Examples of unsupervised ML algorithms are *k*-means clustering [21], principle component analysis (PCA) [26], and independent component analysis (ICA) [27]. The objective of *k*-means clustering is to divide user observations into *k* clusters, where each observation is associated with the adjacent cluster. It uses the center of gravity (centroid) of the cluster, which is the mean value of the observation points within that particular cluster. Continuous iteration of the *k*-means clustering algorithm keeps assigning an agent to the particular cluster in which the centroid is close to the agent based on a similarity metric. This similarity metric is known as Euclidean distance. Further, the in-cluster differences are also minimized until convergence by iteratively updating the cluster centroid is achieved [18]. PCA is used to transform a set of possibly associated parameters into a set of unassociated parameters that are known as the principal components (PCs). The number of PCs is always less than or equal to the number of original parameters/components. The first PC has the largest possible variance, and each subsequent PC has the utmost variance probable under the limitation that it is unassociated with the prior PCs. Basically, the PCs are orthogonal (unassociated) because they are the eigenvectors of the covariance matrix that is symmetric. Unlike PCA, ICA is a statistical method applied to expose unseen elements that inspire sets of haphazard parameters/components within the system [18].

Clustering is one of the common problems in densely deployed wireless networks of IoT-based systems, especially in heterogeneous network environments with diverse cell sizes. In such cases, small cells have to be wisely grouped to avoid interference using coordinated multi-point transmission, whereas the mobile devices are grouped to follow an optimum offloading strategy. The devices are grouped in device-to-device (D2D) wireless networks to attain high energy efficiency, and the WLAN users are grouped to uphold an optimum access point (AP) association. Xia et. al. [28] proposed a hybrid scenario to diminish inclusive wireless traffic by encouraging the exploitation of a high-capacity optical infrastructure. They formulated a mixed-integer programming (MIP) problem to cooperatively optimize both network gateway splitting and the virtual radio channel provision based on typical k-means clustering. Both PCA and ICA are formulated to recover statistically autonomous source signals from their linear combinations using powerful statistical signal processing techniques. One of their key applications is in the area of intrusion detection in wireless networks, which depends on traffic monitoring. Besides, similar issues may also be resolved in the dense wireless communications technologies of IoT-based systems. PCA and ICA can also be invoked to classify user behavior in CRNs. In [29], the authors applied PCA and ICA in a smart grid scenario of IoT systems to improve the concurrent wireless transmissions of smart devices set up in the smart home. The statistical possessions of the received signals were oppressed to blindly isolate them using ICA. Their proposed mechanism enhances transmission capability by evading radio channel assessment and data security by excluding any wideband intrusion.

2.3 Reinforcement learning

Reinforcement learning (RL) is motivated by behaviorist sensibility and a control philosophy, where an agent can achieve its objective by interacting with and learning from its surroundings. In RL, the agent does not have clear information whether it is close to its target objective. However, the agent can observe the environment to augment the aggregate reward in an MDP [30]. RL is an ML technique that learns about the environment, what to do, and how to outline circumstances to current actions to maximize a numerical reward signal. Mostly, the agent is not informed about which actions to perform, and it has to learn which actions will produce maximum reward. In some exciting and inspiring situations, it is possible that actions will affect not only the instant reward but also the following state, and consequently, all succeeding rewards. MDPs offer a precise framework for modeling decision-making in particular circumstances, where the consequences are comparatively haphazard, and the decision-maker partially governs the consequences.

Partially observable MDP (POMDP) [31] and QL [17] are the examples of RL. POMDP might be seen as speculation with MDP, where the agent is inadequate to perceive the original state transitions in a straightforward manner; therefore, it only has constrained information. The agent has to retain the trajectory of the probability distribution of the appropriate states based on a set of annotations, and the probability distribution of both the observation probabilities and the original MDP [32]. QL might be conjured up to discover an optimum strategy for performing action from any finite MDP, particularly when the environment is unknown [18].

The uses of POMDP paradigms create vital tools for supportive decision-making in IoT-based systems, where the IoT devices may be considered agents and the wireless network constitutes the environment. In a POMDP problem, the technique first postulates the environment's state space and the agent's action space. Additionally, it endorses the Markov property among the states. Secondly, it constructs the state transition probabilities formulated as the probability of navigating from one state to another under a specific action. The third and final step is to enumerate both the agent's instant reward and its long-term reward via Bellman's equation [17]. Later, a wisely constructed iterative algorithm may be considered to classify the optimum action in each state. The applications of POMDP comprise the network selection problems of heterogeneous networks, channel sensing, and user access in CRNs. In [32], the authors proposed a mechanism for transmission power control problems of energy-harvesting systems, which were scrutinized with the help of the POMDP model. In their proposed investigation, the battery, channel, data transmission, and data reception states are defined as the state space, and an action by the agent is related to transmitting a packet at a certain transmission power. QL, usually in aggregation with the MDP models, has also been used in applications of heterogeneous networks. Alnwaimi et al. [33] presented a heterogeneous, fully distributed, multi-objective strategy for the optimization of femtocells based on a QL model [33]. Their proposed model solves both the channel resource allocation and interference coordination issues in the downlink of heterogeneous femtocell networks. Their proposed model acquires channel distribution awareness and classifies the accessibility of vacant radio channel slots for the establishment of opportunistic access. Further, it helps choose sub-channels from the vacant spectrum pool.

3 Q-learning-enabled channel access for dense WLANs

As described in the previous section, QL has already been extensively applied in heterogeneous wireless networks [14]. In such a case, the QL paradigm also covers a set of states where an agent can make a decision on an action from a set of available actions. By performing an action in a particular state, the agent collects a reward with the objective of exploiting its collective rewards. A collective reward is illustrated as a Q-function and is updated in an iterative approach after the agent performs an action and attains the subsequent reward [18]. The trade-off between exploration and exploitation is one of the challenges arising in QL but not in other types of ML techniques. To achieve considerable rewards, it is obligatory for a QL agent to choose those actions that it has tried before, and found them to be effective in constructing the reward (exploitation). However, to learn more about the environment, an agent has to try actions that it has not selected before (exploration). In exploitation, the agent has to attain what it has already experienced to optimize the process, and additionally, it must explore the environment to maximize the aggregated reward to make better selections in the future. The quandary is that neither exploration nor exploitation can be pursued exclusively without failing in the other process. The agent must try a diversity of actions and gradually favor those that appear to be the best. It is not possible to both explore and exploit with a particular action selection; therefore, we frequently refer to the "tussle" between these two.

3.1 Q-learning prototype

As aforementioned, QL algorithm utilizes a form of RL to solve MDPs without possessing complete information. In addition to the agent and the environment, a QL system has four main sub-elements: a policy, a reward, a Q-value function, and sometimes a model of the environment as an optional entity [17], as shown in Fig. 4.

3.1.1 Policy

The learning agent's manner of behaving at a particular time is defined as a policy. A policy can be a modest utility or a lookup table; however, it may comprise extensive computations such an exploration process. A policy is fundamental for a QL agent because it alone is adequate to determine the behavior of an agent. Generally, policies might be stochastic. A policy decides which action to perform in which state [17].

3.1.2 Reward

In each iteration, the QL agent receives a particular quantity from the environment known as the reward. The main objective of a QL algorithm is to collect as much reward as possible. An agent's exclusive goal is to exploit the accumulated reward collected over the long run. The reward describes the pleasant and unpleasant events for the agent. Reward signals are the instant and crucial topographies of the problem faced by the agent. The agent decides to change its policy based on the reward. For example, if the current action of the policy is followed by a low reward, then an agent may decide to select other actions in the future [17].

3.1.3 Q-value function

Although the reward specifies what is good at one instant, a Q-value function stipulates what is good in the end. Therefore, the Q-value of a state is the accumulated amount of reward that an agent gains at this state to presume in the future [17]. For example, although a state may continuously produce a low instant reward, it may have a high Q-value owing to being repeatedly trailed by other states that produce high rewards. In a WLAN environment, rewards are similar to a high channel collision probability (unpleased) and a low channel collision probability (pleased), whereas Q-values resemble a more sophisticated and prophetic verdict of how pleased or unpleased the agent is in a particular state (e.g., the back-off stage). If there is no reward, then there will be no Q-value, and the only purpose of estimating the Q-value is to attain additional rewards. An agent is most anxious about the Q-value while giving and assessing verdicts. An agent selects optimum actions based on Q-value findings. It seeks actions that carry states of a maximum Q-value and not a maximum reward because these actions attain the highest amount from the rewards for the agent over the long run.

3.1.4 Environment model

Environment model is an optional element of QL, which imitates the performance of the system to some extent. Typically, it allows drawing inferences to be made about how the environment will perform [17]. For example, given a state and an action, the model might envision the

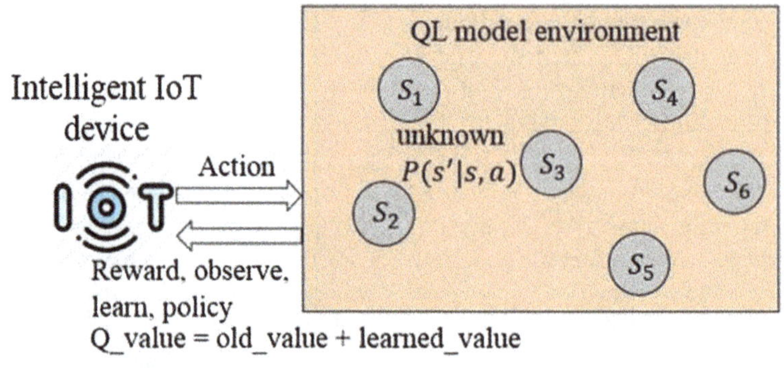

Fig. 4 Q-learning system environment with its sub-elements

subsequent state and the next reward. Environment models are used for planning a method to decide on a sequence of actions by considering latent future situations. In an example of a WLAN system, a device would like to plan its future decisions based on the given state (e.g., the backoff stage) and action, along with its rewards (e.g., channel collision probability).

3.2 Q-learning algorithm

Let S represent a finite set of conceivable states of an environment and A represent a finite set of allowable actions to be performed. At time t, a learner (IoT device) observes the current state (s) of the environment and performs an action (a), i.e., $a_t = a \in A$, based on both the apparent state and its previous experience. The action a_t changes the environmental state from s_t to $s_{t+1} = s^* \in S$; consequently, the agent receives the reward (r) at time t, r_t for the specific action: a_t. The QL algorithm finds an optimal policy for state s that optimizes the rewards over a long period of time. In the QL algorithm, a Q-value function, $Q(s,a)$, estimates the reward as the cumulative discounted reward. An optimal Q-value, i.e., $Q^{\text{opt}}(s,a)$, is determined using the Q-values. The QL algorithm finds the optimal Q-value in a greedy manner. The Q-value is updated as:

$$Q(s,a) = (1-\alpha) \times Q(s,a) + \alpha \times \Delta Q(s,a), \tag{1}$$

where α is the learning rate and takes values such as $0 \leq \alpha \leq 1$. When α is minimum, i.e., zero, the agent does not learn from the environment; therefore, the Q-value is not updated. When α is maximum, i.e., 1, the agent always learns; therefore, learning occurs quickly as seen in the following equation:

$$\Delta Q(s,a) = \left\{r(s,a) + \beta \times \max_{a'} Q\left(s',a'\right)\right\} - Q(s,a), \tag{2}$$

where β ($0 \leq \beta \leq 1$) weighs the immediate rewards more heavily than future rewards, and is known as the discount factor. Over a considerable period of time, $Q(s,a)$ converges into $Q^{\text{opt}}(s,a)$. The simplest policy for action selection is to choose one of the actions with the maximum measured Q-value (i.e., exploitation). If there are more than one greedy actions, then a choice is randomly made among them. This greedy action selection method can be written as:

$$a^{\text{opt}} = \text{argmax}_a Q(s,a), \tag{3}$$

where argmax_a signifies the action a, for which the expression that follows it is exploited. An agent continuously exploits current knowledge to maximize the instant reward. A simple substitute is to perform greedily in most cases; however, sometimes (e.g., with a small probability ε) the agent can randomly select from all the equal probability actions, independent of the Q-value. The method using this greedy and non-greedy action selection rule is known as the ε-greedy method [17]. An advantage of such a technique is that as the number of iterations increases, every action will guarantee that $Q(s,a)$ converges to $Q^{\text{opt}}(s,a)$. This leads to the inference that the probability of choosing the optimum action converges to a value that is larger than $1-\varepsilon$, i.e., to adjacent certainty. In WLANs, for dense IoT-based systems, an agent would choose greedy actions from high-value actions (exploitation) to improve the throughput performance, and would perform a non-greedy action (exploration) to know the dynamicity of the network environment.

3.3 Case study: DCF-based backoff mechanism

The QL-based channel access scheme can be used to guide densely deployed IoT devices and allocate radio resources more efficiently. When an IoT device is deployed in a new environment, usually, no data are available on historical scenarios. Therefore, QL algorithms are the best choice to observe and learn the environment for optimal policy selection. For example, we consider the case study of DCF-based backoff mechanism of dense WLANs in IoT-based systems. In a densely deployed WLAN, channel collision is the most vital issue causing performance degradation. To tackle collision issues at the MAC layer, we propose adopting the QL algorithm. QL finds solutions through interacting and learning with an environment; therefore, we propose using the QL algorithm to model the optimal contention window (CW) in a channel observation-based

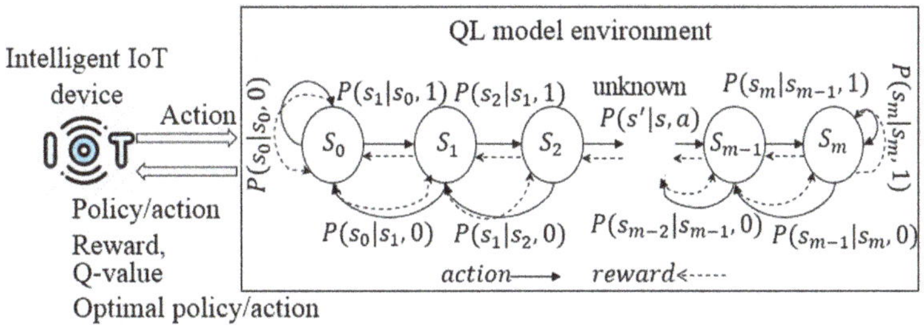

Fig. 5 QL model environment with its elements in the case study of DCF-based backoff mechanism

scaled backoff (COSB) mechanism [34] for dense wireless networks of IoT-based systems. In other words, a station (STA; a WLAN-enabled IoT device referred to as an STA) controls the CW selection intelligently with the aid of the QL-based algorithm.

In COSB [34] protocol, STAs select a random backoff value from the initial CW (CW_{min}) to contend for the wireless medium after observing the channel in an idle state for a distributed inter-frame space (DIFS) period. The period after DIFS is divided into B_{obs} discrete observation time slots. The duration of each discrete time slot is either a constant idle slot time (σ) or a variable busy slot time (owing to successful or collided transmission). In COSB, each STA proficiently measures the channel observation-based collision probability (p_{obs}) as:

$$p_{obs} = 1/B_{obs} \times \sum_{(k=0)}^{(B_{obs}-1)} S_k, \quad (4)$$

where $S_k = 0$ if B_{obs} is observed as idle or if the transmission is successful, whereas $S_k = 1$ if B_{obs} is observed as busy or the transmission has collided [34].

We assume backoff stages of COSB as a set of m states, i.e., $S = \{0, 1, 2, ..., m\}$, where an intelligent IoT device performs an action a from a finite set of permissible actions $A = \{0, 1\}$, where 0 indicates decrement and 1 indicates increment. This is because in COSB, there are two possible actions: increase or decrease the CW size [34]. At time t, the STA collects reward r_t in the response to an action a_t following policy π in a particular state s_t; i.e., $r_t(s_t, a_t)$ with the objective to exploit collective reward $Q(s_t, a_t)$, which is a Q-value function defined in Eqs. (1) and (2). Figure 5 depicts the proposed QL model environment with its elements in a DCF-based backoff mechanism for channel access in WLANs.

The selection of optimal action following π^{opt} is known as a greedy action $\left(a^{\pi^{opt}}\right)$ selection policy that is defined in Eq. (3). A naivest policy can be to exploit in most cases; however, sometimes, the STA explores according to the default policy π, independent of $a^{\pi^{opt}}$. The exploration with probability (ε) and exploitation with probability $(1 - \varepsilon)$ is called ε-greedy method [17]. The ε-greedy technique guarantees the convergence of learning estimate $\Delta Q(s, a)$ with the increase of episodes (instances). In a dense WLAN environment, exploitation can be used to improve throughput performance by an IoT device, and exploration can be used to know the dynamicity of the WLAN environment.

In COSB [34], an STA conducts p_{obs} at every transmission attempt. Therefore, we express p_{obs} as the reward of the action at any specific state. Therefore, reward r_t produced by action a_t taken in state s_t at time t can be described as:

Table 2 MAC layer and PHY layer simulation parameters

Parameter type	Value
Frequency	5 GHz
Channel bandwidth	160/20 MHz
Data rate	1201/11 Mbps
Payload size	1472 bytes
Transmission range	10 m
Simulation time	100/500 s
Propagation loss model	Log distance
Mobility model	Constant position
Rate adaptation models	Constant rate/minsttrel

$$r_t(s_t, a_t) = 1 - p_{obs}. \quad (5)$$

The above equation indicates how pleased the STA was with its action a_t in state s_t.

4 Experimental results and discussion

We used ns3.28 [35] simulator to perform experiments of the proposed *i*QRA mechanism. Some important PHY layer and MAC layer simulation parameters are shown in Table 2. The results in Fig. 6a and b indicate that a small value of α and a large value of β make ΔQ (learning estimate) converge faster. The convergence of ΔQ clearly indicates that there exist optimal values that can be learned and exploited in the future. The throughput performance optimization of COSB using proposed *i*QRA is depicted in Fig. 7a. The performance of *i*QRA may degrade in small networks (i.e., for < 10 contending STAs as shown in Fig. 7a owing to low and irregular rewards).

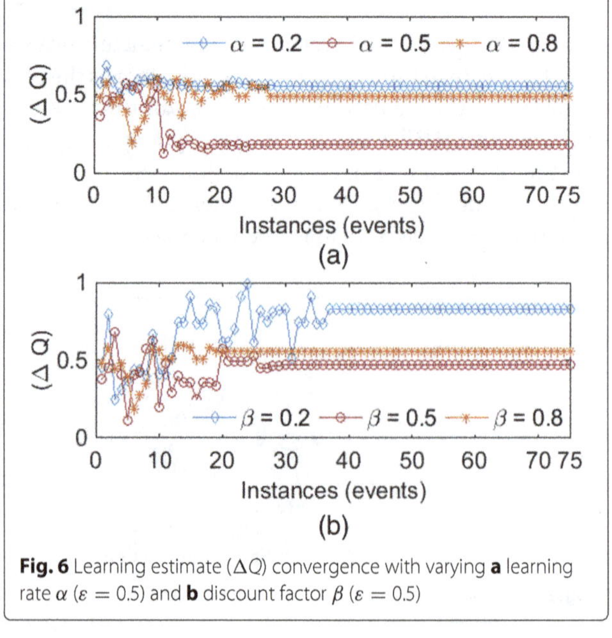

Fig. 6 Learning estimate (ΔQ) convergence with varying **a** learning rate α ($\varepsilon = 0.5$) and **b** discount factor β ($\varepsilon = 0.5$)

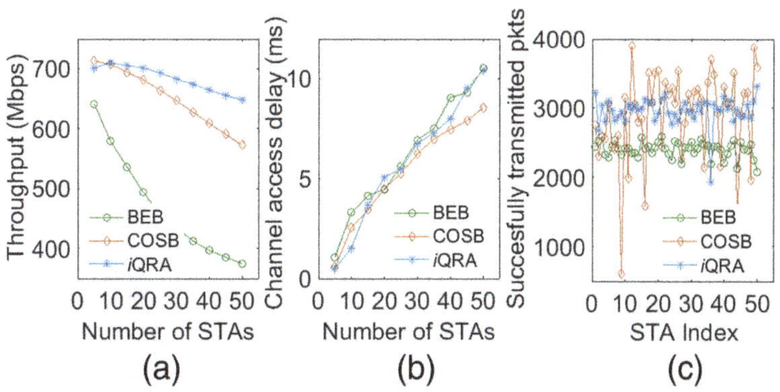

Fig. 7 Comparison of BEB, COSB, and *i*QRA for **a** throughput (Mbps), **b** channel access delay (ms), and **c** successfully transmitted packets (fairness)

Additionally, the channel access delay is also increased for *i*QRA as compared to COSB; this is obvious owing to the environment inference characteristics; however, it remains lower than the conventional binary exponential backoff (BEB) [11] mechanism shown in Fig. 7b. Figure 7c portrays that the proposed *i*QRA also improves the fairness of COSB. The optimized performance of COSB using *i*QRA clearly stipulates that the QL-based proposed mechanism is effective in learning the network environment. Additionally, *i*QRA is essentially intended to intelligently adjust its learning parameters according to the dynamics of the WLAN. Therefore, we simulated a dynamic network environment by increasing the number of contenders by 5 after every 50 s until the number of STAs reached 50. Figure 8 depicts the properties of network dynamics on ΔQ. The figure shows simulation of a 500 s period with 1500 learning instances of a tagged STA. As shown in the figure, with the network dynamics, a tagged STA observes fluctuation in its learning estimate ΔQ, thereby indicating the inference of change in the network. We see that the throughput performance of *i*QRA eventually reaches a steady state in a dynamic network environment, as shown in Fig. 9a. To evaluate the performance of the proposed *i*QRA for moving devices in the network, we simulated a distance-based rate adaptation model. This model changes the transmission rate of the sender device according to the distance between the sender and receiver to achieve the best possible performance. IEEE 802.11a (11 Mbps) WLAN with 10 contending STAs is simulated for distance-based rate adaptation performance evaluation, as shown in Fig. 9b. Contending STAs are placed randomly around the access point (AP) within a distance of 25 m. A tagged STA starts moving away from the AP that is initially placed at a 1-m distance. As the distance from the AP increases, performance of a tagged STA degrades for all the three compared algorithms (BEB, COSB, and *i*QRA), as shown in Fig. 9b. It is observed that the throughput of the BEB algorithm approaches close to zero after the STA reaches a distance of 60 m, and it finally becomes zero after reaching a distance of 80 m. Owing to the observation-based nature of COSB, it achieves higher throughput even after a 60-m distance, compared to BEB. However, the proposed *i*QRA performs optimally, even if the distance reaches 80 m owing to its network inference capability.

5 Conclusion

In this study, we investigated the benefits of ML-based intelligent dense wireless networks for IoT-enabled eHealth systems. We presented the key families of ML algorithms and deliberated their application in the context of dense IoT systems including next-generation wireless networks with massive MIMO; heterogeneous IoT

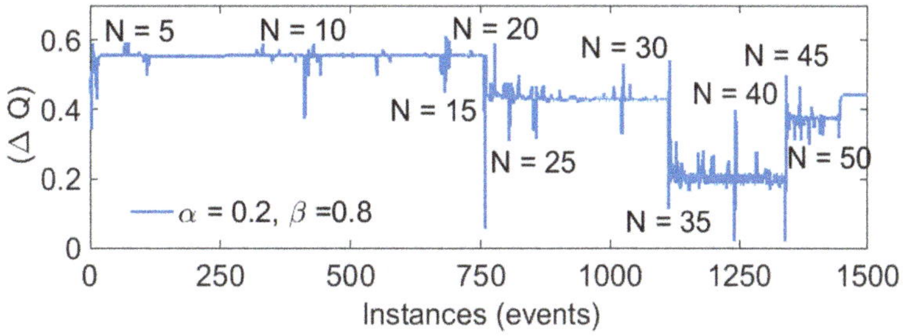

Fig. 8 Convergence of learning estimate (ΔQ) in a dynamic network environment (increasing the number of contenders after every 50 s)

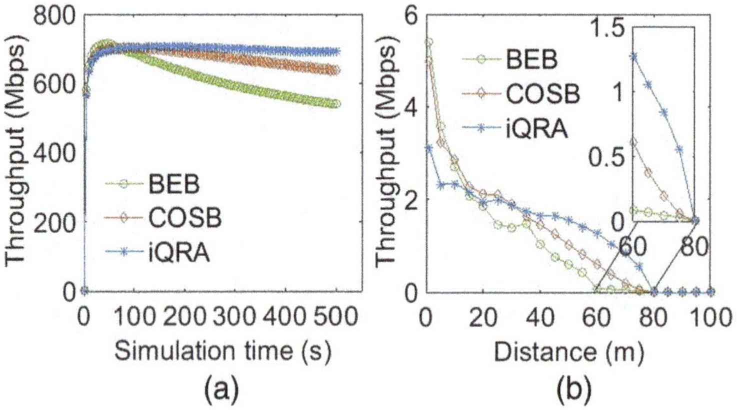

Fig. 9 Throughput comparison of BEB, COSB, and *i*QRA in **a** dynamic network environment with increasing number of contenders after every 50 s and **b** distance-based rate adaptation network environment

networks based on small cells; smart applications, such as the smart grid and smart city; and intelligent cognitive radio. The three well-known categories of ML, supervised learning, unsupervised learning, and RL algorithms, are scrutinized in addition to a consistent sculpting methodology and possible future applications in dense IoT systems. Furthermore, we proposed Q-learning as a promising ML paradigm for MAC layer channel access in dense IoT systems. The proposed paradigm is implemented on a case study of DCF-based backoff mechanism in dense WLANs. We proposed an intelligent Q-learning-based resource allocation (*i*QRA) mechanism to optimize the performance of an existing (COSB) mechanism. The proposed *i*QRA mechanism infers unknown wireless network conditions and exploits rapidly unexpected changes to learn dynamicity in dense WLANs. The experimental results show that *i*QRA significantly enhances the performance of COSB in terms of throughput and fairness. Results reveal the ability of the Q-learning scheme to determine dense wireless network environments in IoT-based systems. In conclusion, ML is a promising area for self-scrutinized intelligence-aided dense wireless network research for IoT-enabled eHealth systems.

In the future, we aim to further investigate the applications of our proposed mechanism in various IoT-based systems such as smart city, smart home, smart grid, and smart industry.

Abbreviations

BLE: Bluetooth Low Energy; CSMA/CA: Carrier sense multiple access with collision avoidance; D2D: Device-to-device; IoT: Internet of Things; KNN: *k*-nearest neighbor; MAC: Medium access control; MIMO: Massive multiple-input multiple-output; ML: Machine learning; QL: Q-learning; RL: Reinforcement learning; SVM: Support vector machine; WBSN: Wireless body sensor network; WHO: World Health Organization; WLAN: Wireless local area network; WMSN: Wireless medical sensor networks

Acknowledgements

This work was supported by the 2019 Yeungnam University Research Grant.

Authors' contributions

RA and YAQ conceived the main idea, designed the algorithm, and proposed the intelligent framework for IoT-based eHealth systems. RA, YBZ, and TU performed the implementation of the proposition in the NS3 simulator. BK and SWK contributed to the structuring, reviewing, and finalizing of the manuscript. All authors read and approved the final manuscript.

Funding

The funding source is the same as described in the acknowledgements.

Competing interests

The authors declare that they have no competing interests.

Author details

[1]Department of Information and Communication Engineering, Yeungnam University, Gyeongsan, 38541, Republic of Korea. [2]Department of Computer Science, COMSATS University Islamabad, Wah, Pakistan. [3]Department of Computer and Information Communication Engineering, Hongik University, Seoul, 04066, Republic of Korea.

References

1. M. Pasha, W. SMShah, Framework for e-health systems in IoT-based environments. Wirel. Commun. Mob. Comput. **2018**(6183732), 1–11 (2018). https://doi.org/10.1155/2018/6183732
2. G. T. Singh, F. Al-Turjman, Learning data delivery paths in QoI-aware information-centric sensor networks. IEEE Internet Things J. **3**(4), 572–580 (2016). https://doi.org/10.1109/JIOT.2015.2504487
3. M. Z. Hasan, F. Al-Turjman, Evaluation of a duty-cycled asynchronous X-MAC protocol for vehicular sensor networks. EURASIP J. Wirel. Commun. Netw. **2017**(1), 95 (2017). https://doi.org/10.1186/s13638-017-0882-7
4. World Health Organization, Global health workforce shortage to reach 12.9 million in coming decades. http://www.who.int/mediacentre/news/releases/2013/health--workforce-shortage/en/. Accessed 10 Dec 2018
5. H. M Alam, M. I. Malik, T. Khan, A. Pardy, Y. L. Kuusik, A. Moullec, Survey on the roles of communication technologies in IoT-based personalized healthcare applications. IEEE Access. **6**, 36611–36631 (2018). https://doi.org/10.1109/ACCESS.2018.2853148
6. M. Faheem, M. Zahid Abbas, G. Tuna, V. C. Gungor, EDHRP: Energy efficient event driven hybrid routing protocol for densely deployed wireless sensor networks. J. Netw. Comput. Appl. **58**, 309–326 (2015). https://doi.org/10.1016/j.jnca.2015.08.002
7. M. Faheem, V. C. Gungor, Energy efficient and QoS-aware routing protocol for wireless sensor network-based smart grid applications in the context of industry 4.0. Appl. Soft Comput. **68**, 910–922 (2018). https://doi.org/10.1016/j.asoc.2017.07.045
8. M. Faheem, R. A. Butt, B. Raza, M. W. Ashraf, S. Begum, Md. A. Ngadi, V. C.

Gungor, in *Transactions on Emerging Telecommunications Technologies*. Bio-inspired routing protocol for WSN-based smart grid applications in the context of Industry 4.0, (2018). https://doi.org/10.1002/ett.3503
9. M. Faheem, V. C. Gungor, Capacity and spectrum-aware communication framework for wireless sensor network-based smart grid applications. Comput. Stand. Interfaces. **53**, 48–58 (2017). https://doi.org/10.1016/j.csi.2017.03.003
10. S. Demir, F. Al-Turjman, Energy scavenging methods for WBAN applications: a review. IEEE Sensors J. **18**(16), 6477–6488 (2018). https://doi.org/10.1109/JSEN.2018.2851187
11. R. Ali, S. W. Kim, B. Kim, Y. Park, Design of MAC layer resource allocation schemes for IEEE 802.11ax: future directions. IETE Tech. Rev. **35**(1), 28–52 (2018). https://doi.org/10.1080/02564602.2016.1242387
12. F. Al-Turjman, E. Ever, H. Zahmatkesh, Small cells in the forthcoming 5G/IoT: traffic modelling and deployment overview. IEEE Commun. Surv. Tutor. **21**(1), 28–65 (2019). https://doi.org/10.1109/COMST.2018.2864779
13. R. Ali, N. Shahin, Y. B. Zikria, B. Kim, S. W. Kim, Deep reinforcement learning paradigm for performance optimization of channel observation-based MAC protocols in dense WLANs. IEEE Access. **7**, 3500–3511 (2019). https://doi.org/10.1109/ACCESS.2018.2886216
14. C. Zhang, P. Patras, H. Haddadi, Deep learning in mobile and wireless networking: a survey. IEEE Commun. Surv. Tutor. Early Access (2019). https://doi.org/10.1109/COMST.2019.2904897
15. Y. Sun, M. Peng, Y. Zhou, Y. Huang, S. Mao, Application of machine learning in wireless networks: key techniques and open issues. ArXiv e-prints (2018). https://arxiv.org/abs/1809.08707
16. E. M. Joo, *Theory and novel applications of machine learning. 12–16.* (IntechOpen, London, 2009). https://doi.org/10.5772/56681
17. R. S. Sutton, A. G. Barto, *Reinforcement learning: an introduction*, Second ed. (MIT Press, Cambridge, 1998). isbn:0262193981
18. E. Alpaydin, *Introduction to machine learning*, Third ed. (MIT Press, Cambridge, 2014). isbn:978-0-262-028189
19. R. Ali, N. Shahin, R. Bajracharya, B. S. Kim, S. W. Kim, A self-scrutinized backoff mechanism for IEEE 802.11ax in 5G unlicensed networks. Sustainability. **10**, 1201 (2018). https://doi.org/10.3390/su10041201
20. Q. H. Abbasi, S. Liaqat, L. Ali, A. Alomainy, in *2013 First International Symposium on Future Information and Communication Technologies for Ubiquitous HealthCare (Ubi-HealthTech)*. An improved radio channel characterisation for ultra wideband on-body communications using regression method, (Jinhua, 2013), pp. 1–4. https://doi.org/10.1109/Ubi-HealthTech.2013.6708063
21. Y. Xu, T. Y. Fu, W. C. Lee, J. Winter, Processing k nearest neighbor queries in location-aware sensor networks. Signal Process. **87**(12), 2861–2881 (2007). https://doi.org/10.1016/j.sigpro.2007.05.013
22. Z. Dong, Y. Zhao, Z. Chen, in *IEEE MTT-S International Wireless Symposium (IWS)*. Support vector machine for channel prediction in high-speed railway communication systems, vol. 2018, (Chengdu, 2018), pp. 1–3. https://doi.org/10.1109/IEEE-IWS.2018.8400912
23. V. S. Feng, S. Y. Chang, Determination of wireless networks parameters through parallel hierarchical support vector machines. IEEE Trans. Parallel Distrib. Syst. **23**(3), 505–12 (2012). https://doi.org/10.1109/TPDS.2011.156
24. C.-K. Wen, S. Jin, K.-K. Wong, J.-C. Chen, P. Ting, Channel estimation for massive MIMO using Gaussian-mixture Bayesian learning. IEEE Trans. Wirel. Commun. **14**(3), 1356–68 (2015). https://doi.org/10.1109/TWC.2014.2365813
25. C.-K. Yu, K.-C. Chen, S.-M. Cheng, Cognitive radio network tomography. IEEE Trans. Veh. Technol. **59**(4), 1980–97 (2010). https://doi.org/10.1109/TVT.2010.2044906
26. M. C. Raja, M. M. A. Rabbani, in *2016 International Conference on Communication and Electronics Systems (ICCES)*. Combined analysis of support vector machine and principle component analysis for IDS, (Coimbatore, 2016), pp. 1–5. https://doi.org/10.1109/CESYS.2016.7889868
27. Z. Luo, C. Li, L. Zhu, Full-duplex cognitive radio using guided independent component analysis and cumulant criterion. IEEE Access. **7**, 27065–27074 (2019). https://doi.org/10.1109/ACCESS.2019.2901815
28. M. Xia, Y. Owada, M. Inoue, H. Harai, Optical and wireless hybrid access networks: design and optimization. IEEE/OSA J. Opt. Commun. Netw. **4**(10), 749–59 (2012). https://doi.org/10.1364/JOCN.4.000749
29. R. C. Qiu, Z. Hu, Z. Chen, N. Guo, R. Ranganathan, S. Hou, G. Zheng, Cognitive radio network for the smart grid: experimental system architecture, control algorithms, security, and micro grid testbed. IEEE Trans. Smart Grid. **2**(4), 724–40 (2011). https://doi.org/10.1109/TSG.2011.2160101
30. R. Li, Z. Zhao, X. Zhou, G. Ding, Y. Chen, Z. Wang, H. Zhang, Intelligent 5G: when cellular networks meet artificial intelligence. IEEE Wirel. Commun. **24**(5), 175–183 (2017). https://doi.org/10.1109/MWC.2017.1600304WC
31. Y. Li, B. Yin, H. Xi, Partially observable Markov decision processes and performance sensitivity analysis. IEEE Trans. Syst Man Cybern. Part B Cybern. **38**(6), 1645–1651 (2008). https://doi.org/10.1109/TSMCB.2008.927711
32. A. Aprem, C. R. Murthy, N. B. Mehta, Transmit power control policies for energy harvesting sensors with retransmissions. IEEE J. Sel. Top. Signal Process. **7**(5), 895–906 (2013). https://doi.org/10.1109/JSTSP.2013.2258656
33. G. Alnwaimi, S. Vahid, K. Moessner, Dynamic heterogeneous learning games for opportunistic access in LTE-based macro/femtocell deployments. IEEE Trans. Wirel. Commun. **14**(4), 2294–2308 (2015). https://doi.org/10.1109/TWC.2014.2384510
34. R. Ali, N. Shahin, Y. T. Kim, B. S. Kim, S. W. Kim, Channel observation-based scaled backoff mechanism for high-efficiency WLANs. Electron. Lett. **54**(10), 663–665 (2018). https://doi.org/10.1049/el.2018.0617
35. The network simulator-ns-3. https://www.nsnam.org/. Accessed 01 Sept 2018

16

On the performance of receiver strategies for cooperative relaying cellular networks with NOMA

Jinjuan Ju[1,2], Guoan Zhang[1*], Qiang Sun[1,3], Li Jin[1] and Wei Duan[1]

Abstract

In this paper, a receiver scheme for a cooperative multi-relay system with non-orthogonal multiple access (CMRS-NOMA) in cellular networks is proposed. In our proposed system, the base station (BS) would like to communicate with multiple users via multiple relays, and direct links between the BS and users are not considered. Specifically, we assume that the relays in the approaching range are allowed to assist the transmission since other relays are out of the cooperation range. In this way, each user and the corresponding relays can be implemented as one group. In addition, the source simultaneously transmits the superposition coded signals to all relay nodes. After receiving the signals, all the relay nodes decode the symbols by applying successive interference cancellation (SIC) and then reconstruct them into new NOMA signals, which will be forwarded to the users and decoded in a linear combination way. Note that the proposed reconstructions at the relay nodes are practical and simple, which leads to an advantageous decoding for the receivers. Moreover, the closed-form expressions in terms of the sum rate (SR) and the outage performance of the system are derived. Qualitative numerical results corroborating our theoretical analysis show that the proposed scheme significantly improves the performance in terms of the SR and outage probability compared with the existing works.

Keywords: Cooperative multi-relay system (CMRS), Non-orthogonal multiple access (NOMA), Sum rate (SR), Outage performance

1 Introduction

Recently, non-orthogonal multiple access (NOMA) has received considerable attention as a promising candidate for 5G systems, to realize an aggregate goal, such as superior spectral efficiency, balanced user fairness, intense connections, and low access latency [1, 2]. In contrast to the traditional orthogonal multiple access (OMA) scheme and recent studies [3–6], the NOMA technique allows multiple users to share time/frequency radio resources and allows distinguishing users by the different power levels [7–9]. In addition, a cooperative relaying system with the NOMA technique (i.e., CRS-NOMA) is proposed in [2], which not only improves spectral efficiency but also reduces the system complexity based on some special approaches. To further improve the system performance in CRS-NOMA, the authors in [10] investigated a novel receiver design, where the performance in terms of the SR and outage probability are significantly improved. Since the distributions have an important impact, a new CRS-NOMA protocol is proposed in [11], in which the applications of simultaneous wireless information and power transfer (SWIPT) to NOMA networks are considered. Moreover, the works focusing on the power allocation schemes and different channel models, such as Rician fading channels and Nakagami-m fading channels in the NOMA technology, have also been studied in [12–14].

Due to the various advantages promised by NOMA technology in existing and future wireless communications systems, NOMA technology has been shown to be compatible with many other potential 5G techniques, such as massive multiple-input multiple-output (MIMO) and millimeter-wave (mmWave), as well as some applications in Internet of Things (IoT), such as vehicle-to-vehicle (V2V) communication scenarios and

*Correspondence: gzhang@ntu.edu.cn
[1]School of Information Science and Technology, Nantong University, Seyuan Road, Nantong, China
Full list of author information is available at the end of the article

machine-type communications (MTC). As an effective means to improve spectral efficiency, MIMO has been widely studied [15–18]. In [19–22], various algorithmic frameworks and performance optimization problems in MIMO-NOMA scenarios were also studied. However, most of the existing NOMA schemes rely on a key assumption that the channel conditions for the users are heterogeneous. To ensure that the potential of NOMA technology can be realized even if the users' channel conditions are similar, a MIMO-NOMA downlink transmission scenario has been considered, with one transmitter sending data to two users, e.g., an access point is serving two IoT devices [23]. Particularly, the authors considered that one user needs to be served quickly for small packet transmission, i.e., with a low targeted data rate, and the other user is to be served with the best effort. However, due to the scarcity of traditional wireless communication frequency resources, mmWave is slowly becoming popular because it has more bandwidth resources in the high-frequency band [24]. The authors in [25] investigated spatial and frequency wideband effects, called dual-wideband effects in massive MIMO systems, from the array signal processing perspective. By taking millimeter-wave-band communications into account, they described the transmission process to address the dual-wideband effects. In [26], to avoid the requirement that the base station (BS) should know all the channel state information (CSI) of the users, the applications of random beamforming to mmWave-NOMA systems were considered. In the application scenarios of IoT, such as V2V communications, since multiple antennas can be accommodated, large-scale MIMO becomes quite attractive [27, 28]. Inspired by the robustness of spatial modulation (SM) against channel correlation and the benefits of NOMA technology, the authors in [29] combined them into NOMA-SM to handle the effects of wireless V2V environments and to support the improvement in bandwidth efficiency.

The studies of NOMA technology and its compatibility with other technologies used in certain scenarios have clearly received considerable attention. For a scenario with multiple receiving nodes, both the transmission strategies and performance analysis are more practical and challenging for wireless communication systems. In addition, to the best of our knowledge, there are no relevant studies that focus on cellular networks with a multi-relay node transmission model with NOMA technology. Although the NOMA systems over Rician fading channels and/or Nakagami-m fading channels are more interesting and challenging, they are beyond the scope of this paper. Based on the above considerations, in this paper, a new transmission model named cooperative multi-relay system using NOMA (CMRS-NOMA) technology over Rayleigh fading channels is studied. To jointly decode the desired signals for each user, the new superposition coded signals are reconstructed and broadcast by the multiple relay nodes. For assisting the multiple relay nodes in completing the transmission of the new superposition signals, a simple and practical decoding scheme with high efficiency and low complexity is investigated.

The implementations and main contributions of our proposed work are summarized as follows:

- A new CMRS-NOMA transmission scenario is proposed, in which the relays within the cooperation range are allowed to assist the transmissions from the BS to the users. Meanwhile, the direct transmission channels are not considered. After receiving signals, the relay nodes reconstruct the received signals into a new superposition code in a practical way, and the coded signal will be simply decoded with linear combinations at the receiver. Moreover, we also introduce a two-stage NOMA power allocation scheme at the BS and relay nodes, which is interesting and challenging.
- We investigate a new decoding method that can be described as a simple algorithm at each user after the relay nodes forward the reconstructed signals from the BS. Thanks to the reconstructions at the relay, each user can decode the desired signals with linear combinations, where the signals are forwarded from the corresponding relays. In addition, we qualitatively analyze a low-complexity, multi-stage power allocation strategy.
- The performance of the cooperative multi-relay NOMA technology in cellular networks is analyzed in terms of the ergodic SR and outage probability. Considering independent Rayleigh fading channels, we first derive the closed-form expressions of the ergodic SR and outage probability for a single-relay case. By extending the single-relay case into the remaining multi-relay node pair cases, a generalized reconstruction strategy is investigated, where all relay nodes follow the same regularity feature. For the remaining multi-relay node pair cases, in this manner, the power allocation factors of the proposed CMRS-NOMA should be changed in a circular manner, which is completely different from the traditional single-relay cooperative networks.
- Compared to the TDMA and CRS-NOMA schemes, the proposed CMRS-NOMA has better performances in terms of outage probability and spectral efficiency in both the analytical and numerical results. It is also observed that NOMA technology offers better fairness when multiple users' quality of services (QoS) are satisfied.

The remainder of this paper is organized as follows. In Section 2, the proposed new system model, power allocation strategies, and decoding scheme of multi-relay cooperative NOMA for cellular networks are described. In Section 3, we analyze the performance in terms of the ergodic SR, outage probability, and some auxiliary indicators. In addition, closed-form expressions for the ergodic SR and outage performance are derived. In Section 4, we present and discuss the simulation results for our proposed works. This paper is concluded in Section 5.

2 System model and the proposed scheme

Consider a cooperative relaying communication system that consists of one source S; multiple relay nodes, i.e., $R_1 \ldots R_M$; and multiple users, $D_1 \ldots D_N$, as shown in Fig. 1. We further consider that the BS wants to communicate with the users via the relay nodes since there are no direct links between them. In addition, the relays in the approaching range are allowed to assist the transmission, while the other relays are beyond the transmission range. It is assumed that all nodes are operating in a half-duplex mode and that each one is equipped with a single antenna. The channels between S-R_i and R_i-D_j are denoted as h_{SR_i} and $h_{R_iD_j}$, respectively, which are assumed to be Rayleigh fading channels with variances β_{SR_i} and $\beta_{R_iD_j}$. Furthermore, the CSI is assumed to be perfectly known to each receiving node.

There are two phases involved in our proposed scheme. In the first phase, the source broadcasts N symbols (s_1, s_2, $\cdots$, s_N) simultaneously to the relays. By adopting the superposition code, the transmitted signal from the source is given as $\sum_{k=1}^{N} \sqrt{a_k P_t} s_k$, where s_k is the symbol for the kth ($1 \leq k \leq N$) user, a_k with $\sum_{k=1}^{N} a_k = 1$ denotes the power allocation coefficient, and P_t is the total transmit power. Similar to the traditional NOMA technology, the power coefficients are determined by the qualities of the S-R_i channels. The observation at the ith($1 \leq i \leq M, M = N$) relay in multiple relay nodes is given by

$$y_{R_i} = h_{SR_i} \sum_{k=1}^{N} \sqrt{a_k P_t} s_k + n_{R_i}, \tag{1}$$

where $n_{R_i} \sim \mathcal{CN}(0, \sigma_{R_i}^2)$ are the additive white Gaussian noises (AWGNs). Without loss of generality, following the NOMA decoding principle, $s_j (1 \leq j \leq N-1)$ is decoded first and allocated with more transmit power, and then s_{j+1} is subsequently decoded. Successive detection will be performed at the ith relay at the end of this phase. Because

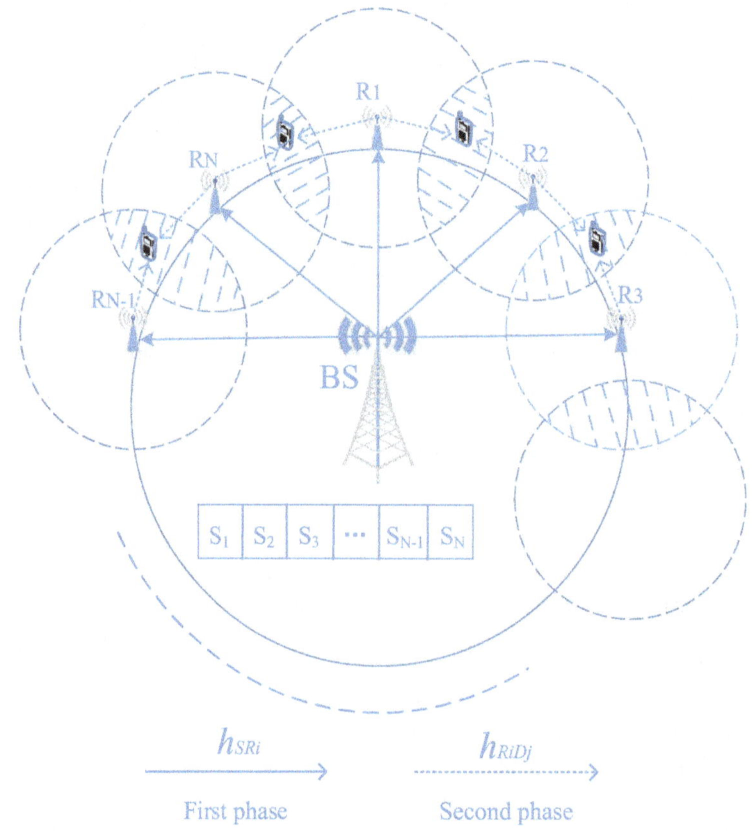

Fig. 1 The proposed multi-relay cooperative system with NOMA (MCRS-NOMA)

of the cellular distribution, all the S-R_i channels are spatially disparate and have distinct transmission qualities at this stage, i.e.,

$$h_{SR_1} \neq h_{SR_2} \neq \cdots \neq h_{SR_N}, \quad (2)$$

In the second phase, with the superposition coding (SC) and successive interference cancellation (SIC), all the relay nodes are available to successively decode the symbols. In this manner, the received signal-to-interference-plus-noise ratio (SINR) at the ith relay node is given by

$$\gamma_{R_i}^{(s_k)} = \frac{\left|h_{SR_i}\right|^2 a_k P_t}{\left|h_{SR_i}\right|^2 \left(\sum\limits_{m=k+1}^{N} a_m\right) P_t + \sigma^2}, \quad (3)$$

and

$$\gamma_{R_i}^{(s_N)} = \frac{\left|h_{SR_N}\right|^2 a_N P_t}{\sigma^2}. \quad (4)$$

To provide a simple and practical decoding, the relays will reconstruct and forward new superposition coded signals, and the user jointly decodes the information symbols from the received signals by employing the linear combination. In fact, only the signals that come from the approaching relays need to be provided for decoding the desired signal. Indeed, when transmitting the reconstructed signals, the power allocation strategy at the relay needs to be designed, which means that a power allocation is still required at each node. As shown in Fig. 1, there are N relay nodes, and the corresponding power allocation coefficients b_{R_i} with $\sum_{i=1}^{N} b_{R_i} = 1$ are considered in the system. At this time, considering the power reallocation at each relay node for transmission, third stage power allocation will occur with coefficients c_k with $\sum_{k=1}^{N} c_k = 1$, which will make the system analysis complex and not suitable for research. For simplicity, the information transmitted at each relay node will no longer be allocated in the relay layer but rather sent out with full power.

2.1 Single-relay scenario

To successfully analyze the multi-relay case, we first consider a single-relay case, which is a simple CRS consisting of only one source, one relay, and one destination. Assume that a direct link between the source and the destination exists, the destination jointly decodes the received signals from both the source and the relay.

At the first time slot, the BS simultaneously transmits the signal $\sqrt{a_1 P_t} s_1 + \sqrt{a_2 P_t} s_2$ to the relay and destination, where s_k denotes the broadcast symbol at the source and a_1 and a_2 with $a_1 + a_2 = 1$ are the power allocation factors. At the second time slot, a new symbol s_r with superposition coding $s_r = \sqrt{b_1 P_t} s_1 - \sqrt{b_2 P_t} s_2$ is forwarded from the relay node to the destination, where b_1 and b_2 with $b_1 + b_2 = 1$ are new power allocation coefficients. Note that the subtraction is necessary, and it can perfectly cancel the interference signals at the destination. The received signals at the destination can be written as

$$y_D^{(1)} = h_{SD}\left(\sqrt{a_1 P_t} s_1 + \sqrt{a_2 P_t} s_2\right) + n_D^{(1)}, \quad (5)$$

$$y_D^{(2)} = h_{RD}\left(\sqrt{b_1 P_t} s_1 - \sqrt{b_2 P_t} s_2\right) + n_D^{(2)}, \quad (6)$$

where $n_D^{(i)}$ is the AWGN at the destination during the ith time slot with zero mean and variance σ^2. To jointly decode s_1 and s_2 at the destination, a linear combination is employed to produce interference-free signals $y_D^{(1)}\sqrt{b_2} h_{RD} + y_D^{(2)}\sqrt{a_2} h_{SD}$ and $y_D^{(1)}\sqrt{b_1} h_{RD} - y_D^{(2)}\sqrt{a_1} h_{SD}$. Therefore, the target signals can be expressed as

$$T_{s_1} = h_{SD} h_{RD} \zeta \sqrt{P_t} s_1 + \sqrt{b_2} h_{RD} n_D^{(1)} + \sqrt{a_2} h_{SD} n_D^{(2)}, \quad (7)$$

and

$$T_{s_2} = h_{SD} h_{RD} \zeta \sqrt{P_t} s_2 + \sqrt{b_1} h_{RD} n_D^{(1)} - \sqrt{a_1} h_{SD} n_D^{(2)}, \quad (8)$$

where $\zeta = \sqrt{a_1 b_2} + \sqrt{a_2 b_1}$. With (7) and (8), the corresponding effective SNRs for s_1 and s_2 are given by

$$\gamma_D^{(s_1)} = \frac{\left|h_{SD}\right|^2 \left|h_{RD}\right|^2 \zeta^2 \rho}{b_2 \left|h_{RD}\right|^2 + a_2 \left|h_{SR}\right|^2}, \quad (9)$$

and

$$\gamma_D^{(s_2)} = \frac{\left|h_{SD}\right|^2 \left|h_{RD}\right|^2 \zeta^2 \rho}{b_1 \left|h_{RD}\right|^2 + a_1 \left|h_{SR}\right|^2}, \quad (10)$$

respectively, where $\rho = \frac{P_t}{\sigma^2}$ denotes the transmit SNR.

2.2 Proposed multi-relay scheme

In this subsection, we extend the single-relay case into the case with multiple users and relays, i.e., CMRS-NOMA. Note that direct links between the BS and the users are not considered since such consideration will lead to extreme complexity for the power allocation. Moreover, in this study, our aim is to demonstrate that our proposed reconstruction at the relay node has advantages compared with the directly forwarding one.

In this scheme, in the first phase, the source broadcasts N symbols $\sum_{k=1}^{N} \sqrt{a_k P_t} s_k$ only to the relay nodes and the received signal at the ith relay in multiple relay nodes, as shown in Eq. (1). By employing SIC, all the relay nodes can decode the symbols successively. In the second phase, the relay nodes reconstruct new superposition coded signals by using linear operations. In contrast to the single-relay case, there are multiple relay nodes, where different relay nodes will reconstruct in a different way by using different linear operations and forward them to the user in a short-range circular communication area. In this process, a two-stage NOMA power allocation scheme at the BS and the relay nodes is introduced. The choice of decoding method at each user after the relay nodes forward new NOMA signals directly affects whether the interference signals are perfectly cancelled. Because the relays in the approaching range are allowed to assist the transmission, each user and the corresponding relay nodes can be implemented as one group. By using simple addition and division, all the other signals except the desired one will be removed.

Our proposed linear operations at the relay nodes can be described as a simple algorithm. The detailed steps of the linear algorithm are as follows:

- *Step 1*: We first choose a relay in the direction of 90° of the coordinate axis and mark it as the number 1 relay; then, along the clockwise direction, we mark the remaining relays as 2,3... until the last relay is marked as N. Assume that multiple relay nodes are distributed in a cellular around the BS with equal distances from the relays to the BS and that the last relay node is connected with the first one.
- *Step 2*: To handle the operational demands in the distributed relay nodes, we place all the adjacent relay node pairs into a node pair list (NPL) and identify the corresponding users within the overlapping area of the relay node pairs at the destinations that are outside of the BS service area. The NPL and CU of step 2 are given in Table 1.
- *Step 3*: For the node pairs that have been served, we obtain each new symbol $s_{R_i} (1 \leq i \leq N)$ from a simple linear combination in sequence and forward them to the area with the full power of a single antenna. The reconstructed signal combinations for each relay are shown in the following group.

$$\begin{cases} s_{R_1} = \sqrt{b_{R_1} a_1 P_t} s_1 + \sqrt{b_{R_1} a_2 P_t} s_2 + \sqrt{b_{R_1} a_3 P_t} s_3 + \cdots + \sqrt{b_{R_1} a_N P_t} s_N & (a) \\ s_{R_2} = \sqrt{b_{R_2} a_1 P_t} s_1 - \sqrt{b_{R_2} a_2 P_t} s_2 - \sqrt{b_{R_2} a_3 P_t} s_3 - \cdots - \sqrt{b_{R_2} a_N P_t} s_N & (b) \\ s_{R_3} = \sqrt{b_{R_3} a_1 P_t} s_1 + \sqrt{b_{R_3} a_2 P_t} s_2 - \sqrt{b_{R_3} a_3 P_t} s_3 - \cdots - \sqrt{b_{R_3} a_N P_t} s_N & (c) \\ s_{R_4} = \sqrt{b_{R_4} a_1 P_t} s_1 + \sqrt{b_{R_4} a_2 P_t} s_2 + \sqrt{b_{R_4} a_3 P_t} s_3 - \cdots - \sqrt{b_{R_4} a_N P_t} s_N & (d) \\ \vdots & \\ s_{R_N} = \sqrt{b_{R_N} a_1 P_t} s_1 + \sqrt{b_{R_N} a_2 P_t} s_2 + \sqrt{b_{R_N} a_3 P_t} s_3 + \cdots - \sqrt{b_{R_N} a_N P_t} s_N & (\cdot) \end{cases} \tag{11}$$

- *Step 4*: After we finish reconstructing the new signals for each relay and send them out, we can jointly decode the received signals from two adjacent relays by employing a simple linear combination method at the users. As shown in the above expressions, for example, signal 1 is obtained by adding $\left(s_{R_1} \cdot \sqrt{b_{R_2}}\right)$ to $\left(s_{R_2} \cdot \sqrt{b_{R_1}}\right)$ and then dividing by $\left(2 \cdot \sqrt{b_{R_1} b_{R_2} a_1 P_t}\right)$. Signal 2 is achieved by subtracting $\left(s_{R_2} \cdot \sqrt{b_{R_3}}\right)$ from $\left(s_{R_3} \cdot \sqrt{b_{R_2}}\right)$ and then dividing by $\left(2 \cdot \sqrt{b_{R_2} b_{R_3} a_2 P_t}\right)$. The remainder can be achieved in the same manner, and the last signal N will be acquired by also deducting the end equation, i.e., $\left(s_{R_N} \cdot \sqrt{b_{R_1}}\right)$, from the first, i.e., $\left(s_{R_1} \cdot \sqrt{b_{R_N}}\right)$, and then dividing by $\left(2 \cdot \sqrt{b_{R_N} b_{R_1} a_N P_t}\right)$. Afterwords, all the signals will have been successfully decoded.

Table 1 RNP and corresponding user

Relay node pair	Corresponding user
(R_1, R_2)	$(D_1, user_1)$
(R_2, R_3)	$(D_2, user_2)$
...	...
(R_{N-1}, R_N)	$(D_{N-1}, user_{N-1})$
(R_N, R_1)	$(D_N, user_N)$

Taking signal 1 as an example, we analyze the received signal of $user_1$ that is cooperated by the number 1 and number 2 relays. Because a_k and b_{R_i} in group (11) are power allocation factors that meet their respective conditions, i.e., $\sum_{k=1}^{N} a_k = 1$ and $\sum_{i=1}^{N} b_{R_i} = 1$, group (11) can be simplified by allowing $c_{ik} = b_{R_i} \times a_k$, and formulae (11-a) and (11-b) can be described as

$$s_{R_1} = \sqrt{c_{11} P_t} s_1 + \sqrt{c_{12} P_t} s_2 + \cdots + \sqrt{c_{1(N-1)} P_t} s_{N-1} + \sqrt{c_{1N} P_t} s_N, \tag{12}$$

and

$$s_{R_2} = \sqrt{c_{21} P_t} s_1 - \sqrt{c_{22} P_t} s_2 - \cdots - \sqrt{c_{2(N-1)} P_t} s_{N-1} - \sqrt{c_{2N} P_t} s_N. \tag{13}$$

In our proposed system, the relay nodes forward new reconstructed symbols s_{R_i} with superposition signals to users as shown in the above formulae. Note that signals from two adjacent relay nodes only need to be processed simply to obtain the desired signal. With (12) and (13), the received signal at $user_1$ can be written as

$$\begin{aligned} y_{D_1}^{(R_1)} &= h_{R_1 D_1} \times s_{R_1} + n_{D_1}^{(R_1)} \\ &= h_{R_1 D_1} \times \left(\sqrt{c_{11} P_t} s_1 + \sum_{k=2}^{N} \sqrt{c_{1k} P_t} s_k \right) + n_{D_1}^{(R_1)}, \end{aligned} \tag{14}$$

$$\begin{aligned} y_{D_1}^{(R_2)} &= h_{R_2 D_1} \times s_{R_2} + n_{D_1}^{(R_2)} \\ &= h_{R_2 D_1} \times \left(\sqrt{c_{21} P_t} s_1 - \sum_{k=2}^{N} \sqrt{c_{2k} P_t} s_k \right) + n_{D_1}^{(R_2)}, \end{aligned} \tag{15}$$

where $n_{D_1}^{(R_i)}$ is the AWGN at $user_1$ from the ith relay node during the second time slot with zero mean and variance σ^2. From the above, we find that there are two signals received at $user_1$, which are $y_{D_1}^{(R_1)}$ and $y_{D_1}^{(R_2)}$. As the single-relay model scheme, $user_1$ jointly decodes s_1. With a simple linear combination, we could produce interference-free signal $y_{D_1}^{(R_1)}\sqrt{b_{R_2}}h_{R_2D_1}+y_{D_1}^{(R_2)}\sqrt{b_{R_1}}h_{R_1D_1}$ for s_1 at user$_1$, with the target signal T_{s_1} as

$$T_{s_1}=h_{R_1D_1}h_{R_2D_1}\varsigma\sqrt{P_t}s_1+\sqrt{b_{R_2}}h_{R_2D_1}n_{D_1}^{(R_1)}+\sqrt{b_{R_1}}h_{R_1D_1}n_{D_1}^{(R_2)}, \tag{16}$$

where $\varsigma=\sqrt{c_{11}b_{R_2}}+\sqrt{c_{21}b_{R_1}}=2\sqrt{b_{R_1}b_{R_2}a_1}$. If we perform symbol-by-symbol detection based on the output signals (16), the corresponding effective SNR for s_1 is given by

$$\gamma_{D_1}^{(s_1)}=\frac{\left|h_{R_1D_1}\right|^2\left|h_{R_2D_1}\right|^2\varsigma^2\rho}{b_{R_2}\left|h_{R_2D_1}\right|^2+b_{R_1}\left|h_{R_1D_1}\right|^2}, \tag{17}$$

where $\rho=\frac{P_t}{\sigma^2}$ denotes the transmit SNR.

Clearly, signal 2 to signal $N-1$ can be obtained in a similar manner, i.e., for $s_k\,(2\le k\le N-1)$ at user i, the interference-free signals $y_{D_i}^{(R_{i+1})}\sqrt{b_{R_i}}h_{R_iD_i}-y_{D_i}^{(R_i)}\sqrt{b_{R_{i+1}}}h_{R_{i+1}D_i}$ could also be generated by employing the same method. As described in step 4 above, the interference-free signal for s_N at user N (the last signal N) will be generated by $y_{D_N}^{(R_1)}\sqrt{b_{R_N}}h_{R_ND_N}-y_{D_N}^{(R_N)}\sqrt{b_{R_1}}h_{R_1D_N}$. Similar to the previous descriptions, the target signal $T_{s_i}(2\le i\le N-1)$ can be expressed as

$$T_{s_i}=h_{R_iD_i}h_{R_{i+1}D_i}\varsigma\sqrt{P_t}s_i+\sqrt{b_{R_{i+1}}}h_{R_{i+1}D_i}n_{D_i}^{(R_i)}+\sqrt{b_{R_i}}h_{R_iD_i}n_{D_i}^{(R_{i+1})}, \tag{18}$$

where $\varsigma=2\sqrt{b_{R_i}b_{R_{i+1}}a_i}$; hence, the expressions of interference-free signal for $T_{s_i}(1\le i\le N-1)$ are the same. However, the last one is slightly different from the ones in front; it is

$$T_{s_N}=h_{R_1D_N}h_{R_ND_N}\varsigma\sqrt{P_t}s_N+\sqrt{b_{R_N}}h_{R_ND_1}n_{D_N}^{(R_1)}+\sqrt{b_{R_1}}h_{R_1D_N}n_{D_N}^{(R_N)}, \tag{19}$$

where $\varsigma=2\sqrt{b_{R_N}b_{R_1}a_N}$. Therefore, the corresponding effective SNRs for $s_i(1\le i\le N-1)$ and s_N are given as

$$\gamma_{D_i}^{(s_i)}=\frac{4\left|h_{R_{i+1}D_i}\right|^2\left|h_{R_iD_i}\right|^2(b_{R_i}b_{R_{i+1}}a_i)\rho}{b_{R_{i+1}}\left|h_{R_{i+1}D_i}\right|^2+b_{R_i}\left|h_{R_iD_i}\right|^2}, \tag{20}$$

and

$$\gamma_{D_N}^{(s_N)}=\frac{4\left|h_{R_1D_N}\right|^2\left|h_{R_ND_N}\right|^2(b_{R_N}b_{R_1}a_N)\rho}{b_{R_N}\left|h_{R_ND_1}\right|^2+b_{R_1}\left|h_{R_1D_N}\right|^2}, \tag{21}$$

where $\rho=\frac{P_t}{\sigma^2}$ denotes the transmit SNR.

3 Performance analysis

In this section, we will analyze the performance of the achievable ergodic SR and the outage probability to characterize the superiority of our proposed scheme.

3.1 Ergodic SR analysis

According to Shannon's theorem, the achievable rate for symbol s_1 can be obtained from $\gamma_{R_1}^{(s_1)}$, $\gamma_{R_2}^{(s_1)}$, and $\gamma_{D}^{(s_1)}$ as

$$C_{s_1}=\frac{1}{2}\log_2\left(1+\min\left\{\gamma_{R_1}^{(s_1)},\gamma_{R_2}^{(s_1)},\gamma_{D_1}^{(s_1)}\right\}\right), \tag{22}$$

where $\frac{1}{2}$ results from the two-hop transmission. Without loss of generality, the received signals corresponding to the relay node pairs can be described as

$$C_{s_i}=\frac{1}{2}\log_2\left(1+\min\left\{\gamma_{R_i}^{(s_i)},\gamma_{R_{i+1}}^{(s_i)},\gamma_{D_i}^{(s_i)}\right\}\right). \tag{23}$$

Therefore, the achievable SR can be expressed as

$$C_{sum}=\sum_{i=1}^{N}C_{s_i}. \tag{24}$$

Denoting $\left|h_{SR_i}\right|^2=\nu_{SR_i}$, $\left|h_{R_iD_j}\right|^2=\nu_{R_iD_j}$, $d_i=b_{R_i}b_{R_{i+1}}a_i$, and $\tau=\sum_{m=i+1}^{N}a_m$, we have

$$\begin{aligned}
&\min\left\{\gamma_{R_i}^{(s_i)},\gamma_{R_{i+1}}^{(s_i)},\gamma_{D_i}^{(s_i)}\right\}\\
&=\min\left\{\frac{\left|h_{SR_i}\right|^2a_i\rho}{\left|h_{SR_i}\right|^2\tau\rho+1},\frac{\left|h_{SR_{i+1}}\right|^2a_i\rho}{\left|h_{SR_{i+1}}\right|^2\tau\rho+1},\frac{4\left|h_{R_iD_i}\right|^2\left|h_{R_{i+1}D_i}\right|^2d_i\rho}{b_{R_{i+1}}\left|h_{R_{i+1}D_i}\right|^2+b_{R_i}\left|h_{R_iD_i}\right|^2}\right\}\\
&=\min\left\{\frac{\nu_{SR_i}a_i\rho}{\nu_{SR_i}\tau\rho+1},\frac{\nu_{SR_{i+1}}a_i\rho}{\nu_{SR_{i+1}}\tau\rho+1},\frac{4\nu_{R_iD_i}\nu_{R_{i+1}D_i}d_i\rho}{b_{R_{i+1}}\nu_{R_{i+1}D_i}+b_{R_i}\nu_{R_iD_i}}\right\},
\end{aligned} \tag{25}$$

where $i\in[1,N-1]$ and $i\neq m$. For the case of $i=N$, the achievable rate C_{s_N} can be written as

$$C_{s_N}=\frac{1}{2}\log_2\left(1+\min\left\{\gamma_{R_N}^{(s_N)},\gamma_{R_1}^{(s_N)},\gamma_{D_N}^{(s_N)}\right\}\right). \tag{26}$$

According to (21), $d_N=b_{R_N}b_{R_1}a_N$, for $i=N$, we have

$$\begin{aligned}
&\min\left\{\gamma_{R_N}^{(s_N)},\gamma_{R_1}^{(s_N)},\gamma_{D_N}^{(s_N)}\right\}\\
&=\min\left\{\left|h_{SR_N}\right|^2a_N\rho,\ \left|h_{SR_1}\right|^2a_N\rho,\ \frac{4\left|h_{R_ND_N}\right|^2\left|h_{R_1D_N}\right|^2d_N\rho}{b_{R_1}\left|h_{R_1D_N}\right|^2+b_{R_N}\left|h_{R_ND_N}\right|^2}\right\}\\
&=\min\left\{\nu_{SR_N}a_N\rho,\ \ \nu_{SR_1}a_N\rho,\ \ \frac{4\nu_{R_ND_N}\nu_{R_1D_N}d_N\rho}{b_{R_1}\nu_{R_1D_N}+b_{R_N}\nu_{R_ND_N}}\right\}.
\end{aligned} \tag{27}$$

Letting $X=\min\{\gamma_{R_i}^{(s_i)},\gamma_{R_{i+1}}^{(s_i)},\gamma_{D_i}^{(s_i)}\}$, from (25) in which $i\neq N$, the complementary cumulative distribution function(CCDF) of X can be formulated as

$$\overline{F}_X(x)=Pr\left\{\frac{\nu_{SR_i}a_i\rho}{\nu_{SR_i}\tau\rho+1}>x,\ \frac{\nu_{SR_{i+1}}a_i\rho}{\nu_{SR_{i+1}}\tau\rho+1}>x,\ \frac{4\nu_{R_iD_i}\nu_{R_{i+1}D_i}d_i\rho}{b_{R_{i+1}}\nu_{R_{i+1}D_i}+b_{R_i}\nu_{R_iD_i}}>x\right\}. \tag{28}$$

Note that the CCDF of $\overline{F}_{\nu(\delta)}=e^{-\frac{x}{\beta_\delta}}$, for $\delta\in\{SR_i,R_iD_j\}$, Eq. (28) is equivalent to (29), as shown at the top of the next page, where $Pr\{A|B\}$ denotes the conditional probability of event A on event B, and $E[\cdot]$ stands for statistical expectation.

$$\begin{aligned}
\overline{F}_X(x) &= \overline{F}_{SR_i}\left(\frac{x}{a_i\rho-\tau\rho x}\right)\overline{F}_{SR_{i+1}}\left(\frac{x}{a_i\rho-\tau\rho x}\right) \\
&\quad\left(E_{\nu_{R_{i+1}D_i}}\left[Pr\left\{\nu_{R_iD_i} > \frac{b_{R_{i+1}}\nu_{R_{i+1}D_i}x}{4\nu_{R_{i+1}D_i}d_i\rho-b_{R_i}x}|\nu_{R_{i+1}D_i} > \frac{b_{R_i}x}{4d_i\rho}\right\}\right]\right. \\
&\quad\left.+E_{\nu_{R_{i+1}D_i}}\left[Pr\left\{\nu_{R_iD_i} < \frac{b_{R_{i+1}}\nu_{R_{i+1}D_i}x}{4\nu_{R_{i+1}D_i}d_i\rho-b_{R_i}x}|\nu_{R_{i+1}D_i} < \frac{b_{R_i}x}{4d_i\rho}\right\}\right]\right) \\
&= \left\{\frac{1}{\beta_{R_{i+1}D_i}}e^{-\frac{x}{[a_i\rho-\tau\rho x]\beta_{SR_i}}-\frac{x}{[a_i\rho-\tau\rho x]\beta_{SR_{i+1}}}-\frac{b_{R_i}x}{4d_i\rho\beta_{R_{i+1}D_i}}}\right. \\
&\quad\left.\int_{\frac{b_{R_i}x}{4d_i\rho}}^{\infty} e^{-\frac{b_{R_{i+1}}ux}{(4d_i\rho u-b_{R_i}x)\beta_{R_iD_i}}-\frac{u}{\beta_{R_{i+1}D_i}}}du\right\} \\
&\quad+\left\{e^{-\frac{x}{[a_i\rho-\tau\rho x]\beta_{SR_i}}-\frac{x}{[a_i\rho-\tau\rho x]\beta_{SR_{i+1}}}}\left(1-e^{-\frac{b_{R_i}x}{4d_i\rho\beta_{R_{i+1}D_i}}}\right)\right. \\
&\quad\left.\int_0^{\frac{b_{R_i}x}{4d_i\rho}} 1-\frac{1}{\beta_{R_{i+1}D_i}}e^{-\frac{b_{R_{i+1}}ux}{(4d_i\rho u-b_{R_i}x)\beta_{R_iD_i}}-\frac{u}{\beta_{R_{i+1}D_i}}}du\right\},
\end{aligned} \tag{29}$$

Since Eq. (29) is too complex to calculate, the high transmit SNR case is considered to simplify the calculation, i.e., $\rho >> 1$, and we have the following approximations

$$\frac{\nu_{SR_i}a_i\rho}{\nu_{SR_i}\tau\rho+1} \sim \frac{a_i}{\tau}, \tag{30}$$

and

$$\frac{\nu_{SR_{i+1}}a_i\rho}{\nu_{SR_{i+1}}\tau\rho+1} \sim \frac{a_i}{\tau}. \tag{31}$$

Letting $t = 4d_i\rho u - b_{R_i}x$ and using $\int_0^\infty e^{-\frac{A}{x}-Bx}dx = 2\sqrt{\frac{A}{B}}K_1(2\sqrt{AB})$ [30]—Eq. (3.324.1), when $x < \frac{a_i}{\tau}$ $\left(\tau = \sum_{m=i+1}^{N} a_m\right)$, (29) can be equivalently written as

$$\begin{aligned}
\overline{F}_X(x) &= \frac{e^{-\frac{x}{4d_i\rho}\left(\frac{b_{R_{i+1}}}{\beta_{R_iD_i}}+\frac{b_{R_i}}{\beta_{R_{i+1}D_i}}\right)}}{4\beta_{R_{i+1}D_i}d_i\rho}\int_0^\infty e^{-\frac{b_{R_i}b_{R_{i+1}}x^2}{4d_i\rho t\beta_{R_iD_i}}-\frac{t}{4d_i\rho\beta_{R_{i+1}D_i}}}dt \\
&= \frac{x}{\xi}e^{-x\phi}K_1\left(\frac{x}{\xi}\right),
\end{aligned} \tag{32}$$

where $\xi = 2d_i\rho\sqrt{\frac{\beta_{R_iD_i}\cdot\beta_{R_{i+1}D_i}}{b_{R_i}\cdot b_{R_{i+1}}}}$, $\phi = \frac{1}{4d_i\rho}\left(\frac{b_{R_{i+1}}}{\beta_{R_iD_i}}+\frac{b_{R_i}}{\beta_{R_{i+1}D_i}}\right)$, and $K_1(\cdot)$ denotes the first-order modified Bessel function of the second kind [31]. For the case $x > \frac{a_i}{\tau}$ $\left(\tau = \sum_{m=i+1}^{N} a_m\right)$, $\overline{F}_X(x) = 0$ always holds due to Eqs. (30) and (31). With (32), the achievable ergodic rate can be calculated as (33).

$$\begin{aligned}
\widetilde{C}_{s_i} &= \int_0^{\frac{a_i}{\tau}}\frac{1}{2}\log_2(1+x)dF_X(x)+\frac{1}{2}\log_2\left(1+\frac{a_i}{\tau}\right)\left(1-F_X\left(\frac{a_i}{\tau}\right)\right) \\
&= \frac{1}{2}\log_2\left(1+\frac{a_i}{\tau}\right)-\frac{1}{2\ln 2}\int_0^{\frac{a_i}{\tau}}\frac{1}{1+x}\left(1-\frac{x}{\xi}e^{-x\phi}K_1\left(\frac{x}{\xi}\right)\right)dx \\
&= \int_0^{\frac{a_i}{\tau}}\frac{1}{2\ln 2(1+x)}\frac{x}{\xi}e^{-x\phi}K_1\left(\frac{x}{\xi}\right)dx.
\end{aligned} \tag{33}$$

Using the equality of $\int_0^\infty \frac{1}{2}\log_2(1+x)f_X(x)dx = \frac{1}{2\ln 2}\int_0^\infty \frac{1-F_X(x)}{1+x}dx$ and $F_X(x) = 1-\overline{F}_X(x)$ and the approximation $K_\nu(x) \approx \frac{\Gamma(\nu)}{2}\left(\frac{2}{x}\right)^\nu$ [32]—Eq. (9.69) in the case of small x, where $\Gamma(\cdot)$ denotes the gamma function, (33) can be approximately rewritten as

$$\begin{aligned}
\widetilde{C}_{s_i} &\sim \frac{1}{2\ln 2}\int_0^{\frac{a_i}{\tau}}\frac{1}{1+x}e^{-x\phi}dx \\
&= \frac{e^\phi}{2\ln 2}\left[Ei\left(-\phi\frac{a_i}{\tau}-\phi\right)-Ei\left(-\phi\right)\right],
\end{aligned} \tag{34}$$

where $\int_0^u \frac{e^{-\mu x}dx}{x+\psi} = e^{\mu\psi}[Ei(\mu u-\mu\psi)-Ei(-\mu\psi)]$ [31]—Eq. (3.352.1) is used, and $Ei(\cdot)$ denotes the exponential integral function. The approximate closed expression obtained from the above analysis only satisfies the case where $i \in [1, N-1]$.

By using a similar analysis method, letting $Y = \min\left\{\gamma_{R_N}^{(s_N)}, \gamma_{R_1}^{(s_N)}, \gamma_{D_N}^{(s_N)}\right\}$, from (27) in which $i = N$, the CCDF of Y can be formulated as

$$\begin{aligned}
\overline{F}_Y(y) &= Pr\left\{\nu_{SR_N}a_N\rho > y,\ \nu_{SR_1}a_N\rho > y,\ \frac{4\nu_{R_ND_N}\nu_{R_1D_N}d_N\rho}{b_{R_1}\nu_{R_1D_N}+b_{R_N}\nu_{R_ND_N}} > y\right\} \\
&= \overline{F}_Y^{(1)}(y)\times\overline{F}_Y^{(2)}(y)\times\overline{F}_Y^{(3)}(y) \\
&= e^{-\frac{y}{a_N\beta_{SR_N}\rho}}\times e^{-\frac{y}{a_N\beta_{SR_1}\rho}}\times\overline{F}_Y^{(3)}(y),
\end{aligned} \tag{35}$$

where $\overline{F}_Y^{(1)}(y) = Pr\{\nu_{SR_N}a_N\rho > y\}$, $\overline{F}_Y^{(2)}(y) = Pr\{\nu_{SR_1}a_N\rho > y\}$ and $\overline{F}_Y^{(3)}(y) = Pr\{\frac{4\nu_{R_ND_N}\nu_{R_1D_N}d_N\rho}{b_{R_1}\nu_{R_1D_N}+b_{R_N}\nu_{R_ND_N}} > y\}$. Letting $\lambda_1 = \frac{2b_{R_N}}{4d_N\rho\beta_{R_1D_N}}$ and $\lambda_2 = \frac{2b_{R_1}}{4d_N\rho\beta_{R_ND_N}}$, since $\overline{F}_Y^{(3)}(y)$ could be written as $\overline{F}_Y^{(3)}(y) = Pr\left\{\frac{2d_N\rho}{b_{R_N}b_{R_1}}\times\frac{2(b_{R_N}\nu_{R_ND_N})(b_{R_1}\nu_{R_1D_N})}{b_{R_N}\nu_{R_ND_N}+b_{R_1}\nu_{R_1D_N}} > y\right\}$, with the help of the cumulative distribution function(CDF) of the harmonic mean of two exponential variables, i.e., assuming that Z_1 and Z_2 are two independent exponential variables with parameters λ_1 and λ_2 respectively, the CDF of $Z = \frac{2Z_1Z_2}{Z_1+Z_2}$, $P_Z(z)$, is given by

$$P_Z(z) = 1 - z\sqrt{\lambda_1\lambda_2}e^{-\frac{z}{2}(\lambda_1+\lambda_2)}K_1(z\sqrt{\lambda_1\lambda_2}). \tag{36}$$

Thus, the CCDF of Z can be written as

$$1 - P_Z(z) = z\sqrt{\lambda_1\lambda_2}e^{-\frac{z}{2}(\lambda_1+\lambda_2)}K_1(z\sqrt{\lambda_1\lambda_2}). \tag{37}$$

Hence, we have

$$\overline{F}_Y^{(3)}(y) = \frac{y}{\xi'}e^{-y\phi'}K_1\left(\frac{y}{\xi'}\right), \tag{38}$$

where $\xi' = 2d_N\rho\sqrt{\frac{\beta_{R_ND_N}\cdot\beta_{R_1D_N}}{b_{R_N}\cdot b_{R_1}}}$ and $\phi' = \frac{1}{4d_N\rho}\left(\frac{b_{R_1}}{\beta_{R_ND_N}}+\frac{b_{R_N}}{\beta_{R_1D_N}}\right)$. Note that the analysis result of the method is consistent with the extended form of (32), which verifies the correctness of our analysis idea. Therefore, by substituting (38) into (35), the CCDF of Y can finally be expressed as

$$\overline{F}_Y(y) = \frac{y}{\xi'}e^{-y\phi'-\frac{y}{a_N\rho}\left(\frac{1}{\beta_{SR_N}}+\frac{1}{\beta_{SR_1}}\right)}K_1\left(\frac{y}{\xi'}\right). \quad (39)$$

Letting $\psi = \phi'+\frac{1}{a_N\rho}\left(\frac{1}{\beta_{SR_N}}+\frac{1}{\beta_{SR_1}}\right)$ and employing [31]—Eq. (3.352.2), the result (the achievable ergodic rate) of the case $i = N$ ($\widetilde{C}_{s_N}$) can be obtained as (40).

$$\begin{aligned}\widetilde{C}_{s_N} &= \frac{1}{2\xi' ln2}\int_0^\infty \frac{ye^{-\psi y}K_1\left(\frac{y}{\xi'}\right)}{1+y}dy \\ &\sim -\frac{e^{\psi}}{2ln2}Ei(-\psi). \end{aligned} \quad (40)$$

Finally, considering all the cases of i where $i \in[1,N]$ together, we can describe the ergodic SR of our proposed system in a closed form as (41).

$$\widetilde{C}_{sum} \sim \frac{e^{\phi}}{2\ln 2}\left[Ei\left(-\phi\frac{a_i}{\tau}-\phi\right)-Ei\left(-\phi\right)\right]-\frac{e^{\psi}}{2ln2}Ei(-\psi). \quad (41)$$

3.2 Outage probability analysis

According to the QoS rates required by users, each user has a predetermined target data rate. When the link capacity cannot meet the required user data rate, communication interruption will occur. Thus, the analysis of outage probability is one of the important performance metrics in our proposed system. In this section, we will analyze the closed-form solutions of the outage probability with its asymptotic expressions. Assume that for s_i the user's target rate is R_{si} and the predefined target rate threshold is $R_{\mathcal{T}_{si}}$, the outage probability can be described as

$$\begin{aligned}P_{out} &= 1 - Pr\left\{R_{s_1} > 2^{2R_{\mathcal{T}_{s1}}}-1,\cdots,R_{s_i} > 2^{2R_{\mathcal{T}_{si}}}-1,\cdots,R_{s_N} > 2^{2R_{\mathcal{T}_{sN}}}-1\right\} \\ &= 1 - Pr\left\{(R_{s_1} > 2^{2R_{\mathcal{T}_{s1}}}-1)\cap\cdots\cdots\cap(R_{s_N} > 2^{2R_{\mathcal{T}_{sN}}}-1)\right\} \\ &= 1-\prod_{i=1}^{N}\underbrace{Pr(R_{s_i} > 2^{2R_{\mathcal{T}_{si}}}-1)}_{u_i}. \end{aligned} \quad (42)$$

Assume that $\omega_i = 2^{2R_{\mathcal{T}_{si}}}-1$ for the sake of simplicity and convenience of analysis, and refer to the above analysis of ergodic SR; therefore, the exact expressions of u_i where $i \in[1,N-1]$ can be described as

$$\begin{aligned}u_i &= Pr\left\{\min(\gamma_{R_i}^{(s_i)},\gamma_{R_{i+1}}^{(s_i)},\gamma_{D_i}^{(s_i)}) > \omega_i\right\} \\ &= Pr\left\{\gamma_{R_i}^{(s_i)} > \omega_i\right\}Pr\left\{\gamma_{R_{i+1}}^{(s_i)} > \omega_i\right\}Pr\left\{\gamma_{D_i}^{(s_i)} > \omega_i\right\} \end{aligned} \quad (43)$$

i.e.,

$$u_i = \begin{cases} e^{-\frac{\omega_i}{4d_i\rho}\left(\frac{b_{R_{i+1}}}{\beta_{R_iD_i}}+\frac{b_{R_i}}{\beta_{R_{i+1}D_i}}\right)} & for \quad \omega_i < \frac{a_i}{\tau} \\ 0 & for \quad \omega_i > \frac{a_i}{\tau} \end{cases}. \quad (44)$$

For the case u_N, it can be written as

$$\begin{aligned}u_N &= Pr\left\{\min(\gamma_{R_N}^{(s_N)},\gamma_{R_1}^{(s_N)},\gamma_{D_N}^{(s_N)}) > \omega_N\right\} \\ &= Pr\left\{\gamma_{R_N}^{(s_N)} > \omega_N\right\}Pr\left\{\gamma_{R_1}^{(s_N)} > \omega_N\right\}Pr\left\{\gamma_{D_N}^{(s_N)} > \omega_N\right\}. \end{aligned} \quad (45)$$

similarly,

$$u_N = \begin{cases} e^{-\frac{\omega_N}{4d_N\rho}\left(\frac{b_{R_1}}{\beta_{R_ND_N}}+\frac{b_{R_N}}{\beta_{R_1D_N}}\right)-\frac{\omega_N}{a_N\rho}\left(\frac{1}{\beta_{SR_N}}+\frac{1}{\beta_{SR_1}}\right)} & for \quad \omega_N < \frac{a_N}{\tau} \\ 0 & for \quad \omega_N > \frac{a_N}{\tau} \end{cases}. \quad (46)$$

By substituting (44) and (46) back into (35), with the condition $\omega_i < \frac{a_i}{\tau}$ where $i \in[1,N]$, the outage probability can be obtained in the closed-form expression as

$$\begin{aligned}P_{\text{out}} &= 1-\prod_{i=1}^{N}\underbrace{Pr(R_{s_i} > 2^{2R_{\mathcal{T}_{si}}}-1)}_{u_i} \\ &= 1-e^{-\left[\sum_{i=1}^{N-1}\frac{\omega_i}{4d_i\rho}\left(\frac{b_{R_{i+1}}}{\beta_{R_iD_i}}+\frac{b_{R_i}}{\beta_{R_{i+1}D_i}}\right)\right]} \\ &\quad e^{-\left[\frac{\omega_N}{4d_N\rho}\left(\frac{b_{R_1}}{\beta_{R_ND_N}}+\frac{b_{R_N}}{\beta_{R_1D_N}}\right)-\frac{\omega_N}{a_N\rho}\left(\frac{1}{\beta_{SR_N}}+\frac{1}{\beta_{SR_1}}\right)\right]}. \end{aligned} \quad (47)$$

4 Numerical results

In this section, we examine the performance of our proposed CMRS-NOMA scheme in terms of ergodic SR and outage probability. Taking the number of signals N as 3 as an example, we study the ergodic rates of three signals (s_1, s_2, and s_3) and the corresponding sum rate through a Monte Carlo simulation, and we compare the obtained analytical results with the simulated ones. Eight rates of both types match better with the increase in transmit SNR, as shown in Figs. 2 and 3. We also investigate the ergodic SR performance with respect to power allocation factors a_i and b_i for our proposed scheme with different transmit SNRs ρ, as seen in Figs. 4, 5, 6, 7, 8, and 9. In addition, we investigate the outage performance of our proposed scheme, as demonstrated in Fig. 10, and also compare the simulations with the analysis results in three cases.

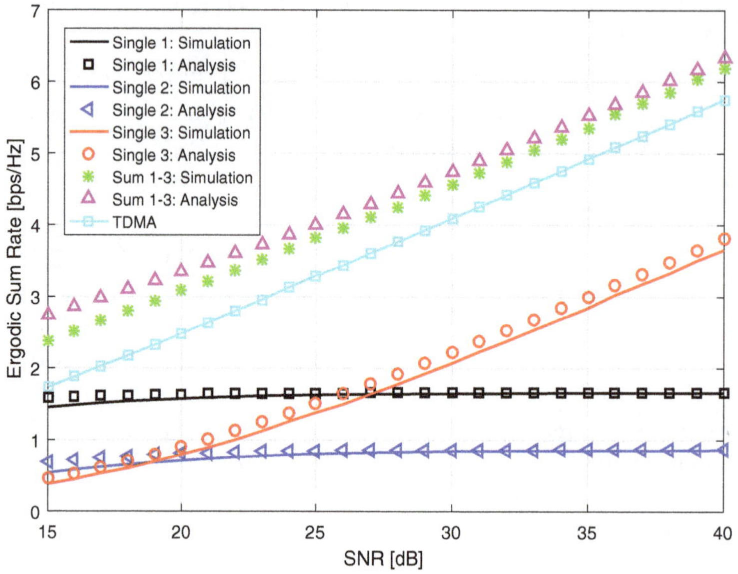

Fig. 2 The ergodic SRs achieved by our proposed MCRS-NOMA scheme with fixed $\nu_{SR1} = 7$, $\nu_{SR2} = 9$, $\nu_{SR3} = 10$, $\nu_{R1D} = 3$, $\nu_{R2D} = 4$, and $\nu_{R3D} = 2$

4.1 Ergodic SR

Figure 2 presents the average rates of the proposed algorithm with fixed ν_{SR_i}, $\nu_{R_iD_j}$ and corresponding β_{SR_i}, $\beta_{R_iD_j}$ for the transmission when $a_1 = 0.9$, $a_2 = 0.07$, $a_3 = 1 - a_1 - a_2$, and $b_i = \frac{1}{3}$ $(i \in [1,3])$. Clearly, due to the approximation of the analytical formula, at low transmit SNR, the theoretical analysis of the exact average rate is not in perfect agreement with the simulation results. However, in the high SNR region, the asymptotic results match the simulation results well, which is consistent with the practical situation. In Fig. 3, the power allocation factors are modified, i.e., $a_1 = 0.88$, $a_2 = 0.09$, and $a_3 = 1 - a_1 - a_2$, and the graphic shows the same trend: the simulated results and analytical results are almost identical. Thus, as shown in the figures, the differences of power allocation factors and signal-to-noise ratio have an effect

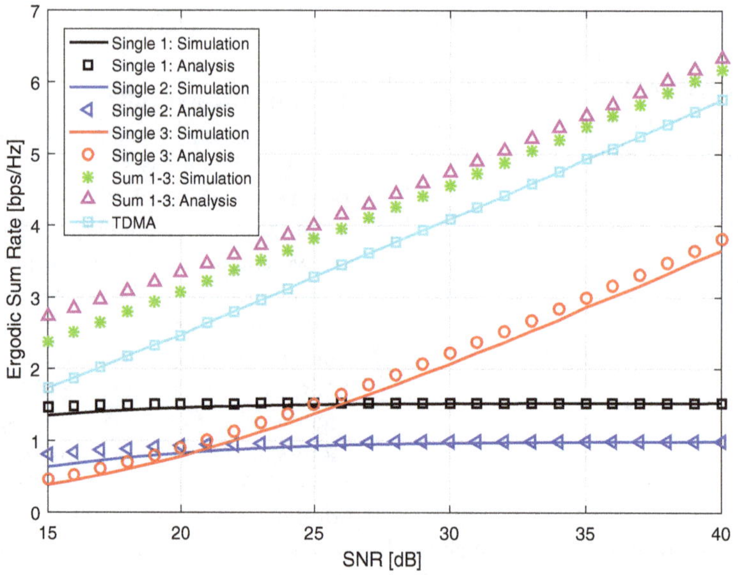

Fig. 3 The ergodic SRs achieved by modifying power allocation factors with fixed $\nu_{SR1} = 7$, $\nu_{SR2} = 9$, $\nu_{SR3} = 10$, $\nu_{R1D} = 3$, $\nu_{R2D} = 4$, and $\nu_{R3D} = 2$

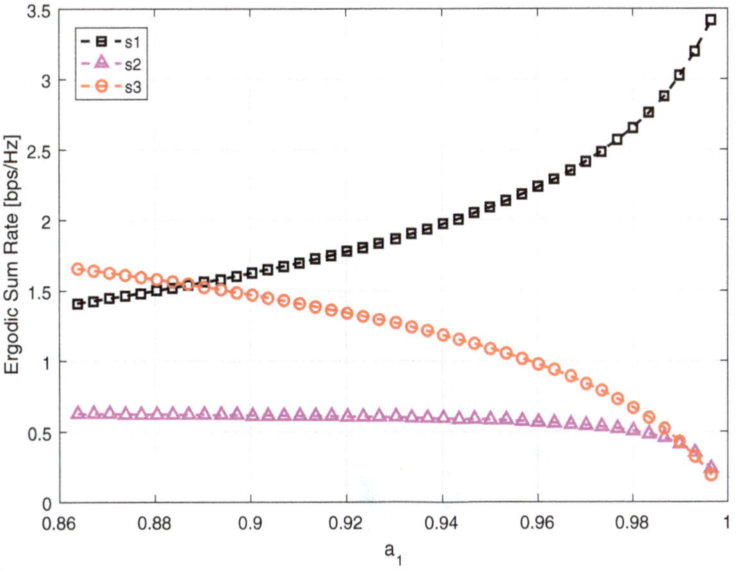

Fig. 4 The ergodic SRs achieved by our proposed scheme versus different power allocation factors a_1 for transmit SNR as $\rho = 25$ dB

on the average rates. In sum, the ergodic SR of our proposed case has an advantage over the TDMA scheme in which the ergodic SR can be described as (48).

$$C_{S_T} = \sum_{i=1}^{N} \frac{1}{2N} \log_2 \left(1 + \frac{1}{2N} \min\left\{ \nu_{SR_i}\rho, \nu_{SR_{i+1}}\rho, \nu_{R_iD_i}\rho + \nu_{R_{i+1}D_i}\rho \right\}\right). \quad (48)$$

Figures 4 and 5 depict the ergodic SR performance with respect to power allocation factor a_1 for our proposed scheme with different transmit SNRs as ρ={25, 30} dB. For both figures, we have set fixed $\nu_{SR_i} = 10$ and $\nu_{R_iD_j} = 2$, and we have a predetermined value range for different power allocation factors, i.e., $a_1 \in [0, 0.997)$, $a_2 = 0.6 * (1 - a_1)$, and $a_3 = 0.4 * (1 - a_1)$. Particularly, to simplify the analysis, we further assume that $b_i = \frac{1}{3}(i \in [1, 3])$, and then $d_i = b_{R_i} * b_{R_{i+1}} * a_i = \frac{1}{9} * a_i (i \in [1, 3])$ will be taken in the simulation. As shown in the figures, an optimal value of a_1 exists that maximizes the ergodic SR. Meanwhile, with the increase in the power ratio a_1 of NOMA symbols,

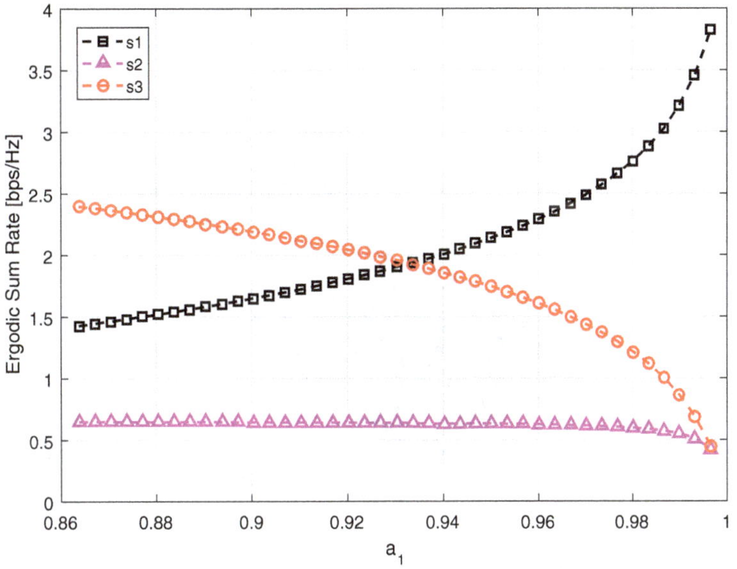

Fig. 5 The ergodic SRs achieved by our proposed scheme versus different power allocation factors a_1 for transmit SNR as $\rho = 30$ dB

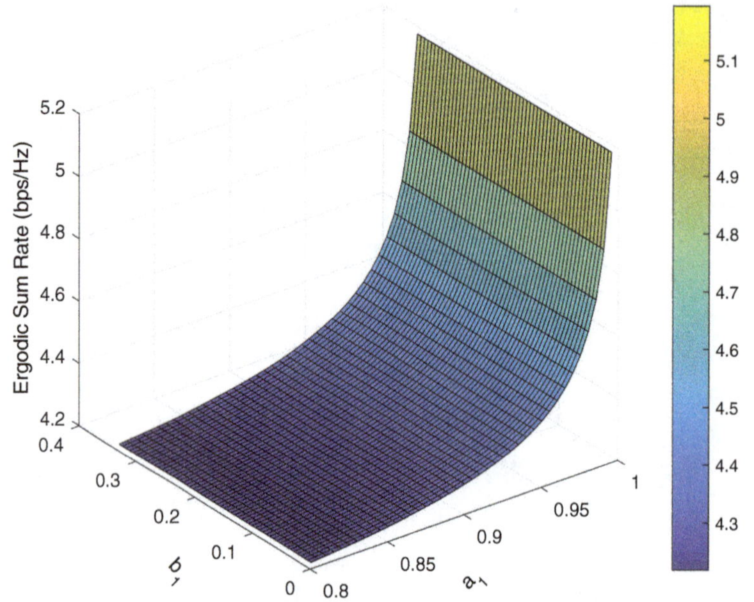

Fig. 6 The ergodic SRs achieved by our proposed scheme versus power allocation factors a_1 and b_1 with fixed $\nu_{SR1} = 3$, $\nu_{SR2} = 5$, $\nu_{SR3} = 6$, $\nu_{R1D} = 10$, $\nu_{R2D} = 18$, $\nu_{R3D} = 25$, and $\rho = 25$ dB

the rate of data symbol s_1 increases while those of data symbols s_2 and s_3 decrease. In addition, comparing these figures, the corresponding a_1 for the optimal ergodic SR will be close to 1 with an increasing SNR.

Figures 6 and 7 show the ergodic SR performance of our proposed scheme versus the power allocation factors a_1 and b_1 with fixed $\nu_{SR1} = 3$, $\nu_{SR2} = 5$, $\nu_{SR3} = 6$, $\nu_{R1D} = 10$, $\nu_{R2D} = 18$, $\nu_{R3D} = 25$, and $\rho = \{25, 35\}$ dB. We have a predetermined value range for power allocation factors a_i, i.e., $a_1 \in (0.801, 0.999)$, $a_2 = 0.8 * a_1$, and $a_3 = 1 - a_1 - a_2$. The maximum results of the ergodic SR are obtained with a_1 approaching to 1. In Fig. 6,

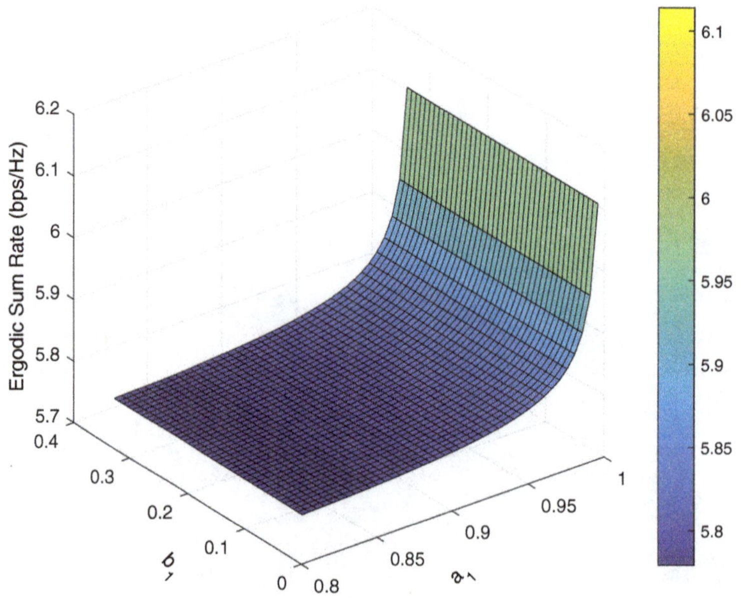

Fig. 7 The ergodic SRs achieved by our proposed scheme versus power allocation factors a_1 and b_1 with fixed $\nu_{SR1} = 3$, $\nu_{SR2} = 5$, $\nu_{SR3} = 6$, $\nu_{R1D} = 10$, $\nu_{R2D} = 18$, $\nu_{R3D} = 25$, and $\rho = 35$ dB

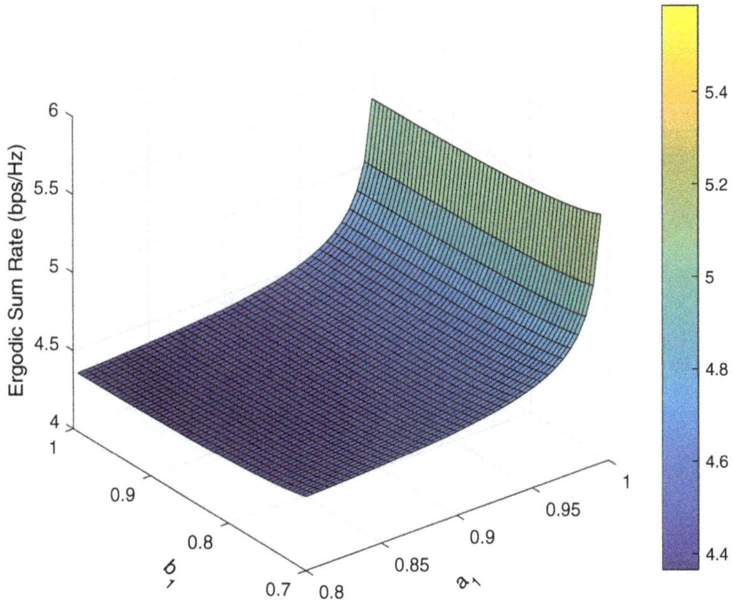

Fig. 8 The ergodic SRs achieved by our proposed scheme versus power allocation factors a_1 and b_1 with fixed $\nu_{SR1} = 3$, $\nu_{SR2} = 5$, $\nu_{SR3} = 6$, $\nu_{R1D} = 10$, $\nu_{R2D} = 18$, $\nu_{R3D} = 25$, and $\rho = 25$ dB

the maximum result is 5.194 bps/Hz with a_1 as 0.996, whereas these values are 6.114 bps/Hz and 0.996, respectively, for Fig. 7. As shown, the surfaces of ergodic SR in the above figures are relatively flat. This phenomenon can be explained by assuming that the value of b_i is uniform. Figures 8 and 9 are the results of non-uniform values of b_i, e.g., $b_1 \in (0.701, 0.999)$, $b_2 = 0.7*b_1$, and $b_3 = 1-b_1-b_2$. Hence, the surface of ergodic SR presents a process of changes in values. Accordingly, the maximum value in Fig. 8 is 5.586 bps/Hz with b_1 as 0.701, which is beyond the range shown in Fig. 6. By using the data cursor in Fig. 9, we can observe that the maximum SR is 6.338 bps/Hz.

4.2 Outage probability

Figure 10 demonstrates the outage performance of our proposed scheme in terms of the simulation and analytical

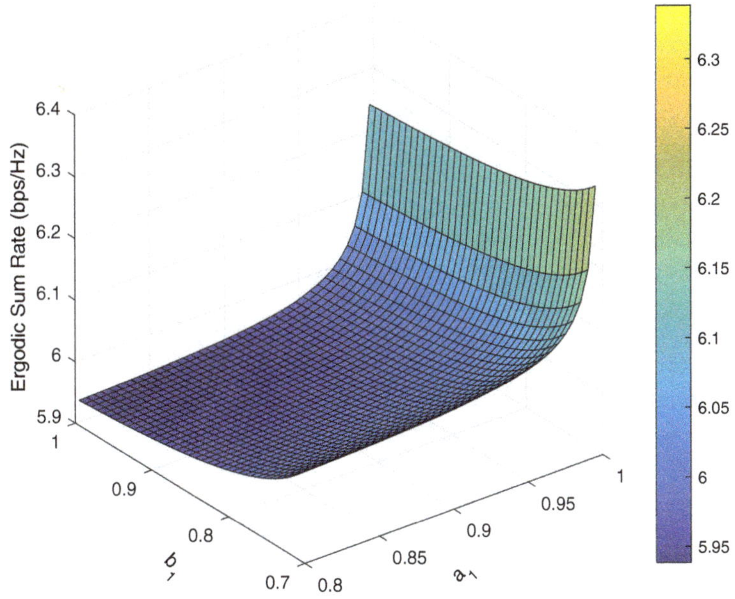

Fig. 9 The ergodic SRs achieved by our proposed scheme versus power allocation factors a_1 and b_1 with fixed $\nu_{SR1} = 3$, $\nu_{SR2} = 5$, $\nu_{SR3} = 6$, $\nu_{R1D} = 10$, $\nu_{R2D} = 18$, $\nu_{R3D} = 25$, and $\rho = 35$ dB

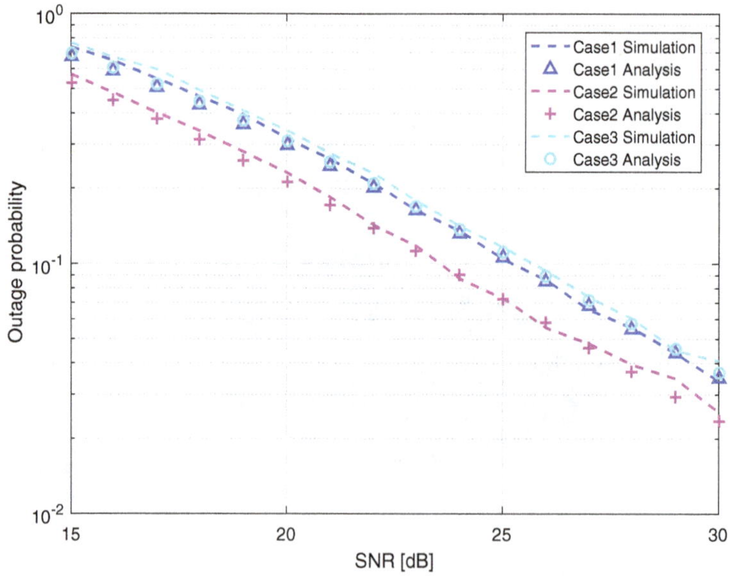

Fig. 10 The outage probability for our proposed scheme with $\rho = \{15, 30\}$ dB in three cases. Case 1 versus the target rate as 0.6 with fixed $\nu_{SR1} = 10$, $\nu_{SR2} = 18$, $\nu_{SR3} = 25$, $\nu_{R1D} = 3$, $\nu_{R2D} = 3$, and $\nu_{R3D} = 3$. Case 2 versus the target rate as 0.6 with fixed $\nu_{SR1} = 10$, $\nu_{SR2} = 18$, $\nu_{SR3} = 25$, $\nu_{R1D} = 5$, $\nu_{R2D} = 5$, and $\nu_{R3D} = 5$. Case 3 versus the target rate as 0.8 with fixed $\nu_{SR1} = 10$, $\nu_{SR2} = 18$, $\nu_{SR3} = 25$, $\nu_{R1D} = 5$, $\nu_{R2D} = 5$, and $\nu_{R3D} = 5$

results with three cases, where the power allocation factors are $a_1 = 0.88$, $a_2 = 0.09$, $a_3 = 1 - a_1 - a_2$, and $b_i = \frac{1}{3}$ ($i \in [1, 3]$). The following system setups are considered: (1) $\omega = 0.6$, $\rho = \{15, 30\}$dB, $\nu_{SR1} = 10$, $\nu_{SR2} = 18$, $\nu_{SR3} = 25$, and $\nu_{R1D} = \nu_{R2D} = \nu_{R3D} = 3$ for case 1; (2) $\omega = 0.6$, $\rho = \{15, 30\}$dB, $\nu_{SR1} = 10$, $\nu_{SR2} = 18$, $\nu_{SR3} = 25$, and $\nu_{R1D} = \nu_{R2D} = \nu_{R3D} = 5$ for case 2; and (3) $\omega = 0.8$, $\rho = \{15, 30\}$dB, $\nu_{SR1} = 10$, $\nu_{SR2} = 18$, $\nu_{SR3} = 25$, and $\nu_{R1D} = \nu_{R2D} = \nu_{R3D} = 5$ for case 3. The results reveal that a good match exists between the simulation and analytical results. The outage probabilities of the above three cases change with the transmission SNR (ρ). For example, the values are respectively {0.2997, 0.2118, 0.3110}, where ρ = 20 dB and {0.1065, 0.0724, 0.1111} where ρ =25 dB. It is clear that the performance of the outage probability improves as the channel gains increase and the threshold decreases. The former can be observed by comparing case 1 with case 2, while the latter can be shown by comparing case 2 with case 3. The values of the previous examples reveal this rule well.

5 Conclusions and future work

In this paper, a new transmission model, CMRS-NOMA, in cellular networks has been proposed. In the proposed scheme, SIC is employed to decode the receptions from the source to multiple relay nodes, and the exact and asymptotic expressions for the achievable SR and outage probability of the proposed system over the independent Rayleigh fading channels are presented. In addition, we qualitatively analyze the complexity of the multi-stage power allocation strategy. It is easy to observe that all the signals can be simply and practically decoded. Numerical results have been presented to corroborate the theoretical analyses, and the results have shown that the performance of the ergodic SR for the proposed CMRS-NOMA case gains a significant improvement and outperforms the TDMA. For a high SNR case, the proposed CMRS-NOMA technology can also provide a greater spectral efficiency. Even that the NOMA systems over Rician fading channels and/or Nakagami-*m* fading channels are more interesting and challenging, our paper considers providing the mentality to study the receiver strategies for cooperative relaying cellular networks. Therefore, investigating different channel models and efficient power allocation methods for the receiving nodes with different QoS requirements are remained as the future works.

Abbreviations

AWGN: Additive white Gaussian noise; BS: Base station; CCDF: Complementary cumulative distribution function; CDF: Cumulative distribution function; CMRS-NOMA: Cooperative multi-relay system with non-orthogonal multiple access; CRS-NOMA: Cooperative relaying system with NOMA; CSI: Channel state information; CU Corresponding user; IoT: Internet of Things, MIMO: Multiple-input multiple-output, mmWave: Millimeter-wave, MTC: Machine-type communications; NPL: Node pair list; OMA: Orthogonal multiple access; QoS: Quality of services; SC: Superposition coding; SIC: Successive interference cancellation; SINR: Signal-to-interference-plus-noise ratio; SM: Spatial modulation; SR: Sum rate; SWIPT: Simultaneous wireless information and power transfer; TDMA: Time-division multiple access; V2V: Vehicle-to-vehicle

Acknowledgements

The author would like to thank the Editor and reviewers for their valuable and insightful comments.The authors would gratefully acknowledge the grants from the National Natural Science Foundation of China (61801249, 61501264 and 61371113), Nantong University-Nantong Joint Research Center for

Intelligent Information Technology (KFKT2016B01, KFKT2017B01), Basic Scientic Research of Nantong Science and Technology Project (JCZ18052), Natural Science Research Program of Nantong Vocational University (18ZK01), the open research fund of National Mobile Communications Research Laboratory, Southeast University (No.2015D02).

Authors' contributions

JJ, WD, and GZ conceived and designed the study. JJ, WD, and QS performed the simulations. JJ and LJ wrote the paper. JJ, GZ, QS, LJ, and WD reviewed and edited the manuscript. All authors read and approved the manuscript.

Competing interests

The authors declare that they have no competing interests.

Author details

[1]School of Information Science and Technology, Nantong University, Seyuan Road, Nantong, China. [2]School of Electronic and Information Engineering, Nantong Vocational University, Qingnian Road, Nantong, China. [3]The Nantong Research Institute for Advanced Communication Technologies, Chongchuan Road, Nantong, China.

References

1. Z. Ding, M. Peng, H. V. Poor, Cooperative non-orthogonal multiple access in 5G systems. IEEE Commun. Lett. **19**, 1462–1465 (2015)
2. J. B. Kim, I. H. Lee, Capacity analysis of cooperative relaying systems using non-orthogonal multiple access. IEEE Commun. Lett. **19**, 1949–1952 (2015)
3. M. Wen, E. Basar, Q. Li, B. Zheng, M. Zhang, Multiple-mode orthogonal frequency division multiplexing with index modulation. IEEE Trans. Commun. **65**, 3892–3906 (2017)
4. M. Wen, B. Ye, E. Basar, Q. Li, F. Ji, Enhanced orthogonal frequency division multiplexing with index modulation. IEEE Trans. Wirel. Commun. **16**, 4786–4801 (2017)
5. M. Wen, X. Cheng, M. Ma, B. Jiao, H. V. Poor, On the achievable rate of OFDM with index modulation. IEEE Trans. Signal Process. **64**, 1919–1932 (2016)
6. J. Li, M. Wen, X. Cheng, Y. Yan, S. Song, M. H. Lee, Generalized precoding-aided quadrature spatial modulation. IEEE Trans. Veh. Technol. **66**, 1881–1886 (2017)
7. F. Cheng, H. Yan, J. Song, V. Leung, Energy-efficient resource allocation for downlink non-orthogonal multiple access network. IEEE Trans. Commun. **64**, 3722–3732 (2016)
8. Y. Sun, D. Ng, Z. Ding, R. Schober, Optimal joint power and subcarrier allocation for full-duplex multicarrier non-orthogonal multiple access systems. IEEE Trans. Commun. **65**, 1077–1091 (2017)
9. Y. Saito, A. Benjebbour, Y. Kishiyama, T. Nakamura, Generalized precoding-aided quadrature spatial modulation. Proc. IEEE Annu. Symp. PIMRC. **66**, 611–615 (2013)
10. M. Xu, M. Wen, F. Ji, W. Duan, Novel receiver design for the cooperative relaying system with non-orthogonal multiple access. IEEE Commun. Lett. **66**, 1679–1682 (2016)
11. Y. Liu, Z. Ding, M. Elkashlan, H. V. Poor, Cooperative non-orthogonal multiple access with simultaneous wireless information and power transfer. IEEE J. Sel. Areas Commun. **34**, 938–953 (2016)
12. Z. Yang, Z. Ding, P. Fan, N. Al-Dhahir, A general power allocation scheme to guarantee quality of service in downlink and uplink noma systems. IEEE Trans. Wirel. Commun. **15**, 7244–7257 (2016)
13. R. Jiao, L. Dai, J. Zhang, R. MacKenzie, M. Hao, On the performance of NOMA-based cooperative relaying systems over Rician fading channels. IEEE Trans. Veh. Technol. **66**, 11409–11413 (2017)
14. X. Yue, Y. Liu, S. Kang, A. Nallanathan, Performance analysis of NOMA with fixed gain relaying over Nakagami-m fading channels. IEEE Access. **5**, 5445–5454 (2017)
15. H. Xie, F. Gao, S. Jin, J. Fang, Y. C. Liang, Channel estimation for TDD/FDD massive MIMO systems with channel covariance computing. IEEE Trans. Wirel. Commun. **17**, 4206–4218 (2018)
16. H. Xie, F. Gao, S. Zhang, S. Jin, A unified transmission strategy for TDD/FDD massive MIMO systems with spatial basis expansion model. IEEE Trans. Veh. Technol. **66**, 3170–3184 (2017)
17. B. Zheng, M. Wen, E. Basar, F. Chen, Multiple-input multiple-output OFDM with index modulation: low-complexity detector design. IEEE Trans. Signal Process. **65**, 2758–2772 (2017)
18. J. Zhang, L. Dai, S. Sun, Z. Wang, On the spectral efficiency of massive MIMO systems with low-resolution ADCs. IEEE Commun. Lett. **20**, 842–845 (2016)
19. J. Choi, Minimum power multicast beamforming with superposition coding for multiresolution broadcast and application to NOMA systems. IEEE Trans. Commun. **63**, 791–800 (2015)
20. M. F. Hanif, Z. Ding, T. Ratnarajah, G. K. Karagiannidis, A minorization-maximization method for optimizing sum rate in the downlink of non-orthogonal multiple access systems. IEEE Trans. Signal Process. **64**, 76–88 (2016)
21. Q. Sun, S. Han, C. L. I, Z. Pan, On the ergodic capacity of MIMO NOMA systems. IEEE Wirel. Commun. Lett. **4**, 405–408 (2015)
22. J. Choi, On the power allocation for MIMO-NOMA systems with layered transmissions. IEEE Trans. Wirel. Commun. **15**, 3226–3237 (2016)
23. Z. Ding, L. Dai, H. V. Poor, Mimo-noma design for small packet transmission in the internet of things. IEEE Access. **4**, 1393–1405 (2016)
24. W. Feng, Y. Wang, D. Lin, N. Ge, J. Lu, S. Li, When mmWave communications meet network densification: a scalable interference coordination perspective. IEEE J. Sel. Areas Commun. **35**, 1459–1471 (2017)
25. B. Wang, F. Gao, S. Jin, H. Lin, G. Y. Li, Spatial- and frequency-wideband effects in millimeter-wave massive MIMO systems. IEEE Trans. Signal Process. **66**, 3393–3406 (2018)
26. Z. Ding, P. Fan, H. V. Poor, Random beamforming in millimeter-wave noma networks. IEEE Access. **5**, 7667–7681 (2017)
27. R. Zhang, Z. Zhong, J. Zhao, B. Li, K. Wang, Channel measurement and packet-level modeling for V2I spatial multiplexing uplinks using massive MIMO. IEEE Trans. Veh. Technol. **65**, 7831–7843 (2016)
28. P. Harris, Performance characterization of a real-time massive MIMO system with LOS mobile channels. IEEE J. Sel. Areas Commun. **35**, 1244–1253 (2017)
29. Y. Chen, L. Wang, Y. Ai, B. Jiao, L. Hanzo, Performance analysis of NOMA-SM in vehicle-to-vehicle massive MIMO channels. IEEE J. Sel. Areas Commun. **35**, 2653–2666 (2017)
30. I. S. Gradshteyn, I. M. Ryzhik, *Table of Integrals, Series, and Products. 7th ed.* (Academic, San Diego, 2007)
31. M. Abramowitz, I. A. Stegun, *Handbook of Mathematical Functions with Formulas, Graphs, and Mathematical Tables. 9th. ed.* (Dover Publications, New York, 1970)
32. A. Milton, I. A. Stegun, *Handbook of Mathematical Functions with Formulas, Graphs, and Mathematical Tables. 10th ed.* (US GPO, New York, 1972)

17

Research on Wi-Fi indoor positioning in a smart exhibition hall based on received signal strength indication

Qing Yang*, Shijue Zheng, Ming Liu and Yawen Zhang

Abstract

To improve the management of science and technology museums, this paper conducts an in-depth study on Wi-Fi (wireless fidelity) indoor positioning based on mobile terminals and applies this technology to the indoor positioning of a science and technology museum. The location fingerprint algorithm is used to study the offline acquisition and online positioning stages. The positioning flow of the location fingerprint algorithm is discussed, and the improvement of the location fingerprint algorithm is emphasized. The raw data of the RSSI (received signal strength indication) is preprocessed, which makes the location fingerprint data more effective and reliable, thus improving the positioning accuracy. Three different improvement strategies are proposed for the nearest neighbor classification algorithm: a balanced joint metric based on distance weighting and a compromise between the two. Then, in the experimental simulation, the positioning results and errors of the traditional KNN (k-nearest neighbor) algorithm and three improvement strategy algorithms are analyzed separately, and the effectiveness of the three improved strategy algorithms is verified by experiments.

Keywords: Indoor positioning, Wi-Fi, Location fingerprint, Nearest neighbor classification, Received signal strength indication

1 Introduction

Science and technology museums are public science education institutions with exhibition education as its main function. There, scientific interest is stimulated, and scientific concepts are enlightened by participation, experience, interactive exhibits and auxiliary displays. For the science and technology museum, it is necessary to know the real-time specific orientation of each visitor in the exhibition hall and the number of visitors and their residence time for each booth. Moreover, according to the number of visitors to each booth, the management of the exhibition center can maintain the better order.

At present, there are a variety of indoor positioning technologies [1], which are broadly divided into the technologies of Wi-Fi positioning [2], radio frequency identification (RFID) positioning [3], ZigBee positioning [4], Bluetooth positioning [5], ultra-wide-band (UWB) positioning [6], infrared positioning [7], ultrasonic positioning [8], optical tracking positioning [9], and computer vision positioning [9]. Among them, the more influential positioning systems are: the active badge positioning system based on infrared positioning technology researched by the Cambridge Laboratory [10]; the RADAR positioning system [11] realized by Wi-Fi positioning technology in wireless LANs and the Active Bat positioning system [12] realized by ultrasonic positioning technically researched by Microsoft Design Institute; the Cricket positioning system [13] researched by MIT which is an improvement on the Active Bat system; AHLos (ad hoc localization system) positioning system [14] researched by UCLA which is an improvement on the Cricket system.

There are many different exhibition halls in science and technology museums. Each exhibition hall has many exhibitions. The managers of science and technology museums want to know the flow of people in different exhibitions at different times. Human flow data are conducive to scientific management. With the development

* Correspondence: yangqing@mail.ccnu.edu.cn
School of Computer, Central China Normal University, Wuhan 430079, People's Republic of China

of the mobile Internet, Wi-Fi hotspots have been covered in most public indoor areas of the city, and Wi-Fi wireless LANs have data communication functions. Wi-Fi wireless networks are easy to deploy and have good scalability. However, accuracy still needs to be further improved. Indoor positioning technology based on the Wi-Fi mobile terminal has a great advantage in positioning in the exhibition hall of science and technology museums. The purpose of this paper is to improve the indoor location algorithm in the environment of science and technology museum, to improve the accuracy of indoor location, and to minimize the complexity of the algorithm, so that it can be used in small- and medium-sized science and technology museums or museums. Based on the literature surveys about the various existing Wi-Fi indoor positioning algorithms and analyses of their advantages and disadvantages, an effective location fingerprint algorithm is proposed. Performance analysis of the proposed algorithm and an evaluation of the algorithm with respect to other existing algorithms are given. This improved algorithm can accurately describe the flow data of people in science and technology museums when it is used for indoor positioning in science and technology museums.

The rest of this paper is organized as follows. Section 2 discusses related work, followed by the improvement of the location fingerprint algorithm designed in Section3. Section 4 shows the simulation experimental results, and Section 5 concludes the paper with a summary and future research directions.

2 Related work

This section provides a brief review of the common Wi-Fi indoor positioning algorithms. After comparing these algorithms, the location fingerprint algorithm is selected for the indoor location of science and technology museum. The architecture of the Wi-Fi location is introduced, and the flow of the location fingerprint algorithm is described.

2.1 Wi-Fi indoor positioning algorithm

Wi-Fi indoor positioning algorithms [15–17] are commonly found as follows: (1) based on the central point algorithm, (2) the propagation model algorithm, and (3) the location fingerprint algorithm.

The location fingerprint algorithm is simpler and easier to implement on the mobile terminal, and the location accuracy based on location fingerprint feature matching is very robust. Wi-Fi fingerprint positioning has become the preferred technology for indoor environment positioning and navigation because of its convenient use of existing Wi-Fi network infrastructure, low cost and strong environmental adaptability. Therefore, the Wi-Fi location fingerprint algorithm based on a mobile terminal is selected to conduct experiments in a science and technology museum.

The important task of signal-based fingerprint location is to select the appropriate location algorithm. There are three typical methods: (1) neighbor method; (2) probability method; (3) neural network method. The number of hidden layers of the neural network method can be one or more. The more hidden layers, the higher the accuracy of data fitting, but the computational complexity and training time will also be improved. Normal servers cannot afford the computational load of the neural network method [18]. This method is not suitable for small- and medium-sized science and technology museums. The main work of this paper is to study the indoor positioning method suitable for small- and medium-sized science and technology museums.

2.2 Wi-Fi positioning architecture

In a Wi-Fi indoor positioning system, the position of the reference point is known, and the position of the target node is usually the one to be tested. The mobile terminal is at the location to be tested. During the positioning process, all the surrounding Wi-Fi signal source APs are perceived, and the metrics are uploaded (the specific metrics may be the time of arrival (TOA) of the signal, or the angle of arrival (AOA) of the signal, the time difference of arrival (TDOA) of the signal and received signal strength RSSP.) to the positioning server background. The server background estimates the position coordinates of the point to be tested by a specific positioning algorithm, and then sends it to the terminal and displays it, as shown in Fig. 1.

In a Wi-Fi wireless network environment, a Wi-Fi signal access point (i.e., an AP) periodically broadcasts a beacon signal frame, and although a mobile terminal device with a Wi-Fi access function module does not establish a connection with any AP, it can also receive at least three important parameter indicators of the transmitting source AP from the AP broadcast signal frame: the Wi-Fi AP's MAC address (BSSID), the Wi-Fi AP name (SSID), and the Wi-Fi signal strength indication (RSSI). These three indicators provide strong support for the model for positioning based on location fingerprint feature matching.

In general, the following factors affect the Wi-Fi positioning effect [19]: the absorption effects of Wi-Fi signals, the reflection and diffraction of Wi-Fi signals, the multipath propagation and shadow attenuation of Wi-Fi signals, the size of indoor areas, the changes in temperature and humidity in indoor environment sand the signal interference from other electronic devices. In addition, due to the lack of uniform rules and optimization of Wi-Fi, the network environment of the Wi-Fi AP is extremely complicated, and the number of manufacturers producing Wi-Fi

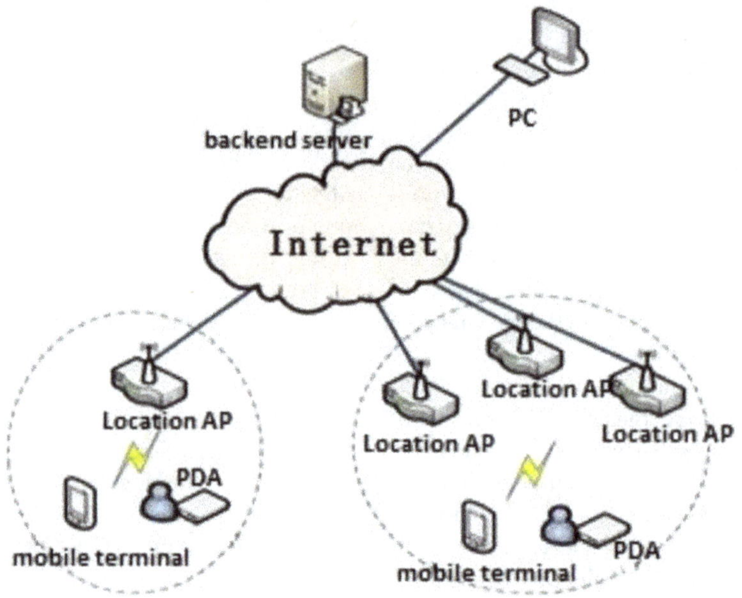

Fig. 1 Overall architecture of Wi-Fi positioning

APs and terminals is large, so the performance difference between devices is very obvious, which means that the various influencing factors must be considered in a specific position to improve the accuracy of positioning.

2.3 The location progress of the position fingerprint algorithm

The position fingerprint algorithm is a feature matching algorithm, and the feature value is the position fingerprint data of the Wi-Fi signal strength.

In the algorithm based on the propagation model, since the Wi-Fi signal is affected by various factors in the indoor environment, the attenuation often does not meet the theoretical value; a position fingerprinting algorithm based on feature matching is proposed. That is, the signal strength information in a group of different Wi-Fi APs is received at each position, and the fingerprint data vector signature sequences of all APs generated by different positions make differences. In this way, specific Wi-Fi position fingerprint information can be obtained at each location. The overall process of position fingerprint algorithm location is shown in Fig. 2.

As Fig. 3 shows, the locating process has two stages: the offline acquisition stage and the online positioning stage. The main task of the offline acquisition stage is to collect the fingerprint data of the reference point position in the indoor environment where the reference point has been deployed, and build a fingerprint database. In the online positioning stage, the fingerprint data collected by the node to be tested is matched with the fingerprint database, and a matching algorithm is used to find a group of data that best matches it, thereby finding the position of the node to be tested.

2.3.1 Offline acquisition stage

The offline acquisition stage generally needs to complete the following work steps (as shown in Fig. 3).

Step 1. The indoor positioning environment establishes a location metric and is divided into several regions. As shown in Fig. 3, the positioning area is divided into 16 small grids, and the position information of each grid is described by Li, i = 1, 2....16.

Step 2. Take signal strength information from different Wi-Fi APs. Figure 3 indicates that the location fingerprint information is being collected at the L7 position. In general, there are two methods for obtaining AP signal strength information at a specific location. (1) Based on the terminal method, the signal strength information of a group of APs is collected one by one at the position of each reference point. (2) Based on the propagation model method, the signal strength information at each reference point is derived based on the position of the AP.

The terminal-based method achieves accurate signal strength and high positioning accuracy. Although it takes requires considerable time, and each new positioning area has to be re-acquired and built a location fingerprint database built, this method is adopted in this paper.

Step 3. Generate location feature fingerprint information at the reference point and store it in

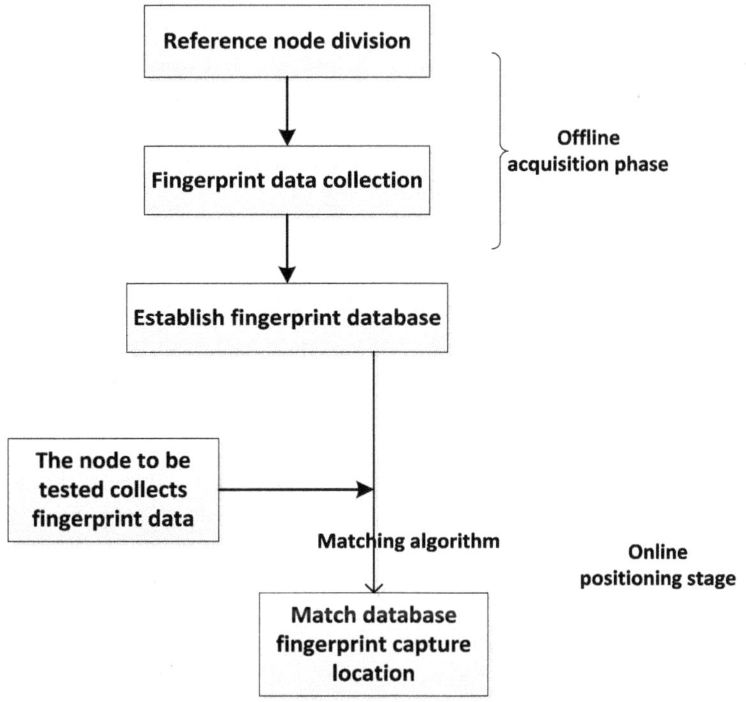

Fig. 2 The overall process of position fingerprint algorithm locating

the location fingerprint database. In Fig. 3, the location fingerprint of the deterministic algorithm is generated; that is, the fingerprint at the location Li is (rssii-1, rssii-2, ssii-3, rssii-4, rssii-5, rssii-6), where rssii is the signal strength indicator; i is the number of the network; 1-6 are the numbers of the APs deployed in Fig. 3.

2.3.2 Online positioning stage

In the matching algorithm used in the positioning phase, the KNN algorithm used in this paper uses some metrics to distinguish the similarity between two different fingerprints, also called the dissimilarity.

1. Fingerprint dissimilarity

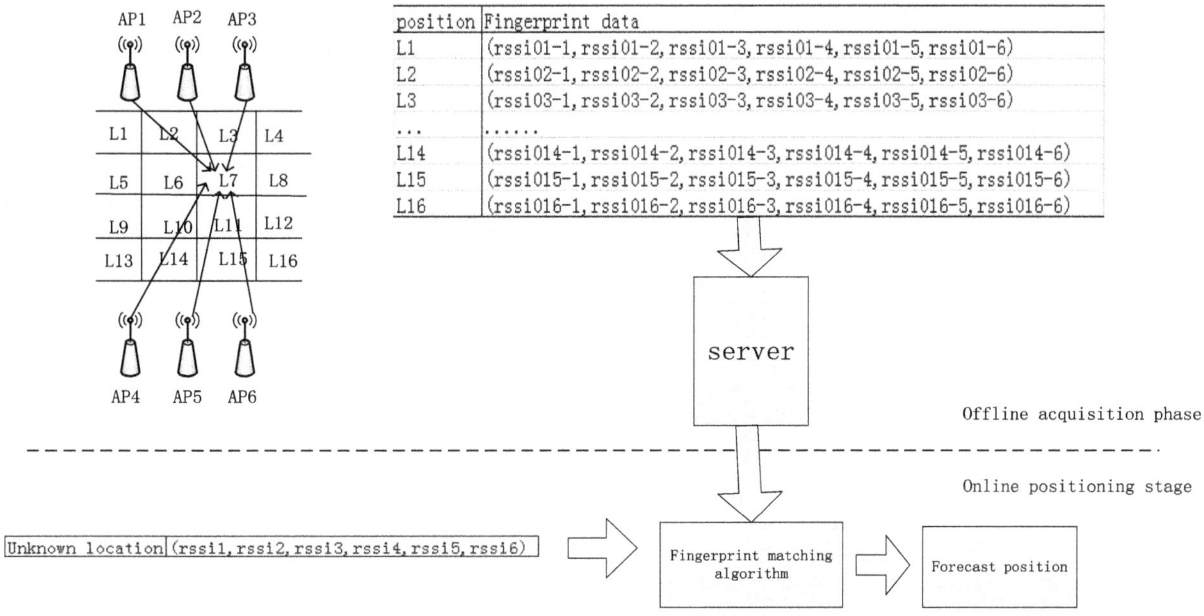

Fig. 3 The schematic diagram of position fingerprint locating algorithm

An informal definition of the degree of dissimilarity between two objects is a numerical measure of the degree of difference between the two objects. The more similar objects are, the lower their dissimilarity becomes. The distance is usually used as a synonym for dissimilarity, where distance is used to indicate the degree of dissimilarity between different nodes (such as the reference node and the point to be measured). Common methods are these, such as Euclidean distance, taxicab geometry, and the Minkowski distance.

2. Locating process

The matching algorithm during the locating process is divided into two types: deterministic positioning algorithm and probability-based positioning algorithm. This paper adopts a probability-based positioning algorithm.

The probability-based positioning algorithm stores the probability distribution model parameters of the collected signal strength information at different positions in the database (information such as histograms and Gaussian probability distributions are stored in the database), and Bayesian probability classification is used in the positioning phase to estimate the user's position. In the actual locating process, as shown in Fig. 3, the user-held mobile terminal collects a set of real-time location fingerprints at the unknown node position to be tested (rssii-1, rssii-2, rssii-3, rssii-4, rssii-5, rssii-6) and then uses a certain matching algorithm (e.g., KNN algorithm), applying the Euclidean distance in the fingerprint database. Measure the difference between the reference node and the node to be tested, and select the value of the K reference point with the smallest difference for test training. Too small or too large will affect accuracy.

A simplified version of the proposed algorithm can be described as follows:

Algorithm 1: k-Nearest Neighbor (KNN)

Input: Node to be tested S, i=1,distance=0

Begin

1. Calculate the size of the fingerprint database
2. Calculate the Euclidean distance between the point in the fingerprint database and the node to be tested
3. Sort by distance
4. Select K reference points with the smallest distance
5. Calculate the coordinates of the node to be tested
6. Average K reference points
7. Repeat steps 2-5 until all points to be measured are calculated.

End

Output: Position of the point to be measured

3 Improvement of the location fingerprint algorithm

The RSSI positioning method [20] is mainly divided into two categories: (1) based on theoretical or empirical propagation models, transforming propagation loss into distance to achieve positioning; (2) based on pattern recognition, positioning by location fingerprint feature matching. Both are based on the intensity variation in the received signal, the difference being in the measurement index and whether the position of the reference node needs to be predicted [21].

3.1 Preprocessing of RSSI data

In the complex exhibition hall of a science and technology museum, the Wi-Fi signal is affected by various interference factors, and the signal attenuation is irregular. The signal strength of the same AP collected by the acquisition terminal at a fixed position is not fixed. In other words, the signal strength fluctuates within a certain range.

In this paper, the method of averaging multiple acquisition and measurement is used to smooth the calculated result to some extent, and thus, reduce the error caused by irregular fluctuations. However, in actual situations, even the fluctuation of different APs collected at the same position has different amplitude changes. Table 1 shows the Wi-Fi signal strength changes in different APs collected 2000 times at a fixed location.

The strength of the signal also changes according to the change in the position distance and is relatively stable within a certain distance range. After a certain range, it starts to increase. The Wi-Fi-based location fingerprint algorithm collects the signal strength feature values. To obtain more reliable location fingerprint feature data, the difference between each reference point and the node to be tested is more specific, which can reduce the positioning error, and improve the positioning accuracy in the positioning matching stage. This section uses the following methods to process the collected RSSI raw information data.

3.1.1 Truncated mean method

When the number of acquisitions is large enough, the arithmetic mean can represent the central trend of the sampling position, which is closer to the actual effective value of the position, and its calculation formula is Formula 1:

$$\bar{rssi} = \frac{\sum_{i=1}^{N} rssi_i}{N} = \frac{rssi_1 + rssi_2 + \cdots + rssi_N}{N} \tag{1}$$

N is the total number of acquisitions. Although the mean after a large number of samples is a useful measure to describe the RSSI data set, it is not the best method; there will be some irregular outliers, resulting in a larger error.

The truncated mean method can be used to eliminate the negative effect of outliers on the valid data. The core

Table 1 Signal amplitude change table for different APs with fixed positions

AP name	Average value	Maximum value	Minimum value	Volatility
TL-WR886N	– 42.58 dBm	– 35 dBm	– 64 dBm	29 dBm
TL-WR742N	– 39.14 dBm	– 29 dBm	– 63 dBm	29 dBm
STM-AUTO	– 57.14 dBm	– 53 dBm	– 66 dBm	13 dBm
Mobile	– 47.29 dBm	– 37 dBm	– 64 dBm	27 dBm
Complex	– 59.43 dBm	– 57 dBm	– 65 dBm	8 dBm
STM	– 61.58 dBm	– 58 dBm	– 70 dBm	12 dBm

idea of this method is to discard the portion of the larger value end and the smaller value end, and then average the remaining data. For example, after sorting the collected data, discard 5% of the larger value end and 5% of the smaller value end, which the ratio must be controlled.

3.1.2 Screening mean method

Some researchers have proposed using the normal distribution model to preprocess the RSSI value. The model is based on the assumption that the RSSI value conforms to the normal distribution, that is, Formula 2 is satisfied.

$$f(x) = \frac{1}{\sqrt{2\pi}\sigma} e^{\frac{-(x-u)^2}{2\sigma^2}} \tag{2}$$

$\mu = \frac{\sum_{i=1}^{N} x_i}{N}$ is the mean value and $\sigma = \sqrt{\frac{\sum_{i=1}^{N} (x_i - \mu)^2}{n-1}}$ is the standard deviation, where N is the number of acquisitions, and x_i is the RSSI value received during acquisition.

The theoretical probability value of the normal distribution in the interval is 0.954, and the theoretical probability value of the interval is 0.997.

Instead of using the normal distribution model directly, the normal distribution model and the trimmed tail estimation method can be comprehensively considered to establish a normal distribution model in the middle part where signal strength is relatively stable (which satisfies the premise), which can eliminate the outliers by the interval, to select the normal RSSI with higher probability.

Although there are many other methods for RSSI data preprocessing, for specific and complex indoor environments, it is necessary to consider the influence of various factors on Wi-Fi signals, and select different data processing methods according to local conditions to eliminate each collection work. The outlier data in the middle effectively improve the validity and reliability of the location fingerprint database, and ultimately improve the accuracy of Wi-Fi indoor positioning.

3.2 Algorithm improvement

The matching algorithm of the nearest neighbor classification algorithm in the Wi-Fi location fingerprint localization algorithm is based on the RSSI location fingerprint feature matching method to achieve position calculation. It compares the dissimilarity (or similarity) of the feature fingerprint information in the Wi-Fi feature fingerprint database with the Wi-Fi feature fingerprint information measured at the location of the unknown node to be determined, and thus determines the result of the location of the unknown node to be tested.

3.2.1 Improved algorithm based on distance weighting

The core idea of the distance-based weighted KNN algorithm is to assign different weighting parameters w_p according to the distance between the reference point and the point to be measured. A high weighting value is assigned to the node closer to the reference point and the reference point, while a low weighting value is assigned to the node far away from them. Therefore, the reference point closer to the point to be measured has a larger influence coefficient, and the farther reference point has a smaller influence coefficient, instead the influence coefficient in the conventional KNN algorithm is k(k is the latest number in the algorithm),and then the estimated position results are corrected to a certain extent, which improves the accuracy of positioning.

$$w_p = \frac{\frac{1}{d_{pj}}}{\sum_{i=1}^{k} \frac{1}{d}} \tag{3}$$

Among them, $\frac{1}{d_{ij}}$ is the reciprocal of the Euclidean distance between the reference point i and the pint j to be measured, and $\frac{1}{d_{pj}}$ is the reciprocal of the Euclidean distance from a reference point p of the k reference points to the point j to be measured.

From Formula 3, the coordinate calculation formula for the position of the point to be measured can be derived as:

$$(x,y)=\sum_{p=1}^{k}\frac{\frac{1}{d_{pj}}}{\sum_{i=1}^{k}\frac{1}{d_{ij}}}\left(x_p,y_p\right)=\frac{\frac{1}{d_{1j}}}{\sum_{i=1}^{k}\frac{1}{d_{ij}}}(x_1,y_1)+\cdots+\frac{\frac{1}{d_{kj}}}{\sum_{i=1}^{k}\frac{1}{d_{ij}}}(x_k,y_k) \quad (4)$$

According to Formula 4, the smaller the d_{pj} value is, the larger the reciprocal value is, and the larger the corresponding w_p value is, which increases the influence of the estimated position coordinate of the near reference point to the measured point.

Algorithm 2:Improved algorithm based on distance weighting

Input: Node to be tested S, i=1,distance=0
Begin
1. Calculate the size of the fingerprint database
2. Calculate the Euclidean distance between the point in the fingerprint database and the node to be tested
3. Sort by distance
4. Different weighting parameters are given according to the distance between the reference point and the point to be measured
5. Calculate the coordinates of the node to be tested according to the weighted distance
6. Average the calculated results
7. Repeat steps 2-6 until all points to be measured are calculated.

End
Output: Position of the point to be measured

Assuming that the quantity of data is expressed in n, the time complexity of the first, second, fourth, fifth and sixth steps in algorithm 2 is O(n). The time complexity of step 3 is O (n^2). Step 7 makes the time complexity of the whole algorithm O (n^3).

3.2.2 Improved algorithm based on cosine similarity

Cosine similarity is a measure of similarity, assuming that x and y are two vectors to be compared, and the similarity function is shown in Formula 5.

$$\text{simality}(\vec{x},\vec{y})=\frac{\vec{x}\cdot\vec{y}}{\|\vec{x}\|\cdot\|\vec{y}\|} \quad (5)$$

$\|\vec{x}\|$ represents the Euclidean norm of vector $\vec{x}=(x_1,x_2,\cdots,x_p)$, which is the length of the vector $\sqrt{x_1+x_2+\cdots+x_p}$. Similarly, $\|\vec{y}\|$ represents the Euclidean norm of the vector. The measure of similarity is the cosine between $\vec{x}$ and $\vec{y}$. A cosine value of 0 means that the vectors are orthogonal and the angle is 90°. There is no similarity. The closer the cosine value is to 1, indicating a smaller angle, the greater the similarity in the vectors. The relationship between distance and cosine is shown in Fig. 4.

In the KNN matching algorithm, the cosine similarity can be selected as the position fingerprint between the reference point and the point to be measured. Calculate the cosine similarity between the point to be measured and each reference node. The following is an artificial pseudo code description for calculating the cosine similarity between two vectors.

Algorithm 3: Improved algorithm based on cosine similarity

Input: Node to be tested S=1, i=1
Begin
1. Calculate the size of the fingerprint database
2. Calculate the cosine similarity between the point in the fingerprint database and the node to be tested
3. Sort by cosine similarity
4. Calculate the coordinates of the node to be tested according to the cosine similarity
5. Average the calculated results
6. Repeat steps 2-5 until all points to be measured are calculated.

End
Output: Position of the point to be measured

The second step in algorithm 3 is to compute cosine similarity, and its time complexity is O (n^2). The time complexity of this step is the highest in the whole algorithm. Step 6 makes the time complexity of the whole algorithm O (n^3).

3.2.3 Improved algorithm for balancing joint metrics

The balanced joint metric algorithm is a trade-off balance algorithm for the above two algorithms. Assume that parameter is α. In the calculation of the nearest k reference points, distance weighting and cosine similarity are used as metrics to estimate the final position results of two points of the same point to be measured. Then, the results are weighted again, as shown in Formula 6.

$$(X_{\text{final}},Y_{\text{final}})=(1-\alpha)\cdot(X_{\text{weight}},Y_{\text{weight}})+\alpha(X_{\text{simality}},Y_{\text{simality}}) \quad (6)$$

where $(X_{\text{final}}, Y_{\text{final}})$ is the final calculated position coordinate result, $(X_{\text{weight}}, Y_{\text{weight}})$ is the result of the position to be measured calculated by distance weighting as the metric, and $(X_{\text{simality}}, Y_{\text{simality}})$ is the cosine similarity as the metric calculated result of the position to be tested. When the value of a is equal to 0.5, it is equivalent to averaging the above two results; the specific values need to be dynamically adjusted according to different indoor environments. The following is the source code description of the simulation core part of cosine similarity.

Algorithm 4: Improved algorithm for balancing joint metrics

Input: Node to be tested S, i=1
Begin
1. Calculate the size of the fingerprint database
2. Calculate the cosine similarity between the point in the fingerprint database and the node to be tested
3. Sort by cosine similarity
4. Calculate your coordinates based on cosine similarity
5. Calculate the distance between the point in the fingerprint database and the node to be tested
6. Sort by distance
7. Different weighting parameters are given according to the distance between the reference point and the point to be measured
8. Calculate the coordinates of the node to be tested according to the weighted distance
9. The result of the calculation is weighted again to obtain the final result
10. Repeat steps 2-9 until all points to be measured are calculated.

End
Output: Position of the point to be measured

The highest time complexity in algorithm 4 is computational cosine similarity. Therefore, the time complexity of the algorithm and the time complexity of algorithm 3 are O(n^3).

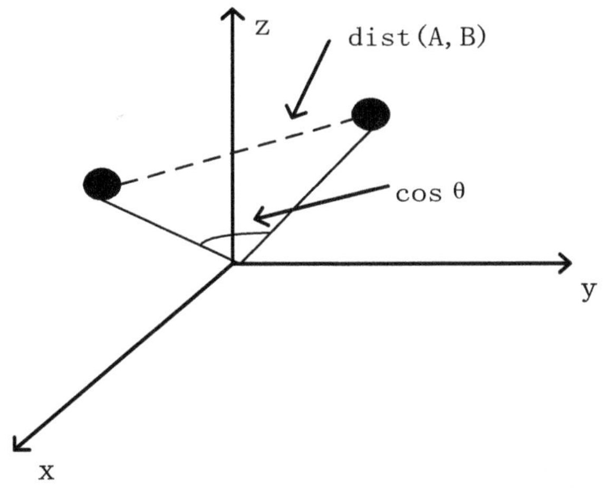

Fig. 4 The schematic diagram of cosine similarity measure

3.3 Summary of design ideas

In order to improve the positioning accuracy, we start from two aspects: data preprocessing and algorithm improvement.

(1) Data preprocessing: Because Wi-Fi signal is affected by various interference factors and its attenuation is irregular, the original data need to be preprocessed. The main methods are truncated mean method and screening mean method. The core idea of truncated mean method is to discard the portion of the larger value end and the smaller value end, and then average the remaining data. The screening mean method establishes a normal distribution model in the middle part of the relatively stable signal strength, and eliminates abnormal values in this way. Truncated mean method and screening mean method make the location fingerprint data more effective and reliable, and provide a guarantee for improving the location accuracy.

(2) Algorithm improvement: The core idea of the distance-based weighted KNN algorithm is to assign different weighting parameters according to the distance between the reference point and the point to be measured, to assign higher weighting values to the nodes whose distance between the reference point and the point to be measured, and to assign lower weighting values to the nodes whose distance is closer, instead of the traditional KNN algorithm where the influence coefficients are all 1/k (k is the number of the nearest neighbors in the algorithm), so as to estimate the location. The results are corrected to improve the positioning accuracy. The idea of the improved algorithm based on cosine similarity is that the cosine value is closer to 1, which indicates that the smaller the angle is, the greater the similarity between vectors. Choosing cosine similarity as the measurement index of fingerprint feature vector between reference point and test point can improve the positioning accuracy. Balancing joint metrics algorithm is a balancing algorithm which combines the above two algorithms. The main idea is to use distance weighting as the measurement index to calculate the result of the position to be measured, cosine similarity as the measurement index to calculate the result of the position to be measured, and then to average the two results, so as to improve the positioning accuracy.
The experiment in Section 4 shows the accuracy of the design idea.

4 Simulation experiment and analysis

4.1 Fingerprint positioning environment

The 10.5 × 18 m^2 areas of a science and technology museum were selected as the indoor positioning experimental environment in Fig. 5. First, the test area is modeled, and the upper left corner of the area is used as the coordinate origin to establish a two-dimensional coordinate system.

4.2 Location fingerprint collection

In the offline sampling measurement phase, an Android terminal is used as a fingerprint information device for collecting Wi-Fi signals, and the mobile phone model is MI-A. In the acquisition stage, set a scan interval of 5 seconds, collect 66 sets of RSSI raw data for all reference points, save them to the local file in txt format, upload them to the server through Wi-Fi (or a mobile communication network), and build on the server side. After the fingerprint data are built on the server side, a mapping relationship between a coordinate point (x, y) and a strong vector of Wi-Fi signals is established to form an offline fingerprint database. In the actual acquisition, there are 6APs with stable Wi-Fi signals for reference, and the others are filtered. Then, at each reference point, 66 sets of fingerprint data of the location feature fingerprint and the point coordinates are generated to complete the

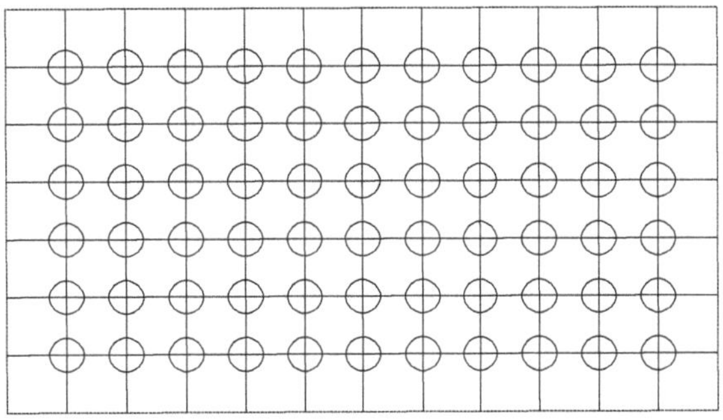
Fig. 5 Regional grid diagram

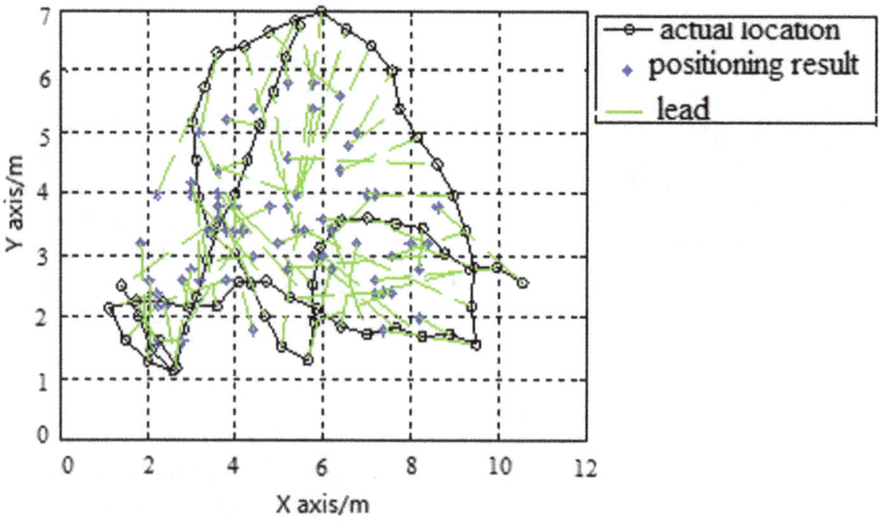

Fig. 6 The result of positioning analysts based on traditional KNN

construction of the fingerprint database. The mapping format is as follows:

<Loc_x, Loc_y, Rssi-1, Rssi-2, Rssi-3, Rssi-4, Rssi-5, Rssi-6>.

4.3 Locating result simulation

This article uses MATLAB R2013b as the simulation environment, and under the premise that 66 reference point offline location fingerprint databases have been constructed, 70 sets of measured points of measured position fingerprint data are selected as analysis indicators. Different improved algorithms are selected as the positioning result analysis.

(1) Positioning analysis based on traditional KNN algorithm

The positioning result of the traditional KNN algorithm using Euclidean distance as a metric is shown in Fig. 6. Is the comparison between the actual position generated and the positioning estimation result with the minimum deviation of average positioning distance. The minimum deviation of average positioning distance is 1.7105 m, and k = 5.

(2) Positioning analysis based on distance weighting improvement

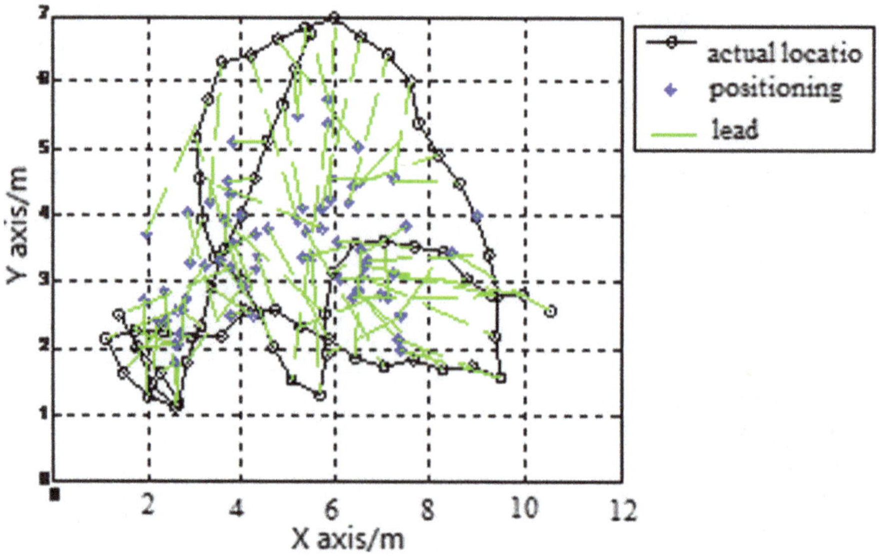

Fig. 7 The result of positioning analysts based on distance weighting KNN

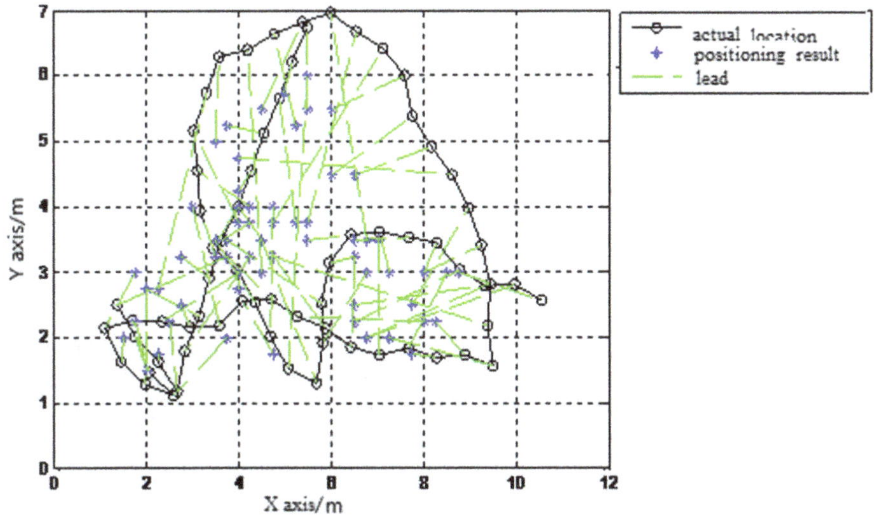

Fig. 8 The result of positioning analysts based on cosine similarity

The result of positioning analysis based on KNN algorithms of distance weighting improvement is shown in Fig. 7. In the figure, the average positioning deviation of the 70 groups of nodes to be tested is the smallest, and the deviation is 1.7105 m, and k = 7. It can be found that the positioning accuracy is improved compared with the traditional Euclidean distance positioning algorithm.

(3) Improved positioning results based on cosine similarity

When the cosine similarity is selected as the metric to improve the algorithm, it is found that while k is taken as 4, the average positioning error of the node to be tested is small, which is 1.6549 m. The comparison of the positioning results at this time is shown in Fig. 8. The positioning accuracy is improved, comparing with the traditional KNN algorithm based on Euclidean distance.

(4) The result of positioning analysts based on balanced joint metric improvement algorithm

The situation of balancing joint metrics is slightly complicated. First, assuming that a takes a value of 0.5, then separately calculate the effect of k taking a values from 1 to 10 on the positioning accuracy. The result is shown in Table 2.

Obviously, the positioning accuracy is higher when k = 4. k takes the value of 4, and the influence of the joint average metric improvement algorithm on the positioning deviation is shown in Table 3.

The joint metric algorithm model shows that when a = 0, it is an algorithm based on distance weighting. When a = 1, it is an algorithm based on similarity metric. Combining Tables 2 and 3, select k = 4, a = 0.7 for the positioning result analysis, as shown in Fig. 9.

(5) Comparison of positioning deviation of three improved algorithms

After completing the analysis of the positioning results of different improved algorithms, different k values are selected here. The traditional positioning algorithm based on Euclidean distance and the positioning deviation between the improved algorithms of three different strategies are comprehensively compared and analyzed. In Fig. 10, it is found that the positioning accuracy improved by the improved algorithm of three different strategies compared with the traditional algorithm. The improved algorithm based on distance weighting and the balanced joint metric algorithm are smoother and more stable, and the joint balance metric algorithm performs better than another, which preliminarily verifies the effectiveness and superiority of the joint balance metric algorithm.

(6) Comparisons of algorithms

Table 2 The influence of the value of k on the positioning deviation of joint metrics

The value of K	Positioning deviation (m)	The value of K	Positioning deviation (m)
1	2.1264	6	1.6915
2	1.9458	7	1.6844
3	1.6473	8	1.7189
4	1.6466	9	1.7269
5	1.6661	10	1.7917

Table 3 The influence of the value of a on the positioning deviation of joint metrics when k = 4

Value of α	Positioning deviation (m)	Value of α	Positioning deviation (m)
0	1.7145	0.5	1.6466
0.1	1.6952	0.6	1.6416
0.2	1.6791	0.7	1.6396
0.3	1.6649	0.8	1.6411
0.4	1.6541	0.9	1.6462

We performed experiments in various regions, such as 20 m × 20 m and 15 m × 24 m. Although the accuracy of the joint balance metric improvement algorithm is not the best, the algorithm is simple and the time complexity is only O (n^3).

The accuracy of the neural network algorithm is higher [18]. Global time complexity of convolutional Neural Networks is: tmes $\sim O(\sum_{l=1}^{D} M_l^2 K_l^2 C_{l-1} C_l)$, Among them: D is the number of convolution layers of the neural network; L is the first convolution layer of the neural network; M is the edge length of the output characteristic graph of each convolution core; K is the edge length of each convolution core; C_l is the number of convolution kernels of the lth layer.

It can be seen that the time complexity of the algorithm we designed is lower. This algorithm can be applied to general science and technology museums.

5 Conclusions

In recent years, with the development of science and technology and the progress of society, the science and technology museums have become increasingly popular. The main concern of the manager is to improve the quality of the science and technology museum's exhibition hall and to understand its popularity. This article uses a phone as the mobile terminal, collecting visitors' trajectories through indoor positioning technology, which is helpful for the scientific management of the science and technology museum. Today's indoor segment distance positioning is still in the stage of exploration and research, and there are no large-scale deployments applied in China. Due to the high popularity of mobile terminals and extensive coverage of Wi-Fi networks, this article chooses Wi-Fi technology based on mobile terminals for the positioning of users of the science and technology museum, improves the location fingerprint algorithm, and illustrates its effectiveness in experiments.

Although this paper proposes some methods for processing the RSSI raw data of the location fingerprint algorithm, provides three improved schemes for the KNN matching algorithm and improves the validity of fingerprint data and optimization of algorithm accuracy, it

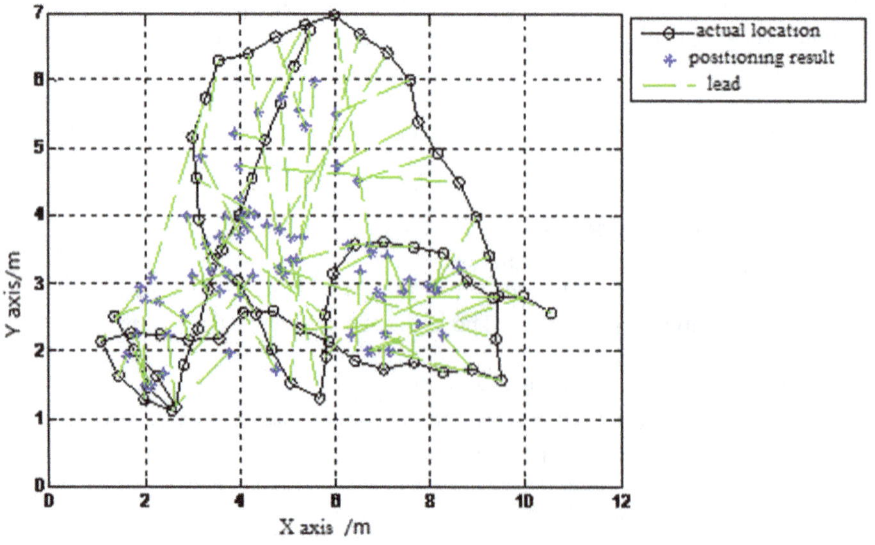

Fig. 9 Positioning result of joint balance metric improvement algorithm

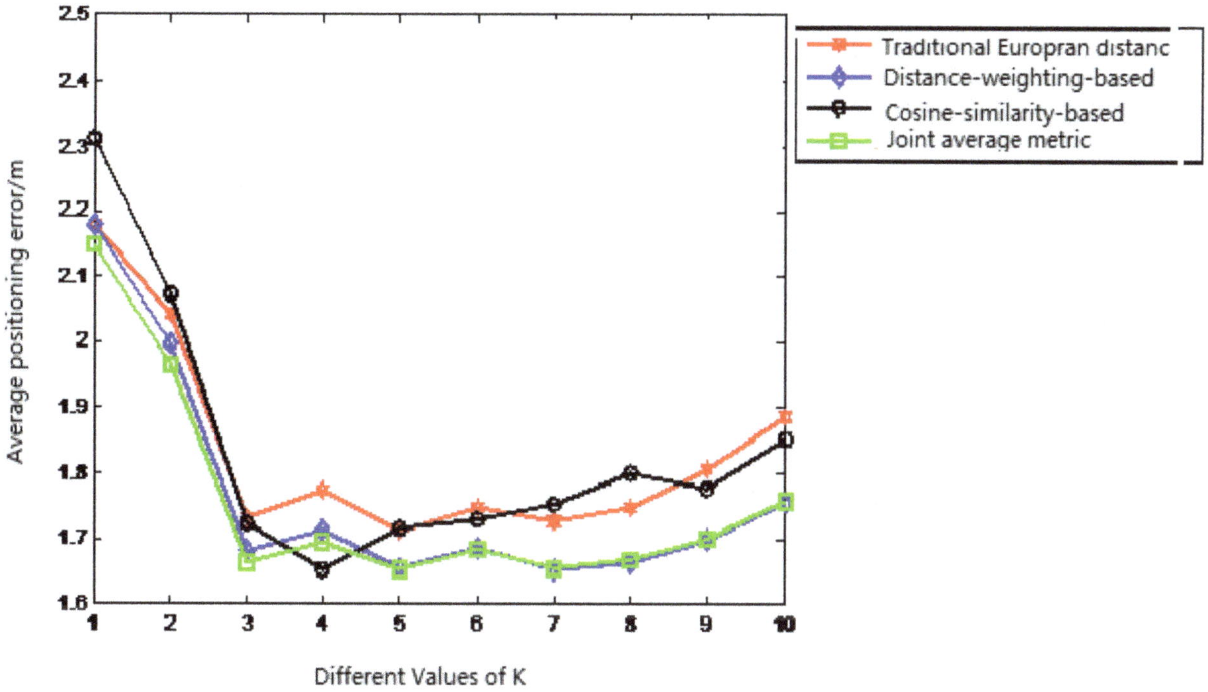

Fig. 10 Deviation map of four positioning algorithms under different k values

only considers the location fingerprint information of two-dimensional space in the simulation experiment to simplify the research work. In fact, the real environment in the room is three dimensional, and the fingerprint information at different heights will be different even in the same position, which is ignored in this article. Therefore, to obtain higher positioning accuracy, it is necessary to consider the construction of the location fingerprint database in three-dimensional space.

Abbreviations

AHLos: Ad hoc localization system; AOA: Angle of arrival; AP: Wireless access point; APs: Approximately 6 wireless access points; BSSID: Basic service set; KNN: k-nearest neighbor; LAN: Local area network; MIT: Massachusetts Institute of Technology; PS: Photoshop; RADAR: Radio detection and ranging; RFID: Radio frequency identification; RSSI: Received signal strength indication; RSSP: Received signal strength; SI: Received; SSID: Service set identifier; TDOA: Time difference of arrival; TOA: Time of arrival; UCLA: University of California, Los Angeles; UWB: Ultra wide band; Wi-Fi: Wireless-fidelity

Acknowledgements

The authors thank the anonymous reviewers and editors for their efforts in valuable comments and suggestions.

Authors' contributions

QY proposes the algorithm design and data analysis. SZ proposes innovation ideas. ML proposes data analysis. YZ carried out experiments. All authors read and approved the final manuscript.

Funding

This paper is supported by the national science and technology support Program. Program number is 2015BAK33B00.

Competing interests

The authors declare that they have no competing interests.

References

1. C. Jin, D. Qiu, Research on indoor positioning technology based on WiFi signal [J]. Bull Survey Mapp **2017**(05), 21–25 (2017)
2. Cheng Y C, Chawathe Y, Lamarca A, Accuracy characterization for metropolitan-scale Wifi localization. 2005 Proceedings of the 3rd International Conference on Mobile System, Applications and Services (MobiSys, Washington, 2005), pp. 233-245. https://dl.acm.org/citation.cfm?doid=1067170.1067195. Accessed Oct 2017.
3. L.M. Ni, Y. Liu, Y.C. Lau, A.P. Patil, LANDMARC: indoor location sensing using active RFID [J]. Wirel Netw **10**(6), 701–710 (2004)
4. Lee J S, Su Y W, Shen C C, A comparative study of wireless protocols: Bluetooth, UWB, ZigBee, and Wi-Fi . IECON 2007-33rd Annual Conference of the IEEE Industrial Electronics Society IEEE. (Taiwan, 2007), pp. 46-51. https://ieeexplore.ieee.org/document/4460126. Accessed Oct 2017.
5. W. Rong, Brief discussion on the popular short-distance wireless communication technology nowadays. Youth Era (14), 199–200 (2015, 2015)
6. Wei-qing Huang, Chang Ding, Si-ye Wang, Junyu Lin, Shao-yi Zhu, Yue Cui, Design and Realization of an Indoor Positioning Algorithm Based on Differential Positioning Method. International Conference on Wireless Algorithms, Systems, and Applications. (WASA. Guilin, 2017), pp. 546-558. https://link.springer.com/chapter/10.1007/978-3-319-60033-8_47. Accessed Oct 2018.
7. Yucel, Edizkan, Ozkir, Yazici, Development of indoor positioning system with ultrasonic and infrared signals. International Symposium on Innovations in Intelligent Systems & Applications. (Trabzon, 2012). https://ieeexplore.ieee.org/document/6246983. Accessed May 2018.
8. Alessio De Angelis, Antonio Moschitta, Paolo Carbone, Massimo Calderini, Stefano Neri, Renato Borgna, Manuelo Peppucci, Design and characterization of an ultrasonic indoor positioning technique . Instrumentation and Measurement Techno-logy Conference (I2MTC) Proceedings 2014 IEEE International. (Montevideo, 2014), pp. 1623-1628. https://ieeexplore.ieee.org/document/6861020. Accessed May 2018.
9. Lingling Zhu, Aolei Yang, Dingbing Wu, Li Liu, Survey of Indoor Positioning

Technologies and Systems. 2014 International Conference on LifeSystem Modeling and Simulation. (Shanghai, 2014), pp. 400-409. https://link.springer.com/chapter/10.1007/978-3-662-45283-7_41. Accessed Oct 2017.
10. Want R, Hopp A, Falcão V, Gibbons J, The active badge location system. Acm Transactions on information Systems. 10(1), 91-102(1992)
11. Bahl P, Padmanabhan V N, RADAR: an in-building RF-based user location and tracking system. Nineteenth Annual Joint Conference of the IEEE Computer and Communications Societies. (IEEE, Tel Aviv, 2000), pp. 775-784. https://ieeexplore.ieee.org/document/832252. Accessed Mar 2018.
12. Woodman O J, Harle R K, Concurrent scheduling in the Active Bat location system. IEEE International Conference on Pervasive Computing and Communications Wordshops. (Mannheim, 2010), pp. 431-437. https://ieeexplore.ieee.org/document/5470631. Accessed Mar 2018.
13. Priyantha N, Chakraborty A, Balakrishnan H, The Cricket location-support system. Proceedings of the 6th annual international conference on Mobile computing and networking. (Boston, 2010), pp. 32-43.
14. Oliveira H A B F, Boukerche A, Nakamura E F, Localization in time and space for wireless sensor networks: an efficient and lightweight algorithm. Performance Evaluation. 66(35), 209-222 (2009)
15. D.J. Coleman, A. Rajabifard, K.W. Kolodziej, Expanding the SDI environment: comparing current spatial data infrastructure with emerging indoor location-based services. Int J Digital Earth **9**(6), 629–647 (2016)
16. Hanhui Yue, Xiao Zheng, Juan Wang, Li Zhu, Chunyan Zeng, Cong Liu, Meng Liu, Research and Implementation of Indoor Positioning Algorithm for Personnel Based on Deep Learning . International Conference on Emerging Internetworking, Data & Web Technologies. (Tirana, 2018), pp. 782-791. https://link.springer.com/content/pdf/10.1007%2F978-3-319-75928-9.pdf. Accessed Feb 2019.
17. He Z ,Xu M , Guo A, Advanced Computational Methods in Life System Modeling and Simulation. Multi-channel feature for pedestrian detection (Nanjing, 2017), pp. 472-480. https://link.springer.com/chapter/10.1007/978-981-10-6370-1_47. Accessed Oct 2018.
18. J. Luo, H. Gao, I. Gondal, Deep belief networks for fingerprinting indoor localization using ultrawideband technology [J]. Int J Distributed Sensor Netwo (2016)
19. Aguilar, W.G., Luna, M.A., Moya, J.F., Abad, V., Ruiz, H., Parra, H., Lopez, W, Cascade classifiers and saliency maps based people detection . International Conference on Augmented Reality, Virtual Reality and Computer Graphics. (Otranto, 2017), pp. 501-510.
20. Jaemin Hong, KyuJin Kim, ChongGun Kim, Comparison of Indoor Positioning System Using Wi-Fi and UWB. Asian Conference on Intelligent Informationand Database Systems. (Dong Hoi City, 2018), pp. 623-632. https://link.springer.com/content/pdf/10.1007%2F978-3-319-75417-8.pdf. Accessed Oct 2018.
21. Jianguo Yu, Zhian Deng, Xin Liu, Juan Chen, Zhenyu Na, WLAN Indoor Positioning Based on DLDA Feature Extraction Algorithm. International Conference in Communications, Signal Processing, and Systems. (Harbin, 2017), pp. 2779-2787.https://link.springer.com/chapter/10.1007/978-981-10-6571-2_336. Accessed Feb 2019.

Permissions

All chapters in this book were first published in EURASIP JWCN, by Springer; hereby published with permission under the Creative Commons Attribution License or equivalent. Every chapter published in this book has been scrutinized by our experts. Their significance has been extensively debated. The topics covered herein carry significant findings which will fuel the growth of the discipline. They may even be implemented as practical applications or may be referred to as a beginning point for another development.

The contributors of this book come from diverse backgrounds, making this book a truly international effort. This book will bring forth new frontiers with its revolutionizing research information and detailed analysis of the nascent developments around the world.

We would like to thank all the contributing authors for lending their expertise to make the book truly unique. They have played a crucial role in the development of this book. Without their invaluable contributions this book wouldn't have been possible. They have made vital efforts to compile up to date information on the varied aspects of this subject to make this book a valuable addition to the collection of many professionals and students.

This book was conceptualized with the vision of imparting up-to-date information and advanced data in this field. To ensure the same, a matchless editorial board was set up. Every individual on the board went through rigorous rounds of assessment to prove their worth. After which they invested a large part of their time researching and compiling the most relevant data for our readers.

The editorial board has been involved in producing this book since its inception. They have spent rigorous hours researching and exploring the diverse topics which have resulted in the successful publishing of this book. They have passed on their knowledge of decades through this book. To expedite this challenging task, the publisher supported the team at every step. A small team of assistant editors was also appointed to further simplify the editing procedure and attain best results for the readers.

Apart from the editorial board, the designing team has also invested a significant amount of their time in understanding the subject and creating the most relevant covers. They scrutinized every image to scout for the most suitable representation of the subject and create an appropriate cover for the book.

The publishing team has been an ardent support to the editorial, designing and production team. Their endless efforts to recruit the best for this project, has resulted in the accomplishment of this book. They are a veteran in the field of academics and their pool of knowledge is as vast as their experience in printing. Their expertise and guidance has proved useful at every step. Their uncompromising quality standards have made this book an exceptional effort. Their encouragement from time to time has been an inspiration for everyone.

The publisher and the editorial board hope that this book will prove to be a valuable piece of knowledge for researchers, students, practitioners and scholars across the globe.

List of Contributors

Esmaeil Nik Maleki and Ghasem Mirjalily
Department of Electrical Engineering, Yazd University, Yazd, Iran

Wei Wu, Xinmeng Shen and Pei Li
College of Communication and Information, Nanjing University of Posts and Telecommunications, Xinmofan Road 66, Nanjing 210003, China

Ping Deng
College of Communication and Information, Nanjing University of Posts and Telecommunications, Xinmofan Road 66, Nanjing 210003, China
College of Automation and College Of Artificial Intelligence, Nanjing University of Posts and Telecommunications, Xinmofan Road 66, Nanjing 210003, China

Baoyun Wang
College of Overseas Education, Nanjing University of Posts and Telecommunications, Xinmofan Road 66, Nanjing 210003, China

Genghua Yu, Zhi Gang Chen, Jia Wu and Jian Wu
School of Computer Science and Engineering, Central South University, Changsha 410083, China

Biao Wang, Yufeng Ge, Cheng He, You Wu and Zhiyu Zhu
School of Electronic and Information, Jiangsu University of Science and Technology, Zhenjiang 212003, China

Jingwen Pan, Jie Cui, Lu Wei, Yan Xu and Hong Zhong
School of Computer Science and Technology, Anhui University, Hefei 230039, Anhui, China

Daniyal Munir, Kae Won Choi, Tae-Jin Lee and Min Young Chung
Department of Electrical and Computer Engineering, Sungkyunkwan University, 2066 Seobu-Ro, Jangan-Gu, 16419 Suwon, Gyeonggi-Do, South Korea

Syed Tariq Shah
Department of Electrical and Computer Engineering, Sungkyunkwan University, 2066 Seobu-Ro, Jangan-Gu, 16419 Suwon, Gyeonggi-Do, South Korea
Department of Telecommunication Engineering, FICT, Balochistan University of Information Technology, Engineering and Management Sciences, Airport Road, Baleli, 87300 Quetta, Pakistan

Xinqi Jiang and Fu-Chun Zheng
School of Electronic and Information Engineering, Harbin Institute of Technology, Shenzhen, China

Zhen Cai and Mangui Liang
Institute of Information Science, Beijing Jiaotong University, Beijing 100044, People's Republic of China

Qiushi Sun
Beijing Key Lab of Traffic Data Analysis and Mining, Beijing Jiaotong University, Beijing 100044, People's Republic of China

Solmaz Nobahary and Amir Masoud Rahmani
Department of Computer Engineering, Science and Research Branch, Islamic Azad University, Tehran, Iran

Hossein Gharaee Garakani and Ahmad Khademzadeh
Iran Telecom Research Center (ITRC), Tehran, Iran

Avishek Mukherjee and Zhenghao Zhang
Department of Computer Science, Florida State University, Tallahassee, FL, USA

Jing Zhang
School of Economics and Management, Beijing Jiaotong University, Beijing 100044, China

Jessica Mendoza, David Palacios, Isabel de-la-Bandera, Eduardo Baena, Emil J Khatib and Raquel Barco
Department of Communication Engineering, University of Malaga, Andalucía Tech., 29071 Malaga, Spain

Min Wang and Zhongbo Wu
Computer Engineering School, Hubei University of Arts and Science, Xiangyang, China

Hang Qin
Computer School, Yangtze University, Jingzhou, China

Xuefei Zhang, Yinjun Liu and Yue Wang
National Engineering Laboratory for Mobile Network Technologies, Beijing University of Posts and Telecommunications, 100876 Beijing, China

Juan Bai
Air Force Engineering University, 710051 Xi'an, China

Rashid Ali, Yazdan Ahmad Qadri, Yousaf Bin Zikria and Sung Won Kim
Department of Information and Communication Engineering, Yeungnam University, Gyeongsan, 38541, Republic of Korea

Tariq Umer
Department of Computer Science, COMSATS University Islamabad, Wah, Pakistan

Byung-Seo Kim
Department of Computer and Information Communication Engineering, Hongik University, Seoul, 04066, Republic of Korea

Guoan Zhang, Li Jin and Wei Duan
School of Information Science and Technology, Nantong University, Seyuan Road, Nantong, China

Jinjuan Ju
School of Information Science and Technology, Nantong University, Seyuan Road, Nantong, China
School of Electronic and Information Engineering, Nantong Vocational University, Qingnian Road, Nantong, China

Qiang Sun
School of Information Science and Technology, Nantong University, Seyuan Road, Nantong, China
The Nantong Research Institute for Advanced Communication Technologies, Chongchuan Road, Nantong, China

Qing Yang, Shijue Zheng, Ming Liu and Yawen Zhang
School of Computer, Central China Normal University, Wuhan 430079, People's Republic of China

Index

www.ingramcontent.com/pod-product-compliance
Lightning Source LLC
Chambersburg PA
CBHW061244190326
41458CB00011B/3577
* 9 7 8 1 6 3 9 8 7 0 2 4 0 *